Techniques of Control for Energy Optimization in Actuators, Motors and Power Generation Systems

Techniques of Control for Energy Optimization in Actuators, Motors and Power Generation Systems

Guest Editor

Paolo Mercorelli

MDPI

Basel • Beijing • Wuhan • Barcelona • Belgrade • Novi Sad • Cluj • Manchester

Guest Editor
Paolo Mercorelli
Institute for Production
Technology and Systems
Leuphana University of
Lueneburg
Lueneburg
Germany

Editorial Office
MDPI AG
Grosspeteranlage 5
4052 Basel, Switzerland

This is a reprint of the Special Issue, published open access by the journal *Energies* (ISSN 1996-1073), freely accessible at: https://www.mdpi.com/journal/energies/special_issues/energy_optimization.

For citation purposes, cite each article independently as indicated on the article page online and as indicated below:

Lastname, A.A.; Lastname, B.B. Article Title. *Journal Name* **Year**, *Volume Number*, Page Range.

ISBN 978-3-7258-4977-2 (Hbk)
ISBN 978-3-7258-4978-9 (PDF)
https://doi.org/10.3390/books978-3-7258-4978-9

Contents

About the Editor

Paolo Mercorelli

Paolo Mercorelli (Member, IEEE) received his PhD degree in Systems Engineering from the University of Bologna, Bologna, Italy, in 1998. In 1997, he was a Visiting Researcher for one year at the Department of Mechanical and Environmental Engineering, University of California, Santa Barbara, CA, USA. From 1998 to 2001, he was a Postdoctoral Researcher with Asea Brown Boveri Corporate Research, Heidelberg, Germany, and his research activity there earned him three patents. From 2002 to 2005, he was a Senior Researcher at the Institute of Automation and Informatics, Wernigerode, Germany, where he was the leader of a control group. From 2005 to 2011, he was an Associate Professor of Process Informatics at Ostfalia University of Applied Sciences, Wolfsburg, Germany. Since 2012, he has served as Full Professor and the Chair of Control and Drive Systems at the Institute for Production Technology and Systems, Leuphana University of Lueneburg, Lueneburg, Germany. Dr. Mercorelli was the recipient of a three-year scholarship, extending from 1998 to 2001, from the Marie Curie Actions Research Fellowship Program, which is one of the most competitive and prestigious European awards, sponsored by the European Commission. He has also received seven Best International Conference Paper awards—IECON 2013, IECON 2014, CoDIT 2014, ICCC 2017, FedCSIS 2019, ACD 2019, ICCC 2020, ICSTCC 2023, and IECON 2024—and given thirteen plenary lectures at international conferences. In the years 2019, 2020, 2021, 2022, 2023, and 2024, he was included in the list of the top 2% of scientists by the Elsevier Database and (METRICS) University of Stanford (USA) (Ioannidis, J.; Boyack, K. & Baas, J., 2020, Updated science-wide author databases of standardized citation indicators, PLOS, October 16). Since 2022, he has been Editor-in-Chief for the "Engineering Mathematics" section in *Mathematics*, an Open Access Journal published by MDPI, Basel, Switzerland.

Preface

It is increasingly evident that the intelligent administration of energy resources is not only a technological imperative, but an ethical responsibility. As we navigate the transition toward more sustainable and efficient energy systems, the role of advanced control strategies and optimization algorithms becomes ever more crucial. Intelligent control is no longer a niche field; it is at the heart of contemporary energy challenges, enabling systems to achieve higher efficiency, reduce losses, and adapt dynamically to varying operating conditions.

This Reprint of the Special Issue of *Energies* titled "Techniques of Control for Energy Optimization in Actuators, Motors and Power Generation Systems" brings together cutting-edge research that addresses energy optimization through the lens of control engineering. The contributions featured herein span a wide range of applications, from smart grids and renewable energy integration to electric and hybrid vehicles, traditional engines, storage systems, and the diverse world of actuators, including electrical, hydraulic, and pneumatic types.

While large-scale power systems often dominate discussions on energy efficiency, it is essential to recognize the pivotal role played by smaller components such as actuators and motors. Their ubiquity in industrial, automotive, and consumer applications means that even marginal improvements in their performance can lead to significant global energy savings. As such, the development and implementation of intelligent control techniques in these systems are not just beneficial, but essential.

This Reprint is a collection of forward-looking studies and practical innovations aimed at optimizing energy use through control. It provides valuable insights into methodologies that improve energy conversion efficiency, enhance system responsiveness, and reduce operational costs. Whether through model predictive control, machine learning, adaptive systems, or novel algorithmic frameworks, the studies presented here share a common goal: to contribute to a smarter, more sustainable future in energy production.

We trust that this Reprint will serve as a valuable resource for researchers, practitioners, and policymakers interested in the intersection of energy optimization and intelligent control systems.

Paolo Mercorelli
Guest Editor

energies

MDPI

Editorial

Techniques of Control for Energy Optimization in Actuators, Motors, and Power Generation Systems

Paolo Mercorelli

Institute for Production Technology and Systems (IPTS), Leuphana Universität Lüneburg, 21335 Lüneburg, Germany; paolo.mercorelli@leuphana.de

1. Introduction

It is common sense that an intelligent administration of energy resources represents a unique way to respect the environment and equally redistribute and reutilize resources. In today's energy systems, control and optimization techniques and algorithms play a decisive role. In fact, energy optimization represents a crucial issue in smart-grid/micro-grid energy flow, renewable energy, electrical and hybrid vehicles, as well as in traditional engines, energy storage devices, electrical motors, and in any kind of actuator, including alternative actuators.

In terms of energy, all of these systems, from the biggest to the smallest, play an extremely important role. Any kinds of small motors and actuators (electrical, hydraulic, pneumatic, etc.), which are devoted to the transformation of energy into movement because of their widespread application, should be considered in this process of global, intelligent energy optimization.

This Special Issue of Energies will explore the latest developments in intelligent control algorithms and techniques dedicated to the optimal management of energy in the systems listed above, or in any kind of system in which an algorithm or a technique plays a decisive role in terms of energy optimization and performance. This Special Issue brings together cutting-edge research on advanced control techniques aimed at improving the energy efficiency of actuators, motors, and power generation systems. These technologies are crucial to reducing energy consumption, enhancing system performance, and minimizing environmental impact.

Recent announcements, such as the European Commission's new Green Deal initiatives and updated emissions targets for 2030, underscore the urgency of developing smart energy solutions. Likewise, the U.S. Department of Energy's latest funding round for AI-driven grid optimization tools highlights the growing investment in intelligent systems that enhance energy reliability and efficiency.

The issue covers a diverse range of topics, including, but not limited to, model-based control strategies, adaptive control, smart grid integration, and the application of artificial intelligence in optimizing power output.

By providing insights into both theoretical advancements and practical implementation, this Special Issue aims to offer a comprehensive view of the latest developments in the field. It is intended to foster collaboration among researchers, engineers, and practitioners working towards the common goal of creating more energy-efficient, reliable, and sustainable systems.

2. Presentation of the Research Papers

Contribution 1 explores a low-cost silver-coated polyamide (PA6) as a thermoresponsive metal-polymer hybrid actuator. PA6's properties are influenced by humidity and water absorption, affecting mass, resistance, and surface morphology. The material's resistance dynamics are modeled using hysteresis, and a switching logic algorithm is introduced to control the actuator force.

In 2, a multi-objective optimization model for large-scale wind power generation planning is proposed, considering power flow, investment costs, and network loss. The improved differential evolution (IDE) algorithm is used to calculate Pareto optimal solutions, ranked by entropy weights and finalized with TOPSIS. The model is tested on China's 2020 power grid, achieving optimized wind power integration.

Contribution 3 proposes a fault-tolerant control (FTC) scheme for a distributed high-speed generator system to maintain operation despite four simultaneous generator faults. The FTC consists of a baseline controller and a lower layer for fault diagnosis, tested through simulations with various actuator faults.

Contribution 4 focuses on a model for wind turbines developed with direct-drive permanent magnet synchronous generators, accounting for active power losses in power system simulations. Analytical assessments and simulations validate the model, emphasizing the importance of active power losses in wind turbine control.

The applicability of tie-line bias control (TBC) in multi-area load frequency control (LFC) systems is re-examined in 5. A new LFC mode is proposed for ring networks where TBC is unsuitable, guiding future LFC designs for complex topologies.

Contribution 6 introduces a new power control scheme for multi-terminal low-frequency AC transmission systems using cycloconverters and virtual synchronous generator (VSG) control. The scheme eliminates the need for communication links for frequency synchronization, and is tested with a two-phase LFAC system.

Contribution 7 shows a hybrid propulsion system to reduce fuel consumption and emissions in off-highway applications. A causal optimization-based power management algorithm is introduced to optimize power split in hybrid powertrains. Simulations show the algorithm's potential to reduce fuel consumption in a diesel multiple-unit train.

In 8, a new fault-tolerant control strategy for a five-phase permanent-magnet vernier (FT-PMV) machine under short-circuit faults is proposed. The strategy minimizes torque ripple and compensates for faults using remedial currents. Simulations and experiments validate its effectiveness.

Contribution 9 defines a method for determining reserve margins in power systems based on short-term reliability of balance. Cost analysis compares this method with conventional approaches, showing greater reliability and efficiency, particularly for wind power generation.

The compound-structure permanent-magnet (CSPM) motor for hybrid electric vehicles (HEVs) is studied in 10. Magnetic coupling and design parameters are optimized using finite element analysis, leading to a proposed decoupling scheme for improved HEV performance.

In 11, a new power system operation model incorporates the output characteristics of wind and solar power generation. The model, tested on China's Northwest power grid, optimizes the integration of large-scale new energy power generation (NEPG).

A feedback linearization controller for a doubly-fed induction generator-based wind turbine is proposed in 12. The controller ensures system stability and track-desired states, and is validated by simulation results.

In Contribution 13, the impact of curtailment on the capacity factors of wind and solar power in Germany's renewable energy system is studied. Simulation results show that curtailment significantly reduces capacity factors, emphasizing the need for enhanced the transmission infrastructure.

In Contribution 14, an adaptive variable-rated sliding mode controller (ASMC) is designed for Bearingless Induction Motors (BIM) to improve performance under large disturbances. Simulations and experiments demonstrate reduced chattering and improved dynamic performance.

In Contribution 15, two optimization frameworks for hybrid vehicle design are compared: nested optimization and enhanced iterative optimization. The iterative approach, applied to a hybrid electric bus, is more computationally efficient and robust compared to varying parameters.

Contribution 16 presents a 3D blade redesign for a five-stage axial compressor using multi-stage CFD methods. The redesign improves stall margin by 13% while maintaining efficiency, which is validated through aerodynamic analysis.

In Contribution 17, a steady-state identification method and optimization scheme for electric variable-pitch propellers in quadrotors are proposed. This method improves endurance and positioning accuracy, and is validated through experimental results.

In 18 a new framework for designing offshore wind farms (OWFs) optimizes layout and electrical infrastructure simultaneously, considering trade-offs like wake losses and collection system length. The framework, applied to the Dutch Borssele area, shows potential for cost savings and more efficient designs.

3. Conclusions

In conclusion, the efficient and intelligent management of energy resources is integral for achieving sustainability and optimizing the performance of modern systems. As we continue to integrate diverse technologies such as smart grids, renewable energy, electric vehicles, and various actuators, the role of advanced control and optimization algorithms becomes increasingly critical. By enhancing energy utilization across a wide spectrum of applications—from small motors to large-scale systems—we can reduce waste, improve efficiency, and foster a more sustainable future. This Special Issue aims to highlight the cutting-edge research and innovations in intelligent energy-efficient systems that benefit both the environment and society. Moreover, it presents a wide variety of research papers that highlight the central importance of energy-related issues in defining control in all possible systems.

Conflicts of Interest: The author declares no conflicts of interest.

List of Contributions:

1. Schimmack, M.; Feistauer, E.E.; Amancio-Filho, S.T.; Mercorelli, P. Hysteresis Analysis and Control of a Metal-Polymer Hybrid Soft Actuator. *Energies* **2017**, *10*, 508. https://doi.org/10.3390/en10040508.
2. Li, H.; Li, G.; Liu, S.; Wang, Y.; Wang, Z.; Wang, J.; Zhang, N. Research on Optimal Planning of Access Location and Access Capacity of Large-Scale Integrated Wind Power Plants. *Energies* **2017**, *10*, 442. https://doi.org/10.3390/en10040442.
3. Giger, U.; Kühne, P.; Schulte, H. Fault Tolerant and Optimal Control of Wind Turbines with Distributed High-Speed Generators. *Energies* **2017**, *10*, 149. https://doi.org/10.3390/en10020149.
4. Bonfiglio, A.; Delfino, F.; Invernizzi, M.; Procopio, R. Modeling and Maximum Power Point Tracking Control of Wind Generating Units Equipped with Permanent Magnet Synchronous Generators in Presence of Losses. *Energies* **2017**, *10*, 102. https://doi.org/10.3390/en10010102.
5. Chen, C.; Zhang, K.; Yuan, K.; Teng, X. Tie-Line Bias Control Applicability to Load Frequency Control for Multi-Area Interconnected Power Systems of Complex Topology. *Energies* **2017**, *10*, 78. https://doi.org/10.3390/en10010078.

6. Pichetjamroen, A.; Ise, T. Power Control of Low Frequency AC Transmission Systems Using Cycloconverters with Virtual Synchronous Generator Control. *Energies* **2016**, *10*, 34. https://doi.org/10.3390/en10010034.
7. Schalk, J.; Aschemann, H. A Causal and Real-Time Capable Power Management Algorithm for Off-Highway Hybrid Propulsion Systems. *Energies* **2016**, *10*, 10. https://doi.org/10.3390/en10010010.
8. Gu, C.; Zhao, W.; Zhang, B. Simplified Minimum Copper Loss Remedial Control of a Five-Phase Fault-Tolerant Permanent-Magnet Vernier Machine under Short-Circuit Fault. *Energies* **2016**, *9*, 860. https://doi.org/10.3390/en9110860.
9. Kwon, K.b.; Park, H.; Lyu, J.K.; Park, J.K. Cost Analysis Method for Estimating Dynamic Reserve Considering Uncertainties in Supply and Demand. *Energies* **2016**, *9*, 845. https://doi.org/10.3390/en9100845.
10. Xu, Q.; Sun, J.; Luo, L.; Cui, S.; Zhang, Q. A Study on Magnetic Decoupling of Compound-Structure Permanent-Magnet Motor for HEVs Application. *Energies* **2016**, *9*, 819. https://doi.org/10.3390/en9100819.
11. Li, H.; Li, G.; Wu, Y.; Wang, Z.; Wang, J. Operation Modeling of Power Systems Integrated with Large-Scale New Energy Power Sources. *Energies* **2016**, *9*, 810. https://doi.org/10.3390/en9100810.
12. Alrifai, M.; Zribi, M.; Rayan, M. Feedback Linearization Controller for a Wind Energy Power System. *Energies* **2016**, *9*, 771. https://doi.org/10.3390/en9100771.
13. Kies, A.; Schyska, B.; Von Bremen, L. Curtailment in a Highly Renewable Power System and Its Effect on Capacity Factors. *Energies* **2016**, *9*, 510. https://doi.org/10.3390/en9070510.
14. Yang, Z.; Wan, L.; Sun, X.; Li, F.; Chen, L. Sliding Mode Variable Structure Control of a Bearingless Induction Motor Based on a Novel Reaching Law. *Energies* **2016**, *9*, 452. https://doi.org/10.3390/en9060452.
15. Zhuang, W.; Zhang, X.; Peng, H.; Wang, L. Simultaneous Optimization of Topology and Component Sizes for Double Planetary Gear Hybrid Powertrains. *Energies* **2016**, *9*, 411. https://doi.org/10.3390/en9060411.
16. Ning, T.; Gu, C.W.; Ni, W.D.; Li, X.T.; Liu, T.Q. Aerodynamic Analysis and Three-Dimensional Redesign of a Multi-Stage Axial Flow Compressor. *Energies* **2016**, *9*, 296. https://doi.org/10.3390/en9040296.
17. Sheng, S.; Sun, C. Control and Optimization of a Variable-Pitch Quadrotor with Minimum Power Consumption. *Energies* **2016**, *9*, 232. https://doi.org/10.3390/en9040232.
18. Rodrigues, S.; Restrepo, C.; Katsouris, G.; Teixeira Pinto, R.; Soleimanzadeh, M.; Bosman, P.; Bauer, P. A Multi-Objective Optimization Framework for Offshore Wind Farm Layouts and Electric Infrastructures. *Energies* **2016**, *9*, 216. https://doi.org/10.3390/en9030216.

energies

MDPI

Article

Hysteresis Analysis and Control of a Metal-Polymer Hybrid Soft Actuator

Manuel Schimmack [1], Eduardo E. Feistauer [2], Sergio T. Amancio-Filho [2,3] and Paolo Mercorelli [1,*]

1 Institute of Product and Processinnovation, Leuphana University of Lueneburg, Volgershall 1, D-21339 Lueneburg, Germany; schimmack@uni-leuphana.de
2 Institute of Materials Research, Materials Mechanics, Solid State Joining Processes, Centre for Materials and Coastal Research, Helmholtz-Zentrum Geesthacht, Max-Planck-Str. 1, D-21502 Geesthacht, Germany; feistauer@hzg.de (E.E.F.); amancio@hzg.de (S.T.A.-F.)
3 Institute of Polymer Composites, Hamburg University of Technology, Denicke Str. 15, D-21073 Hamburg, Germany; sergio.amancio@tuhh.de
* Correspondence: mercorelli@uni-leuphana.de; Tel.: +49-4131-677-5571; Fax: +49-4131-677-5300

Academic Editor: Chunhua Liu
Received: 31 October 2016; Accepted: 13 March 2017; Published: 8 April 2017

Abstract: The number of applications of stimulus-responsive polymers is growing at an impressive rate. The motivation of this contribution is to use a commercially available low-budget silver-coated polyamide (PA6) as a thermo-responsive metal-polymer hybrid soft actuator. Polyamide is a hygroscopic polymer; therefore, its mechanical and physical-chemical properties are affected by exposition to humidity or immersion in water. The effect of water absorption content on the PA6 and silver-coated PA6 monofilament properties, such as mass change and resistance, were evaluated. Moreover, the influence of swelling and shrinking effects on the surface morphology, caused by variations of moisture and water immersion, was investigated. Based on these variations, the dynamics of the resistance of the hybrid material were analyzed in the context of the proposed hysteresis model. An identification procedure of the hysteresis is presented along with an approximation of the upper and lower bound based on a constrained least square approach. A switching logic algorithm for this hybrid dynamic system is introduced, which makes it possible to structure the non-linear function in a switching mode. Finally, a non-linear integral sliding manifold is proposed and tested to control the resulting force of the actuator.

Keywords: soft actuator; metal-polymer hybrid material; switching system

1. Introduction

Stimulus-responsive polymers are relevant for the realization of smart actuators. The term smart, in this context, indicates the actuator ability to both sense and actuate. Polymer fiber-based actuators are also called soft actuators because of their low rigidity. The conductive properties that polymer fibers offer after coating make them suitable for a huge range of applications in biomedical engineering, such as conductors in sensors or bio-sensing textiles for healthcare [1,2]. Polymers that respond to different types of stimuli have become available, such as heat (thermo-responsive materials) [3], stress and pressure (mechano-responsive materials), electric current or voltage (electro-responsive materials) [4], magnetic fields (magneto-responsive materials), changes in pH or moisture (chemo-responsive materials) and light (photo-responsive materials) [5]. The number of proposed applications is growing at an impressive rate. Recently, applications have been proposed where soft actuators are able to solve even complex problems in strategic sectors, such as bio-inspired robotics. For example, a soft actuator mimicking human esophageal peristalsis for a swallowing robot is presented in [6], where the use of soft material gives an opportunity to imitate the soft-bodied esophagus and its biological function. The deformation arising from the non-linear material behavior, the structure and the actuation system causes difficulties in modeling and control. The same type of actuators, pneumatic artificial muscles (PAMs), are used in [7]. Shape-memory alloys (SMA), such as Nitinol in the form of wires, are the most well-known artificial muscles. They undergo a phase change with temperature (or stress) that is responsible for actuation. In reference [8] the electroactive polymer actuators (EAPs) are discussed as an alternative. EAPs, as a class of smart and soft actuators, are ideal candidates for bioinspired mechatronic applications due to their physiological compliance and built-in actuation ability [9]. The complete development and analysis of a soft robotic platform based on antagonistic arrangement of circular and longitudinal muscle groups is presented in [10]. Furthermore, nanotubes drawn from a carbon multi-wall nanotube (MWNT) as artificial muscles are of practical interest, as presented in [11]. The authors showed repeatable contractions of these super-coiled polymer (SCP) actuators constructed from MWNT.

Metal-polymer hybrid fibers are the fundamental core for wearable technology, an integral part of smart textiles and have possible applications in aerospace, sports, military and energy fields [12]. Commercially produced synthetic polymers like polyethylene (PE) and polyamide (PA6.6) fibers are important artificial muscle precursors. They combine the ability of reversible contraction, large volumetric thermal expansion and large anisotropy in thermally-induced dimension changes to provide enhanced muscle stroke [13,14]. High performance robotic muscles from conductive PA6.6 sewing thread and the thermo-mechanical and thermo-electric models of the SCP actuators are presented in [15].

In reference [13], artificial muscles from fishing lines and sewing threads are presented as a possible substitute of commercial NiTi shape-memory metal wires, which are expensive and have strongly pronounced hysteresis. This type of current-voltage hysteresis makes them difficult to control because of the variability of its resistance. In particular, using soft actuators, the kinematic and dynamic model of the actuator is typically affected by uncertainties in the parameters, which are difficult to identify. Therefore, they need to be controlled in a robust way. Moreover, because of the current-voltage hysteresis effect, the resistance is a non-linear function, and the control of the system requires specific and more advanced methods.

Traditionally, quantitative feedback theory (QFT) and H_∞ methods are used to achieve robust control of linear systems. Recently, these two methods were also applied in the context of non-linear control through some adaptations. An approach in the context of H_∞, along with a linear model adaption around an operating point of a shape memory alloy actuator, is presented in [16]. A dielectric elastomer diaphragm actuator to replicate human pulse signals is controlled with H_∞ techniques in [17]. The flexible soft actuator proposed in this work is controlled using a sliding mode control (SMC) approach. The system model is formulated as a switching system. Through the introduced thickness δ of the proposed sliding surface, the switching nature of this approach is moderated. Thanks to its robust property, the sliding mode approach is one of the principal techniques used today for non-linear systems in the presence of disturbances and uncertainties. During the last few years, considerable efforts have been made in this direction. In [18], a sliding surface is adopted. This technique is based on a combination between a sliding surface and an adaptive pre-action. In [19], the authors describe a comparison of two stator-flux and speed observers for a variable structure in a machine drive. More recent literature considers many related aspects with control, monitoring and disturbances in terms of interference in actuators, which are minimized by using sliding mode control. In particular, in [20] a new monitoring system for removing non-linearities and unmodeled disturbances in piezo electric actuators is proposed. In [21], a high-gain sliding mode observer is designed to reduce the effect of disturbances and uncertainties. The suggested algorithm has the capacity to ensure a vanishing average error in estimation in a finite time, even in the presence of model uncertainties.

The structure of the work is the following. Section 2 shows some morphological features of the soft actuator facilitated by scanning electron microscopy (SEM). The influence of swelling and shrinking effects on the surface morphology, caused by variations of moisture and water immersion, is investigated. Based on these variations, the dynamics of the resistance of the silver-coated hybrid material were analyzed in the context of the proposed hysteresis model. The parameter identification and modeling of the soft actuator and its hysteresis are presented in Section 3. The approximation of the hysteresis curve is realized using a constrained least squares method. Based on the results, a switching logic algorithm for the hybrid system is proposed. Section 3.2 proposes a non-linear integral sliding approach. Section 4 provides a concise and precise description of the experimental results, with conclusions resolving the paper. The main goal of the research is to use a commercial low budget silver-coated polyamide (PA6) monofilament as a thermo-responsive metal-polymer hybrid soft actuator, which is stimulated with a defined power.

2. Experimental Procedure

2.1. Physical-Chemical Analysis

In this work, a thermally-expandable silver-coated polyamide 6 (PA6) monofilament (STATEX GmbH, Bremen, Germany) was selected as a metal-polymer hybrid soft actuator.

2.1.1. Surface Morphology Analysis

The monofilament morphology was observed by scanning electron microscopy (SEM). The SEM analyses were carried out with Quanta^{TM} FEG 650 equipment (FEI Company, Hillsboro, OR, USA) with a voltage of 10 kV and a working distance of approximately 15 mm. Before SEM analysis, the non-conductive PA6 monofilaments were sputtered with gold for 30 s with a current of 65 mA using Q150R ES equipment (Quorum Technologies Ltd., Singapore). As a result, a nanometric layer of gold is formed on the surface of the sample, making them conductive. Thus, it facilitates their imaging by SEM and the resolution of small surface features. In the case of silver-coated PA6 monofilament, due to the high conductivity of the silver coating, further gold coating was not required. Nevertheless, a lower voltage was set during SEM measurements in order to prevent charging of defective coating regions, where the PA6 was eventually exposed to the electron beam.

2.1.2. Water Immersion Test

Polyamide is a hygroscopic polymer; therefore, its mechanical and physical-chemical properties are affected by exposition to humidity or immersion in water. The effect of water absorption content on the PA6 monofilament properties, such as mass change and resistance, were evaluated by the immersion test. The test was based on the DIN EN ISO 62:2008.8 [22].

Before the water immersion test, the uncoated PA6 monofilament and the silver-coated monofilament were heated in an oven at 50.0 ± 2.0 °C for 24 ± 1 h to remove remaining absorbed water from the delivered samples. In this manner, both evaluated monofilaments have a water-free reference, where the mechanical and physical-chemical properties are not affected by the moisture content. After the drying process, the monofilaments cool down to the ambient temperature of 23.0 ± 2.0 °C with a relative humidity of $50.0\% \pm 5.0\%$. The initial mass m_1 was measured with a Sartorius scale (Model CPA225D-0CE, measurement precision of 10 µg). The next step involves immersing the monofilaments in a beaker filled with 300 mL distilled H_2O. After each of the immersion cycles, the samples were manually dried in a paper tissue to remove excess water. Following this step, the monofilaments were weighted and returned to the distilled H_2O. After 24 h, the H_2O concentration change (c_1) in relation to the initial mass (m_1) after one hour was calculated as:

$$c_1 = \frac{m_2 - m_1}{m_1} \cdot 100\% \tag{1}$$

and for the subsequent cycles follows as:

$$c_n = \frac{m_n - m_1}{m_1} \cdot 100\% \tag{2}$$

For the water absorption investigation, three replicates of the uncoated PA6 monofilaments with average initial mass of $m = 103.98\ \text{mg} \pm 0.66\ \text{mg}$ and three replicates of the silver-coated monofilaments with average initial mass of $m = 102.91\ \text{mg} \pm 1.03\ \text{mg}$ were used.

2.1.3. Thermogravimetric Analysis

Thermogravimetric analysis (TGA) was performed to study the changes in the physical and chemical properties of the monofilaments and to assess the maximum absorbed water content as a function of temperature with a constant heating rate. A TGA instrument, TG 209 F3 Tarsus (Netzsch GmbH, Selb, Germany), was used. After 24 h of water immersion, samples with masses of 21–26 mg were cut with a scalpel from the uncoated PA6 and from the silver-coated monofilaments. Heating temperature in the interval of 20–800 °C under a continuous nitrogen flux of 20 $\frac{\text{mL}}{\text{min.}}$ and a constant heating rate of 20$\frac{^\circ\text{C}}{\text{min.}}$ were applied and the change in the mass was continuously recorded.

2.2. Parameter Identification

2.2.1. Resistance Analysis

An important question to be answered in order to better understand the soft actuator response is related to the correlation between water content and resistance. For this analysis, a monofilament with an initial length of $l = 100$ mm was selected and submitted to three different experimental conditions, as shown in Table 1. The resistance was measured with a precision digital multimeter HP 34401A (Agilent Technologies, Santa Clara, CA, USA).

Table 1. Overview of the defined conditions to analyze the correlation between the water content and the resistance.

State 1	indicates the dried monofilaments, which were maintained in an oven at 50.0 ± 2.0 °C for 24 ± 1 h
State 2	indicates the "as received" monofilaments with stabilized ambient temperature at 23.0 ± 2.0 °C and a relative humidity of 50% ± 5%
State 3	indicates the monofilaments immersed in a beaker filled with 300 mL distilled H_2O 24 ± 1 h, maximum water absorption as previously described for the water immersion test

2.2.2. Hysteresis Analysis of Metal-Polymer Hybrid Material

A low noise B2961A precision source/measure unit (SMU, Agilent Technologies, Santa Clara, CA, USA) was used for the stimulation and characterization of the metal-polymer hybrid soft actuator. It consists of a DC current source, a DC voltage source and an arbitrary waveform generator, which presents 100 nV/10 fA resolution. The SMU allows a voltage-current correct measurement. The low noise measurements are guaranteed by triaxial 4-wire Kelvin probes, which eliminate the measurement error caused by a residual lead resistance. Considering that the electrical power is defined as:

$$P \equiv \frac{1}{T}\int_{t=0}^{t=T} p\ \mathrm{d}t = \frac{1}{T}\int_{t=0}^{t=T} i^2 R\ \mathrm{d}t \tag{3}$$

where the influence of the previously-analyzed resistor R of the monofilament is linear and the controlled current i is squared. The metal-polymer hybrid monofilament is connected to the triaxial 4-wire Kelvin probes and stimulated by the current $i(t)$. The soft actuator is loaded with a weight of mass m = 2.13 g while the stimulation process is carried out.

3. Metal-Polymer Hybrid Soft Actuator as a Hybrid Dynamic System

3.1. Switching Logic Algorithm

In this section, a switching logic algorithm for a hybrid dynamic system is proposed. Figure 1 shows the hysteresis curve with the upper bound voltage trajectory u_{upbo_m} and lower bound voltage trajectory u_{lobo_m} of the soft actuator in combination with the limitation of the input current i_{max} and the maximum input voltage u_{max} to protect the soft actuator.

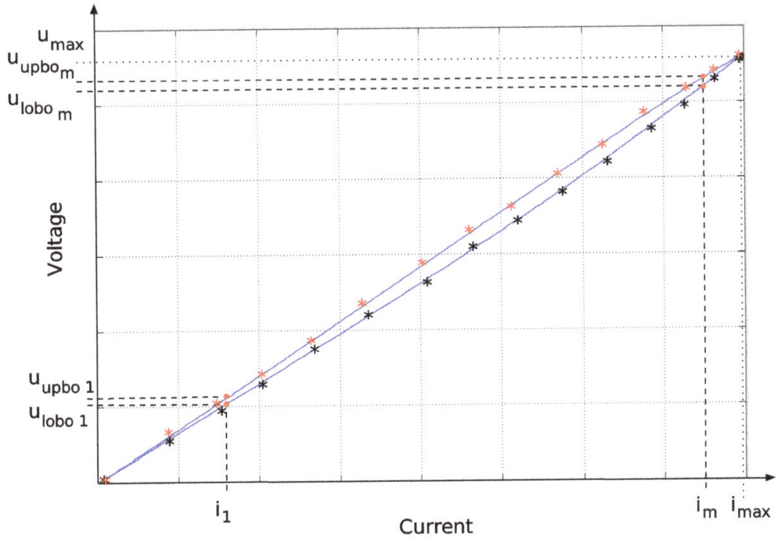

Figure 1. Hysteresis curve after fitting of the upper and lower bound of the metal-polymer hybrid soft actuator with the maximum input voltage u_{max} and the maximum input current i_{max}.

To find an approximate hysteresis curve using measured data, a fifth order polynomial is used. It represents, in the given interval, the upper and lower bound of input current $i(t)$ and the system voltage $u(t)$. Thus:

$$u(t) = a_6 + a_5 i(t) + a_4 i^2(t) + a_3 i^3(t) + a_2 i^4(t) + a_1 i^5(t) \tag{4}$$

where $a_6 = 0$ guarantees $u(t) = 0$ if $i(t) = 0$. The following constraint guarantees that i_{max} and u_{max} meet the approximating curve, which is depicted in Figure 1. The first coefficient of the polynomial holds:

$$a_1 = \frac{1}{i_{max}^5}\left(u_{max} - a_4 i_{max}^2 - a_3 i_{max}^3 - a_2 i_{max}^4 - a_5 i_{max}\right) \tag{5}$$

Inserting the coefficient a_1 in (4), it follows for the voltage that:

$$u(t) = a_5 i(t) + a_4 i^2 + a_3 i^3(t) + a_2 i^4(t) + \frac{i^5(t)}{i_{max}^5}\left(u_{max} - a_4 i_{max}^2 - a_3 i_{max}^3 - a_2 i_{max}^4 - a_5 i_{max}\right) \tag{6}$$

and:

$$u(t) = a_5\Big(1 - \frac{i^4(t)}{i^4_{\max}}\Big)i(t) + a_4\Big(1 - \frac{i^3(t)}{i^3_{\max}}\Big)i^2(t) + a_3\Big(1 - \frac{i^2(t)}{i^2_{\max}}\Big)i^3(t) + a_2\Big(1 - \frac{i(t)}{i_{\max}}\Big)i^4(t) + \frac{u_{\max} i^5(t)}{i^5_{\max}} \tag{7}$$

Considering the already discussed constrained least squares method in the frame of m measurement data, the following matrix structure is derived in the general way for the lower and for the upper bound:

$$\begin{bmatrix} u_1 - \frac{u_{\max} i_1^5}{i^5_{\max}} \\ u_2 - \frac{u_{\max} i_2^5}{i^5_{\max}} \\ \vdots \\ u_m - \frac{u_{\max} i_m^5}{i^5_{\max}} \end{bmatrix} = \begin{bmatrix} \left(1 - \frac{i_1}{i_{\max}}\right) i_1^4 & \left(1 - \frac{i_1^2}{i^2_{\max}}\right) i_1^3 & \left(1 - \frac{i_1^3}{i^3_{\max}}\right) i_1^2 & \left(1 - \frac{i_1^4}{i^4_{\max}}\right) i_1 \\ \left(1 - \frac{i_2}{i_{\max}}\right) i_2^4 & \left(1 - \frac{i_2^2}{i^2_{\max}}\right) i_2^3 & \left(1 - \frac{i_2^3}{i^3_{\max}}\right) i_2^2 & \left(1 - \frac{i_2^4}{i^4_{\max}}\right) i_2 \\ \vdots & \vdots & \vdots & \vdots \\ \left(1 - \frac{i_m}{i_{\max}}\right) i_m^4 & \left(1 - \frac{i_m^2}{i^2_{\max}}\right) i_m^3 & \left(1 - \frac{i_m^3}{i^3_{\max}}\right) i_m^2 & \left(1 - \frac{i_m^4}{i^4_{\max}}\right) i_m \end{bmatrix} \begin{bmatrix} a_2 \\ a_3 \\ a_4 \\ a_5 \end{bmatrix} \tag{8}$$

The main idea for the implementation of the switching logic algorithm is to prevent the system from going into saturation. Figure 2 shows the general block diagram of the switching logic algorithm for hybrid systems and its conditions for the lower and upper bound.

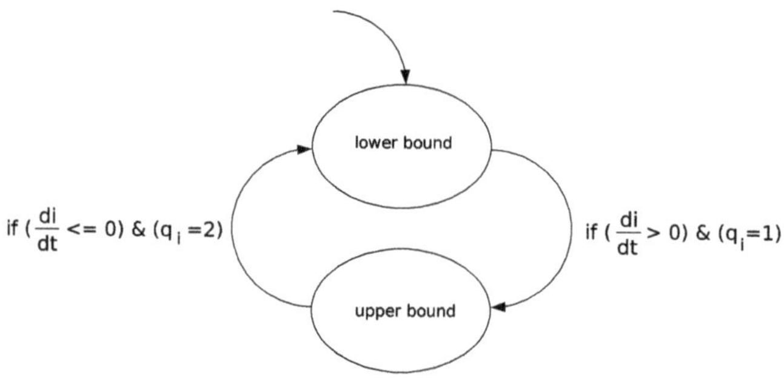

Figure 2. General block diagram of the switching logic algorithm for hybrid systems and its conditions.

3.2. *A Constructive Sufficient Condition for Stability*

Consider the current dynamic of the system as follows:

$$\frac{\mathrm{d}i(t)}{\mathrm{d}t} = \frac{-f_i\left(i(t), \frac{\mathrm{d}i(t)}{\mathrm{d}t}, t\right) + u(t) + \mathrm{d}(t)}{L} \tag{9}$$

and the resulting force dynamic of the system as:

$$F(t) = f_f\left(i(t), \frac{\mathrm{d}i(t)}{\mathrm{d}t}, t\right) \tag{10}$$

where $i(t)$ represents the current as the state of the system, the voltage $u(t)$ is a controlled input and $d(t)$ a bounded external voltage disturbance $|d(t)| < D$ with $D \in \mathbb{R}$.

The model consist of two cascade hysteresis f_i and f_f of (9) and (10). The two algebraic non-linear functions are structured in a switching mode, with a current-voltage switching parameter q_i as follows:

$$f_i\left(i(t), \frac{di(t)}{dt}, t\right) = \begin{cases} f_{i_{\text{upper}}}\left(i(t), \frac{di(t)}{dt}, t\right) & if \left(\frac{di}{dt} > 0\right) \& (q_i = 1) \\ f_{i_{\text{lower}}}\left(i(t), \frac{di(t)}{dt}, t\right) & if \left(\frac{di}{dt} <= 0\right) \& (q_i = 2) \end{cases} \tag{11}$$

and:

$$f_f\left(i(t), \frac{di(t)}{dt}, t\right) = \begin{cases} f_{f_{\text{upper}}}\left(i(t), \frac{di(t)}{dt}\right) & if \ q_i = 1 \\ f_{f_{\text{lower}}}\left(i(t), \frac{di(t)}{dt}\right) & if \ q_i = 2 \end{cases} \tag{12}$$

The effect of the current-force hysteresis is directly connected with the current-voltage hysteresis. With this approach, a non-linear integral sliding control is proposed to compensate this effect using a double integral surface:

$$S(t) = k_1 e(t) + k_2 \int_0^t e(\tau)d\tau + k_0 \int_0^t \left(\int_0^\tau e(\eta)d\eta\right) d\tau \tag{13}$$

where $e(t) = i_d(t) - i(t)$ with the desired current $i_d(t)$ and the obtained current $i(t)$. The constants k_0, k_1 and k_2 are positive.

Definition 1. *Define the following control law:*

$$u(t) = \hat{u}(t) + k_3 \cdot \text{sat}\left(\frac{S(t)}{\delta}\right) \tag{14}$$

where $\hat{u}(t)$ is the sliding equivalent input signal and δ is a constant that defines the thickness of the boundary layer and with which the chattering phenomenon can be limited. It follows for the saturation function that:

$$\text{sat}\left(\frac{S(t)}{\delta}\right) = \begin{cases} 1 & if \ \frac{S(t)}{\delta} \geq 1 \\ \frac{S(t)}{\delta} & if \ -1 < \frac{S(t)}{\delta} < 1 \\ -1 & if \ \frac{S(t)}{\delta} \leq -1 \end{cases} \tag{15}$$

Proposition 1. *Consider the system described in (9); when $|S(t)| > \delta$, then the following control law:*

$$u(t) = \hat{u}(t) + k_3 \cdot \text{sgn}(S(t)) \tag{16}$$

with the sliding equivalent input signal $\hat{u}(t)$ and the sign function sgn$(\cdot)$, *(16) stabilizes the sliding surface defined by (13), if the condition $k_3 < -|d(t)|$ holds.*

Proof 1. *Define the following candidate function:*

$$V(t) = \frac{1}{2}S^2(t) \tag{17}$$

considering that:

$$\dot{S}(t) = k_1 \cdot \dot{e}(t) + k_2 \cdot e(t) + k_0 \int_0^t e(\tau)d\tau \tag{18}$$

$$\begin{aligned}\dot{V}(t) &= S(t)\left(k_1 \cdot \dot{e}(t) + k_2 \cdot e(t) + k_0 \int_0^t e(\tau)d\tau\right) \\ &= S(t)\left(k_1\left(\frac{\mathrm{d}i_\mathrm{d}(t)}{\mathrm{d}t} - \frac{-f_i(\mathbf{x}(t),t) + u(t) + d(t)}{L}\right) + k_2 \cdot e(t) + k_0 \int_0^t e(\tau)\mathrm{d}\tau\right)\end{aligned} \tag{19}$$

With (16), the following expression is obtained:

$$S(t)\left(k_1\left(\frac{\mathrm{d}i_d(t)}{\mathrm{d}t} - \frac{-f_i(\mathbf{x}(t),t) + \hat{u}(t) + k_3 \cdot \mathrm{sgn}(S(t)) + d(t)}{L}\right) + k_2 \cdot e(t) + k_0 \int_0^t e(\tau)\mathrm{d}\tau\right) \tag{20}$$

Choosing:

$$\hat{u}(t) = f_i(\mathbf{x}(t),t) + L\frac{\mathrm{d}i_\mathrm{d}(t)}{\mathrm{d}t} + L\frac{k_2}{k_1}e(t) + L\frac{k_0}{k_1}\int_0^t e(\tau)\mathrm{d}\tau \tag{21}$$

it follows:

$$\dot{V}(t) = S(t)\frac{1}{L}\left(k_3 \cdot \mathrm{sgn}(S(t)) + d(t)\right) = \frac{1}{L}\left(k_3|S(t)| + S(t) \cdot d(t)\right) = \frac{1}{L}\left(|S(t)|k_3 + S(t) \cdot d(t)\right) \tag{22}$$

If $k_3 < -|d(t)|$, *then* $\dot{V}(t) < \frac{k_3}{L}|S(t)| < 0$.

4. Results and Discussion

In this section, the influence of swelling and shrinking effects on the surface morphology, caused by variations of moisture and water immersion, is investigated. Moreover, based on these variations, the dynamics of the resistance of the hybrid material were analyzed in the context of the proposed hysteresis model.

4.1. Physical-Chemical Analysis

4.1.1. Surface Morphology Analysis

Figure 3a shows the uncoated PA6 monofilament (State 1, Table 1), and Figure 3b depicts a detailed SEM image of the uncoated fiber surface morphology in the same condition. It is possible to see extrusion defects (marked with white arrows in Figure 3a) and surface ripples (marked with white arrows in Figure 3b). The silver-coated PA6 monofilament is depicted in Figure 4a and the details of its surface in Figure 4b. Defects of the surface finishing are visible (marked with white arrows in Figure 4a), and a line direction based on the manufacturing process is also seen (marked with red arrows in Figure 4a). Furthermore, in the detailed surface, cracks and rough surface ripples are present (marked with white arrows in Figure 4b). An overview of the morphology of the silver-coated PA6 monofilament in State 1 is depicted in Figure 5a and the details of the silver-coating surface in Figure 5b after the stimulation. Defects of the surface finishing are visible (marked with white arrows). It is possible to see, based on the orientation of the marked spots (marked with white arrows), that there is no visible changing of the surface finishing after the stimulation of the fiber. Similar results were obtained for the monofilament stimulated in State 3, as shown in Figure 6. The stimulation process does not change the morphological features of the monofilament, as observed through the comparison of defect orientation and distribution (marked with white arrows) before and after the stimulation (see Figure 6a,b in the same sequence). Therefore, it is suggested that defects of the surface finishing are visible (marked with white arrows). It is possible to see, based on the orientation of the marked spots (marked with white arrows), that there is also no visible changing of the surface finishing after the stimulation of the fiber. The dark area in the picture is the distilled water.

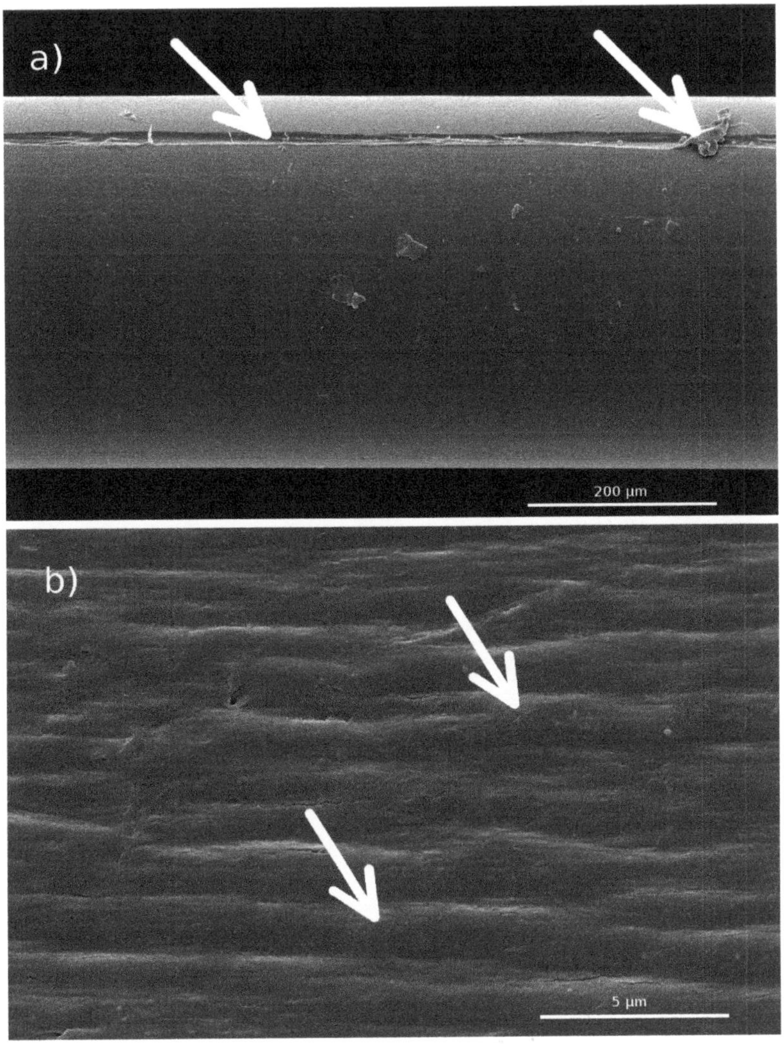

Figure 3. (**a**) Overview of the morphology of the uncoated PA6 monofilament in State 2 and (**b**) details of the monofilament surface morphology presented in (**a**).

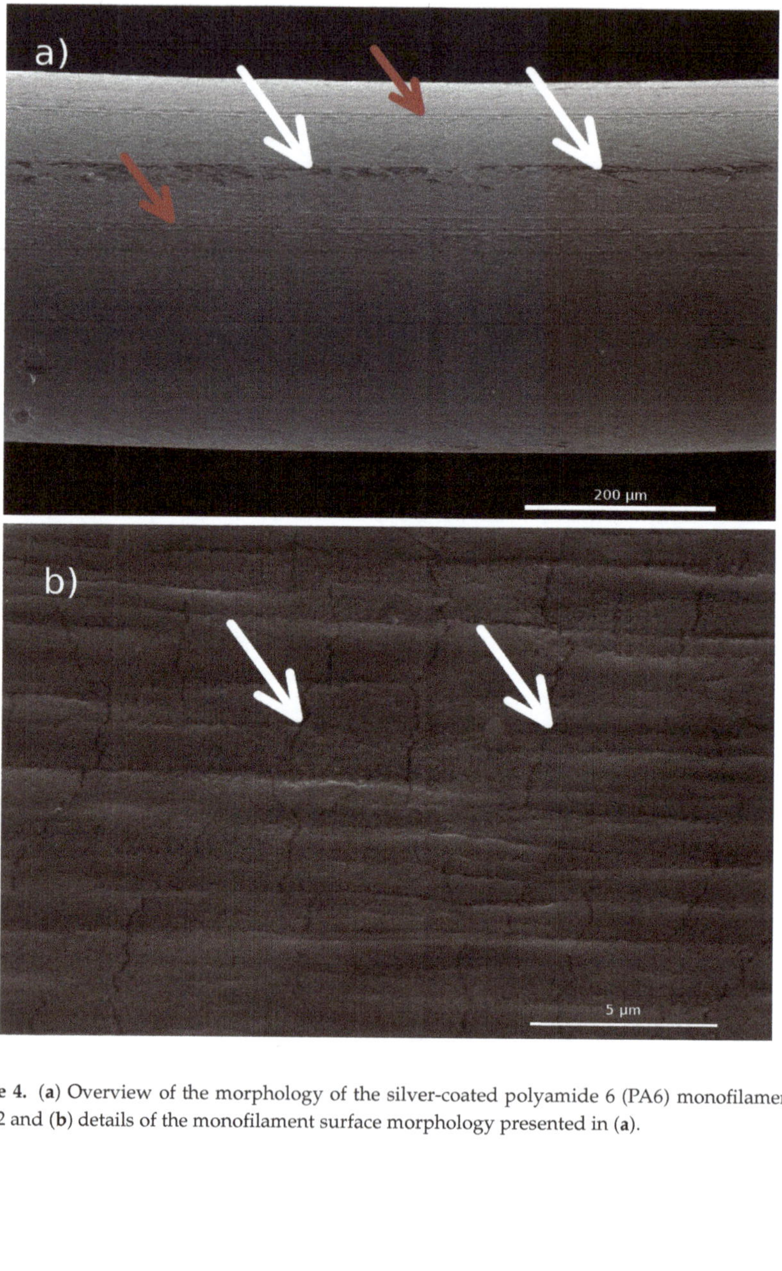

Figure 4. (**a**) Overview of the morphology of the silver-coated polyamide 6 (PA6) monofilament in State 2 and (**b**) details of the monofilament surface morphology presented in (**a**).

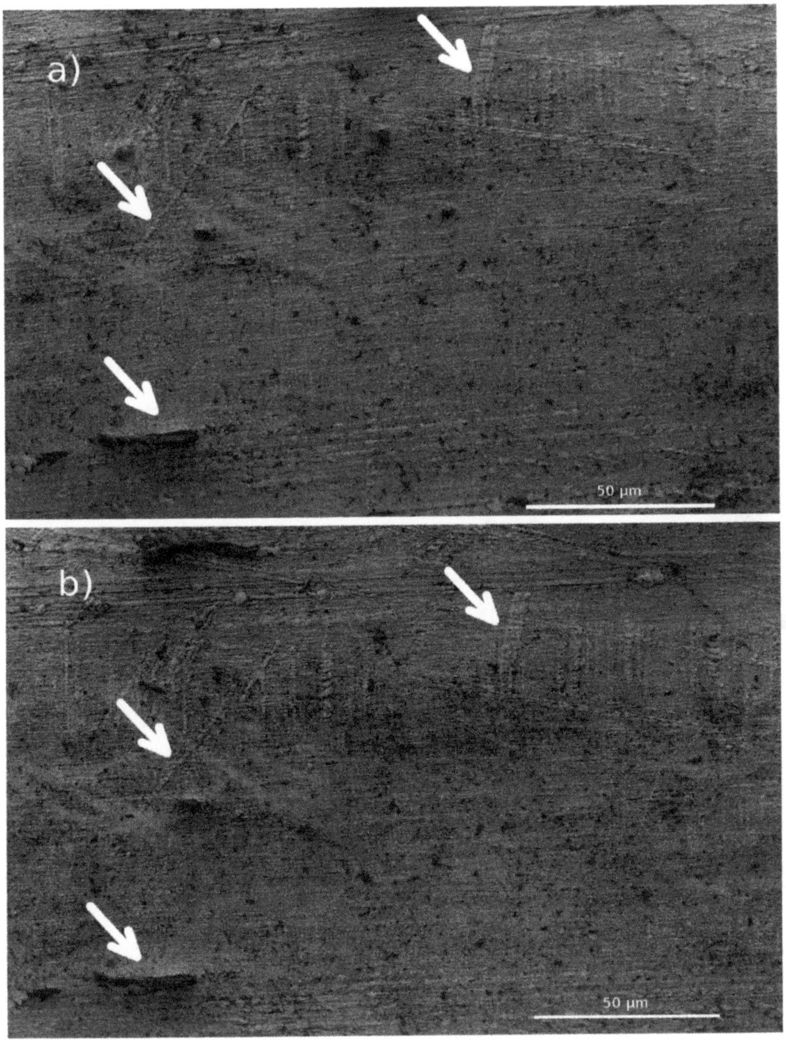

Figure 5. (**a**) Overview of the morphology of the silver-coated PA6 monofilament in State 1 and (**b**) details of the monofilament surface morphology presented in (**a**) after the stimulation.

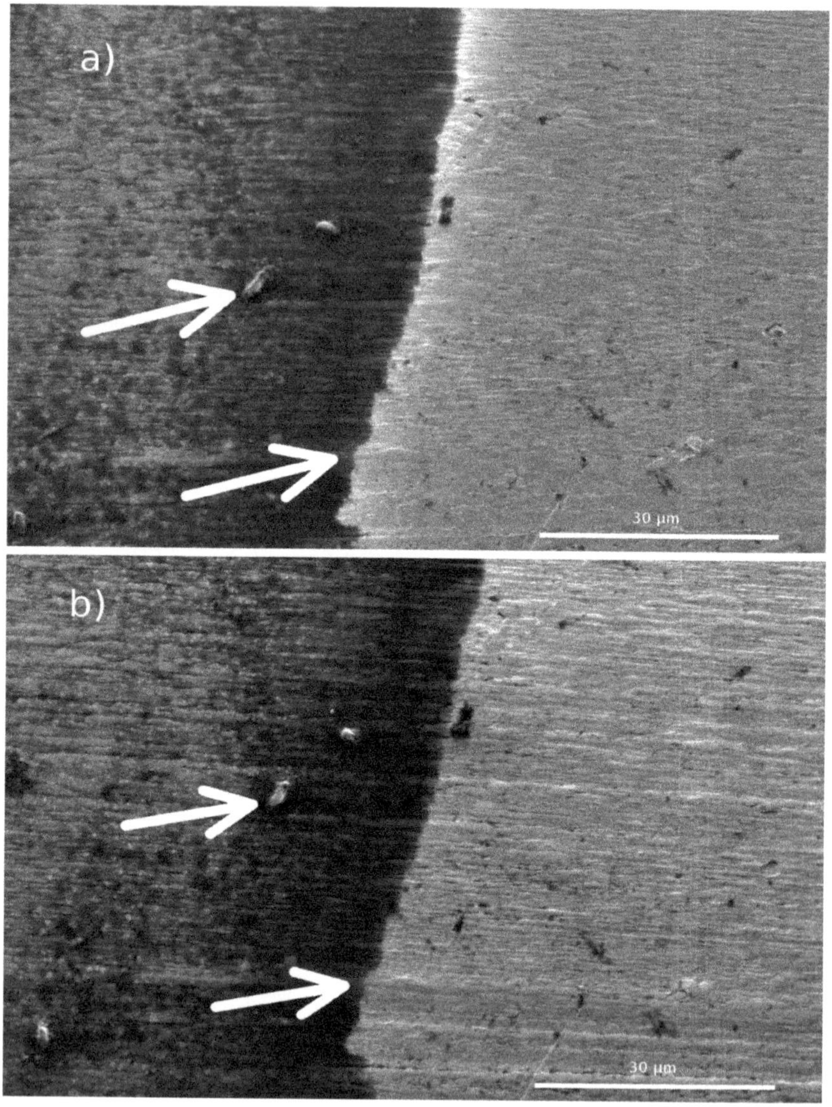

Figure 6. (**a**) Overview of the morphology of the silver-coated PA6 monofilament in State 3 and (**b**) details of the monofilament surface morphology presented in (**a**) after the stimulation.

4.1.2. Water Immersion Test

Figure 7 shows a graph of the measured water absorption over time. The water absorption saturation point was observed after 20 h. The uncoated polyamide monofilament absorbed $5.97\% \pm 0.10\%$ of water, while the silver-coated monofilament $5.93\% \pm 0.15\%$ (measurement at 24 h). Therefore, based on the measurements of absorption and its evaluation together with the graphical representations, it can be concluded that the filaments behave very similarly regarding the absorption of water, and the silver-coated monofilament has no effect (or not a detectable one with this experimental setup).

4.1.3. Thermogravimetric Analysis

Figure 8 shows the thermogravimetric curves of the uncoated PA6 and silver-coated PA6 monofilament after 24 h of water immersion test (State 3). The curve is characterized by the occurrence of two weight loss steps. The first step, between 20 and 200 °C, is formed due to the water evaporation. The measured percent weight losses for both filaments were similar, $\Delta m = 6.28\%$ and $\Delta m = 6.34\%$ for the uncoated PA6 and silver-coated PA6 monofilament, respectively. These values are in agreement with the previous presented water absorption test and correspond to the maximum water content absorbed by the filaments. Moreover, as previously observed, the silver coating does not affect the water absorption mechanism of the PA6 filaments. The second step observed in the TGA thermogram, with the onset at 240 °C, is a result of the final thermo-degradation of PA6, and at this stage, the remaining mass of polymer is lost.

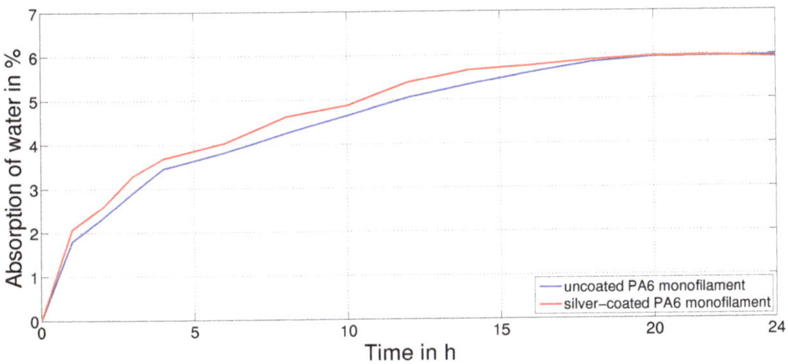

Figure 7. Water concentration after immersion-testing in distilled water of the uncoated PA6 monofilament and the silver-coated PA6 monofilament.

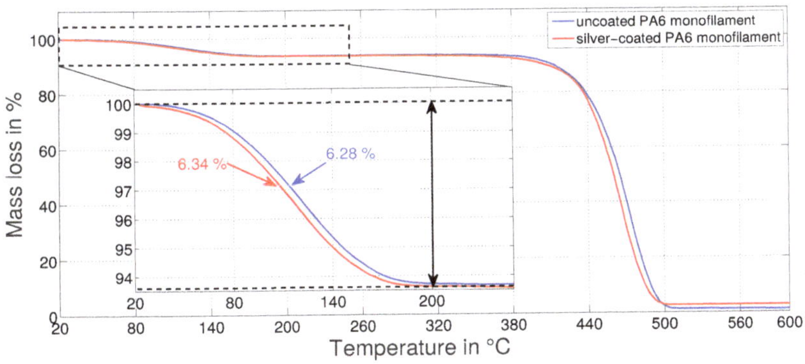

Figure 8. Results of the thermogravimetric analysis (TGA) of the uncoated PA6 in State 3.

4.2. Parameter Identification

4.2.1. Analysis of the Resistance

Figure 9a shows the measured resistance in the case of States 1, 2 and 3 with the standard deviation (σ_x) of the metal-polymer hybrid monofilament with a defined length (l = 100 mm). Figure 9b shows the results of the weight measurements in the same states. As expected, by drying the filaments, a significant decrease in resistance was observed in comparison with States 2 and 3. Therefore, the water content must be taken into consideration when using it as a soft actuator. Ongoing studies are being carried out to better understand this phenomenon. Moreover, one can observe an increase in the mass from State 1 to 3, which can be explained by the water absorbed during the immersion test. Finally, it can be observed that the swelling effect based on the water absorption of the polymer still has no relevant influence on the resistance.

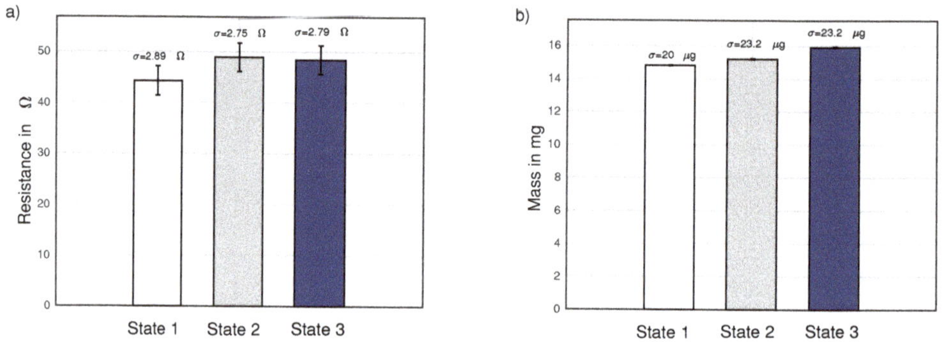

Figure 9. Graphical representation of the measurements of (**a**) the resistance R_n and (**b**) the mass m_n in States 1 to 3 of the soft actuator.

Figure 10a shows the results of the resistance measurements in the case of State 1 after the stimulation process of the samples (which is done by applying an electrical current to heat the metal-polymer hybrid actuator). Figure 10b represents the results of the weight measurements in State 2 and the aforementioned stimulation. It is possible to observe that there is a relevant decrease of about 16.6% in the resistance for the coated monofilament. However, no variations in mass were observed.

In the previous experimental setup, the effect of the stimulation of State 2 samples was analyzed. The initial state is an intact coating of the monofilament before drying or stimulating. Figure 11a shows the compiled results of the resistance measurements in the mentioned states. The results of the weight measurements under the same states are depicted in Figure 11b. Once more, an accentuated decrease for State 1 samples of 16.4% in the resistance after stimulation of the coated monofilament (Figure 11a) is detected, while no variation of the mass happened.

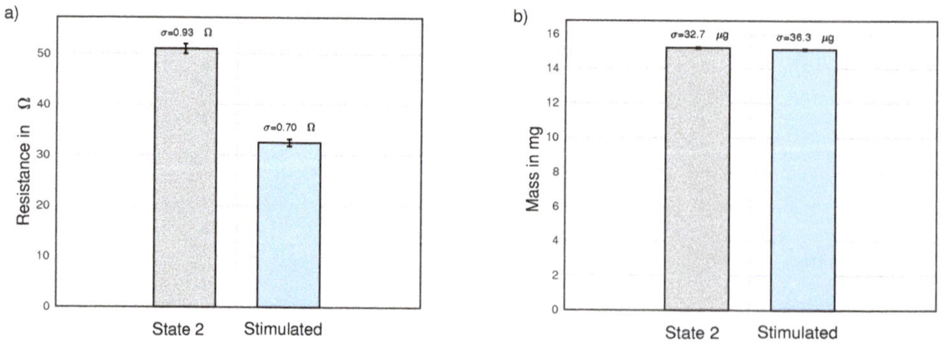

Figure 10. Graphical representation of the measurements of (**a**) the resistance R_n and (**b**) the mass m_n in State 2 and after stimulation of the soft actuator.

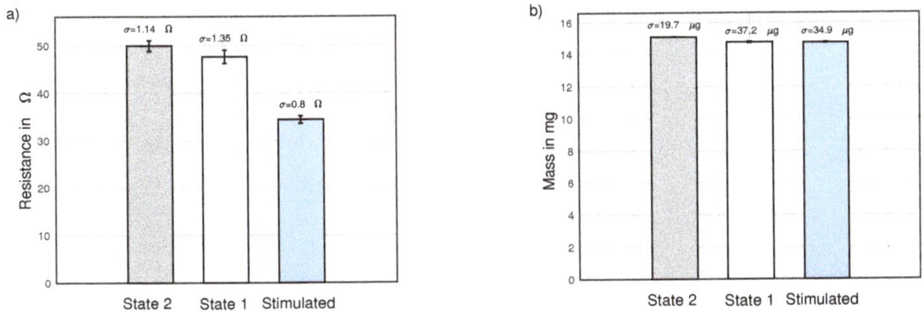

Figure 11. Measurements of (**a**) the resistance R_n and (**b**) the mass m_n in case of the initial States 2 to 1 and after the stimulation process of the replicates.

4.2.2. Hysteresis Analysis of Metal-Polymer Hybrid Material

Figure 12a shows the applied input current $i(t)$ as the stimulation signal. The results of the measured power $P(R)$ by the precision SMU are presented in Figure 12b. The metal-polymer hybrid monofilament is connected to the triaxial four-wire Kelvin probes and stimulated by the current $i(t)$. The soft actuator is loaded with a weight of mass m = 2.13 g, while the stimulation process is carried out. Figure 13 shows a considerable variation of the current-voltage hysteresis after the first cycle. This hysteresis curve is stabilized and remains unchanged during the next stimulation cycles. Moreover, the fiber does not return to its original (twisted) configuration.

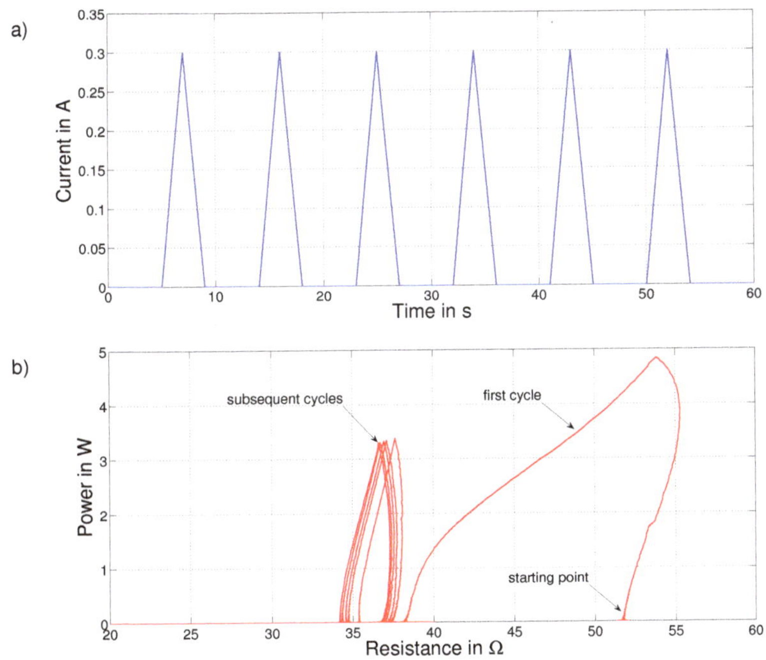

Figure 12. (**a**) The input current $i(t)$ as stimulation signal and (**b**) the power $P(R)$ measured by the precision source/measure unit (SMU).

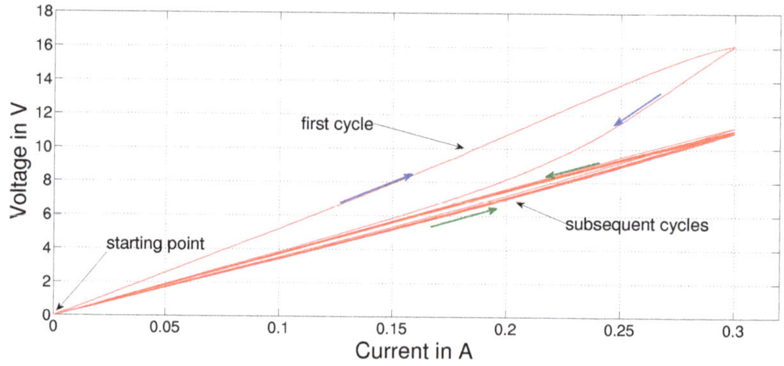

Figure 13. Graphical representation of the measured current-voltage hysteresis of the soft actuator and its upper and lower bounds.

4.3. Switching Logic Algorithm for Hybrid Dynamic System and a Non-Linear Integral Sliding Mode

A switching logic algorithm was developed and successfully implemented, based on the previous results of the parameter identification, which makes it possible to structure the non-linear function in switching mode. Finally, the proposed non-linear integral sliding manifold is implemented and tested to control the resulting force of the soft actuator. Figure 14 shows the disturbance $d(t)$ with a peak-to-peak amplitude of 100 mV. Figure 15 shows the result of the desired and obtained current in the presence of the disturbance $d(t)$. It is possible to see that the dynamics of the obtained current consist of peaks in the amplitude around 10 mA. This is due to the switching control signal of the sliding approach; see Figure 16. The thickness of the boundary layer of the saturation function was chosen as ($\delta = 0.1$) to guarantee a limit of the chattering phenomenon. In general, thanks to the parameter δ, a compromise between robustness against disturbances (uncertainties) and chattering suppression can be found. The result of the desired and obtained force $f(t)$ of the soft actuator is shown in Figure 17.

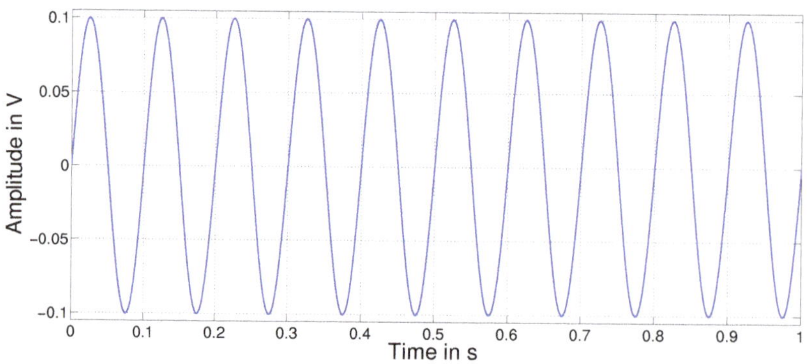

Figure 14. Graphical representation of the voltage disturbance $d(t)$.

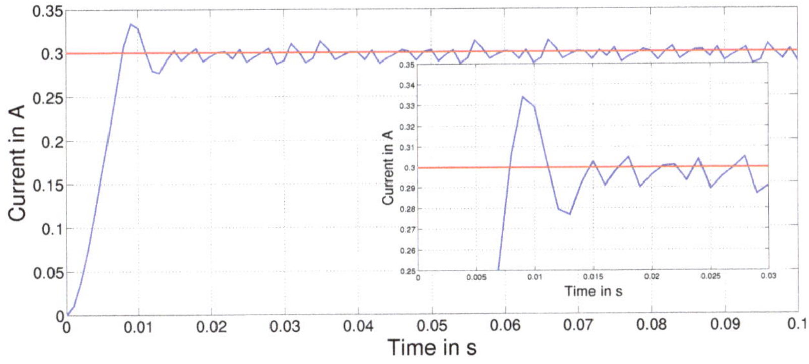

Figure 15. Graphical representation of the desired and obtained current.

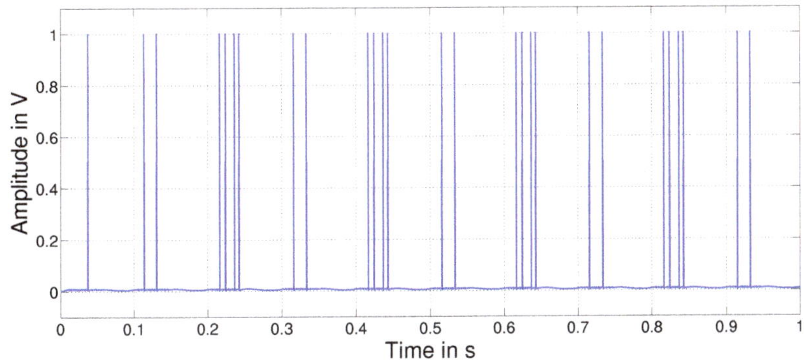

Figure 16. Graphical representation filtered switching variable of the sliding approach.

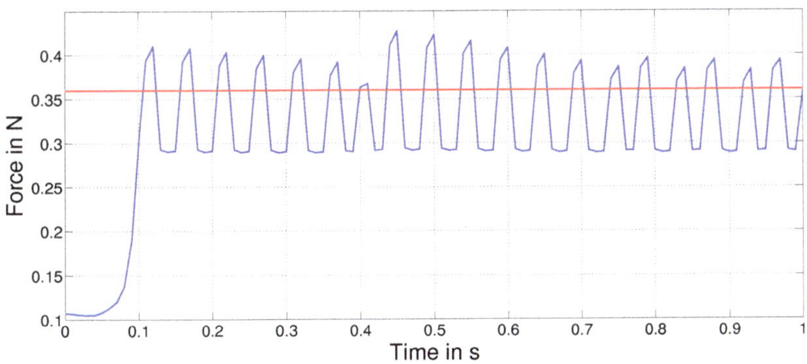

Figure 17. Graphical representation of the controlled force f.

5. Conclusions

In this work, a thermally-expandable silver-coated PA6 monofilament was selected as a metal-polymer hybrid soft actuator. The main goal of the research was to use a commercial low budget silver-coated PA6 as a thermo-responsive metal-polymer hybrid soft actuator, which was stimulated with a defined power. Polyamide is a hygroscopic polymer, and the mentioned monofilament was examined under different environmental conditions, more precisely its interaction with water. Through an analysis of the water absorption and of both the coated and uncoated monofilament, it was determined whether or not a coated filament is able to store water. Moreover, the influence of swelling and shrinking effects on the surface morphology, caused by variations of moisture and water immersion, was investigated. Results indicate that the silver coating has no effect (or not a detectable one with the experimental setup) in water desorption.

As expected, the drying process decreased considerably the resistance of the silver-coated filament in comparison with the "as received" (State 2) and immersed in distilled water (State 3). Therefore, the water content must be taken into consideration when applying this PA6 polymer-metal-based soft actuators. Moreover, one can observe an increase in the mass from State 1 to 3, which can be explained by the water absorbed during the immersion test. Finally, it was observed that the swelling effect based on the water content of the polymer still has no relevant influence on the resistance. Based on these variations, the dynamics of the resistance of the hybrid material were analyzed in the context of the proposed hysteresis model.

An identification procedure of the hysteresis was presented along with an approximation of the upper and lower bound based on the constrained least square approach. A switching logic algorithm for this hybrid dynamic system was developed and successfully implemented, which makes it possible to structure the non-linear function in the switching mode. Finally, a non-linear integral sliding manifold was proposed and tested to control the resulting force of the actuator.

Acknowledgments: The authors would like to acknowledge the financial support provided to Eduardo E. Feistauer by the *Conselho Nacional de Desenvolvimento Científico e Tecnológico* – CNPq (Brazil) and the financial support of the Helmholtz Association through the Young Investigator Group, "Advanced Polymer Metal Hybrid Structures" (Grant no. VH-NG-626). The authors thank Michel L. da Costa for his support in preparing the simulations.

Author Contributions: Paolo Mercorelli conceived the control strategies and the structure of the manuscript; Manuel Schimmack conceived the switching model of the system and the logic algorithm. Together with Paolo Mercorelli he worked on the constructive sufficient condition to obtain the convergence constraints for the control strategy. He designed and performed most of the experiments, analyzed all experimental data and at the end he wrote the manuscript; Eduardo E. Feistauer and Sergio T. Amancio-Filho contributed with polymer-metal fiber characterization (SEM, TGA, water content); All the authors contributed in the preparation, discussions of experimental results and review of the manuscript.

Conflicts of Interest: The authors declare no conflict of interest.

Abbreviations

SEM	Scanning electron microscope
SMU	Source/measure unit
TGA	Thermogravimetric analysis
EAP	Electro-active polymers
PE	Polyethylene
SMA	Shape-memory alloy

Nomenclature

a_m	Polynomial coefficient
$e(t)$	Error
$d(t)$	Voltage disturbance
q_i	Switching parameter
k_0	Constant
k_1	Constant
k_2	Constant
k_3	Constant
$i(t)$	Input current
$i_{\mathrm{d}}(t)$	Desired current
i_{max}	Limited current
$u(t)$	Input voltage
$\hat{u}(t)$	Sliding equivalent input signal
$u_{\mathrm{upbo_m}}$	Upper bound voltage trajectory
$u_{\mathrm{lobo_m}}$	Lower bound voltage trajectory
R_n	Resistance
L	Inductance
S(t)	Sliding surface
L	Liter
l	Length
m_x	Mass
δ	Thickness of the boundary layer
μ_x	Mean value
σ_x	Standard deviation

References

1. Carpi, F.; De Rossi, D. Electroactive polymer-based devices for e-textiles in biomedicine. *IEEE Trans. Inf. Technol. Biomed.* **2005**, *9*, 295–318.
2. Schimmack, M.; Hand, A.; Mercorelli, P.; Georgiadis, A. Using a seminorm for wavelet denoising of sEMG signals for monitoring during rehabilitation with embedded orthosis system. In Proceedings of the IEEE MeMeA—International Symposium on Medical Measurements and Applications, Torino, Italy, 7–9 May 2015; pp. 1773–1778.
3. Schimmack, M.; Mercorelli, P. *An Extended Kalman Filter as an Observer in a Sliding Mode Controller for a Metal-Polymer Composite Actuator*; Springer: Cham, Switzerland, 2005; Volume 427, pp. 305–314.
4. Kruusamäe, K.; Punning, A.; Aabloo, A.; Asaka, K. Self-sensing ionic polymer actuators: A review. *Actuators* **2015**, *4*, 17.
5. Hilber, W. Stimulus-active polymer actuators for next-generation microfluidic devices. *Appl. Phys. A* **2016**, *122*, 751.
6. Chen, F.J.; Dirven, S.; Xu, W.L.; Li, X.N. Soft actuator mimicking human Esophageal peristalsis for a swallowing robot. *IEEE/ASME Trans. Mech.* **2014**, *19*, 1300–1308.
7. Kodama, T.; Okabe, A.; Kogiso, K. Simultaneous estimation of contraction ratio and parameter of McKibben pneumatic artificial muscle model using log-normalized unscented Kalman Filter. In Proceedings of the 2016 IEEE 4th International Conference on Cyber-Physical Systems, Networks, and Applications (CPSNA), Aichi, Japan, 6–7 October 2016; pp. 44–48.
8. Smela, E. Conjugated polymer actuators for biomedical applications. *Adv. Mater.* **2003**, *15*, 481–494.
9. Mutlu, R.; Alici, G.; Li, W. A soft mechatronic microstage mechanism based on electroactive polymer actuators. *IEEE/ASME Trans. Mech.* **2016**, *21*, 1467–1478.
10. Seok, S.; Onal, C.D.; Cho, K.J.; Wood, R.J.; Rus, D.; Kim, S. Meshworm: A peristaltic soft robot with antagonistic nickel titanium coil actuators. *IEEE/ASME Trans. Mech.* **2013**, *18*, 1485–1497.
11. Lima, M.D.; Li, N.; de Andrade, M.J.; Fang, S.; Oh, J.; Spinks, G.M.; Kozlov, M.E.; Haines, C.S.; Suh, D.; Foroughi, J.; et al. Electrically, chemically, and photonically powered torsional and tensile actuation of hybrid carbon nanotube yarn muscles. *Science* **2012**, *338*, 928–932.
12. Schimmack, M.; McGaw, D.; Mercorelli, P. Wavelet based fault detection and RLS parameter estimation of conductive fibers with a simultaneous estimation of time-varying disturbance. *IFAC Symp. Inf. Control Probl. Manuf.* **2015**, *48*, 1773–1778.

13. Haines, C.; Lima, M.; Li, N.; Spinks, G.M.; Foroughi, J.; Madden, J.W.; Kim, S.H.; Fang, S.; de Andrade, M.J.; Göktepe, F.; et al. Artificial muscles from fishing line and sewing thread. *Science* **2014**, *343*, 868–872.
14. Ramazani, S.; Morshed, M.; Ghane, M. Effect of service temperature on structure and mechanical properties of polyamide 6 & 66 tyre cords. *J. Polym. Res.* **2011**, *18*, 781–792.
15. Yip, M.C.; Niemeyer, G. High-performance robotic muscles from conductive nylon sewing thread. In Proceedings of the 2015 IEEE International Conference on Robotics and Automation (ICRA), Seattle, WA, USA, 26–30 May 2015; pp. 2313–2318.
16. Jayender, J.; Patel, R.V.; Nikumb, S.; Ostojic, M. Modeling and control of shape memory alloy actuators. *IEEE Trans. Control Syst. Technol.* **2008**, *16*, 279–287.
17. Ye, Z.; Chen, Z.; Kong, K.W.; Chan, H. Robust control of dielectric elastomer diaphragm actuator for replicating human pulse. In Proceedings of the 2016 IEEE International Conference on Automation Science and Engineering (CASE), Fort Worth, TX, USA, 21–25 August 2016; pp. 188–193.
18. Mercorelli, P. An anti-saturating adaptive pre-action and a slide surface to achieve soft landing control for electromagnetic actuators. *IEEE/ASME Trans. Mech.* **2012**, *17*, 76–85.
19. Xu, Z.; Rahman, M.F. Comparison of a sliding observer and a Kalman filter for direct-torque-controlled IPM synchronous motor drives. *IEEE Trans. Ind. Electron.* **2012**, *59*, 4179–4188.
20. Xu, Q. Digital sliding-mode control of piezoelectric micropositioning system based on input-output model. *IEEE Trans. Ind. Electron.* **2014**, *61*, 5517–5526.
21. Mercorelli, P. A two-stage sliding-mode high-gain observer to reduce uncertainties and disturbances effects for sensorless control in automotive applications. *IEEE Trans. Ind. Electron.* **2015**, *62*, 5929–5940.
22. *ISO 62:2008 Plastics—Determination of Water Absorption*; International Organization for Standardization (ISO): Geneva, Switzerland, **2008**; pp. 1–17.

energies

MDPI

Article

Research on Optimal Planning of Access Location and Access Capacity of Large-Scale Integrated Wind Power Plants

Hui Li [1,2], Gengyin Li [1], Siwei Liu [2,*], Yuning Wang [3], Zhidong Wang [2], Jiaming Wang [2] and Ning Zhang [4]

1 State Key Laboratory of Alternate Electrical Power System with Renewable Energy Sources, North China Electric Power University Changping District, Beijing 102206, China; lihui@chinasperi.sgcc.com.cn (H.L.); ligy@ncepu.edu.cn (G.L.)
2 State Power Economic Research Institute, Changping District, Beijing 102209, China; wangzhidong@chinasperi.sgcc.com.cn (Z.W.); wangjiaming@chinasperi.sgcc.com.cn (J.W.)
3 State Grid Materials Co. Ltd., Xicheng District, Beijing 100120, China; wangyuning@sgm.sgcc.com.cn
4 State Key Lab of Control and Simulation of Power Systems and Generation Equipment, Department of Electrical Engineering, Tsinghua University, Haidian District, Beijing 100084, China; ningzhang@tsinghua.edu.cn
* Correspondence: liuswmax@163.com; Tel.: +86-138-1121-9687

Academic Editor: Paolo Mercorelli
Received: 13 December 2016; Accepted: 21 March 2017; Published: 1 April 2017

Abstract: This paper proposes a multi-objective optimal planning model of access location and access capacity for large-scale integrated wind power generation considering the mutual restriction between the planning of large-scale wind power plants and the planning of power system network. In this model, the power flow equilibrium degree, investment costs and active network loss are taken as the optimization goals. The improved differential evolution (IDE) algorithm is applied to calculate the Pareto optimal solution set of wind power's access planning. With the solution results described by the Pareto pattern, all the alternative solutions are then ranked based on the entropy weight method and the final compromised solution is selected by the method of technique for order preference by similarity to ideal (TOPSIS). And the proposed optimal planning model is tested based on a practical planning need of large-scale integrated wind power generation in an actual power grid of China in 2020. The simulation results show that applied with the proposed optimization model and matching algorithm, the planning scheme of large-scale wind power's access location and access capacity under complex and practical power system circumstances has been successfully optimized.

Keywords: large-scale wind power generation; power plant planning; power flow equilibrium degree; economic efficiency; Pareto optimal solution set; improved differential evolution (IDE) algorithm; technique for order preference by similarity to ideal (TOPSIS) based non-poor sorting

1. Introduction

In the northwest and northeast regions of China, which are rich in wind energy resources, the wind power generation has experienced rapid development in recent years, which brought a need of optimal planning of the access location and access capacity of large-scale integrated wind power plants to improve wind power consumption and to ensure the power system safely operation [1,2].

Power plant planning aims to solve the issue about when and where to build what kind of power plants with an overall consideration of power grid conditions, such as grid network structure, load development, power system reliability, economic efficiency, operation efficiency, and so on [3]. The common optimal planning of conventional power plants considers the smallest system economic cost as the optimization objective with the load demand supplied and the power system operating safely. The system's total economic cost usually comprises of two parts: the power plants' investment costs, which are related to the installed capacity, and the operating costs, which are related to the actual generated capacity.

Regarding the issue of optimal planning of wind power plants, according to wind power's generation characteristics, the peaking ability of other types of power generation will be influenced by the integrated large-scale wind power generation. In solving the access location and access voltage level of integrated wind power plants, the total capacity of integrated wind power, the constraints of grid network structure, and the power system's reliability and economy need to be taken into account [4–6]. In [7], based on a specific power operation mode, the capable access capacity of wind power generation was solved by the interior-point method based on the affine scaling theory. In [8], considering the constraints of the transmission lines' limited capacity, system's spinning reserve requirements, and regular power units' output limits, the wind farm's capable access capacity was calculated using the chance-constrained method. The literature [9] proposed an optimal method based on a market constraints optimization algorithm to solve multi-wind farms' access locations and access capacities considering quarterly load demand and wind speed distribution characteristics. The literature [10] also analyzed the transmission network's reliability with integration of large-scale wind power, based on which of the relationships between wind farms' installed capacity and the transmission system's expansion planning, as well as the balance between system reliability and investment benefit, were investigated. In [11], based on the evaluation of annual electricity demand, average wind speed, and illumination intensity, an optimization model using a genetic algorithm was proposed to calculate the total access capacity, the optimal capacity proportion, and the installation position of a wind-photovoltaic power hybrid system.

Up to now, among the studies focused on the optimal planning of wind power plants at home and abroad, most research are carried out based on distribution networks [12–15]. Though abundant achievements have been made, the existing methods show limitations facing the optimal planning of large-scale wind power plants which are centrally integrated into the transmission network. In China, according to the regional distribution characteristic of wind resource, wind power based generation is mostly exploited in a large scale more than hundred-MW-level in the northwest and northeast areas, however, which are a thousand kilometers far away from China's east load center. In the meantime, China's current new energy development policy encourages wind power energy to be exploited vigorously under the condition that the installed wind power can be effectively consumed. In this situation, there may be dozens or hundreds of wind farms integrated in a central area, and the power generated by wind farms need to be sent to distant load centers through the main transmission network. Hence, the optimal planning of access location and access capacity of wind power plants group is facing the challenges of large amount of total access capacity, widely available exploited areas, and variously-capable access locations.

In this paper, considering the mutual restriction between the planning of large-scale wind power plants and the planning of the power system's network, an optimal planning model of access location and access capacity of large-scale wind power plants is put forward. The optimization objective takes the power flow equilibrium degree, investment costs, and active network loss as multi-optimization goals, and the proposed model is solved by the improved differential evolution (IDE) algorithm. With the solution results described by the Pareto pattern, the weights of the multi-objective attributes of each solution are calculated based on the entropy weight method, according to which all the alternative solutions are ranked and the final compromised solution is selected by the method of technique for order preference by similarity to ideal (TOPSIS). Based on a practical planning need of large-scale integrated wind power in an actual power grid in China in 2020, the best wind power access solution is successfully identified under the complex actual grid circumstances using the proposed model, which verifies the model's effectiveness.

2. Optimal Planning Model of Large-Scale Wind Power Generation

The plan of access location and access capacity of newly planned wind power plants influences the corresponding expansion of power grid networks, the operation reliability, and the economic efficiency of the power system. To avoid the probable investment losses caused by wind farm construction and network construction, and to reach the optimal overall social benefits, the planning of integrated wind farms' access locations and access capacities is better carried out with the consideration of the planning of network expansion. However, in this situation, the issue about optimal planning of wind power plants turns to be a multi-objective and nonlinear planning problem.

2.1. Objective Function

(1) Power flow equilibrium degree

With large-scale wind power integrated into the power system, if the wind power's access has not been well planned, the power flow disequilibrium in different transmission lines may happen due to the volatility and randomness of wind power output characteristics. When one transmission line's power flow reaches its capacity limit while the other lines are still lightly loaded, the power system's security and reliability declines. However, the balance degree of system power flow on different transmission lines can be improved by better general planning schemes of wind farms and system network expansion. Hence, the power flow equilibrium degree can be an effective index indicating the effectiveness of the planning scheme of wind power and network expansion.

The power flow equilibrium degree can be described by the standard deviation of different line's power flow ratios, as written as below:

$$\min\sqrt{\frac{1}{l}\sum(r_{ij}-\overline{r})^2},\ r_{ij}=\frac{|P_{ij}|}{P_{ij\max}} \tag{1}$$

where l is the amount of the transmission lines in the system, r_{ij} is the power flow ratio between the practical power flow and the rated capacity of line from bus i to bus j, $|P_{ij}|$ is the actual power flow, $P_{ij\max}$ is the rated capacity of line, and $\overline{r}$ is the average of r_{ij}.

(2) Economic efficiency

On this issue about the optimal planning of large-scale centrally integrated wind power plants, the economic efficiency is mainly considered in two parts: the investment costs and the operation costs. Investment costs includes the expenses of wind farms' construction, corresponding connection lines' construction, and main power grid network's expansion. Operation costs is mainly caused by system operation loss which changes for different planning schemes since the power flow distribution will be different. In the composition of system operation losses, the active network loss is mainly considered in this issue.

Taking the minimum sum of the investment costs and the active network loss as the economic optimization objective, the economic efficiency is described as:

$$\min f(n_{ij},P_{\rm loss})=\left(\sum c_{ij}n_{ij}+C_{\rm loss}P_{\rm loss}\right) \tag{2}$$

where n_{ij} is the amount of planning lines between bus i and bus j, and c_{ij} is the price of each line, $P_{\rm loss}$ is the active network loss of power system, $C_{\rm loss}$ is the cost coefficient of $P_{\rm loss}$, determined by the electricity price and annual hours of maximum load loss.

2.2. Constraints

The constraints of the proposed model mainly include the wind power access capacity constraint, the power and electricity balance constraint, the power system operation constraint, and the power grid network constraint, etc.

(1) Constraint of wind power access capacity:

$$\sum_{i=1}^{N} \lambda_i W_i \leq W_{\text{ins}} \tag{3}$$

where W_{ins} is the limit of total capacity of integrated wind power.

(2) Power flow constraint:

$$\widetilde{S}_G + \lambda \widetilde{S}_w - \widetilde{S}_L = U[Y_0 + \Delta Y]^* U^* \tag{4}$$

where Y_0 is the initial node admittance matrix, ΔY is the change of node admittance matrix with the planning lines. y_{ij} is the admittance of each planning line.

ΔY can be expressed by:

$$\begin{cases} \Delta Y_{ii} = \sum_j n_{ij} \times y_{ij}, & i \neq j \\ \Delta Y_{ij} = \Delta Y_{ji} = -n_{ij} \times y_{ij}, & i \neq j \end{cases} \tag{5}$$

(3) Constraint of conventional power plants' output:

$$P_{Gmin} \leq P_G \leq P_{Gmax} \tag{6}$$

(4) Node voltage constraint:

$$U_{min} \leq U \leq U_{max} \tag{7}$$

(5) Constraint of transmission line power flow:

$$|P_{ij}| \leq \left(n_{ij}^0 + n_{ij}\right) P_{ijmax} \tag{8}$$

(6) Constraint of network expansion:

$$0 \leq n_{ij}^0 + n_{ij} \leq \overline{n}_{ij} \tag{9}$$

where, n_{ij}^0 and $\overline{n}_{ij}$ are the amounts of existed and planned transmission lines between bus i and bus j, respectively.

(7) Constraint of wind power distribution at booster station:

$$0 \leq W_i \leq W_{i,\max} \tag{10}$$

where $W_{i,\max}$ is the maximum limit of bus i's capable access capacity.

(8) Security and stability constraint

The planning of large-scale wind power access locations and capacity should satisfy the requirement in the standard *Guidelines for Power System Security and Stability* [16] written by the State Grid Corporation of China. This standard stipulates the reliability requirement for power systems under different stability levels including: (i) *N-1* condition, when the power system loses any one operating electric equipment; (ii) *N-2* condition, when the system loses any two operating electric equipment; and (iii) serious fault conditions.

3. Multi-Objective Optimization Process

3.1. Pareto-Based Optimal Solution Set

Since the proposed optimization model of large-scale wind power's access location and access capacity is a multi-objective and nonlinear programming problem, the improved differential evolution (IDE) algorithm [17] is adopted here to solve the Pareto optimal solutions. The basic process of the DE algorithm consists three main steps: (i) Mutation: for two random vectors in the population, their difference is set as the variable source to generate a new vector; (ii) Crossover: based on mutation, a new generation is evolved from the current population each round; (iii) Choice: the population generation evolves until the evolutionary population contains the optimal solution.

Considering the calculation complexity and the DE algorithm's characteristics, the original DE algorithm's process is improved as below:

(1) Improved mutation process

For two random vectors $\boldsymbol{p}$ and $\boldsymbol{q}$ in the original population, if every attribute in vector $\boldsymbol{p}$ is not poor and at least one attribute is better than those in vector $\boldsymbol{q}$, it is called that $\boldsymbol{p}$ and $\boldsymbol{q}$ obey Pareto dominance and $\boldsymbol{p}$ is the dominated vector. In the G_{th} mutation process, a new temporary population will be composed by the way that vector $u_{i,G+1}$ begins to compare with its parent vector $x_{i,G}$: (i) If the two vectors satisfy the Pareto dominance principle, the dominated vector is chosen as the new temporary population; (ii) If not, the two vectors are both chosen as the new temporary population. In this competing way, the amount of vectors in the temporary population is between N_P and $2N_P$, and the former N_P vectors become the next population generation by non-dominated sorting and crowded distance sorting based on the improved non-dominated sorting genetic algorithm [18].

(2) Adaptive adjustment of control parameters

Using the IDE algorithm, the convergence and robustness of the solving process are directly affected by the control parameters F and CR. In the standard DE algorithm, F and CR are constants parameters, and it always takes a long time to find the most appropriate value. Therefore, the adaptive adjustment is adopted to calculate the value of F and CR as the population evolves, which can eliminate the inconvenience of manual calculation and increase the diversity of the population. The adaptive adjustment strategies of F and CR are described as follows [19]:

$$F_i^{G+1} = \begin{cases} F_{\min} + r_1 F_{\max}, & r_2 < \tau_1 \\ F_i^G, & \text{other} \end{cases} \tag{11}$$

$$CR_i^{G+1} = \begin{cases} r_3, & r_4 < \tau_2 \\ CR_i^G, & \text{other} \end{cases} \tag{12}$$

where r_1, r_2, r_3, r_4 are random numbers in [0,1] obeying the uniform distribution. τ_1 and τ_2 are the adaptive adjustment possibility of F and CR, respectively. In this work, the values of $F_{\min}$, $F_{\max}$, τ_1, and τ_2 are set as 0.1, 0.9, 0.1, and 0.1, respectively.

3.2. Acquisition of the Final Compromised Solution

With the Pareto optimal solution set solved by the IDE algorithm, the attribute weights of optimization objectives are then calculated, and the weighted Pareto decision-making matrix is composed using the information entropy method [20]. In the weighted Pareto decision-making matrix, all the alternate solutions are sorted using the TOPSIS [21], based on the distances from each alternative solution to the ideal solution (IS) and the negative solution (NS), which are both calculated. After the comparison, the solution closest to the IS and at the mean time farthest away from NS, is chosen as the compromised Solution. The decision-making steps based on the TOPSIS method are as follows:

(1) Establishment of the weighted-standardized decision matrix

Firstly, establish the standardized decision matrix $\boldsymbol{R}$, of which the square sum of a same attribute of all solutions adds up to 1:

$$r_{ij} = \frac{x_{ij}}{\sqrt{\sum\limits_{i=1}^{m} x_{ij}^2}} \tag{13}$$

Based on Equation (13), the weighted-standardized decision matrix $\boldsymbol{V} = \{v_{ij}\}$ is then established through the way that each column of matrix $\boldsymbol{R}$ multiply by its corresponding weight w_j.

(2) Calculation of IS and NS

Second, provided that a lower value of the objective attribute indicates a better solution, then the ideal solution A^+ and the negative solution A^- can be expressed as:

$$\begin{aligned} A^+ &= \left\{ \min_i v_{ij}, i = 1,2,\ldots,m \right\} = \left\{ v_1^+, v_2^+, \ldots, v_n^+ \right\} \\ A^- &= \left\{ \max_i v_{ij}, i = 1,2,\ldots,m \right\} = \left\{ v_1^-, v_2^-, \ldots, v_n^- \right\} \end{aligned} \tag{14}$$

(3) Calculation of distance

Then the distances of every alternative solution to IS and NS are calculated based on N-dimensional Euclid Norm, expressed as:

$$\begin{aligned} S_{i^+} &= \sqrt{\sum\limits_{j=1}^{n} \left(v_{ij} - v_j^+\right)^2} \\ S_{i^-} &= \sqrt{\sum\limits_{j=1}^{n} \left(v_{ij} - v_j^-\right)^2} \end{aligned} \tag{15}$$

(4) Sorting of alternative solutions and decision-making

Finally, the approximate degree of each solution to the ideal solution is described by:

$$C_i = \frac{S_{i^-}}{S_{i^+} + S_{i^-}} \tag{16}$$

According to Equation (16), it satisfies $0 \le C_i \le 1$, and if v_i is the IS, $C_i = 1$, if v_i is the NS, $C_i = 0$. Since the value of C_i directly indicates the approximate degree of each alternative solution to the IS, all the solutions in Pareto optimal solution set can be sorted according to the value of C_i, and the one which has largest value of C_i is chosen as the final compromised solution.

3.3. Overall Optimization Process

Matched with the proposed optimal planning model of large-scale wind power, the access location and access capacity of wind farms are calculated and optimized in the following steps:

(1) Initialization of the calculation model. With the original system data, parameters, and algorithms input, the original generation of wind farm planning solutions are generated. Then the system power flow for each original solution is calculated by the Newton-Raphson method, based on which of the adaptive degrees of each solution is fetched.
(2) Acquisition of the first test generation of wind power plants planning solutions. Based on the results of the adaptive degree calculation of each original solution, the adaptive adjustment is applied calculating the value of F and CR as the population evolves. After mutation and crossover

processes, the first test generation of the wind power plants planning solutions is generated, and the adaptive degree of each solution is calculated as well.

(3) Calculation of the Pareto-based optimal solution set. The IDE algorithm is applied to calculate the Pareto optimal solution set, as described in Section 3.1.

(4) Acquisition of the final compromised solution. The TOPSIS method is applied to rank all the alternate solutions and reveal the final compromised solution out as described in Section 3.2. The overall optimization process is shown in Figure 1.

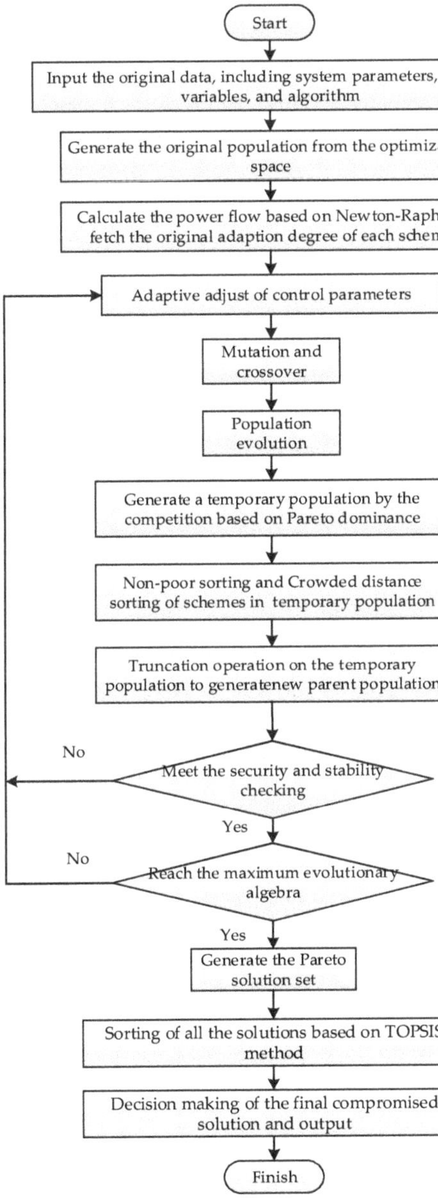

Figure 1. Diagram of the overall optimization process.

4. Simulation Analysis

4.1. Simulation Models

4.1.1. Power Grid Conditions

The simulation analysis is carried out based on a practical regional power grid in China which already contains 7300 MW first-phase integrated wind power generation. A second-phase access project of 4350 MW wind power generation in the year 2020 is ready for planning. The power system circumstances in 2020 are given in Table 1 in which the data are given according to the future forecasting of the constructions of power plants, substations, transmission lines, and the development of load demands based on the current data, and the corresponding diagram is shown in Figure 2.

Table 1. Power system circumstances in 2020.

Substation	Voltage Level/kV	Installed Capacity of Conventional Power Plants/MW	Installed Capacity of Wind Farms/MW	Loads/MW	Transmitted Power through HVDC/MW	Equivalent Loads/MW
T1	330	132	550	353	-	353
T2	330	-	-	353	-	353
T3-1	750	-	-	-	−5000	−5000
T3-2	330	-	3500	-	-	-
T4	330	-	800	16	-	16
T5-1	750	6000	-	-	8000	8000
T5-2	330	-	2200	-	-	-
T6	330	224	1110	176	-	176
T7	330	167	-	337	-	337
T8	330	1722	-	978	-	978
T9-1	750	3200	-	-	2000	2000
T9-2	330	660	-	403	-	403
T10	330	408	-	1109	-	1109

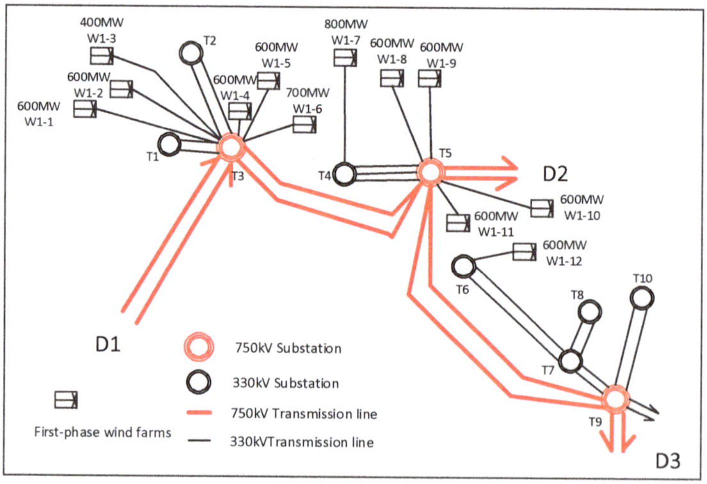

Figure 2. Diagram of the test system network in 2020.

As shown in Figure 2, the basic information and constraints of this power gird are as follows:

(1) This regional power system consists of three 750 kV substations: T3, T5, and T9, of which each 750 kV substation is able to connect up to sixteen 330 kV lines, and this regional system consists of seven 330 kV substations: T1, T2, T4, T6, T7, T8, and T10, of which each 330 kV substation is able to connect up to eight 330 kV lines.

(2) Wind farms in the first-phase and second-phase are all integrated at 330 kV substation. The economic transport capacity of 330 kV transmission line is taken as 309 MW and the limited continuous transport capacity to be 782 MW. The power output of each wind farm from its booster substation to 330 kV transmission line is assumed as 600 MW.
(3) Through HVDC (high voltage direct current transmission) lines, this regional power system transmits power with other regional power grids with amount of −5000 MW, 8000 MW, and 2000 MW to District 1, District 2, and District 3, respectively.
(4) The locations of wind farms booster substations has been determined, and the capable exploited capacity of wind power around each booster substations is also certain. The network structure of wind farms' internal collection and the corresponding network losses are not taken into consideration in this simulation.

4.1.2. Wind Farms Access Conditions and Boundary Conditions

As shown in Figure 2, the detailed information of installed capacity and the transmission scheme of the first-phase 7300 MW wind power plants are given in Table 2.

Table 2. Data of the first-phase wind farms and the booster substations.

Wind Farm's Booster Substation	Installed Capacity/MW	Accessed 330 kV Substation	Transmission Line's Length/km
W1-1	600	T3	49
W1-2	600	T3	36
W1-3	400	T3	50
W1-4	600	T3	24
W1-5	600	T3	39
W1-6	700	T3	44
W1-7	800	T4	36
W1-8	600	T5	28
W1-9	600	T5	44
W1-10	600	T5	84
W1-11	600	T5	35
W1-12	600	T6	10

Affected by the wind resources distribution limit, there are only eight alternate wind farm's booster substations, named by W2-1–W2-8, capable for integration of second-phase 4350 MW wind power generation. The 330 kV substations capable for wind power integration are the substations T1–T6. The transmission line's length from the capable wind-power booster stations W2-1–W2-8 to the capable 330 kV substation T1–T6 are given in Table 3.

Table 3. Transmission line's length from W2-1–W2-8 to T1–T6/km.

Alternate Wind Farm's Booster Substations	T1	T2	T3	T4	T5	T6
W2-1	50	56	20	62	110	200
W2-2	56	56	30	20	110	205
W2-3	60	60	36	50	100	200
W2-4	72	72	56	48	90	190
W2-5	160	170	150	50	15	110
W2-6	160	200	170	90	69	140
W2-7	170	190	160	80	56	130
W2-8	205	210	170	110	72	90

4.2. Optimal Planning of the Second-Phase Wind Farms

In optimization process of the access location and access capacity planning of the second-phase wind power generation, some key variables are set as: (1) the unit cost of 750 kV line and 330 kV line are 2.44 million RMB and 1.5 million RMB per kilometer, respectively; (2) the cost coefficient of active network loss is 1.5 million RMB per MW; (3) the scale amount of the original population is 200; and (4) the maximum number of calculation iteration rounds is 1000.

In the early stage of the practical optimization process, the power system's power flow equilibrium degree changes only in a small fluctuation degree, which is far less than the variation scale of the economic efficiency. It is suggested that the economy of transmission line's cost plays a leading role among the optimization objectives in the early stage of optimization process. Hence, the optimal access locations of wind power booster substations and the corresponding transmission lines' expansion plan are preliminarily selected.

In the later stage of the optimization process, since the transmission lines' costs are the same based on the acquired wind farm access location solution, the capacity distribution scheme of wind farms are optimized according to the optimization objectives of power flow equilibrium degree and the active network loss.

Eventually the Pareto optimal solution set is generated with 30 solutions regarding wind farms' access locations and access capacity combinations, of which the target cost of transmission lines is 504 million RMB, the target distribution of power flow equilibrium degree is between 0.3578 and 0.3725, and the target distribution of economic efficiency is between 711.0~863.3 million RMB. The distribution trend of all the 30 solutions in the Pareto optimal solution set are shown in Figure 3. The optimization objective's attribute values of each solution are given in Table 4.

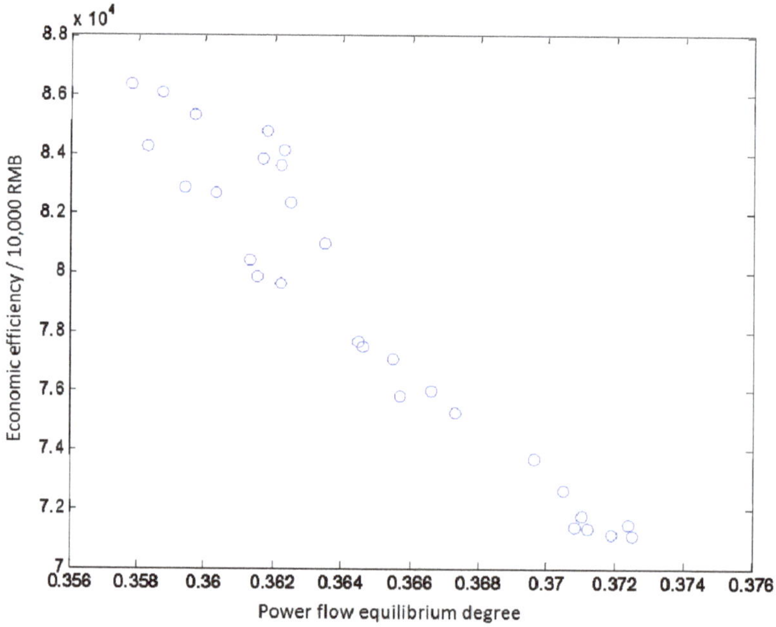

Figure 3. Distribution trend of the Pareto optimal solution set.

Table 4. Optimization objective attribute values of the Pareto optimal solution set.

Pareto optimal Solution Set	Power flow Equilibrium Degree	Economic Efficiency Deviation from the Most Economical Solution (%)
Solution 1	0.3578	21.4%
Solution 2	0.3666	6.8%
Solution 3	0.3725	0.0%
Solution 4	0.3587	21.0%
Solution 5	0.3712	0.3%
Solution 6	0.3719	0.1%
Solution 7	0.3724	0.5%
Solution 8	0.3708	0.4%
Solution 9	0.3594	16.5%
Solution 10	0.3705	2.2%
Solution 11	0.3597	20.0%
Solution 12	0.3719	0.1%
Solution 13	0.3655	8.4%
Solution 14	0.3696	3.7%
Solution 15	0.3613	13.1%
Solution 16	0.3646	9.0%
Solution 17	0.3615	12.3%
Solution 18	0.3625	15.9%
Solution 19	0.3710	0.9%
Solution 20	0.3673	5.8%
Solution 21	0.3617	17.9%
Solution 22	0.3622	17.6%
Solution 23	0.3603	16.3%
Solution 24	0.3657	6.6%
Solution 25	0.3583	18.5%
Solution 26	0.3618	19.2%
Solution 27	0.3623	18.3%
Solution 28	0.3622	12.0%
Solution 29	0.3635	13.8%
Solution 30	0.3645	9.2%

Based on the Pareto optimal solution set, as shown in Figure 3 and Table 4, the attribute weights of the power plow equilibrium degree and active network loss are then, respectively, calculated as 0.0357 and 0.9643 by the information entropy method. This means that the economic efficiency plays a more decisive role in this optimization planning process of the second-phase wind farms' access location and access capacity.

Based on the TOPSIS method, all of the alternate solutions in the Pareto optimal solution set are sorted, and Solution 3 turns to be the optimal compromise solution which has the smallest distance to the ideal solution and the longest distance to the negative ideal solution. In Solution 3, (1) the power plow equilibrium degree is 0.3725; (2) the economic efficiency is 711 million RMB. The 750 kV network in this regional power system does not need strengthening, the total length of the planned transmission lines from second-phase wind farms to the 330 kV substations is up to 336 km with a total investment of 504 million RMB; (3) the active network loss is 138 MW; and (4) Solution 3 also satisfies the power flow check that no transmission line's power flow exceeds the rated capacity.

The detailed access location and access capacity information of the second-phase wind farms in Solution 3 are given in Figure 4 and Table 5.

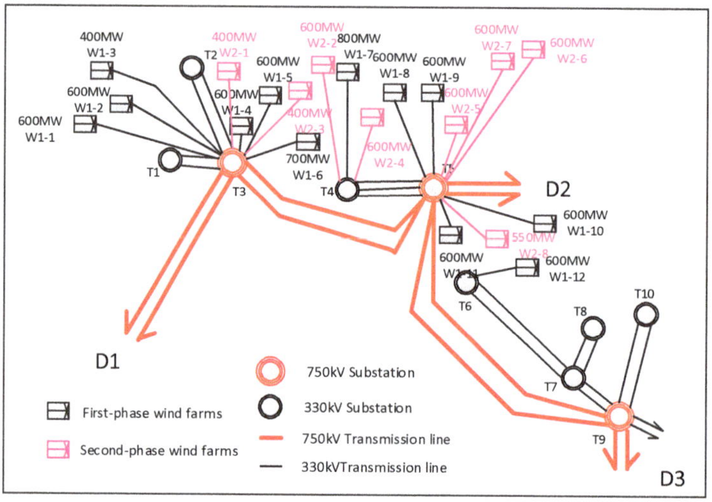

Figure 4. Diagram test system network in 2020 with the second-phase wind farms.

Table 5. The detailed access location and capacity of the second-phase wind farms of Solution 3.

Planned Wind Farm Booster Substation	Planned Wind Farm Access Capacity/MW	Accessed 330 kV Substation	Transmission Line's Length/km	Transmission Line's Cost/Million RMB
W2-1	400	T3	20	30
W2-2	600	T4	20	30
W2-3	400	T3	36	54
W2-4	600	T4	48	72
W2-5	600	T5	15	22.5
W2-6	600	T5	69	10.35
W2-7	600	T5	56	84
W2-8	550	T5	72	10.8

5. Conclusions

This paper proposed a multi-objective optimization planning model of large-scale wind power's access location and access capacity considering the mutual restriction between wind power's access planning and the power system's network planning. Based on a practical planning need of large-scale wind power access in a regional power grid in China in 2020, the proposed optimization model is tested. The power flow equilibrium degree, investment costs, and active network loss are taken as the optimization goals. Applied with the improved differential evolution (IDE) algorithm, the optimal access locations of wind power booster substations and the corresponding transmission lines' expansion plan are preliminarily solved in the early stage of the optimization process, and the capacity distribution plans of integrated wind farms are calculated in a later stage. With the Pareto optimal solution set solved, all of the alternative solutions are sorted by the TOPSIS (technique for order preference by similarity to ideal) method and the final compromised solution is successfully selected. The result verifies the model's effectiveness under complex practical power system circumstances.

Acknowledgments: This research work was supported by the National Key Research and Development Program of China: Topology Evolution Model and Power Forecast Method of the Future Electrical Power Systems (No. 2016YFB0900101).

Author Contributions: Hui Li and Gengyin Li conceived and designed the experiments; Hui Li, Siwei Liu, Yuning Wang, and Zhidong Wang performed the experiments; Hui Li, Yuning Wang, and Ning Zhang analyzed the data; Jiaming Wang contributed analysis tools; Hui Li and Siwei Liu wrote the paper.

Conflicts of Interest: The authors declare no conflict of interest.

References

1. Xiao, C.Y.; Wang, N.B.; Zhi, J. Power characteristics of Jiuquan wind power base. *Autom. Electr. Power Syst.* **2010**, *37*, 64–67.
2. Wang, Z.M.; Su, A.L.; Lu, S. Analysis on capacity of wind power integrated into Liaoning power grid based on power balance. *Autom. Electr. Power Syst.* **2010**, *34*, 86–90.
3. Cheng, H.Z.; Zhang, Y. *Power System Planning*; China Electric Power Press: Beijing, China, 2008.
4. Wiik, J.; Gjerde, J.O.; Gjengedal, T.; Gustafsson, M. Steady state power system issues when planning large wind farms. In Proceedings of the IEEE Power Engineering Society Winter Meeting, New York, NY, USA, 27–31 January 2002.
5. Billinton, R.; Yi, G.; Karki, R. Composite system adequacy assessment incorporating large-scale wind energy conversion systems considering wind speed correlation. *IEEE Trans. Power Syst.* **2009**, *24*, 1375–1382. [CrossRef]
6. Xu, M.; Zhuan, X. Optimal planning for wind power capacity in an electric power system. *Renew. Energy* **2013**, *53*, 280–286. [CrossRef]
7. Lei, Y.Z.; Wang, W.S.; Yin, Y.H. An optimization method for determining wind power penetration limit in power system under static security constraints. *Proc. CSEE* **2001**, *21*, 25–28.
8. Lei, Y.Z.; Wang, W.S.; Yin, Y.H. Wind power penetration limit calculation based on chance constrained programming. *Proc. CSEE* **2002**, *22*, 32–35.
9. Roy, S. Market Constrained Optimal Planning for wind energy conversion systems over multiple installation. *IEEE Trans. Energy Convers.* **2003**, *17*, 124–129. [CrossRef]
10. Billinton, R.; Wangdee, W. Reliability-based transmission reinforcement planning associated with large-scale wind farms. *IEEE Trans. Power Syst.* **2007**, *22*, 34–41. [CrossRef]
11. Masoum, M.A.S.; Mousavi Badejani, S.M.; Kalantar, M. Optimal placement of hybrid PV-wind systems using genetic algorithm. In Proceedings of the 2010 Innovative Smart Grid Technologies (ISGT), Gaithersburg, MD, USA, 19–21 January 2010; pp. 1–5.
12. Qiu, X.Y.; Xia, L.L.; Li, X.Y. Planning of distributed generation in construction of smart grid. *Power Syst. Technol.* **2010**, *34*, 7–10.
13. Wang, J.; Li, X.Y.; Qiu, X.Y. Improved cluster analysis based Ultra-short term load forecasting method. *Autom. Electr. Power Syst.* **2005**, *29*, 90–97.
14. Wang, C.S.; Cheng, K.; Xie, Y.H. Siting and sizing of distributed generation distribution network expansion planning. *Autom. Electr. Power Syst.* **2006**, *30*, 38–43.
15. Hsiao, Y.T.; Chen, C.H.; Chien, C.C. Optimal capacitor placement in distribution systems using a combination fuzzy-GA method. *Electr. Power Energy Syst.* **2004**, *26*, 501–508. [CrossRef]
16. State Grid Corporation of China. *Guide on Security and Stability for Power System*; State Grid Corporation Company: Beijing, China, 2001.
17. Cheng, H.Z.; Gao, X.W.; Ma, Z.L. General optimized model of multi-objective electric power network planning. *Autom. Electr. Power Syst.* **2004**, *38*, 1229–1232.
18. Liu, L.; Cheng, H.X.; Ma, Z.L.; Bazargan, M. Multi-objective transmission expansion planning considering life cycle cost. *Proc. CSEE* **2012**, *32*, 46–54.
19. Hongyun, M.; Xiaohua, Z.; Sanyang, L. A differential evolution based on double populations for constrained multi-objective optimization problem. *Chin. J. Comput.* **2008**, *31*, 228–235.
20. Guoyin, W.; Hong, Y.; Dachun, Y. Decision table reduction based on conditional information entropy. *Chin. J. Comput.* **2002**, *25*, 759–766.
21. Chen, C.T. Extensions of the TOPSIS for group decision-making under fuzzy environment. *Fuzzy Sets Syst.* **2000**, *114*, 1–9. [CrossRef]

energies

MDPI

Article

Fault Tolerant and Optimal Control of Wind Turbines with Distributed High-Speed Generators

Urs Giger [1] Patrick Kühne [2] and Horst Schulte [2,*]

[1] GGS GmbH, Gotthardstrasse 37, 6490 Andermatt, Switzerland; info@ggs-falcon.com
[2] Control Engineering Group, School of Engineering—Energy and Information, University of Applied Sciences (HTW) Berlin, 12459 Berlin, Germany; PatrickKuehne173@googlemail.com
* Correspondence: schulte@htw-berlin.de

Academic Editor: Frede Blaabjerg
Received: 03 October 2016; Accepted: 11 January 2017; Published: 24 January 2017

Abstract: In this paper, the control scheme of a distributed high-speed generator system with a total amount of 12 generators and nominal generator speed of 7000 min^{-1} is studied. Specifically, a fault tolerant control (FTC) scheme is proposed to keep the turbine in operation in the presence of up to four simultaneous generator faults. The proposed controller structure consists of two layers: The upper layer is the baseline controller, which is separated into a partial load region with the generator torque as an actuating signal and the full-load operation region with the collective pitch angle as the other actuating signal. In addition, the lower layer is responsible for the fault diagnosis and FTC characteristics of the distributed generator drive train. The fault reconstruction and fault tolerant control strategy are tested in simulations with several actuator faults of different types.

Keywords: fault tolerant systems; power system control; wind energy system; actuators

1. Introduction

The electrical system of the state-of-the-art variable speed wind turbines (WT) consists of a single generator, a power electronic interface and a transformer, which is connected to the grid. In the partial load region, the single generator adjusts the rotor speed to an optimal tip speed ratio. However, a single generator concept is not an optimal solution for future large-scale wind turbines regarding power density, low downtimes and in particular maintenance with reduced downtimes.

Recently, the multi-generator drive train topology, also called distributed generator drive train, already presented on the market—see [1,2]—has been revitalized and extended to high-speed generators. In particular, the high-speed concept proposed in [3] with a gear box ratio of $i_g = 438$ and the extensive use of flex pin planets originally introduced by Raymond Hicks 1965 leads to a considerable load reduction in the gear drive and a mass reduction of the nacelle. To increase the availability and efficiency despite the increasing amount of components (due to the additional transmission stages) advanced control concepts are necessary. They can take into account the power loss characteristics of the multi-generator system and are able to detect and isolate the faulty behavior of each generator. Moreover, a fault tolerant control (FTC) strategy must be capable enough to avoid overspeed of the rotor and should mitigate induced loads caused by abrupt torque changes. One systematic way to achieve this is to consider the influence of occurred faults with a formal description of the whole system and a deduced model based FTC strategy with fault accommodation or control configuration.

In recent years, some model-based FTC approaches are proposed for wind turbines with *conventional single generator drive trains*: in [4], passive and active fault-tolerant controllers are designed and considered with regard to accommodating altered actuator dynamics in the pitch system model. In [5], a bank of unknown-input observers is used for fault diagnosis in the rotor and generator speed sensors of the fault detection isolation (FDI) benchmark model presented in [6]. In [7,8], active fault-tolerant control is achieved in the partial load region of wind turbines by means of a sensor fault hiding approach. The fault-tolerant control (FTC) strategy uses a multiple integral observer and a fast adaptive fuzzy estimator, where the observer design is based on a nonlinear Takagi-Sugeno (TS) model. In [9], a passive sensor fault-tolerant control strategy is implemented using a sliding mode controller for the partial load region that tolerates generator speed sensor faults and generator torque offset faults. In [10], an FTC strategy using Linear Parameter Varying (LPV) virtual sensors is proposed and applied to the benchmark model [6]. Instead of hiding the fault, the virtual sensor is used to expand the set of available sensors before the state observer is designed.

In [11], an FTC scheme based on adaptive filters obtained via the nonlinear geometric approach is applied to the wind turbine benchmark model. It is shown that the proposed approach allows us to obtain an interesting decoupling property with respect to uncertainty affecting the wind turbine system. A fuzzy modeling and identification method for fault detection and FTC is applied in [12]. The proposed fuzzy gain-scheduled fault-tolerant control system is evaluated by a series of simulations on a wind turbine benchmark in the presence of different fault scenarios. In [13], the method of [12] is compared with a fuzzy model-reference adaptive control in which a fuzzy inference mechanism is used for parameter adaptation without any explicit knowledge of the system faults.

In [14], a Takagi-Sugeno [15] sliding mode observer (TS SMO) is used to reconstruct actuator and sensor faults in wind turbines with conventional drive trains. Here, the proposed FTC strategy is based on the modification of control inputs in the presence of actuator faults and on the active-fault compensation of the sensor output signal in the presence of sensor faults. Both strategies serve a behavior similar to the fault-free case.

In this paper, an FTC scheme for distributed high-speed generator systems with a total amount of 12 generators and a nominal generator speed of 7000 min^{-1} is presented. The main contribution is as follows. First, existing FTC strategies for wind turbines with single-generators such as [9–11] do not take into account the design freedom of multi-generator systems. In this paper, the fault accommodation is achieved by adaption of the division of the demanded torque between the healthy generators and the rotor speed dependent transition from the partial load (torque control) to the full-load region (pitch control). Second, no FDI and FTC schemes have been used so far in distributed generator-drive train concepts on the market [1,2].

This paper is organized as follows: Section 2 presents the reduced-order wind turbine model with distributed generators, the baseline control laws for the partial/full load region and the optimal controller design for the full-load region. In Section 3 the scheme of the reference signal adaptation for FTC is proposed and validated using the NREL (National Renewable Energy Laboratory) of the U.S. Department of Energy benchmark wind turbine [16]. Section 4 deals with the transient behavior of the previously introduced FTC scheme. Further, a classic control design is used for the active damping of weakly damped torsional vibrations. Finally, the conclusion and open problems are presented in Section 5.

2. Wind Turbine Model with Distributed Generators

2.1. Drive Train Concept and Reduced-Order Model

The wind turbine (WT) drive train shown in Figure 1 consists of two stage epicyclical split gears, a spur gear for power distribution and twelve epicyclic split gears directly connected to each generator. Each single generator feeds its own AC-DC converter (see the rectangular devices in Figure 1). All converters are connected via DC bus to a common grid-side DC-AC converter (not shown in Figure 1).

Figure 1. Wind turbine multi-generator drive train with 12 single generators.

The overall ratio i_g from the low speed shaft to the high-speed shafts connected to the generators is $i_g = 438$. From a control-oriented point of view, an ideal gearbox with ratio 1 is considered, which simplifies the equations of motion for the drive train. For the purpose of comparing simulation results of the reduced-order model with more complex rigid body models, a gearbox ratio can be implicitly considered in the proposed control approach.

In this paper, a WT model with four degrees-of-freedom (4-DOF) is used as a *baseline model*. Hereby, the tower top deflection, the drive train rotation and especially the blade tip deflection are assumed to be sensitive to wind speed variations. The complete model consists of several submodels for the mechanical structure, the aerodynamics, as well as the dynamics of the pitch system and the multi-generator/converter system. Note that the main sources of uncertainty are caused by the unmodeled rotor blade deflection in an edge-wise direction and of each individual rotor blade in a flap-wise direction. Therefore, the DOFs of the model are the collective horizontal tip displacements of the rotor blade tips in a flap-wise direction y_B, the displacement of tower top in the wind direction denoted by y_T and the rotor and generator rotational angle θ_r and θ_g. The drive train is modeled by two rigid bodies joined with a torsionally elastic coupling, as described in [17]. The equations of motion of the four DOF yields four coupled differential equations. The state-space model of WT is given as follows

$$\dot{\boldsymbol{x}} = \boldsymbol{A}\,\boldsymbol{x} + \boldsymbol{B}\,\boldsymbol{u} + \boldsymbol{g}(\boldsymbol{x}, v) \tag{1}$$

with the state vector

$$\boldsymbol{x} = \begin{pmatrix} y_T & y_B & \theta_s & \dot{y}_T & \dot{y}_B & \omega_r & \omega_g & \beta & T_g \end{pmatrix}^T, \tag{2}$$

where $\theta_s = \theta_r - \theta_g$ denotes the shaft torsion angle, β denotes the pitch angle and T_g denotes the collective generator torque. The input vector is

$$\boldsymbol{u} = \begin{pmatrix} \beta_d & T_{g,d} \end{pmatrix}^T, \tag{3}$$

where β_d denotes the demanded pitch angle and $T_{g,d}$ the demanded collective generator torque. The system matrix of (1) is

$$A = \begin{pmatrix} 0 & 0 & 0 & 1 & 0 & 0 & 0 & 0 & 0 \\ 0 & 0 & 0 & 0 & 1 & 0 & 0 & 0 & 0 \\ 0 & 0 & 0 & 0 & 0 & 1 & -1 & 0 & 0 \\ -\frac{k_T}{m_T} & \frac{Nk_B}{m_T} & 0 & -\frac{d_T}{m_T} & \frac{Nd_B}{m_T} & 0 & 0 & 0 & 0 \\ \frac{k_T}{m_T} & -\frac{k_B}{m_B} - \frac{Nk_B}{m_T} & 0 & \frac{d_T}{m_T} & -\frac{d_B}{m_B} - \frac{Nd_B}{m_T} & 0 & 0 & 0 & 0 \\ 0 & 0 & -\frac{k_S}{J_r} & 0 & 0 & -\frac{d_S}{J_r} & \frac{d_S}{J_r} & 0 & 0 \\ 0 & 0 & \frac{k_S}{J_g} & 0 & 0 & \frac{d_S}{J_g} & -\frac{d_S}{J_g} & 0 & -\frac{1}{J_g} \\ 0 & 0 & 0 & 0 & 0 & 0 & 0 & -\frac{1}{\tau} & 0 \\ 0 & 0 & 0 & 0 & 0 & 0 & 0 & 0 & -\frac{1}{\tau_g} \end{pmatrix}$$

and the input matrix is as follows

$$B = \begin{pmatrix} \mathbf{0}_{7\times1} & \mathbf{0}_{7\times1} \\ \frac{1}{\tau} & 0 \\ 0 & \frac{1}{\tau_g} \end{pmatrix},$$

where N denotes the number of blades, k_B denotes the effective flap-wise stiffness parameter, k_S the drive train stiffness parameter, k_T the effective tower fore-aft stiffness parameter, m_T the effective mass of nacelle-tower motion, J_r the rotor inertia and J_g the total inertia of the generators, τ denotes the delay time constant of a reduced-order model of the pitch-dynamics and τ_g denotes a common delay time constant of the torque dynamics of the multi-generator system. The WT model (1) has constrained controllable inputs since both the generators and the pitch drive mechanism are limited in the control range

$$0 \le T_g \le T_{g,max}, \quad 0^\circ \le \beta \le 90^\circ, \quad |\dot{\beta}| \le 10\text{m/s} \tag{4}$$

The vector $\boldsymbol{g}$ in Equation (1) nonlinearly depends on the rotor speed and the pitch angle, which are states of the system and the undisturbed wind speed v far in front of the rotor

$$\boldsymbol{g}(\boldsymbol{x}, v) = \begin{pmatrix} \mathbf{0}_{4\times1} \\ \frac{1}{Nm_B} F_T(\boldsymbol{x}, v) \\ \frac{1}{J_r} T_r(\boldsymbol{x}, v) \\ \mathbf{0}_{3\times1} \end{pmatrix}, \tag{5}$$

where F_T denotes the thrust force acting on the rotor and the aerodynamic rotor torque T_r. The rotor torque

$$T_r = \frac{1}{2} \rho_{air} \pi R^3 v^2 C_Q(\lambda, \beta) \tag{6}$$

and also the thrust force

$$F_T = \frac{1}{2} \rho_{air} \pi R^2 v^2 C_T(\lambda, \beta) \tag{7}$$

depend on the pitch angle β, the wind speed v and the aerodynamic torque coefficient C_Q or aerodynamic thrust coefficient C_T. Both depend on the tip speed ratio ratio $\lambda = \frac{\omega_r R}{v}$, where ρ_{air} denotes the air density and R denotes the rotor radius. The rotor torque and thrust force are therefore nonlinearly dependent on the wind speed, the rotor speed and the pitch angle.

Note that (5) is derived directly (without any linearization) by the aerodynamic rotor torque (6) and thrust force (7). This form is well suited for a transformation into a Takagi-Sugeno (TS) form [15] proposed in [17]. Here, the nonlinear term $\boldsymbol{g}(\boldsymbol{x}, v)$ is first written as a product of a matrix and the state vector $\boldsymbol{x}$ and then reformulated into a TS structure using the sector nonlinearity approach [18]. This approach yields an exact representation of the nonlinear functions defined on a compact set and is used as an control-oriented model for the optimal controller design proposed in [19] and applied in Section 2.3.

The multi-generator/converter dynamics is modeled by a bank of first order delay systems illustrated in Figure 2. The baseline model (1) and the bank of first order systems can be combined to an augmented switched system with the state vector

$$\boldsymbol{x_a} = \begin{pmatrix} y_T & y_B & \theta_s & \dot{y}_T & \dot{y}_B & \omega_r & \omega_g & \beta & T_{g,1} & \ldots & T_{g,12} \end{pmatrix}^T . \tag{8}$$

The switch position (on/off) in Figure 2 represents the current status of each generator (healthy/total failure). Here, first it is assumed that the multi-generator is healthy or the i'th generator has totally failed, where up to four faults can simultaneously occur. Second, the fault detection and isolation (FDI) problem is not considered in this work. Instead it is assumed that the on-board electronics of the generator/converter unit are able to detect and isolate a generator fault. To accomplish this, an observer-based FDI scheme is proposed in [14] and experimentallx validated in [20] to detect and isolate a speed sensor and actuator fault in wind turbine drive trains with a single generator.

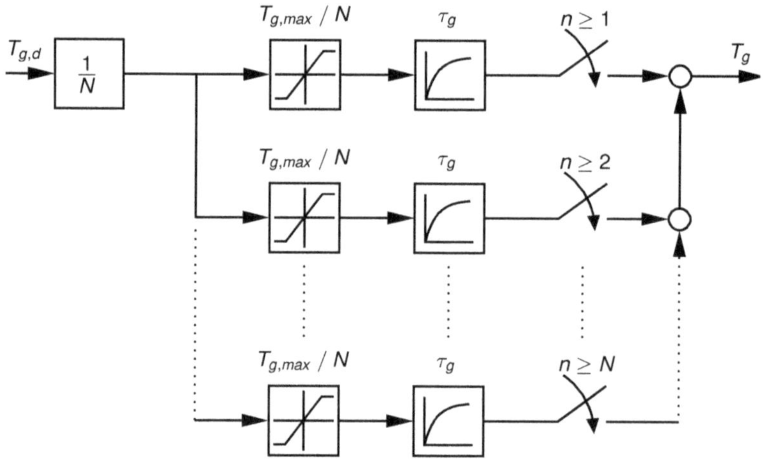

Figure 2. Reduced-order model of multi-generator/converter (Type 1).

An alternative reduced-model is illustrated in Figure 3 and formalized in Equation (9). It is equivalent only if $n \in \{1, \ldots, N\}$ with $n = N - k$ and $k = 0, \ldots, k_{max}$, wherein a maximum number of $k_{max} = 4$ partial generator failures are taken into consideration:

$$\begin{aligned} \dot{T}_{g,single} &= -\frac{1}{\tau_g} T_{g,single} + \frac{1}{\tau_g} \operatorname{sat}\left(\frac{T_{g,d}}{n}\right), \\ T_g &= n \cdot T_{g,single} \end{aligned} \tag{9}$$

with

$$\operatorname{sat}(x) = \begin{cases} x & \text{if} \quad x \leq \frac{T_{g,max}}{N} \\ \frac{T_{g,max}}{N} & \text{otherwise} \end{cases},$$

where $T_{g,single}$ denotes the torque of a single generator unit, N denotes the total number of generator units, n is the number of faulty units and τ_g represents the common delay time constant of each generator unit.

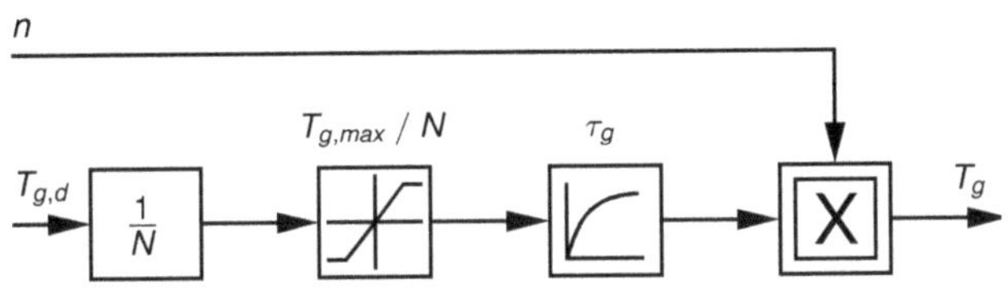

Figure 3. Reduced-order model of multi-generator/converter (Type 2).

2.2. Baseline Controller

The baseline controller of a wind turbine is usually designed separately for the different load regions (partial load region, transition respectively upper partial load region, full load region) [21], see Figure 6. In the partial load region below the rated wind speed, the pitch angle is kept at the fine pitch angle (usually $\beta = 0°$), where maximum energy extraction is possible, while the total torque of the generators is adjusted such that the turbine is operating around the optimal tip speed ratio and thereby around the maximum power coefficient. The standard approach is a deterministic squared-law for the generator torque with dependence on the rotor speed

$$T_{g,d} = k_{opt}\, \omega_r^2 \tag{10}$$

with the coefficient k_{opt} following the optimal torque, such that the rotor is constantly running around the optimal tip speed ratio. In the whole partial load region, the pitch angle is kept at a constant value of $\beta = 0°$.

In the transition region, the generator torque is raised to the maximum value using a fast controller

$$\begin{aligned} T_{g,d} &= T_{g,opt,SP,trq} + \Delta T_{g,d}\,, \\ \Delta T_{g,d} &= -k_{\omega_{r,trq}} \Delta\omega_r - k_{I,trq} \int \Delta\omega_r\, dt\,, \end{aligned} \tag{11}$$

where $T_{g,opt,SP,trq}$ denotes the desired rated collective (total) generator torque, $k_{\omega_{r,trq}}$ and $k_{I,trq}$ denotes the controller coefficient. In the full load region above the rated wind speed, only the pitch angle is adjusted to control the rotor speed around the desired rated rotor speed set point $\omega_{r,SP}$ where $T_{g,d} = T_{g,max}$. Increasing the pitch angle reduces the aerodynamic lift force at the rotor blades and thereby leads to a reduced rotor torque. For the pitch control design in the full load region, the wind turbine state space model (3) is thus considered without an explicitly demanded generator torque and only the pitch angle is used as an input. The control law is given as follows

$$\beta_d = -\sum_{i=1}^{N_r} h_i(\beta)\ \ (\boldsymbol{k}_i^T \Delta\boldsymbol{x}_i + k_{I,i} x_I) \tag{12}$$

with $\Delta\boldsymbol{x}_i = \boldsymbol{x} - \boldsymbol{x}_i$ where $\boldsymbol{x}_i$ denotes the states of the i'te stationarity point and $h_i(\beta)$ are the membership function, which fulfil the convex sum condition

$$0 \le h_i(\beta) \le 1\,, \qquad \sum_{i=1}^{N_r} h_i(\beta) = 1\,. \tag{13}$$

2.3. *Optimal Controller Design for the Full-Load Region*

The details of the Linear–quadratic regulator (LQR) design for wind turbine models in Takagi-Sugeno form and a formal stability analysis are proposed in [19]. In the following, the necessary steps are presented for the practical controller design. In this section, the matrices $\tilde{\mathbf{k}}_i^T = \begin{pmatrix} \mathbf{k}_i^T & k_{I,i} \end{pmatrix}$ of the control law (12) are designed separately for each linear submodel by means of the LQR design with the augmented state vector

$$\boldsymbol{x_a} = \begin{pmatrix} y_T & y_B & \theta_s & \dot{y}_T & \dot{y}_B & \omega_r & \omega_g & \beta & x_I \end{pmatrix}^T \tag{14}$$

where

$$x_I := \int \left(\omega_r - \omega_{r,SP}\right) dt\,. \tag{15}$$

The LQR controller design minimizes the functional

$$J = \int_0^\infty \left(\tilde{\mathbf{x}}^T \mathbf{Q} \tilde{\mathbf{x}} + u^T R u\right) d\tau\,, \tag{16}$$

with the weighting matrices $\mathbf{Q}$ and R. In order for these to be determined, the maximal absolute values of the contributing states first have to be estimated. For the wind turbine model, the following estimates are taken: $y_{T,\max} = 1.5\,\mathrm{m}$, $y_{B,\max} = 10\,\mathrm{m}$, $\theta_{s,\max} = 0.002\,\mathrm{rad}$, $\dot{y}_{T,\max} = 0.2\,\frac{\mathrm{m}}{\mathrm{s}}$, $\dot{y}_{B,\max} = 2\,\frac{\mathrm{m}}{\mathrm{s}}$, $\omega_{g,\max} = \omega_{r,\max} = 1.57\,\frac{\mathrm{rad}}{\mathrm{s}}$, $\beta_{\max} = 1.57\,\mathrm{rad}$

Subsequently, the individual states are weighted with relative weights, where the state $x_6 = \omega_r$ is weighted most since it is the primary quantity to be controlled: $Q_1 = 0$, $Q_2 = 0$, $Q_3 = 2 \times 10^{-9}$, $Q_4 = 0.02$, $Q_5 = 0.1$, $Q_6 = 1.57$, $Q_7 = 0$, $Q_8 = 1.5706 \times 10^{-6}$, $Q_I = 0.05$.

In the final weighting matrix, the relative weights are divided by the squares of estimated maximum absolute values to take account the magnitudes of the individual states:

$$\mathbf{Q} = \mathrm{diag}\begin{pmatrix} \frac{Q_1}{y_{T,\max}^2} & \frac{Q_2}{y_{B,\max}^2} & \frac{Q_3}{\theta_{s,\max}^2} & \frac{Q_4}{\dot{y}_{T,\max}^2} & \frac{Q_5}{\dot{y}_{B,\max}^2} & \frac{Q_6}{\omega_{r,\max}^2} & \frac{Q_7}{\omega_{g,\max}^2} & \frac{Q_8}{\beta_{\max}^2} & Q_I \end{pmatrix}. \tag{17}$$

The weight of the control input is chosen as $R = 1$.

3. Fault Tolerant Control Using Adaption of Reference Signals

In this section, the impact on the closed loop dynamics of the total generator faults (up to four) that occurred is considered. First, an FTC scheme with an adaptation of the reference signal is proposed. Second, the improvement is illustrated by means of steady state operation in the partial and full load region and the impact on the closed-loop dynamics is shown by rotor acceleration tests.

3.1. *Fault Dependent Adaptation of the Reference Signals*

The working principle of the reference signal adaptation is illustrated by the block diagram in Figure 4.

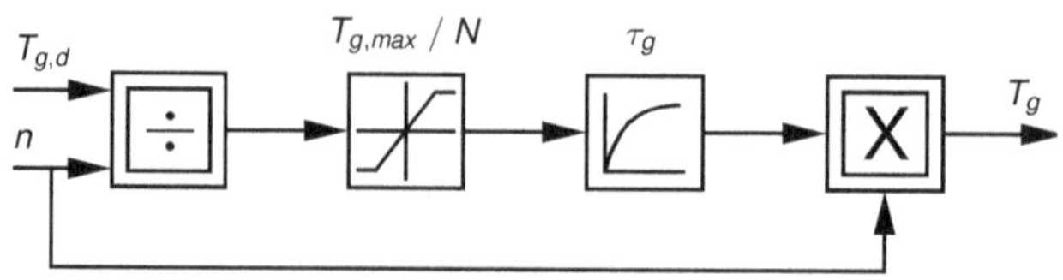

Figure 4. Reference signal adaptation for demanded generator torque.

In contrast to an equidistant nominal distribution of N generator torques (see the first left block in Figures 2 and 3), the adaptation scheme exploits the known number n of healthy generators. Note: In the last section it was assumed that the number of failures is known. If a generator failure occurs, the reference signals of the remaining generators increases. This works as long as the saturation has not been reached. The effect on the overall system is investigated in the next subsections.

3.2. Fault Tolerant Interaction between the Partial and Full Load Region

In Figure 5, the different control zones for the fault free case 12/12 (also shown in Figure 6) and the faulty cases $12/n$ with $n = 12 - k$ for $k = 1, ..., 4$ generator faults as total single generator failures are illustrated by plotting the torque versus the rotor speed. Typically, there is a cut-in rotor speed $\omega_{r,1} = 0.72$ (refers to the NREL benchmark WT [22]) which is shown in Figure 6. Due to a fault in $k = 1, ..., 4$ single generators, the maximum controllable torque gradually decreases. The multi-generator drive train is fault-tolerant as long as the demanded torque is less than or equal to the cumulated generator saturation of the remaining fault-free generators. This has implications in the full load region ($\omega_r > \omega_{r,SP}$, see Figure 6) and the pitch control loop due to the lower rated power. This is made clear in Figure 7, where the curve of stationary pitch angle value vs. the wind speed's upward shift is dependent on increasing generator torque losses. This means that for $k = 1, ..., 3$ generator fault(s), a reduction of the maximum controllable torque is caused in the upper partial load region defined by $\omega_{r,SP,trq} \leq \omega_r \leq \omega_{r,SP}$ and for $k = 4$ a reduction of the maximum controllable torque in the lower partial load region $\omega_{r,2} \leq \omega_r < \omega_{r,SP,trq}$ is seen, see Figure 6. That results in an acceleration of the rotor and if the rotor speed reaches a value above the nominal speed, the pitch controller for the full load operation is activated. The pitch controller has the speed error (difference of the current rotor speed and its nominal speed) as an input and produces the collective reference pitch angle for the pitch drives. An increasing pitch angle greater than zero causes a reduction of the aerodynamic torque in such a way that this is adapted to the limited maximum torque. The following pseudo-code illustrates the overall behavior of the generator torque and pitch control in the partial and full load region.

Algorithm 1: FTC hybrid wind turbine control scheme

if $\omega_{r,2} \leq \omega_r(t) < \omega_{r,SP,trq}$ **then**
    /* lower partial load region */;
    $T_{g,d}(t) = k_{opt}\ \omega_r^2(t)$ /* Equation (10) */;
    $\beta_d(t) = 0°$;
else
    /* upper partial load region */;
    if $\omega_{r,SP,trq} \leq \omega_r(t) \leq \omega_{r,SP}$ **then**
        $T_{g,d}(t) =$ TS-cntrl-Tg$[e(t)]$ with $e(t) = \omega_r(t) - \omega_{r,SP}$ /* Equation (11) */;
        $\beta_d(t) = 0°$;
    else
        /* full load region */;
        $T_{g,d}(t) = T_g$;
        $\beta_d(t) =$ TS-cntrl-beta$[e(t), \mathbf{x}(t)]$ with $e(t) = \omega_r(t) - \omega_{r,SP}$ /* Equation (12) */;
    end
end
$T_{g,d,FTC}(t) = T_g(t) =$FT-cntrl$[T_{g,d}(t)]$ /* Equation (9) */;

In the above algorithm, PI-cntrl-Tg$[e(t)]$ denotes the PI generator torque controller (11) in the upper partial load, TS-cntrl-beta$[e(t), \mathbf{x}(t)]$ denotes the pitch controller (12) in the full load region and FT-cntrl$[T_{g,d}(t)]$ represents the model-based fault tolerant control scheme using (9).

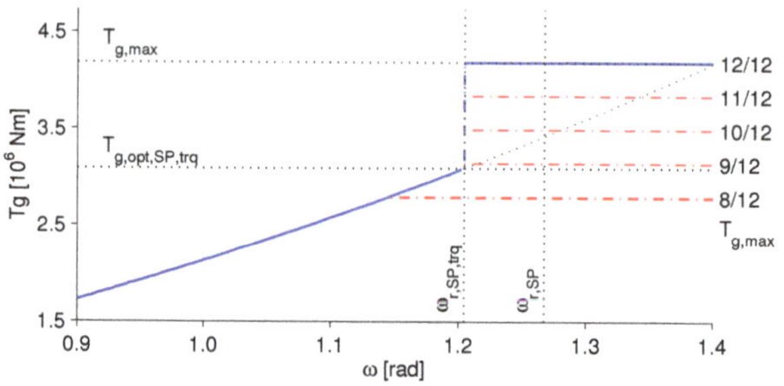

Figure 5. Regions of control and demanded generator torque dependent upon number of $k = 0, 1, ..., 4$ faulty generators.

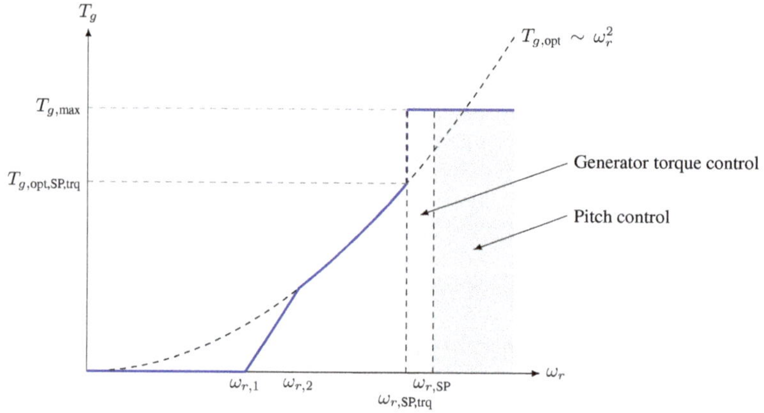

Figure 6. Regions of wind turbine control; Total Generator torque for the fault-free case dependent upon the rotor angular speed.

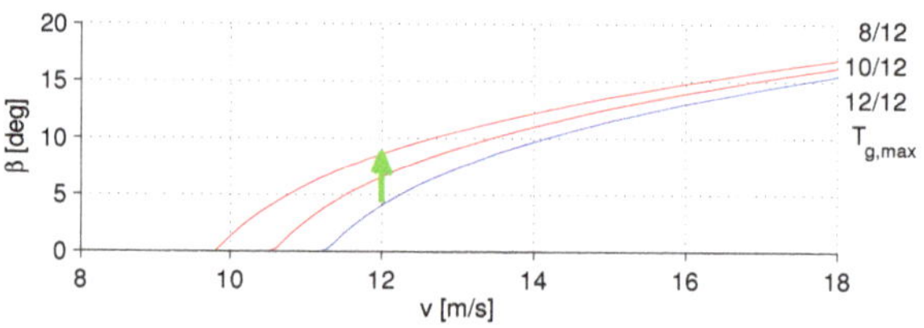

Figure 7. Stationary pitch angle values in the full load region dependent upon number of faults.

3.3. Rotor Acceleration Test of the Multi-Generator Drive Train

Figure 8 shows a comparison between the fault-free case ($n = N$) and faulty case ($n = 8$) during rotor acceleration from cut-in to medium strong wind speed. It can be seen as in the previous stationary consideration, but now more clearly the *fault tolerant control characteristics* of the baseline controller in combination with an fault dependent adaptation of the demanded generator torque can be observed:

- partial load: remaining healthy generators compensate the occurred generator faults
- full load: previous engagement of pitch control compensates for the decreased maximum total torque

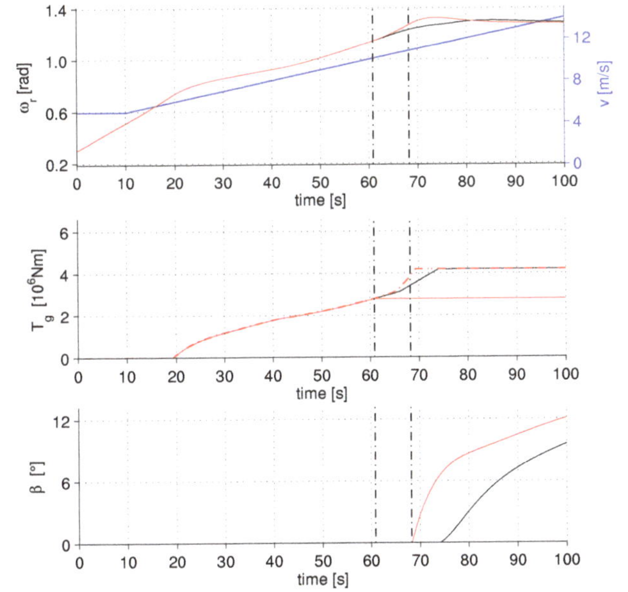

Figure 8. Rotor acceleration test in partial, transition and full load region: fault-free case, $n = N$ (black line) and faulty case $n = 8$ (red line), wind speed (blue line).

4. Transient Behavior of Fault Tolerant Control Scheme

In this section the transient behavior in the presence of abrupt generator faults in the partial and full load region is considered. Finally, a classic control design is used for the active damping of weakly damped torsional vibrations.

4.1. Transient Behavior in the Partial Load Region

Figure 9 shows the impact of an abrupt fault on the system dynamics. Here, the controller is able to reach the demanded generator torque in a sufficient time. Unfortunately, the excited oscillation of the torsion angle cannot be compensated for by the same strategy. For this purpose, an active damping is designed in Section 4.3.

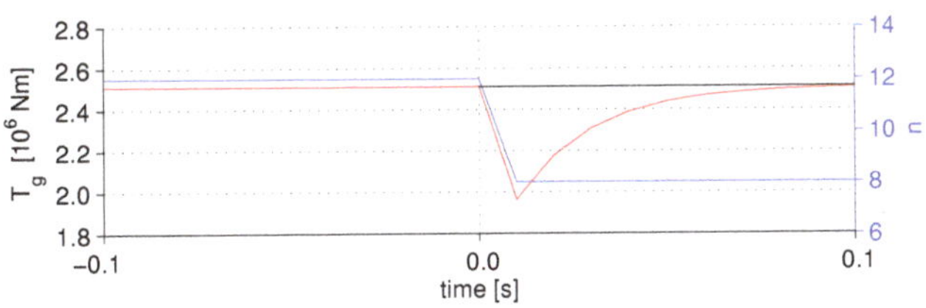

(a)

Figure 9. *Cont.*

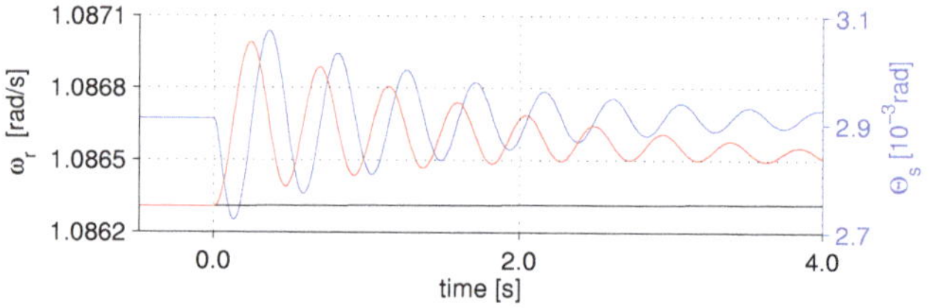

(b)

Figure 9. Generator failure in the partial load region at constant wind speed 8 m/s. (**a**) otal generator torque T_g (red line), demanded torque $T_{g,d}$ (black line), and $n \in \mathbb{N}^+$ denotes the number of healthy generators (blue line); (**b**) rotor speed ω_r (red), demanded rotor speed ω_r (black), shaft torsion angle Θ_s (blue).

4.2. Transient Behavior in the Full Load Region

The transient behavior in the full load region is illustrated in Figure 10. Here, by lowering the full load region the pitch controller is able to reach the demanded rotor speed in a sufficient time. As before, the excited oscillation cannot be compensated for by the same strategy. For this purpose, an active damping is designed in the next subsection.

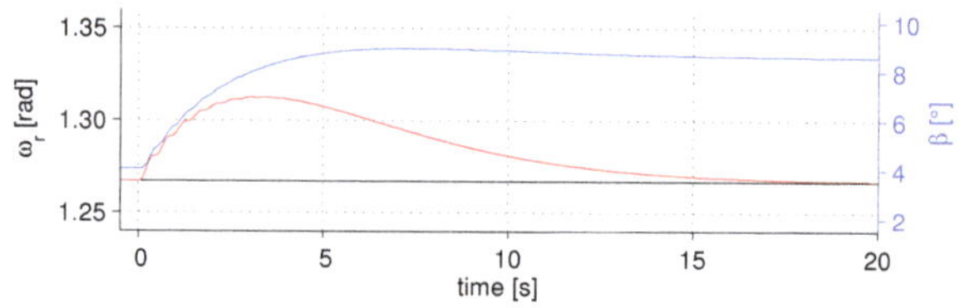

(a)

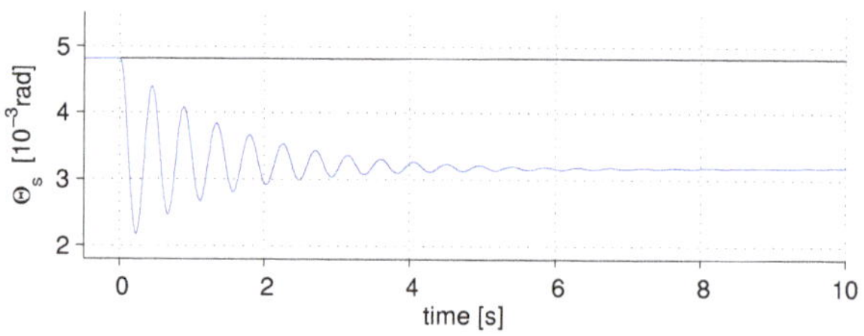

(b)

Figure 10. Generator failure in the full load region at constant wind speed 12 m/s. (**a**) Total generator torque T_g (red line), demanded torque $T_{g,d}$ (black line), pitch angle β (blue line); (**b**) shaft torsion angle Θ_s (bue), demanded angle Θ_s (black).

4.3. Active Vibration Damping

The active damping based on the superposition of the demanded generator torque $T_{g,d}$ calculated with (10) or (11) by

$$\tilde{T}_{g,d} = T_{g,d} + \Delta T_{g,d} \tag{18}$$

with the control law

$$\Delta T_{g,d} = k_{damp} \, \dot{\theta}_s \, . \tag{19}$$

The control design follows the classic approach in [23] and is based on the state space model (3). In Figure 11, two different cases ($n = 11, n = 8$) are illustrated by plotting the shaft torsion angle without and with active damping by (19).

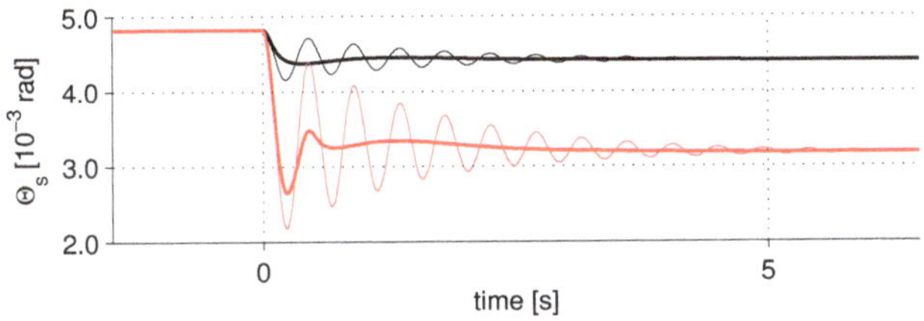

Figure 11. Control performance of active damping systems: Comparison between one (black line) and four total generator failures (red line).

5. Conclusions and Future Works

In this paper, a distributed generator drive train is studied under fault conditions. A control-oriented model is proposed and a controller structure with two layers is studied. The upper layer is the baseline controller, which is separated into a partial load region with generator torque as the actuating signal and the full-load operation region with a collective pitch angle as the actuating signal. The lower layer is responsible for FTC characteristics of the distributed generator drive train, which is based on an adaptation of the demanded generator torque. The practicability is illustrated by means of steady state and rotor acceleration tests operation in the partial and full load region.

An open problem is the delayed detection of generator faults in real systems. The delayed detection causes a mismatch of the distribution of the generator torques, which leads to an intermittent increasing of the total generator torque and a decreasing of the rotor speed. It must be studied in detail whether a torque controller in the upper partial load range is able to avoid significant drop of the rotor speed. Further, the controller robustness for different load cases has to be tested with a much more detailed wind turbine model, which is i.a. implemented in the aero-elastic simulation package FAST (Fatigue, Aerodynamics, Structures, and Turbulence) by the National Renewable Energy Laboratory of the U.S. Department of Energy (NREL).

References

1. Cohen, J.; Schweizer, T.; Laxson, A.; Butterfield, S.; Schreck, S.; Fingersh, L.; Veers, P.; Ashwill, T. *Technology Improvement Opportunities for Low Wind Speed Turbines and Implications for Cost of Energy Reduction*; NREL/TP-500-41036; NREL: Golden, CO, USA, 2008.
2. Robb, D. The return of the Clipper Liberty wind turbine. *Power* **2008**, *152*, 74.
3. Giger, U.; Arnaudov, K. New drive train design for ultra large 15 MW wind turbines. In Proceedings of the International Conference on Gears, Garching, Germany, 7–9 October 2013.
4. Sloth, C.; Esbensen, T.; Stoustrup, J. Robust and fault-tolerant linear parameter-varying control of wind turbines. *Mechatronics* **2011**, *21*, 645–659.
5. Odgaard, P.F.; Stoustrup, J. Fault tolerant control of wind turbines using unknown input observers. In Proceedings of the IFAC Symposium on Fault Detection, Supervision and Safety of Technical Processes, Mexico City, Mexico, 29–31 August 2012; pp. 313–318.
6. Odgaard, P.F.; Stoustrup, J.; Kinnaert, M. Fault Tolerant control of wind turbines—A benchmark model. In Proceedings of the IFAC Symposium on Fault Detection, Supervision and Safety of Technical Processes, Barcelona, Spain, 30 June–3 July 2009; pp. 155–160.
7. Sami, M.; Patton, R.J. An FTC approach to wind turbine power maximisation via T-S fuzzy modelling and control. In Proceedings of the IFAC Symposium on Fault Detection, Supervision and Safety of Technical Processes, Mexico City, Mexico, 29–31 August 2012; pp. 349–354.

8. Sami, M.; Patton, R.J. Active sensor fault tolerant output feedback tracking control for wind turbine systems via T-S model. *Eng. Appl. Artif. Intell.* **2014**, *34* , 1–12.
9. Sami, M.; Patton, R.J. Fault tolerant adaptive sliding mode controller for wind turbine power maximisation. In Proceedings of the 7th IFAC Symposium on Robust Control Design, Aalborg, Denmark, 20–22 June 2012; pp. 499–504.
10. Rotondo, D.; Nejjari, F.; Puig, V.; Blesa, J. Fault tolerant control of the wind turbine benchmark using virtual sensors/actuators. In Proceedings of the IFAC Symposium on Fault Detection, Supervision and Safety of Technical Processes, Mexico City, Mexico, 29–31 August 2012; pp. 114–119.
11. Simani, S.; Castaldi, P. Active actuator fault-tolerant control of a wind turbine benchmark model. *Int. J. Robust Nonlinear Control* **2014**, *24*, 1283–1303.
12. Badihi, H.; Zhang, Y.; Hong, H. Fuzzy gain-scheduled active fault tolerant control of a wind turbine. *J. Frankl. Inst.* **2014**, *351*, 3677–3706.
13. Badihi, H.; Zhang, Y.; Hong, H. Wind turbine fault diagnosis and fault-tolerant torque load control against actuator faults. *IEEE Trans. Control Syst. Technol.* **2015**, *23*, 1351–1372.
14. Georg, S.; Schulte, H. Actuator fault diagnosis and fault-tolerant control of wind turbines using a Takagi-Sugeno sliding mode observer. In Proceedings of the International Conference on Control and Fault-Tolerant Systems (Sys-Tol), Nice, France, 9–11 October 2013; pp. 516–522.
15. Takagi, T.; Sugeno, M. Fuzzy identification of systems and its application to modeling and control. *IEEE Trans. Syst. Man Cybern.* **1985**, *15*, 116–132.
16. Jonkman, J.M.; Buhl, M L., Jr. *FAST User's Guide*; Technical Report NREL/EL-500-38230; National Renewable Energy Laboratory: Golden, CO, USA, 2005.
17. Georg, S.; Schulte, H.; Aschemann, H. Control-oriented modelling of wind turbines using a Takagi-Sugeno model structure. In Proceedings of the IEEE International Conference on Fuzzy Systems, Brisbane, Australia, 11–15 June 2012; pp. 1737–1744.
18. Tanaka, K.; Sano, M. A robust stabilization problem of fuzzy control systems and its application to backing up control of a truck-trailer. *IEEE Trans. Fuzzy Syst.* **1994**, *2*, 119–134.
19. Georg, S. Fault Diagnosis and Fault-Tolerant Control of Wind Turbines Nonlinear Takagi-Sugeno and Sliding Mode Techniques. Ph.D. Thesis, Fakultät für Maschinenbau und Schiffstechnik, University of Rostock, Rostock, Germany, 2014.
20. Schulte, H.; Georg, S. Hardware-in-the-Loop Test-Bed for benchmarking of fault tolerant control schemes for wind turbines. In Proceedings of the 9th IFAC Symposium on Fault Detection, Supervision and Safety of Technical Processes, Paris, France, 2–4 September 2015.
21. Burton, T.; Jenkins, N.; Sharpe, D.; Bossanyi, E. *Wind Energy Handbook*, 2nd ed.; John Wiley & Sons Ltd.: New York, NY, USA, 2011.
22. Odgaard, P.F.; Johnson, K. Wind turbine fault detection and fault tolerant control—An enhanced benchmark challenge. In Proceedings of the American Control Conference, Washington, DC, USA, 17–19 June 2013; pp. 4447–4452.
23. van der Hooft, E.L.; Schaak, P.; van Engelen, T.G. *Wind Turbine Control Algorithms; Technical Report DOWEC-F1W1-EH-03-094/0*; Energy Research Center of the Netherlands (ECN): Delft, The Netherlands, 2003.

energies

MDPI

Article

Modeling and Maximum Power Point Tracking Control of Wind Generating Units Equipped with Permanent Magnet Synchronous Generators in Presence of Losses

Andrea Bonfiglio *, Federico Delfino, Marco Invernizzi and Renato Procopio

Department of Electrical, Electronic, Telecommunications Engineering and Naval Architecture, University of Genoa, Via Opera Pia 11a, I-16145 Genoa, Italy; federico.delfino@unige.it (F.D.); marco.invernizzi@unige.it (M.I.); renato.procopio@unige.it (R.P.)

* Correspondence: a.bonfiglio@unige.it; Tel.: +39-010-353-2730

Academic Editor: Paolo Mercorelli

Received: 27 September 2016; Accepted: 3 January 2017; Published: 15 January 2017

Abstract: This paper focuses on the modeling of wind turbines equipped with direct drive permanent magnet synchronous generators for fundamental frequency power system simulations. Specifically, a procedure accounting for the system active power losses to initialize the simulation starting from the load flow results is proposed. Moreover, some analytical assessments are detailed on typical control schemes for fully rated wind turbine generators, thereby highlighting how active power losses play a fundamental role in the effectiveness of the wind generator control algorithm. Finally, the paper proposes analytical criteria to design the structure and the parameters of the regulators of the wind generator control scheme. Simulations performed with Digsilent Power Factory validated the proposed procedure, highlighting the impact of active power losses on the characterization of the initial steady state and that the simplifying assumptions done in order to synthesize the controllers are consistent with the complete modeling performed by the aforementioned power system simulator.

Keywords: power systems modeling; renewable power generation; wind power

---

1. Introduction

The evolution of the electricity production scenario changed the operating conditions of the electricity system itself providing, on one side, benefits related to distributed generation from renewables [1,2] but, on the other, issues in terms of quality, security and reliability of power generation and delivery related to the stochastic behavior of the prime energy source [3–5]. The strong presence of power electronic devices in renewable energy sources (RES) like photovoltaic or wind power plants makes them suitable and flexible enough to act as supporters for the electricity system operation [6,7], as they can provide some of the ancillary services that in the past needed the installation of dedicated devices [8,9]. For these reasons, numerous grid code requirements of various electricity system operators all over the world are beginning to require large renewable plants to act as providers of ancillary services. Examples of this can be found in the Italian, German, French and Canadian technical regulations [10–13]. As a consequence, both the increasing number of RES installations integrated into large-scale electrical power systems and the necessity of enhancing their performances in order to meet the standard requirements illustrate the need for the development of adequate modeling of such units and effective control systems [14].

From this point of view, wind turbine generator (WTG) controllers are much more demanding, due to the presence of many different devices (turbine, gearbox, shaft, electric machine, power electronic converters, controllers and so on) and configurations.

From a power system modeling point of view, to evaluate the impact of high wind power penetrations on electrical power systems, there is the necessity to incorporate WTG models into already existing software packages that perform all the typical power system analyses (e.g., load flow calculation, root mean square (RMS) transients, electromagnetic (EMT) transients, short circuit analysis and so on). To reach this goal, many different papers have been published in the literature specifying the assumptions that have to be made in order to make such models effective with respect to the results one wants to obtain and consistent with the modeling of the remaining part of the power system (see [15–17] for details). To this extent, the first problem to be solved is related to the calculation of the initial conditions for a dynamic simulation starting from the load flow results of the power system. This aspect has been addressed in [18], under the hypothesis of a lossless system, but, to the best of our knowledge, the initialization problem is still unsolved if active power losses occur. The losses issue has been addressed in [19] and more recently in [20]. In [19], a control scheme for a back-to-back supplied induction generator is proposed in order to minimize the overall system losses, whereas in [20], the authors focused on evaluating how the maximum power point moves if one accounts for them and proposed a new maximum power point tracking (MPPT) iterative method.

However, neither [19] nor [20] discussed the problem of the dynamic system initial condition calculation in the presence of active power losses.

From the control system standpoint, different solutions have been studied in the literature accounting for both traditional and innovative approaches. In this last case, two of the most widely employed nonlinear control techniques, i.e., feedback linearization (FBL) and sliding modes (SM) have been employed respectively in [21–23]. In [22], the attention is concentrated on the control of the machine side converter (MSC) and the application of the particle swarm optimization for the MPPT, while in [21] a unified control scheme for both the machine and grid side converter (GSC) is presented. The problem of model and parametric uncertainties typical of model-based control techniques like FBL is overcome in [23] with a method able to handle disturbances. However, whereas references [21,22] suffer from robustness problems, the control strategy proposed in [23] is very demanding in terms of the design of the WTG controller.

For this reason, in many applications, WTGs are still equipped with controllers designed on the basis of classical, proportional integral derivative (PID)-based controllers. In particular, modeling and control concepts for the doubly fed induction generator (DFIG) are presented in [24,25]; both of them propose a nested control structure where each controller consists of a classical proportional integral (PI) structure, but they do not give evidence of the impact of the losses in the control scheme. Furthermore, while reference [24] does not provide any guidelines on how to choose the controller parameters, reference [25] states that an analytical derivation of the transfer function required to assign the system closed loop poles is not practical and so moves toward a reduced order model.

The other widely employed wind turbine concept is the direct drive synchronous generator (DDSG) consisting of either a permanent magnet synchronous generator (PMSG) or an electrically excited one. Such a structure is addressed in [26–28], but, again, no reference is made to the losses problem both from the steady-state modeling and from the controller design point of view. Moreover, reference [26] focuses on the supervisory reactive power control that produces the reference signals for the lower hierarchical level controller, reference [27] provides a small signal stability analysis and reference [28] presents three possible control strategies according to three different control aims, i.e., the unity power factor, the constant stator voltage and the maximum torque. However, in all the cases, a criterion to choose the controller parameters is missing. This aspect has been addressed for the only case of the GSC controller in [29] and some numerical tests appear in [30], but without a comprehensive dissertation related to the PMSG model.

As a consequence of this very complex and varied literature framework, the aim of the present paper is that of filling the following gaps:

- Defining a methodology to manage the initialization of WTG models in the presence of power system losses;

- Proposing a control scheme that allows tracking the maximum power point even in the presence of losses;
- Providing a detailed analyses on the structure of the controllers and analytical criteria to choose their parameters according to the system's desired dynamic behavior.

The paper is organized as follows. Section 2 proposes a review of the power system model of a fully rated WTG equipped with a PMSG, while Section 3 is dedicated to the problem of the system initialization in the presence of active power losses. Section 4 proposes an application of the procedure defined in Section 3 and provides some analytical results to highlight the error committed by neglecting active power losses. In Section 5, a detailed description of the overall machine control scheme is proposed, while in Section 6 criteria to properly define the parameters of the different unit controllers are detailed. Finally, Section 7 is dedicated to the validation of the procedure described in Section 6, followed by some conclusive remarks in Section 8.

2. Power System Model

One of the two most employed strategies to integrate the power coming from a wind plant into the electric system is that of using a PMSG connected to the grid by means of two power electronic converters (as depicted in Figure 1) [15,16].

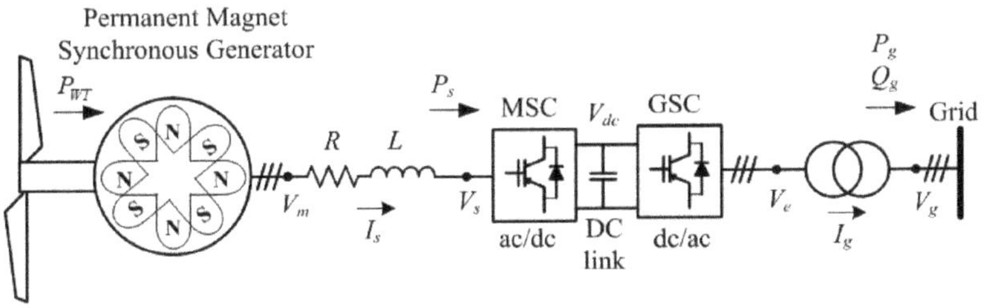

Figure 1. Direct drive synchronous generator (DDSG) wind turbine layout. MSC: machine side converter; and GSC: grid side converter.

The first one (from now on MSC) is typically adopted in order to extract the maximum power from the wind turbine and to control the voltage (V_m) at the machine terminals. The second one (from now on GSC) controls the voltage on the DC link (v_{DC}) and the reactive power (Q_g) delivered to the distribution network.

In order to define a suitable control architecture to follow the abovementioned goals, it is necessary to derive a mathematical model of the system to be controlled.

2.1. Wind Turbine Model

As is well known, in power system simulations wherein the electrical behavior of the wind system is the main point of interest, a simplified modeling of the wind turbine rotor is normally used. Such a model consists of the following algebraic relation between wind speed and the extracted mechanical power:

$$P_{WT}(\omega, v_w, q) = \frac{1}{2S_b}\rho A c_p[\lambda(\omega, v_w), \theta]v_w^3 \quad (1)$$

where P_{WT} is the power extracted from the wind in p.u. on the electric machine base S_b; ρ is the air density (kg/m^3); A is the area covered by the rotor (m^2), θ is the pitch angle of rotor blades (deg), v_w is the wind speed at hub height upstream the rotor (m/s); and ω is the turbine angular speed in p.u. on the base of the rated machine speed ω_n. c_p is the performance coefficient or power coefficient depending on the pitch angle θ and on the tip speed ratio, defined as:

$$\lambda = \frac{R\omega}{v_w} \tag{2}$$

where R is the rotor radius. The performance coefficient's typical shape is depicted in Figure 2 [31].

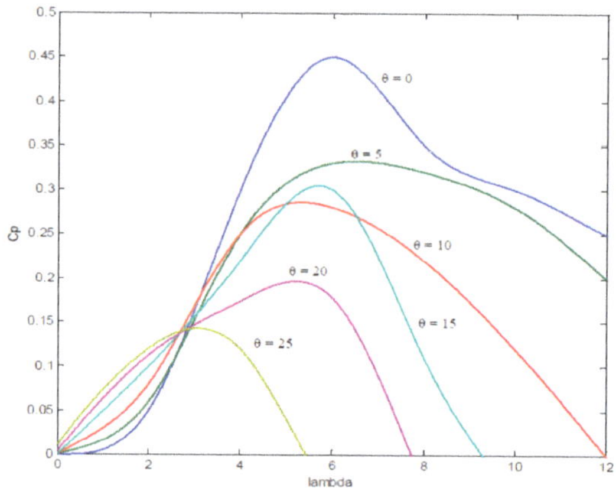

Figure 2. Performance coefficient behavior in accordance with the pitch angle and the tip speed ratio.

From Figure 2 it can be observed that for an assigned value of the pitch angle, there exists only one value of the tip speed ratio that maximizes the performance coefficient. As a consequence, for any wind speed, there exists only one turbine angular speed ω^* that allows extracting the maximum power $P_{WT}{}^*$ from the wind. Therefore, to maximize the unit power production, the following control curve (MPPT curve) can be defined as [32]:

$$P_{WT}* = \begin{cases} 0 & \omega < \omega_{min} \\ \frac{\rho A c_{p,opt} R}{2 S_b \lambda_{p,opt}} \omega^{*3} & \omega_{min} \leq \omega < \omega_{max} \\ P_{max} & \omega \geq \omega_{max} \end{cases} \tag{3}$$

where $\lambda_{p,opt}$ is the optimal tip speed ration at null pitch angle corresponding to the maximum performance coefficient ($c_{p,opt}$).

Moreover, increasing the pitch angle will result in a reduction of the power extracted from the wind. As a consequence, this feature is typically used to avoid over-synchronous speeds and is implemented in the so-called pitch angle controller (see [15,16] for details). The working principle of the pitch controller is the following: when the electric machine angular speed exceeds a specified threshold, the controller changes the value of the pitch angle in order to reduce the power extracted from the wind. Since the pitch control works only for over-synchronous speeds, an anti-wind up limiter locks the pitch angle to 0 for sub-synchronous speeds.

The machine stator voltage amplitude, V_m, is set according to the desired machine speed in the so-called "constant V/f" regulation mode. In other words, the per unit stator voltage amplitude is equal to the per unit machine speed until the stator rated voltage (1 p.u.). For speed values higher than 1 p.u., the stator voltage reference is kept to 1 p.u. To summarize, the machine stator voltage reference can be defined as:

$$V_m^* = \min\{\omega^*, 1\} \tag{4}$$

In accordance with Equations (3) and (4), for a given wind speed the machine angular speed, the power extracted from the primary source and the machine voltage amplitude are uniquely defined.

2.2. Permanent Magnet Synchronous Generator Model

Describing the machine in a Park-oriented frame [33] with the passive sign convention from a mechanical point of view and the active sign convention from the electrical one, the stator electric equations (in p.u. on the machine basis) are as follows [3]:

$$\begin{cases} v_{md} = -R_s i_{sd} - \frac{1}{\omega_n}\frac{d\phi_{sd}}{dt} - \omega\phi_{sq} \\ v_{mq} = -R_s i_{sq} - \frac{1}{\omega_n}\frac{d\phi_{sq}}{dt} + \omega\phi_{sd} \end{cases} \quad (5)$$

where R_s is the stator resistance, $v_{md(q)}$ is the direct (quadrature) axis component of the machine voltage, $i_{sd(q)}$ is the direct (quadrature) axis component of the current outgoing from the machine (see Figure 1) and $\phi_{sd(q)}$ is the direct (quadrature) axis component of the machine linkage. The magnetic stator equations are the following:

$$\begin{cases} \phi_{sd} = x_d i_{sd} + \psi \\ \phi_{sq} = x_q i_{sq} \end{cases} \quad (6)$$

where x_d, x_q and ψ are the stator direct and quadrature axis reactances at rated frequency and the permanent magnet flux respectively. Finally, the electromagnetic torque (T_e) is given by:

$$T_e = -i_{sd}\phi_{sq} + i_{sq}\phi_{sd} = (x_d - x_q)i_{sd}i_{sq} + \psi i_{sq} \quad (7)$$

and the motion equation states that:

$$\frac{P_{WT}}{\omega} - T_e = 2H\frac{d\omega}{dt} \quad (8)$$

where H is the WTG inertia constant.

2.3. Machine Side Converter Model

The MSC can be considered as a voltage source V_s, whose axis components v_{sd} and v_{sq} can be adjusted acting on the pulse width modulation (PWM) inverter parameters. This voltage is related to the machine one according to the following relationships:

$$\begin{cases} v_{sd} = v_{md} - Ri_{sd} - \left(\frac{L}{\omega_n}\right)\frac{di_{sd}}{dt} - \omega L i_{sq} \\ v_{sq} = v_{mq} - Ri_{sq} - \left(\frac{L}{\omega_n}\right)\frac{di_{sq}}{dt} + \omega L i_{sd} \end{cases} \quad (9)$$

where R and L are the connection cable resistance and inductance. Finally, the PWM modulating index m_a and phase angle δ can be obtained as follows:

$$\begin{cases} m_a = \frac{2}{V_{DC}}\sqrt{v_{sd}^2 + v_{sq}^2} \\ \delta = arctg\left(\frac{v_{sq}}{v_{sd}}\right) \end{cases} \quad (10)$$

where V_{DC} is the DC link voltage.

2.4. Grid Side Converter Model

Assuming that the GSC can be modeled as an ideal voltage source (i.e., with no delays and no switching and state losses) and that the connection between the converter and the main network is represented as an *R-L* series impedance (accounting for cables, transformer and the series section of the filters), then, the system portion between the GSC and the network can be represented with the equivalent circuit of Figure 3.

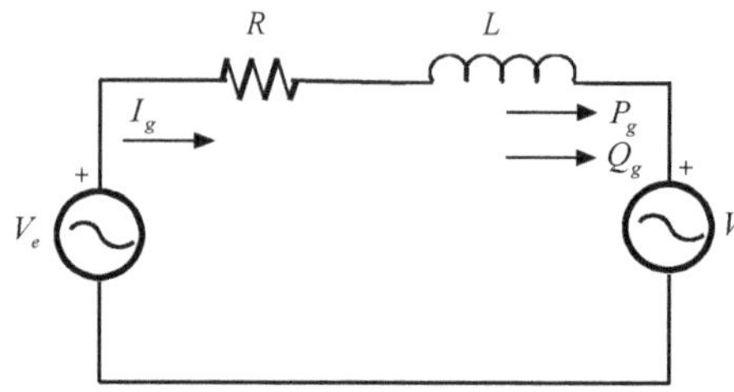

Figure 3. Equivalent circuit of the grid side converter, the connection system and the main network.

Describing the system in a Park reference frame [33] rotating at the network angular frequency ω_e (expressed in p.u. on the rated angular frequency ω_g), the Kirchhoff voltage law applied to the mesh depicted in Figure 3 states that:

$$\begin{cases} v_{gd} = v_{ed} - R_T i_{gd} - \frac{L_T}{\omega_g}\frac{d}{dt} i_{gd} - \omega_e L_T i_{gq} \\ v_{gq} = v_{eq} - R_T i_{gq} - \frac{L_T}{\omega_g}\frac{d}{dt} i_{gq} + \omega_e L_T i_{gd} \end{cases} \tag{11}$$

where (in p.u. on the machine basis):

- R_T and L_T are the connection resistance and the inductance, respectively;
- $v_{ed(q)}$ is the direct (quadrature) axis components of the voltage at the AC terminals of the GSC;
- $v_{gd(q)}$ is the direct (quadrature) axis components of the grid voltage and
- $i_{gd(q)}$ is the direct (quadrature) axis components of the current flowing in the connection.

3. Initialization of Fundamental Frequency Simulations Starting from Load-Flow Calculation

All the fundamental frequency simulations start from an equilibrium point deriving from a load flow calculation. As a consequence, it is necessary to derive a procedure that allows finding out the system equilibria starting from the load flow output, i.e., to initialize the variables of a wind power plant. Such a procedure has been studied in [18] for a lossless system.

Calculating the initial conditions of a dynamic model from load flow data means obtaining the inputs and the states of the system starting from the system outputs. Therefore, the description and modeling of the components of the wind power plant have to start from the grid connection point. Moreover, when calculating the initial conditions, the system is assumed to be at steady state. Therefore, the equations presented in the previous sections will be used for zeroing all the time derivatives. However, the procedure derived in [18] assumes that the system losses (i.e., electrical machine, transformer, cables and so on) are negligible. This means that if one applies this algorithm to a system where such losses are different from zero, the calculated solution will not result in an equilibrium point from which the simulation can start.

The present section proposes a modification of the initialization procedure described in [18] in order to account for the system losses.

With reference to the notations of Figure 3, the load flow calculation exposes the initial values (indicated with the subscript "0") of the grid active power P_{g0}, of the reactive power Q_{g0} and of the voltage phasor V_{g0} at the machine bus. Starting from these values, the initialization problem can be split into two parts regarding the portion of the system between the AC terminals of the GSC and the external network connection (from now on network side portion) and the one between the wind turbine and the AC terminals of the MSC (from now on machine side portion).

3.1. Network Side Portion

For the network side portion, the initialization problem consists of finding the values of v_{gd0}, i_{gd0}, i_{gq0}, v_{ed0}, v_{eq0}. Choosing the Park initial angle such that the network voltage V_g has only a direct axis component [33], it thus follows that:

$$\begin{cases} v_{gd0} = V_{g0} \\ v_{gq0} = 0 \end{cases} \quad (12)$$

and:

$$\begin{cases} P_{g0} = v_{gd0} i_{gd0} \\ Q_{g0} = v_{gd0} i_{gq0} \end{cases} \quad (13)$$

which allows finding the initial values of the current axis components, as:

$$\begin{cases} i_{gd0} = \frac{P_{g0}}{V_g} \\ i_{gq0} = \frac{Q_{g0}}{V_g} \end{cases} \quad (14)$$

Inserting now Equations (12) and (14) into Equation (11) and nullifying the time derivatives, one can finally calculate the initial values of the axis components of the voltage at the AC terminals of the GSC.

3.2. Machine Side Portion

For the machine side portion, the initialization problem consists of finding the values of v_{sd0}, v_{sq0}, i_{sq0}, i_{sd0}, ω_0, P_{WT0}, v_{w0}. Neglecting the two converter losses, the active power P_{s0} delivered at the AC terminals of the MSC can be calculated as:

$$P_{s0} = [P_{g0} + R_T({i_{gd0}}^2 + {i_{gq0}}^2)] \quad (15)$$

It should be observed that, if one wants to account also for the converter losses, an empirical correction to the resistances R and R_T can be done without affecting the procedure (see [34] for details). The power P_{s0} can be expressed as:

$$P_{s0} = v_{sd0} i_{sd0} + v_{sq0} i_{sq0} \quad (16)$$

and, if one inserts Equation (6) into Equation (5) and Equation (5) into Equation (9) and nullifies all the time derivatives (to account for the steady-state condition), it is possible to write the following steady state equations:

$$\begin{cases} v_{sd0} = -(R_S + R) i_{sd0} - \omega_0 x_q i_{sq0} \\ v_{sq0} = -(R_S + R) i_{sq0} + \omega_0 x_d i_{sd0} + \omega_0 \psi \end{cases} \quad (17)$$

Moreover, assuming that the initial point is characterized by a speed belonging to the range $[\omega_{min}, \omega_{max}]$, the MPPT curve coincides with the second of Equation (3), therefore:

$$P_{WT0} = \frac{\rho A c_{p,opt} R}{2 S_b \lambda_{p,opt}} \omega_0^3 \quad (18)$$

The active power balance between the wind turbine and the AC terminals of the MSC allows writing:

$$P_{WT0} - (R + R_S)({i_{sd0}}^2 + {i_{sq0}}^2) = P_{s0} \quad (19)$$

Finally, after some algebraic manipulations, Equation (4) becomes:

$${V_{m0}}^2 = {v_{md0}}^2 + {v_{mq0}}^2 = (v_{sd0} + R i_{sd0} + \omega_0 L i_{sq0})^2 + (v_{sq0} + R i_{sq0} - \omega_0 L i_{sd0})^2 = \min\{{\omega_0}^2, 1\} \quad (20)$$

Equations (16), (17), (19) and (20) originate an algebraic nonlinear system of five equations in five unknown variables ($v_{sd0}, v_{sq0}, i_{sd0}, i_{sq0}, \omega_0$) that can be solved numerically.

Finally, the application of Equation (18) provides the initial value of the wind turbine power, while the corresponding wind speed can be defined by solving Equation (1). If the initial value of the machine angular speed falls out of the range $[\omega_{min}, \omega_{max}]$, then Equation (18) is no longer valid; as a consequence, the procedure has to be modified setting P_{WT0} equal to P_{max} and assigning v_{w0} independently (as described in [18]).

In this case, the set of equations that allows the discovery of the solution is Equations (16), (17), (19) and (20) simply by substituting P_{WT0} with P_{max}.

4. Application of the Proposed Initialization Procedure

In order to provide an application of the proposed methodology and to highlight the impact of power losses on the system equilibrium point, an initial condition calculation is proposed on two MVA wind generating units. The electrical machine main parameters are reported in Table 1 [20].

Table 1. Permanent magnet synchronous generator (PMSG) data.

Description	Value	Unit
Rated Power	2	MVA
Rated voltage	690	V
Stator resistance	0.042	p.u.
Direct axis reactance	1.05	p.u.
Quadrature axis reactance	0.75	p.u.
Permanent magnet flux	1.25	p.u.

Moreover, the following numerical data are assumed (see Figure 1):

- The connection between the machine and the machine side converter is characterized by resistance $R = 0.05$ p.u. and an inductance $L = 0.05$ p.u. (on the machine basis)
- The connection between the GSC and the external network has an equivalent resistance $R_t = 0.005$ p.u. and an equivalent inductance $L_t = 0.05$ p.u. (on the machine basis).

The power system is assumed to be operated in a load flow condition characterized by unitary voltage magnitude on the grid side bus-bar, a power production equal to 1.6 MW (0.8 p.u. on machine power base) and null reactive power production at the point of interconnection with the grid (i.e., $V_g = 1$ p.u., $P_g = 0.8$ p.u. and $Q_g = 0$ p.u.). The results obtained applying the proposed methodology to the present test case are summarized in Table 2. The comparison of the proposed approach with the one described in [18] is presented in Table 2 which also reports the results of the lossless initial conditions calculations.

As one can notice, active power losses affect all the variables characterizing the system steady state. In particular, the error on the wind speed evaluation is around 3%, while the error on the wind power is around 6.9%. Moreover, the most important result is that the loss-accounted system is initialized with a wind speed equal to 8.66 m/s (obtained by the lossless initialization), and the maximum power produced by the wind turbine is equal to 0.8 p.u., which is not consistent with the load flow results, as the active power delivered at the grid bus is going to be lower than 0.8 p.u. due to the losses. This means that using a lossless initialization as the starting point for a dynamic simulation will result in an undesired initial transient.

Table 2. Comparison of the initialization results for the traditional lossless initialization and the proposed one (that accounts for the WTG losses).

Section	Variable	Initial Value with Losses	Initial Value without Losses
GSC	v_{gd0}	1.00 p.u.	1.00 p.u.
	i_{gd0}	0.80 p.u.	0.80 p.u.
	i_{gq0}	0.00 p.u.	0.00 p.u.
	v_{ed0}	1.004 p.u.	1.00 p.u.
	v_{eq0}	−0.04 p.u.	−0.04 p.u.
MSC	v_{sd0}	−0.58 p.u.	−0.57 p.u.
	v_{sq0}	0.76 p.u.	0.81 p.u.
	i_{sq0}	0.68 p.u.	0.64 p.u.
	i_{sd0}	−0.48 p.u.	−0.47 p.u.
Wind	P_{wind0}	0.86 p.u.	0.80 p.u.
	v_{w0}	8.90 p.u.	8.66 p.u.
	ω_0	1.14 p.u.	1.11 p.u.

5. Fully Rated Converter Wind Generator Unit Control Scheme

As outlined in [28], the controllers of both MSC and GSC normally have a nested-loop structure with a fast inner current loop, controlling the stator *d*- or *q*-axis currents, combined with outer slower loops that pursue the control aims described in the previous section (i.e., extracting the maximum power from the wind turbine and controlling the voltage at the machine terminals for the MSC controller and regulating the voltage on the DC link and the reactive power delivered to the main network for the GSC one). The aim of this section is to describe the overall wind generator architecture that accounts for the system power losses before detailing an effective regulator synthesis in Section 6.

5.1. Machine Side Converter Controller

The voltage channel takes its reference signal from Equation (4) and typically implements a structure of the kind depicted in Figure 4, where the output of the machine voltage regulator R_{Vm} becomes the reference signal for the direct axis stator current regulator (the block MIN is the minimum selector representing Equation (4)).

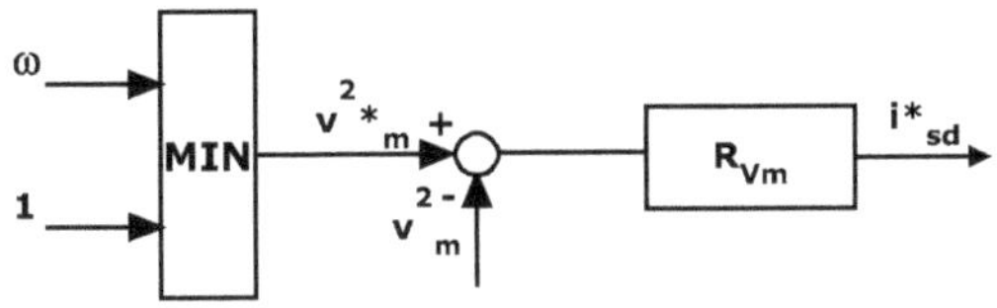

Figure 4. Voltage regulator block diagram. MIN: minimum.

The torque channel aims at extracting the maximum power from the wind source. This is achieved by means of the MPPT function described in Equation (3); it should be noted that, once that curve is available, there is no need to measure the wind speed in order to know the optimal power that can be extracted from the wind for each turbine speed. In the literature, two main possible uses for this curve have been proposed:

- Type I: the MPPT curve input is the actual turbine speed and produces the reference signal for the power to be produced by the PMSG [24–26,30–34] (see Figure 6).
- Type II: The control accounts for the inverse MPPT curve input and feeds it with the PMSG active power producing the reference signal for the speed controller R_W (see e.g., [28,35]), as depicted in Figure 5.

In Type II configuration, the output of the speed controller feeds an active power controller R_P; in both cases, the output of the power controller becomes the reference signal for the quadrature axis current regulator.

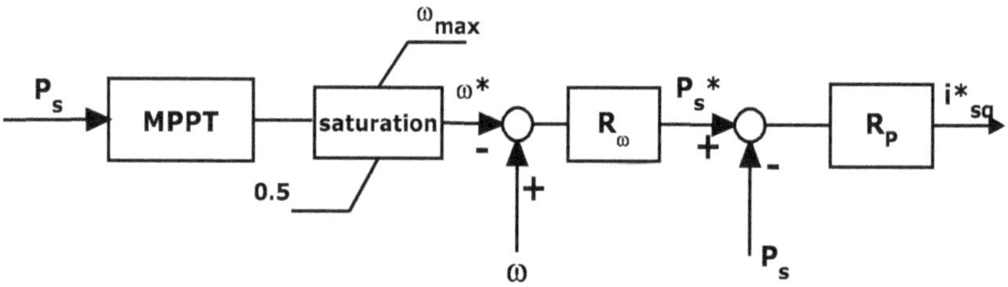

Figure 5. Power controller block diagram (Type II).

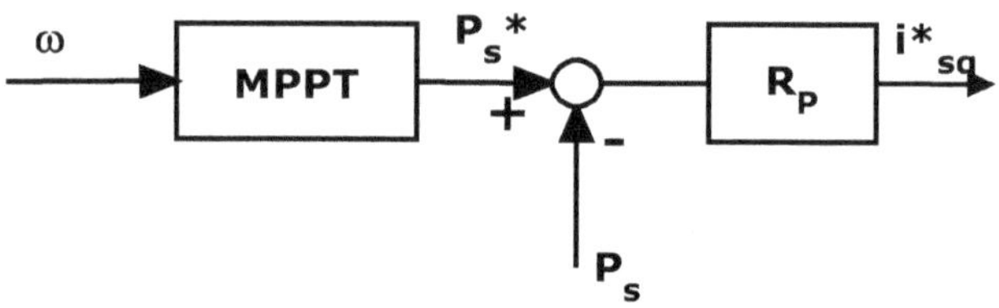

Figure 6. Power controller block diagram (Type I). MPPT: maximum power point tracking.

Nevertheless, in both Type I and Type II configurations, the system is obliged to work at a point whereby the active power P_s produced by the PMSG follows the MPPT curve, which is not feasible, unless one neglects the losses in the electric machine. This section details how to account for power system losses by introducing some corrections in the classical control scheme. Inserting now Equation (6) into Equation (5), and Equation (5) into Equation (9), one obtains:

$$\begin{cases} v_{sd} = -(R_s + R)i_{sd} - \frac{L+x_d}{\omega_n}\frac{di_{sd}}{dt} - \omega(L + x_q)i_{sq} \\ v_{sq} = -(R_s + R)i_{sq} - \frac{L+x_q}{\omega_n}\frac{di_{sq}}{dt} + \omega(L + x_d)i_{sd} + \omega\psi \end{cases} \tag{21}$$

Then, the following power balance can be easily obtained:

$$P_s = v_{sd}i_{sd} + v_{sq}i_{sq} = -(R_s + R)({i_{sd}}^2 + {i_{sq}}^2) - \frac{(L+x_d)}{\omega_n}i_{sd}\frac{di_{sd}}{dt} - \frac{(L+x_q)}{\omega_n}i_{sq}\frac{di_{sq}}{dt} + \omega i_{sq}\psi + \omega i_{sq}i_{sd}(x_d - x_q) \tag{22}$$

This balance can be written at a steady state, neglecting the rotor saliency as:

$$\begin{aligned} P_s &= v_{sd}i_{sd} + v_{sq}i_{sq} = -R_{tot}({i_{sd}}^2 + {i_{sq}}^2) + \omega i_{sq}\psi = -(R_s + R)({i_{sd}}^2 + {i_{sq}}^2) + \omega T_e \\ &= P_{WT} - (R_s + R)({i_{sd}}^2 + {i_{sq}}^2) \end{aligned} \tag{23}$$

Equation (23) means that the power coming from the wind equals the active power available at the machine terminals plus the losses in the stator windings and in the cable connection. Both the control schemes appearing in Figures 5 and 6 imply that P_s follows the MPPT curve; thus, assuming that the desired working point is in the range in which the MPPT curve is increasing (i.e., the pitch controller is not active), with the same notations used in Section 2, it is possible to write the following power balance:

$$\frac{\rho A c_{p,opt} R}{2 S_b \lambda_{p,opt}}\omega^3 = P_{WT}(\omega, v_w, 0) - (R_s + R)\left({i_{sd}}^2 + {i_{sq}}^2\right) \tag{24}$$

The only solution for Equation (24) is in the case that the term $(R_s + R)$ is equal to zero since the meaning of the left-hand side of Equation (24) is, by definition, the maximum value of the wind power P_{WT}.

From a physical point of view, this means that it is not possible to deliver to the AC terminals of the electric machine the maximum power that can be extracted from the wind, as this would imply that the original source (the wind) generates a greater amount of power in order to compensate for the losses.

A possible solution to this problem consists in modifying the control system to account also for power losses, as depicted in Figure 7 (a Type II scheme is presented as an example). However, it is not possible to know exactly the loss amount, due to possible errors in the current measurements, imperfect knowledge of the machine stator winding and cable resistances and impossibility to estimate properly eventual losses due to friction or air resistance. The meaning of Equation (24) suggests that what is necessary is to provide an upper bound for such losses, in order to avoid the situation in which the controller requires the wind turbine to produce more than what is physically possible.

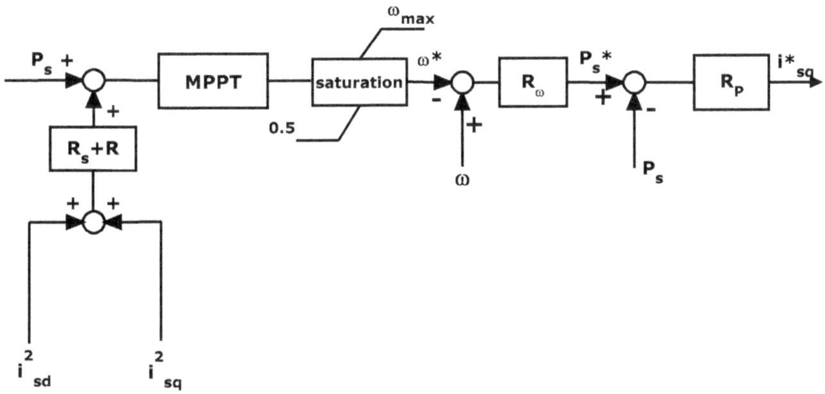

Figure 7. Correction in the power controller block diagram (Type II).

5.2. Grid Side Converter Controller

As far as the GSC is concerned, it is necessary to characterize the DC link transients. The DC link voltage v_{DC} dynamic equation states that:

$$\frac{C}{2}\frac{dv_{DC}^2}{dt} = P_s - R_T({i_{gd}}^2 + {i_{gq}}^2) - P_g \tag{25}$$

As the power coming from the PMSG tracks, the maximum power point, Equation (25), suggests the possibility of using the power injected into the grid to regulate the DC link voltage. As a consequence, one can set up a regulator that accepts as input the error on this voltage and produces the reference signal for the power $P_g{}^*$. Now, recalling Equation (13), this signal can be translated into a request for the direct axis current $i_{gd}{}^*$ simply by dividing by the amplitude of the point of interconnection voltage V_g. The reactive power channel is much simpler as the set point for the reactive power to be delivered to the grid; it can be translated into a request for the quadrature axis current $i_{gq}{}^*$, recalling again Equation (13). The architecture of the GSC regulators is depicted in Figure 8.

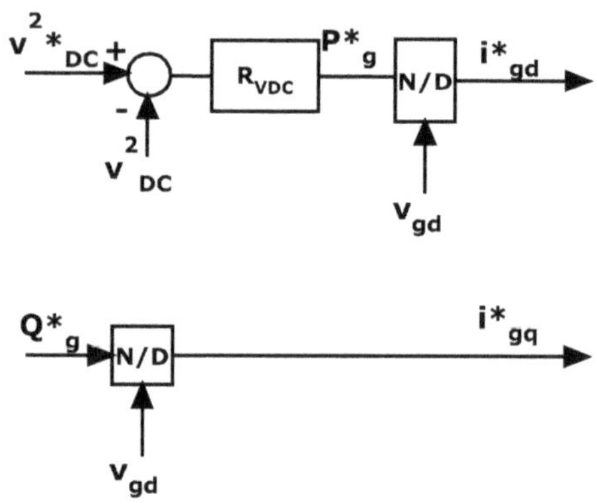

Figure 8. GSC controller architecture.

6. Criteria for the Synthesis of the Wind Turbine Generator Regulators

In this section, some guidelines will be provided in order to support the choice of the structure and parameters of the controllers for both the MSC (Section 6.1) and the GSC (Section 6.2). From now on, the Laplace transform of any time domain quantity will be indicated with the same symbol as the time domain function for the sake of readability.

6.1. Machine Side Converter Regulator Synthesis

As outlined in the previous section, the MSC controller presents a nested structure, whose inner (and faster) loop is the current one, which consists of two regulators accepting as input the error on the direct (quadrature) axis current and producing as output the suitable values of the MSC AC side voltages. In order to derive structure and parameters of such regulators, it is possible recalling Equation (21); defining now two fictitious inputs w_{sd} and w_{sq} as follows:

$$\begin{cases} w_{sd} = -v_{sd} - \omega x_q i_{sq} \\ w_{sq} = -v_{sq} + \omega x_d i_d + \omega \psi \end{cases} \tag{26}$$

This implies that two decoupled relationships hold between such inputs and the axis currents:

$$\begin{cases} w_{sd} = (R_s + R) i_{sd} + \frac{L + x_d}{\omega_n} \frac{di_{sd}}{dt} \\ w_{sq} = (R_s + R) i_{sq} + \frac{L + x_q}{\omega_n} \frac{di_{sq}}{dt} \end{cases} \tag{27}$$

This way it is possible to construct two scalar regulators R_d and R_q that accept as input the error on the direct and quadrature axis current, respectively, and produce as output the signals w_{sd} and w_{sq}, as follows:

$$\begin{cases} w_{sd}(t) = k_P(i_{sd}^*(t) - i_{sd}(t)) + k_I \int_0^t (i_{sd}^*(\tau) - i_{sd}(\tau)) d\tau \\ w_{sq}(t) = k_P(i_{sq}^*(t) - i_{sq}(t)) + k_I \int_0^t (i_{sq}^*(\tau) - i_{sq}(\tau)) d\tau \end{cases} \tag{28}$$

where k_P and k_I are their proportional and integral gains, respectively. As a consequence, the dynamic properties of the resulting system can be investigated by examining the following transfer function:

$$F_{d(q)}(s) = \frac{\left(k_P + \frac{k_I}{s}\right) \frac{1}{R_s + R + \frac{L + x_{d(q)}}{\omega_n} s}}{1 + \left(k_P + \frac{k_I}{s}\right) \frac{1}{R_s + R + \frac{L + x_{d(q)}}{\omega_n} s}} \tag{29}$$

whose poles are the solution of:

$$\frac{L + x_{d(q)}}{\omega_n} s^2 + (R_s + R + k_P)s + k_I = 0 \tag{30}$$

The solution of Equation (30) belongs to the left half plane for any positive value of k_P and k_I, thus guaranteeing the stability of the current loop. Moreover, the solution of Equation (30) allows to choose k_P and k_I according to the desired transient behavior of the axis currents (i.e., placing the poles). The real inputs v_{sd} and v_{sq} can be obtained starting from w_{sd} and w_{sq}, thereby solving Equation (26), as shown in Figure 9.

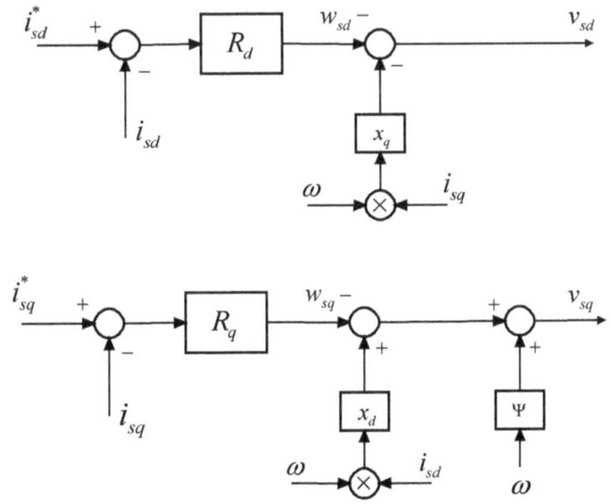

Figure 9. Current control loops.

The reference signal for the quadrature axis current is provided by an active power regulator R_P, as shown in Figures 5 and 6; in order to define an effective structure for such a regulator, one has to consider the scheme depicted in Figure 10.

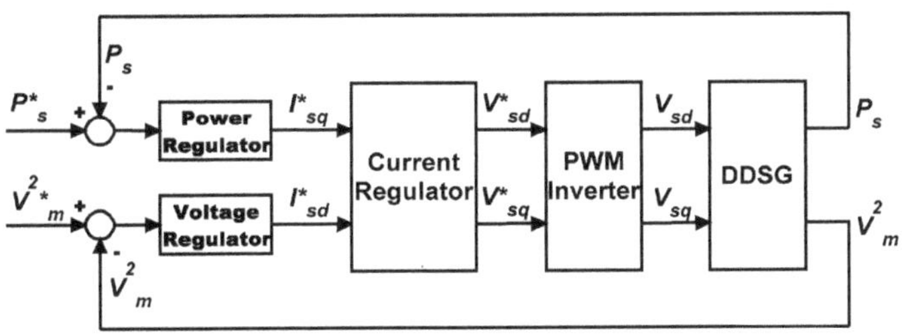

Figure 10. Active power loop.

As the inner current control loop is typically faster, to synthesize a possible active power regulator, one can assume that the action of the quadrature axis current controller is instantaneous, i.e.,

$$i_{sq}(t) = i_{sq}{}^*(t) \tag{31}$$

Moreover, neglecting the rotor saliency and recalling Equation (7), $T_e = \psi i_{sq}$, in the hypothesis of no active power losses in Equations (16) and (17), the active power P_s can be written as:

$$P_s = \omega\psi i_{sq} \tag{32}$$

Finally, supposing that the angular speed transient is much slower than the active power one, the scheme depicted in Figure 10 can be simplified into the one of Figure 11; choosing a PI structure for the active power controller, the closed loop transfer function becomes:

$$F_p(s) = \frac{\omega\psi[k_{PP}s + k_{PI}]}{s + \omega\psi(k_{PP}s + k_{PI})} \tag{33}$$

where k_{PP} and k_{PI} are the proportional and integral gains of the active power controller, respectively. In accordance with Equation (33), any positive value of k_{PP} and k_{PI} will lead the system to be asymptotically stable. It is also possible to place the pole of Equation (33) acting on the proportional and integral gains.

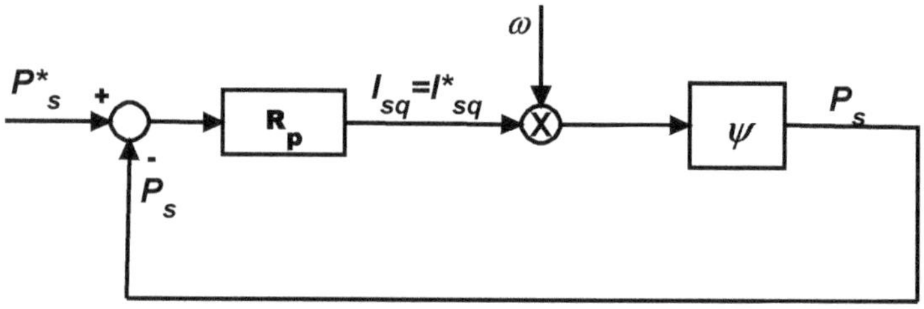

Figure 11. Simplified active power control loop for the synthesis of the regulator.

The reference signal for the active power can be provided either by the MPPT block (Type I) or by a speed controller (Type II); in this latter case, the synthesis of such a controller can be achieved assuming negligible active power losses and instantaneous action of the active power controller, i.e.,

$$P_s(t) = P_s^*(t) \tag{34}$$

In this case, the control scheme depicted in Figure 5 becomes the one of Figure 12, whose closed loop transfer function is:

$$\omega(s) = \frac{k_{Pw}s + k_{Iw}}{2Hs^2 + k_{Pw}s + k_{Iw}}\omega^*(s) + \frac{s}{2Hs^2 + k_{Pw}s + k_{Iw}}P_{WT}(s) \tag{35}$$

where k_{Pw} and k_{Iw} are the proportional and integral gains of the PI speed controller respectively. Once again, any positive value of k_{Pw} and k_{Iw} will provide an asymptotically stable closed loop system, while the prescribed transient behavior can be achieved by tuning k_{Pw} and k_{Iw} in order to obtain the desired values for the poles of Equation (35).

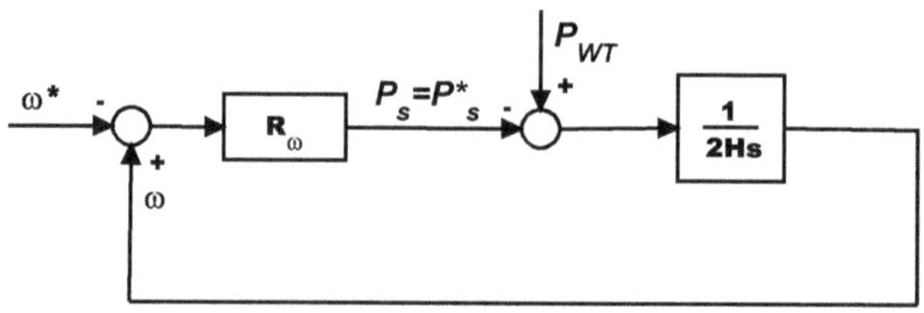

Figure 12. Simplified speed control loop for the synthesis of the regulator.

Let us now summarize the procedure to define the control parameters of the MSC regulator: (i) choose the values of k_P and k_I using Equation (30); (ii) choose the values of k_{PP} and k_{PI} by means of Equation (33) accounting for assumption Equation (31) (i.e., the amplitude of the pole of Equation (33) must be much smaller than the amplitudes of the poles of Equation (30); (iii) define the values of k_{Pw} and k_{Iw} in accordance with Equation (35), ensuring that the dynamics of the power loop are faster than the dynamics of the speed one (to meet condition of Equation (34)).

The reference signal for the direct axis current is provided by a voltage regulator R_{Vm}, as shown in Figure 4; in order to define an effective structure for this regulator, one has to consider the scheme depicted in Figure 10. The amplitude of the machine voltage is given by:

$$v_m{}^2 = v_{md}{}^2 + v_{mq}{}^2 \tag{36}$$

where the axis components are related to the axis currents by means of Equation (5). Nevertheless, Equation (36) defines a nonlinear function, it is possible to consider a first-order Taylor expansion centered in the initial balance point defined in the initialization procedure. To this extent, in the following for any quantity y the variation Δy with respect to the initial equilibrium point y_0 will be considered. As a consequence, Equation (36) becomes:

$$\Delta v_m{}^2 = 2v_{md0}\Delta v_{md} + 2v_{mq0}\Delta v_{mq} \tag{37}$$

and, recalling Equations (5) and (6), it is possible to obtain:

$$\begin{cases} \Delta v_{md} = -R_s\Delta i_{sd} - \frac{x_d}{\omega_n}\frac{d\Delta i_{sd}}{dt} - \omega_0 x_q \Delta i_{sq} - x_q i_{sq0}\Delta\omega \\ \Delta v_{mq} = -R_s\Delta i_{sq} - \frac{x_q}{\omega_n}\frac{d\Delta i_{sq}}{dt} + \omega_0 x_d \Delta i_{sd} + x_d i_{sd0}\Delta\omega + \psi\Delta\omega \end{cases} \tag{38}$$

which, in the Laplace domain, becomes:

$$\Delta v_m{}^2(s) = G_d(s)\Delta i_{sd}(s) + G_q(s)\Delta i_{sq}(s) + G_w(s)\Delta\omega(s) \tag{39}$$

having defined:

$$\begin{cases} G_d(s) = [2v_{mq0}\omega_0 x_d - 2v_{md0}(R_s + \frac{x_d s}{\omega_n})] \\ G_q(s) = [-2v_{mq0}(R_s + \frac{x_q s}{\omega_n}) - 2v_{md0}\omega_0 x_q] \\ G_w(s) = [2v_{mq0}(x_d i_{sd0} + \psi) - 2v_{md0} i_{sq0} x_q] \end{cases} \tag{40}$$

The equivalent block scheme of the simplified machine voltage dynamics is depicted in Figure 13.

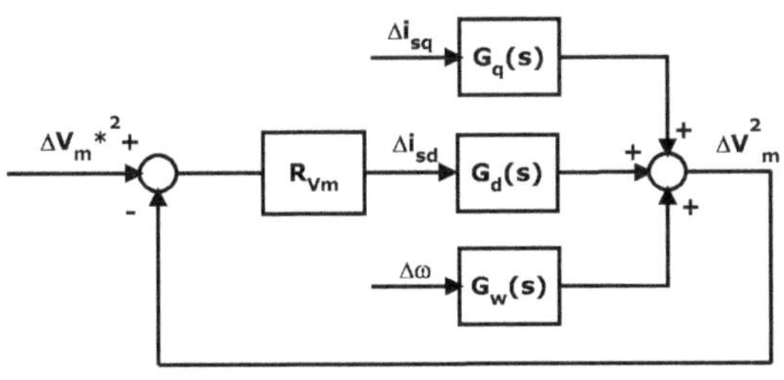

Figure 13. Simplified speed control loop for the synthesis of the regulator.

Supposing that the direct axis current loop is instantaneous, (i.e., $i_{sd}(t) = i_{sd}{}^*(t)$) it is possible to derive:

$$\Delta v_m{}^2(s) = \frac{G_d(s)R_{vm}(s)}{1+G_d(s)R_{vm}(s)}\Delta v_m^{*\,2}(s) + \frac{G_q(s)}{1+G_d(s)R_{vm}(s)}\Delta i_{sq}(s) + \frac{G_w(s)}{1+G_d(s)R_{vm}(s)}\Delta\omega(s) \tag{41}$$

Now choosing $R_{vm}(s) = \frac{sk_{Pvm}+k_{Ivm}}{s}$, the closed loop transfer function poles are the solutions of:

$$2v_{md0}\frac{x_d}{\omega_n}k_{Pvm}s^2 + [k_{Ivm}2v_{md0}\frac{x_d}{\omega_n} + k_{Pvm}(2v_{md0}R_s - 2v_{mq0}\omega_0 x_d) - 1]s + k_{Ivm}(2v_{md0}R_s - 2v_{mq0}\omega_0 x_d) = 0 \quad (42)$$

The parameters of the machine voltage regulator have to be defined in order to guarantee the asymptotic stability of the system for any initial working point. It should be observed that, given the wind power P_{WT0}, the solution of the nonlinear algebraic system consisting of Equations (16), (17), (19) and (20) allows to uniquely define the initial working point; as a consequence, the following algorithm can be set up:

- Choose a couple of parameters k_{Pvm} and k_{Ivm};
- For any value of P_{WT0} belonging to the range $[P_{WTmin}, P_{WTmax}]$, determine the corresponding initial working point;
- Evaluate v_{md0} and v_{mq0} with Equations (5) and (6) having nullified the time derivatives;
- Solve Equation (42);
- Adjust the values of k_{Pvm} and k_{Ivm} s.t. for all the values of P_{WT0}: (i) the solutions of Equation (42) have a negative real part; (ii) the solution of Equation (42) with smaller amplitude (i.e., the one that dictates the dynamic behavior of the voltage loop) is (in modulus) sufficiently smaller than the roots of Equation (30) to guarantee that the current loop is much faster than the voltage one.

To illustrate the proposed procedure, one can refer again to the machine with the data appearing in Table 1. Choosing $k_P = 0.1$ and $k_I = 10$, the current loop has two complex conjugate poles $s_{1,2} = -27 \pm 45j$. Now, selecting $k_{Pvm} = 5$ and $k_{Ivm} = 2.5$, the real part of the voltage loop poles are plotted in Figure 14 as a function of P_{WT}, highlighting both the stability of the system and the different timeframe of the two loops' action.

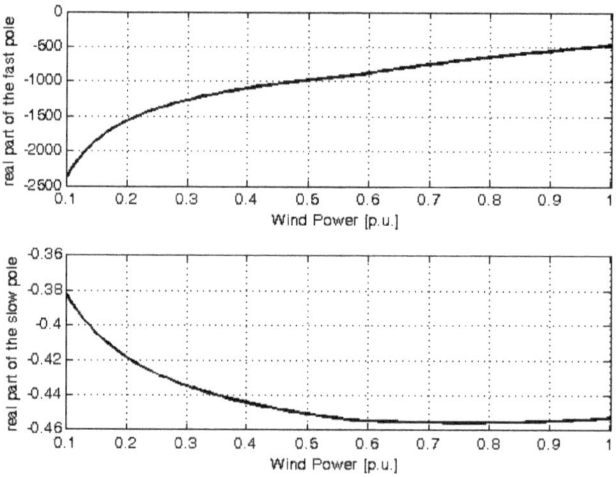

Figure 14. Machine voltage loop poles.

6.2. Grid Side Converter Regulator Synthesis

Also, the GSC controller present a nested structure with an inner current control loop whose regulators can be designed in the same way as for the MSC.

Recalling Equation (11), two scalar regulators can be defined, one for i_{gd} and one for i_{gq}; furthermore, defining the following two quantities:

$$x_{gd} = v_{ed} - v_{gd} - \omega_e L_T i_{gq} \quad (43)$$

and:

$$x_{gq} = v_{eq} - v_{gq} + \omega_e L_T i_{gd} \tag{44}$$

and indicating with x_{gd} and x_{gq} the regulator outputs, the system to be controlled is described by the following decoupled equations:

$$x_{gd} = R_T i_{gd} + \frac{L_T}{\omega_g}\frac{d}{dt} i_{gd} \tag{45}$$

and:

$$x_{gq} = R_T i_{gq} + \frac{L_T}{\omega_g}\frac{d}{dt} i_{gq} \tag{46}$$

This way it is possible to construct two PI regulators R_{gd} and R_{gq} that accept as input the error on the direct and quadrature axis current, respectively, and produce as output the signals x_{sd} and x_{sq}, as for the MSC controller. As a consequence, indicating with k_{Pg} and k_{Ig} the proportional and integral gains of R_{gd} and R_{gq}, the dynamic properties of the resulting system can be investigated by examining the following transfer function:

$$F_{gd(q)}(s) = \frac{\left(k_{Pg} + \frac{k_{Ig}}{s}\right)\frac{1}{R_T + \frac{L_T}{\omega_g}s}}{1 + \left(k_{Pg} + \frac{k_{Ig}}{s}\right)\frac{1}{R_T + \frac{L_T}{\omega_g}s}} \tag{47}$$

whose poles have a negative real part for any positive value of k_{Pg} and k_{Ig}. The reactive power control can be addressed simply by recalling Equation (14), which allows to transform the reference signal for the reactive power exchange into a request of quadrature axis current (see Figure 8).

The DC link voltage v_{DC} regulator synthesis can be done recalling Equation (25); in neglecting the active power losses, it is clear that for any change in wind speed, and therefore in the power P_s, it is possible to keep the DC link voltage constant acting on the power P_g. Completing the scheme of Figure 8 it is possible to achieve the control scheme of Figure 15.

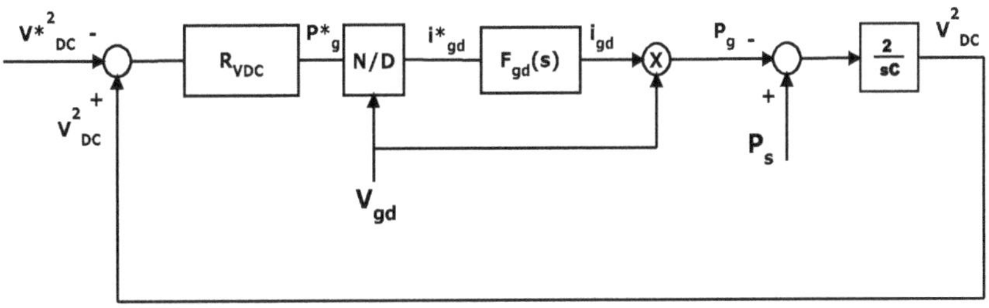

Figure 15. DC voltage control loop.

Supposing that the current loop is instantaneous and a PI controller for R_{VDC} is chosen, it is possible to obtain the following transfer function:

$$\Delta V_{DC}{}^2(s) = \frac{\frac{2}{Cs}\Delta P_s(s) + \frac{2}{Cs}(k_{PDC} + \frac{k_{IDC}}{s})\Delta V^*_{DC}{}^2(s)}{1 + (k_{PDC} + \frac{k_{IDC}}{s})\frac{2}{Cs}} \tag{48}$$

where k_{PDC} and k_{IDC} are respectively the proportional and integral gains of the DC voltage PI controller. The examination of Equation (48) leads to the conclusion that for any step variation of P_s, it is possible to keep the DC voltage constant for any value of the gains, as:

$$\lim_{s \to 0} \frac{\frac{2}{Cs}}{1 + (k_{PDC} + \frac{k_{IDC}}{s})\frac{2}{Cs}} = 0 \tag{49}$$

Once again, it is possible to choose the parameters of the current controllers assigning the poles of Equation (47) and then the parameters of the DC voltage regulator using Equation (49), thus ensuring that the voltage loop is sufficiently slower than the current one.

6.3. Overall Control System Structure and Summary of the Developed Criteria

The WTG overall control scheme is depicted in Figure 16, also including the pitch angle controller and the shaft model (see [15,16] for details). The GSC and its controller are depicted in the block "inverter DC/AC including V_{DC} control and Q control", whose detailed structure appears in Figure 8. As far as the MSC controller is concerned, details on the voltage and power controllers are provided respectively in Figures 4 and 7, while the inner current control loops are depicted in Figure 9.

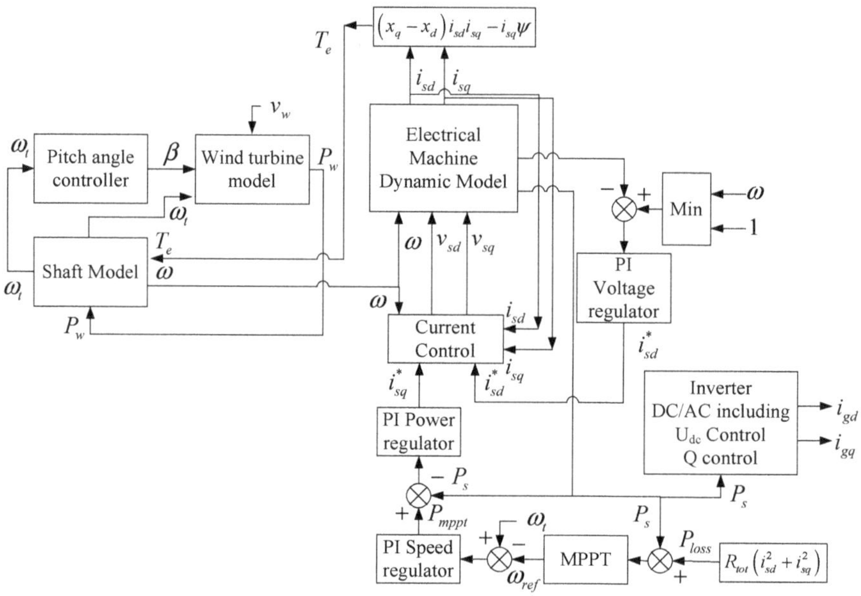

Figure 16. Final control scheme.

The power channel of the MSC controller can be designed as follows:

- Place the poles of the inner control loop according to Equation (30);
- Place the poles of the active power control loop zeroing the denominator of Equation (33) such that the power dynamics are sufficiently slower than the current one in order to meet Equation (31);
- Place the poles of the speed control loop zeroing the denominator of Equation (35) such that the speed dynamics are sufficiently slower than the active power one in order to meet Equation (34).

The voltage channel of the MSC controller can be designed as follows:

- Place the poles of the inner control loop according to Equation (30);
- Place the pole voltage control loop with the procedure described in Section 6.1.

The GSC controller parameters can be chosen as follows:

- Place the poles of the inner control loop zeroing the denominator of Equation (47).
- Place the poles of the DC voltage control loop zeroing the denominator of Equation (49) such that the DC voltage dynamics are sufficiently slower than the current one.
- The choice of the Park reference frame such that the network voltage V_g has only a direct axis component [33] creates an algebraic link between the reactive power and the quadrature axis current; therefore, the reactive power dynamics are the same as the quadrature axis current.

7. Simulations to Validate the Controller Synthesis and the Proposed Simplified Schemes

The aim of the present section is to provide some simulative support to the regulator design provided in the previous sections. In order to account for a detailed and reliable representation of the power system, the PMSG model has been implemented in Digsilent Power Factory® (14.1, DigSILENT Gmbh, Gomaringen, Germany) [35]. For this purpose, the dynamic models of the MSC and GSC controllers have been implemented in Power Factory, in accordance with the equations detailed in Section 5. Therefore, the validation is performed by comparing the results provided by the Digsilent Power Factory full model and the ones obtained applying the simplified schemes that have allowed the derivation of analytical criteria for the choice of the controller parameters. The simplified linear scheme has also been implemented in Power Factory and its run supplied by the same references of the complete model.

Table 3 summarizes the parameters used for the regulators, defined on the basis of the criteria described in the previous section.

Table 3. Controller parameters.

Regulator	Symbol	Value
DC Voltage Regulator	k_{PDC} k_{IDC}	10.0 2.0 s^{-1}
Active Power regulator	k_{pp} k_{ip}	4.0 8.0 s^{-1}
Speed regulator	k_{Pw} k_{Iw}	1.0 p.u. 0.1 s^{-1}

The numerical values of the machine voltage and current controllers have been specified in the previous section, while the parameters of the grid side current controllers are the same as the machine current ones. This choice means the following:

- The poles of the machine currents loop are $-27 \pm 45j$; while the poles of the active power and of the speed loops are respectively -1.6 and $-0.017 \pm 0.05j$. With these values, the dynamics of the three loops act on different time frames, thus meeting the requirements of Equations (31) and (34).
- The poles of the grid current loops are -544 and -115, while the poles of the DC voltage one are $-0.02 \pm 0.02j$, which ensures that its dynamics are much slower than the current one.

The model is initialized with a wind speed corresponding to an active power production equal to 0.55 p.u. (1.1 MW). Then, two consecutive wind speed variations are provided (Figure 17): the first one is triggered at 5 s and increases the wind speed up to 8.4 m/s, while the second one occurs at 300 s and brings the wind speed to 15 m/s, activating the pitch controller.

In accordance with this wind speed profile, the machine undertakes a dynamic that involves all the regulators designed in the previous section. The wind speed variation provokes a variation of the rotational speed of the turbine with a corresponding activation of the speed regulator. The speed is changed acting on the power regulator that correctly provides an increase of the wind generator active power production. Finally, the power dynamics on the MSC activate the DC voltage regulator that allows for an increase in the active power delivered to the grid. The speed dynamic is reported in

Figure 18, where the speed reference variation is provided by the MPPT block of the system. The blue curve is the speed profile obtained by the complete model of the system, while the red one is the one provided by the simplified block scheme proposed in Figure 12.

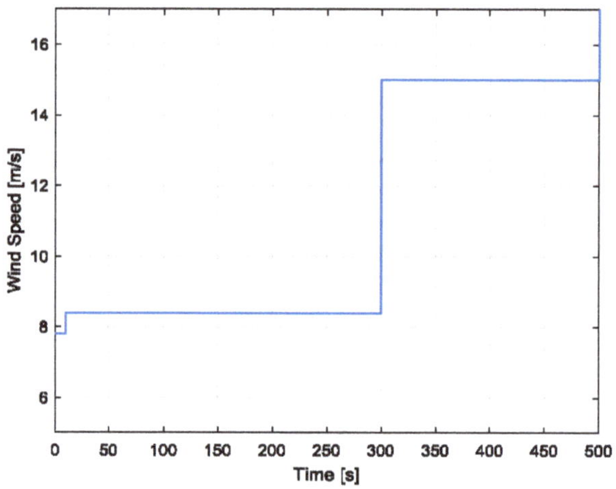

Figure 17. Wind speed time profile.

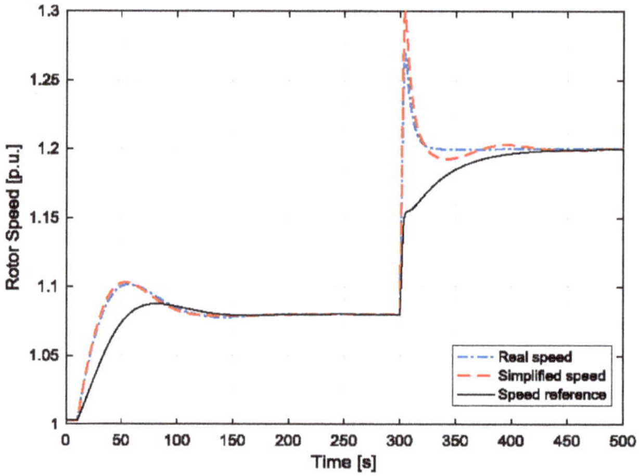

Figure 18. Rotor speed time profile.

The first transient is characterized by a final steady state within the limits of the MPPT speed curve, while the second is associated with the activation of the speed limitation at 1.2 p.u. As one can notice, the simplified dynamics are very close to the real one in both conditions, stating that the control scheme of Figure 12 can be used to synthesize the speed regulator parameters. Figure 19 compares the dynamics of active power P_s considering the complete control scheme, blue line, the simplified one of Figure 11, red line, and the time evolution of the power reference signal as dictated by the speed controller, black line (the three curves are basically overlapping so that it is almost impossible to distinguish them). The fact that the blue line is coincident with the red one leads to the conclusion that the equivalent scheme of Figure 11 is a good approximation of the real model. Moreover, as it is not possible to distinguish between the power time profile and its reference signal in a timeframe of about 500 s, one can conclude that the active power loop dynamics are much faster than the speed one. This can be confirmed through examining Figure 18 where the speed dynamics are much slower. This is consistent with previous assumptions. The current waveform is not reproduced here since it is much faster than the power one.

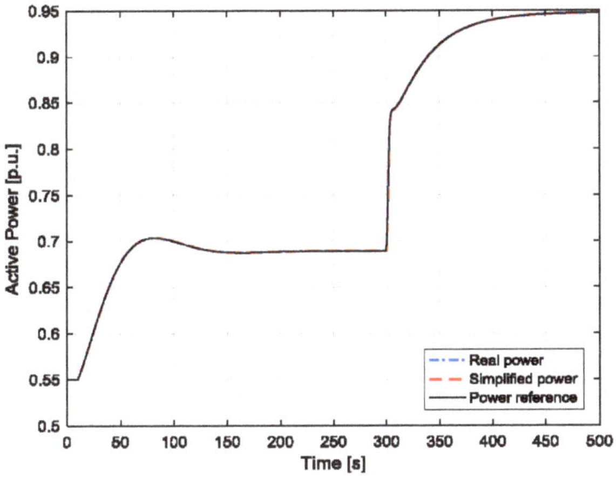

Figure 19. Active power time profile.

Another verification that has been performed is related to the machine voltage regulation loop, comparing the time profile obtained by the simplified linear scheme of Figure 13 with the actual machine voltage time profile obtained by the complete model. Figure 20 depicts the simplified and real machine voltage profile as well as the machine voltage reference provided to the controllers. As one can notice, the two profiles are slightly different but similar enough to provide a good approximation to apply the criteria for the regulator synthesis.

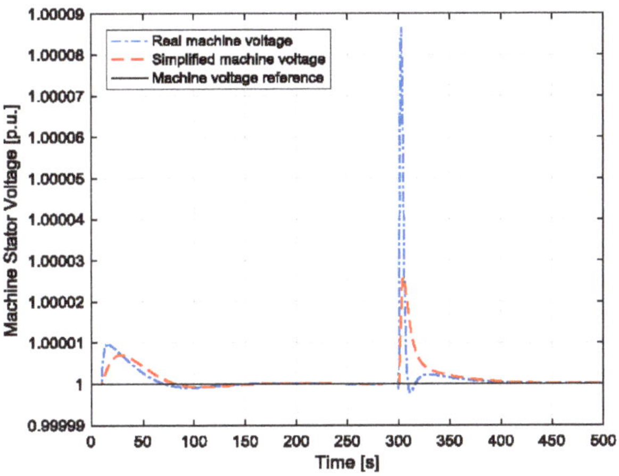

Figure 20. Machine voltage time profile.

Finally, Figure 21 reports the dynamics of the DC link voltage obtained with the full model and with the simplified linear block scheme of Figure 12. Once again, there are some deviations between the simplified dynamic and the real one; nevertheless, the settling time and the overall behavior are acceptable for considering the simplified scheme as valuable in providing the criteria for the regulator synthesis.

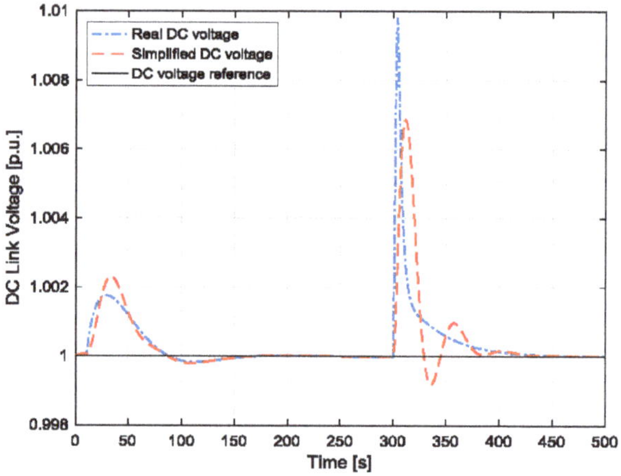

Figure 21. DC voltage time profile.

8. Conclusions

The aim of the present study was to provide an analytical procedure for the calculation of the initial condition of a wind generator unit provided with a permanent magnet DDSG accounting for power system losses and to define some analytical criteria for the definition of the various parameters of the wind generator regulators necessary for its proper operation. In the literature, the problem of the initialization of a wind generating unit is discussed and developed under the assumption of a lossless system, since power losses increase the complexity of the resulting algebraic system. Nevertheless, power losses can have a sensible impact on the system initial conditions and also on the design of an effective control strategy. In the paper, a detailed model of the wind generator has been proposed and the solution of the initial condition problem has been fully described. The impact of power system losses on the results obtained were evaluated in comparison with a traditional lossless approach. Furthermore, the problem of providing analytical criteria to choose both the structures and the parameters of the system controllers has not been widely discussed in the literature. For this reason, the present paper has proposed a detailed characterization of the control strategy of a WTG equipped with a PMSG and provided a simplified representation of the system useful to design the parameters of the various system regulators.

Simulations performed with Digsilent Power Factory software highlighted the good response of the defined regulation system, thus assessing the effectiveness of the approximations done to synthesize the controllers. Further development of this research will include the possible application of nonlinear and decoupled control techniques to the WTG in order to provide a more effective and robust control of the system also capable to integrate frequency support logics (such as synthetic inertial controllers) and active power curtailment.

Author Contributions: Marco Invernizzi and Renato Procopio developed the theoretical approaches Section 3 implemented and tested by Andrea Bonfiglio in Section 3. Federico Delfino and Renato Procopio made the definition of the control scheme of Section 5 and together with Andrea Bonfiglio defined the simplified structure for the regulator design. Dynamic simulation were carried out by Andrea Bonfiglio.

Conflicts of Interest: The authors declare no conflict of interest.

References

1. Chiradeja, P. Benefit of distributed generation: A line loss reduction analysis. In Proceedings of the 2005 IEEE/PES Transmission and Distribution Conference and Exhibition: Asia and Pacific, Dalian, China, 15–18 August 2005; pp. 1–5.
2. Momoh, J.A.; Xia, Y.; Boswell, G.D. An approach to determine distributed generation (DG) benefits in power networks. In Proceedings of the 40th North American Power Symposium, NAPS '08, Calgary, AB, Canada, 28–29 September 2008; pp. 1–7.
3. Conka, Z.; Hocko, P.; Novak, M.; Kolcun, M.; Gyorgy, M. Impact of renewable energy sources on stability of ewis transmission system. In Proceedings of the 14th International Conference on Environmental and Electrical Engineering, Kraków, Poland, 10–12 May 2014; pp. 75–79.
4. Amarasekara, H.W.K.M.; Meegahapola, L.; Agalgaonkar, A.P.; Perera, S. Impact of renewable power integration on VQ stability margin. In Proceedings of the 22nd Australasian Universities Power Engineering Conference, AUPEC 2013, Hobart, Australia, 29 September–3 October 2013; pp. 1–6.
5. Davda, A.T.; Parekh, B.R. System impact analysis of renewable distributed generation on an existing radial distribution network. In Proceedings of the 2012 IEEE Electrical Power and Energy Conference (EPEC), Halifax, NS, Canada, 10–12 October 2012; pp. 128–132.
6. Delfino, F.; Denegri, G.B.; Invernizzi, M.; Procopio, R. Feedback linearisation oriented approach to Q-V control of grid connected photovoltaic units. *IET Renew. Power Gener.* **2012**, *6*, 324–339. [CrossRef]
7. Bonfiglio, A.; Brignone, M.; Delfino, F.; Invernizzi, M.; Pampararo, F.; Procopio, R. A technique for the optimal control and operation of grid-connected photovoltaic production units. In Proceedings of the 2012 47th International Universities Power Engineering Conference (UPEC), London, UK, 4–7 September 2012.
8. Fornari, F.; Procopio, R.; Bollen, M.H.J. SSC compensation capability of unbalanced voltage sags. *IEEE Trans. Power Deliv.* **2005**, *20*, 2030–2037. [CrossRef]
9. Delfino, B.; Fornari, F.; Procopio, R. An effective SSC control scheme for voltage sag compensation. *IEEE Trans. Power Deliv.* **2005**, *20*, 2100–2107. [CrossRef]
10. *CEI 0-16 Reference Technical Rules for the Connection of Active and Passive Consumers to the HV and MV Electrical Networks of Distribution Company*; Comitato Elettrotecnico Italiano (CEI): Milano, Italy, 2012.
11. *BWE Technical Guidelines—Generating Plants Connected to the Medium Voltage Network*; Bundesverband der Energie- und Wasserwirtschaft (BDEW): Berlin, Germany, 2008.
12. RTE Requirements Regulatory Text (Decree 2008-386 dated 23 April 2008). Available online: https://www.legifrance.gouv.fr/affichTexte.do?cidTexte=JORFTEXT000018698004 (accessed on 27 September 2016).
13. *Power to Ontario on Demand Market Rules—Chapter 4 Grid Connection Requirements*; Independent Electricity System Operator (IESO): Toronto, ON, Canada, 2013.
14. Gonzalez-Longatt, F.; Bonfiglio, A.; Procopio, R.; Verduci, B. Evaluation of inertial response controllers for full-rated power converter wind turbine (Type 4). In Proceedings of the IEEE PES General Meeting 2016, Boston, MA, USA, 7–21 July 2016.
15. Slootweg, J.G.; de Haan, S.W.H.; Polinder, H.; Kling, W.L. General model for representing variable speed wind turbines in power system dynamics simulations. *IEEE Trans. Power Syst.* **2003**, *18*, 144–151. [CrossRef]
16. Slootweg, J.G.; Polinder, H.; Kling, W.L. Representing Wind turbine electrical generating systems in fundamental frequency simulations. *IEEE Trans. Energy Convers.* **2003**, *18*, 516–524. [CrossRef]
17. Margaris, I.D.; Hatziargyriou, N.D. direct drive synchronous generator wind turbine models for power system studies. In Proceedings of the 7th Mediterranean Conference and Exhibition Power Generation, Transmission, Distribution and Energy Conversion (MedPower 2010), Agia Napa, Cyprus, 7–10 November 2010; pp. 1–7.
18. Slootweg, J.G.; Polinder, H.; Kling, W.L. Initialization of wind turbine models in power system dynamics simulations. In Proceedings of the 2001 IEEE Porto Power Tech Proceedings (Cat. No.01EX502), Porto, Portugal, 10–13 September 2001; Volume 4.
19. Abo-Khalil, A.G.; Kim, H.-G.; Lee, D.-C.; Seok, J.-K. Maximum output power control of wind generation system considering loss minimization of machines. In Proceedings of the 30th Annual Conference of IEEE Industrial Electronics Society, IECON 2004, Busan, Korea, 2–6 November 2004; Volume 2, pp. 1676–1681.
20. De Kooning, J.D.M.; Vandoorn, T.L.; van de Vyver, J.; Meersman, B.; Vandevelde, L. Displacement of the maximum power point caused by losses in wind turbine systems. *Renew. Energy* **2016**, *85*, 273–280. [CrossRef]

21. Delfino, F.; Pampararo, F.; Procopio, R.; Rossi, M. A feedback linearization control scheme for the integration of wind energy conversion systems into distribution grids. *IEEE Syst. J.* **2012**, *6*, 85–93. [CrossRef]
22. Soufi, Y.; Kahla, S.; Bechouat, M. Feedback linearization control based particle swarm optimization for maximum power point tracking of wind turbine equipped by PMSG connected to the grid. *Int. J. Hydrog. Energy* **2016**, *41*, 20950–20955. [CrossRef]
23. Mozayan, S.M.; Saad, M.; Vahedi, H.; Fortin-Blanchette, H.; Soltani, M. Sliding mode control of PMSG wind turbine based on enhanced exponential reaching law. *IEEE Trans. Ind. Electron.* **2016**, *63*, 6148–6159. [CrossRef]
24. Tapia, A.; Tapia, G.; Ostolaza, J.X.; Saenz, J.R. Modeling and control of a wind turbine driven doubly fed induction generator. *IEEE Trans. Energy Convers.* **2003**, *18*, 194–204. [CrossRef]
25. Ko, H.S.; Yoon, G.G.; Kyung, N.H.; Hong, W.P. Modeling and control of DFIG-based variable-speed wind-turbine. *Electr. Power Syst. Res.* **2008**, *78*, 1841–1849. [CrossRef]
26. Kim, H.-W.; Kim, S.-S.; Ko, H.-S. Modeling and control of PMSG-based variable-speed wind turbine. *Electr. Power Syst. Res.* **2010**, *80*, 46–52. [CrossRef]
27. Wu, F.; Zhang, X.P.; Ju, P. Small signal stability analysis and control of the wind turbine with the direct-drive permanent magnet generator integrated to the grid. *Electr. Power Syst. Res.* **2009**, *79*, 1661–1667. [CrossRef]
28. Li, S.; Haskew, T.A.; Xu, L. Conventional and novel control designs for direct driven PMSG wind turbines. *Electr. Power Syst. Res.* **2010**, *80*, 328–338. [CrossRef]
29. Constantin, A. Advanced Modeling and Control of Wind Power Systems. Master's Thesis, Aalborg University, Aalborg, Denmark, June 2010.
30. Hansen, M.H.; Hansen, A.; Larsen, T.J.; Øye, S.; Sørensen, P.; Fuglsang, P. *Control Design for a Pitch-Regulated, Variable Speed Wind Turbine*; Risø National Laboratory: Roskilde, Denmark, 2005.
31. Liu, H. Grid Integration of Offshore Wind Farms via VSC-HVDC—Dynamic Stability Study. Ph.D. Thesis, Aalborg University, Aalborg, Denmark, June 2014.
32. Milano, F. *Power System Modeling and Scripting*; Springer: London, UK, 2010.
33. Kundur, P. *Power System Stability and Control*; McGraw-Hill: New York, NY, USA, 1994.
34. Delfino, F.; Procopio, R.; Rossi, M.; Ronda, G. Integration of large-size photovoltaic systems into the distribution grids: A P–Q chart approach to assess reactive support capability. *IET Renew. Power Gener.* **2010**, *4*, 329. [CrossRef]
35. *DIgSILENT PowerFactory Application Guide*; DIgSILENT GmbH: Gomaringen, Germany, 2016.

energies

Article

Tie-Line Bias Control Applicability to Load Frequency Control for Multi-Area Interconnected Power Systems of Complex Topology

Chunyu Chen [1], Kaifeng Zhang [1,*], Kun Yuan [1] and Xianliang Teng [2]

[1] Key Laboratory of Measurement and Control of Complex Systems of Engineering, Ministry of Education, School of Automation, Southeast University, Nanjing 210096, China; 230159105@seu.edu.cn (C.C.); kyuan@seu.edu.cn (K.Y.)

[2] State Key Laboratory of Smart Grid Protection and Control, NARI Group Corporation, State Grid Electric Power Research Institute, Nanjing 211106, China; tengxianliang@sgepri.sgcc.com.cn

* Correspondence: kaifengzhang@seu.edu.cn; Tel.: +86-138-1385-4412

Academic Editor: Paolo Mercorelli

Received: 14 October 2016; Accepted: 28 December 2016; Published: 11 January 2017

Abstract: The tie-line bias control (TBC) method has been widely used in the load frequency control (LFC) of multi-area interconnected systems. However, it should be questioned whether the conventional TBC can still apply to LFC when considering the complication of structures of power systems. LFC, in essence, is to stabilize system frequency/tie-line power by controlling controlled outputs' area control error (ACE). In this paper, relations between LFC control variables and controlled outputs are expressed as a system of equations, based on which an exemplary ring network is studied. Sufficient and necessary conditions for TBC applicability is presented, and a novel LFC mode is proposed for a general ring network where TBC cannot work. Finally, TBC applicability to multi-area systems with general topology is studied, and a general LFC mode is proposed for systems where TBC is not definitely applicable, thus rendering routines that may guide LFC design of future power systems with more complex topologies.

Keywords: load frequency control; tie-line bias control (TBC) applicability; ring network; topology

1. Introduction

Interconnection of electrical power systems has been the main trend in modern power grid construction [1–3]. By interconnection, distributed power systems can assist each other in case of emergencies due to their varied load requirement and capacity outage [4]. Interconnection also offers advantages such as overall cost reduction of reserve provision, frequency maintenance and voltage collapse avoidance, etc., thus enhancing system stability and security [5]. Modern power grids are becoming more and more complex in respect to their constitutions/configurations and interconnection manners [6]. Quantitatively, interconnected multi-area systems have evolved from systems of few interconnected areas to power systems with a large number of areas. Topologically, the interconnection manners are becoming more diversified, and ring topology is among the emerging structures. For example, Chinese power systems are in the process of transforming from simple structures to more complex ones like the Three-China Grid (a three-area ring synchronous network connecting the North China, Central China and East China Grid into an annular structure). A multi-area ring network is also under construction in the East China Grid [7,8].

Hierarchical control was introduced into power system control when Schweppe analyzed basic aspects of hierarchical systems and related them to real-time electric power control problems [9]. The hierarchical structure leads to controller localization, which means controllers are distributed among corresponding control areas divided based on the hierarchy. As an important means of power system stability regulation, active power control (real power or frequency control) aims at balancing power generation and consumption [10,11]. On an area-control level, active power control mainly focuses on two indexes: (1) system frequency and (2) tie-line interchange power. By regulating frequency and tie-line interchange power around their nominal values, power balance and system stability can thus be achieved. Active power control on an area-control level is usually termed load frequency control (LFC) [12], and area control error (ACE), a linear weighted sum of frequency and tie-line interchange power, is used as the area control goal (controlled output), which is conventionally achieved by tie-line bias control (TBC).

Thus far, TBC-based LFC schemes have been studied for two-area systems [13–15], three-area systems [16,17] and multi-area systems [18–20]. However, to the best of the authors' knowledge, no research has ever been made specifically on how the topology influences the LFC performance or whether the existing TBC mode can still satisfy the LFC requirement of modern power systems of more complex topologies. As mentioned in the first paragraph, the ring network has emerged as a new mode of transmission, and it is inevitable to face LFC problems of a ring network. TBC-based LFC schemes have briefly addressed ring networks in the simulation studies and seemingly achieve the asymptotical stability of tie-line interchange power [21]. Nevertheless, in this paper, we prove that TBC cannot regulate tie-line interchange power of ring networks back to nominal values except for some specially constructed systems.

One of the most essential problems of LFC is neglected in conventional TBC-based control schemes—namely, whether or not TBC applies for modern power systems of new topologies. By analysis, it is learned that TBC cannot be applicable for multi-area systems of specific network topologies, of which ring is the most typical. Especially when considering the tie-line interchange power stability in control goals, TBC fails to guarantee that. The reason that TBC fails in these circumstances is that the number of control inputs is less than that of control variables. That is to say, frequency and tie-line interchange power deviation variables outnumber the controllers corresponding to ACE of respective areas. In this paper, a hybrid control mode is proposed for LFC of ring networks. The composite control mode is designed by adding an additional controller to existing ones and modifying TBC mode of some areas with pure interchange power control. Thus, both system frequency and tie-line interchange power deviations can return to nominal values. Furthermore, multi-area systems of more general topologies are studied and LFC schemes are presented for them, thus offering routines that may guide LFC design of future power grids of different topologies.

The remaining part of this paper is organized as follows. In Section 2, TBC-based LFC schemes are briefly discussed and applicability of TBC is analyzed. In Section 3, two typical types of multi-area systems are studied: (1) cascade and (2) ring. For a general ring network, the sufficient and necessary conditions of TBC applicability for LFC is given, and, for those rings, when this condition is unsatisfied, a hybrid LFC scheme is presented. LFC of multi-area systems with general topologies are discussed in Section 4, and corresponding control schemes are presented. In Section 5, simulations are implemented for a three-area ring network with the aid of the proposed hybrid scheme. Concluding remarks are given in Section 6.

2. Tie-Line Bias Control-Based Load Frequency Control Scheme and Applicability to Multi-Area Interconnected Systems

In this section, TBC applicability to multi-area interconnected systems is analyzed. A TBC based LFC scheme is the most common method in dealing with system frequency and tie-line interchange power stability of multi-area interconnected systems caused by power imbalance of control areas. ACE is used as the control goal. By controlling ACE to specific ranges based on the control standards, system frequency and tie-line interchange power can go back around nominal values.

ACE under TBC is written by:

$$ACE_i = 10B_i\left(f_{ai} - f_0\right) + \left(P_{ai} - P_0\right), \tag{1}$$

where B_i is the bias coefficient; and f_0 (f_{ai}) and P_0 (P_{ai}) are the nominal (actual) values of system frequency and interchange power.

Remark 1. *Throughout this paper, all tie-lines between each pair of interconnected area (i, j) are equivalent to one lumpy tie-line.*

Remark 2. *Based on the restrictions of ACE, the control standard can be categorized into A1/A2, and CPS (Control Performance Standard) standard, etc., In this paper, all analyses are based on a simple A1/A2 standard ($ACE = 0$).*

Remark 3. *In stable operating conditions, all area frequencies Δf_i are equal to COI (Center of Inertia: the weighted average of frequencies considering inertia of generators [22]) frequency of the whole system Δf_0.*

From the expressions of ACE in TBC in Equation (1), TBC can apparently satisfy the LFC requirements of both regulating frequency and total interchange power. However, by analysis, it will be proved that TBC is only suitable for interchange power per tie-line control under certain circumstances.

It can be found that the area's total interchange power deviation $\Delta P_{tie(i)}$ is exactly the interchange power deviation per tie-line $\Delta P_{tie(i\_j)}$ for two-area systems (since there exists only one lumpy line for a two-area system). However, under those circumstances where certain areas are interconnected by more than one area, $\Delta P_{tie(i)}$ control is not equal to $\Delta P_{tie(i\_j)}$ control, which could lead to undesired tie-line interchange power deviation though total interchange power is restored.

Assume there is an m-area system connected by n tie-lines. The control equation can be written as follows:

$$Ax = y, \tag{2}$$

where x is an $(n+1) \times 1$ control variable vector containing COI frequency Δf_0 and n tie-line interchange power deviations $\Delta P_{tie(i\_j)}$; y is an $m \times 1$ control goal vector (a vector composed of 0 with appropriate dimensions) which is composed of ACE_i of Area i; A is the corresponding $m \times (n+1)$ coefficient matrix that is determined by the interconnection manner/topology of the system.

Based on linear algebra, if A satisfies $n + 1 = m$, there exists a unique zero solution if and only if $rank(A) = m$.

If $n + 1 > m$, the solutions of Equation (2) only have the following two situations:

(a) Each control equation of the control area is independent $rank(A) = m$, and then Equation (2) is insoluble, which means there exists no frequency or interchange power deviation solution under TBC in this condition.
(b) Each control equation of the control area is non-independent $rank(A) < m$, and then Equation (2) has infinitely many solutions, which means there exists more than one frequency or interchange power deviation value under TBC in this condition.

Remark 4. *It is assumed that there exists no isolated power system, namely, for each area, there is at least one tie-line which connects it with other areas. Therefore, the circumstance where $n + 1 < m$ does not exist.*

From the analysis above, it is known that for a general multi-area interconnected system, frequency and tie-line interchange power stability, namely, a unique null solution of Equation 2) cannot be guaranteed under TBC. The reason that TBC is applicable to a two-area system is due to the fact that the number of control variables is equal to that of control inputs (both are three) and $rank(A) = 3$. However, this special condition cannot apply for general systems. In Section 3, multi-variable systems of typical topologies will be given as examples to further illustrate applicability of TBC. Correspondingly, the hybrid LFC scheme is presented to solve the LFC problem insoluble by TBC.

3. Load Frequency Control of Multi-Area Interconnected Systems of Typical Topology

As it is known from TBC applicability analysis in Section 2, the interconnection manner/topology of multi-area systems plays an important role in determining the LFC control mode. With the development of ultra high voltage transmission techniques, the Three-China ("North China-Central China-East China") synchronous grid is under design for large-scale and high efficiencies of power transmission [8]. The diagram of the Three-China Grid is as shown in Figure 1.

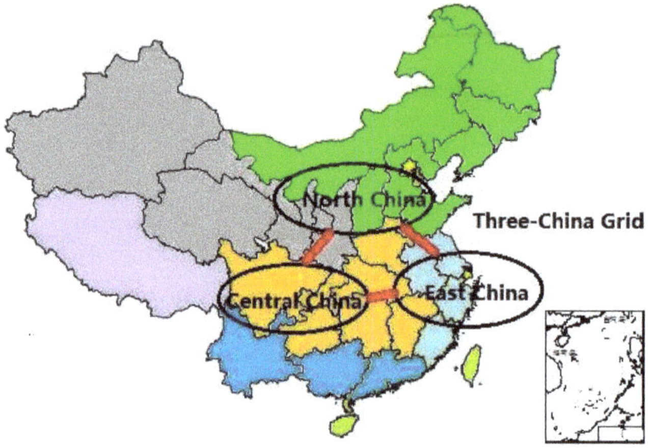

Figure 1. Diagram of the Three-China synchronous 1000 kV grid.

In this section, LFC of multi-area systems with the most prevalent cascade interconnection manner is first addressed considering the emergence of ring networks (e.g., Three-China grid shown in Figure 1) in area interconnection, and then LFC of multi-area systems of ring interconnection is studied.

3.1. Load Frequency Control for Cascade Network under Tie-Line Bias Control

A cascade network is formed by all areas interconnected in series, and the schematic diagram of an m-area cascade network is shown in Figure 2.

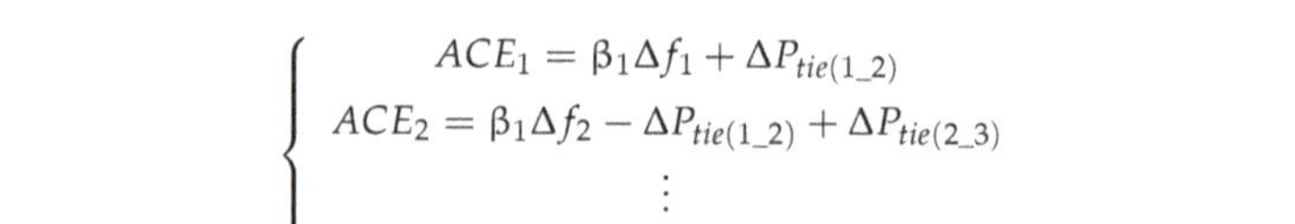

Figure 2. Diagram of the m-area cascade network.

Under TBC mode, control equations of LFC can be expressed by:

$$\begin{cases} ACE_1 = \beta_1 \Delta f_1 + \Delta P_{tie(1\_2)} \\ ACE_2 = \beta_1 \Delta f_2 - \Delta P_{tie(1\_2)} + \Delta P_{tie(2\_3)} \\ \vdots \\ ACE_m = \beta_m \Delta f_m + \Delta P_{tie(m-1\_m)}. \end{cases} \quad (3)$$

Since control inputs guarantee that ACE_i is zero, it can be obtained from Equation (3) that $\Delta f_i = 0$ and $\Delta P_{tie(i\_j)} = 0$, which means TBC applies for LFC of cascade network.

3.2. Load Frequency Control for Ring Network under Tie-Line Bias Control

A typical 5-area ring network is shown in Figure 3.

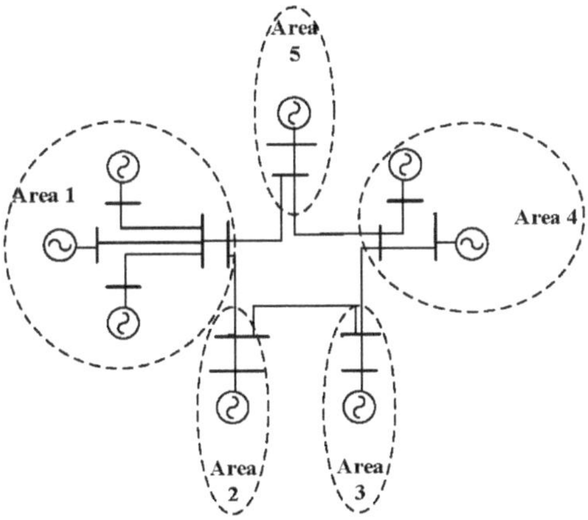

Figure 3. Circuit diagram of a 5-area ring network.

A general m-area ring network (Figure 4) is formed by all areas interconnected in an end-to-end manner.

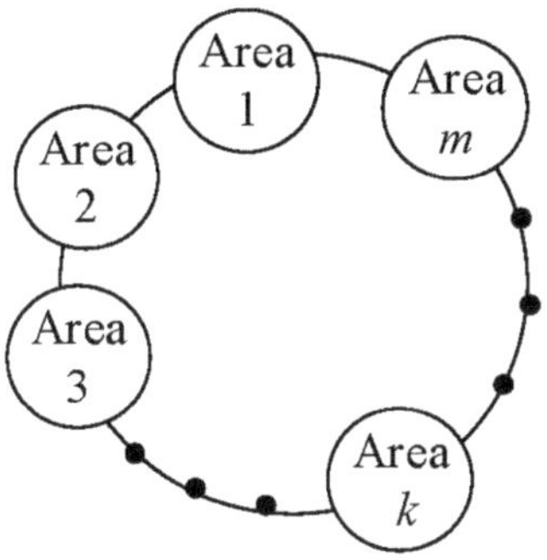

Figure 4. Diagram of a general m-area ring network.

Control equations under TBC are expressed by:

$$\begin{cases} ACE_1 = \beta_1 \Delta f_1 - \Delta P_{tie(m\_1)} + \Delta P_{tie(1\_2)} \\ ACE_2 = \beta_2 \Delta f_2 - \Delta P_{tie(1\_2)} + \Delta P_{tie(2\_3)} \\ \vdots \\ ACE_m = \beta_m \Delta f_m - \Delta P_{tie(m-1\_m)} + \Delta P_{tie(m\_1)}. \end{cases} \quad (4)$$

By summing both sides of Equation (4), it can be obtained that:

$$\sum_{i=1}^{m} ACE_i = \sum_{i=1}^{m} \beta_i \Delta f_i = \Delta f_o \sum_{i=1}^{m} \beta_i. \quad (5)$$

Since $ACE_i = 0$ under control inputs, it can be learned that $\Delta f_1 = \Delta f_2 = \cdots \Delta f_m = \Delta f_o = 0$. However, $\Delta P_{tie(i\_j)}$ cannot be guaranteed to return to zero. Actually, any $\Delta P_{tie(i\_j)} = const$ can satisfy Equation (4), implying the existence of circulating power flow, which is undesirable and extremely dangerous for power system security.

Notice that $\Delta P_{tie(i\_j)} = 0$ is also a solution of Equation (4). Next, the sufficient and necessary conditions will be given for the circumstances under which TBC can apply for LFC of ring networks. Assume that interchange power between Area i and Area j is denoted by P_{i_j}. The tie-line model is equivalent to connecting two voltage sources $U_i \angle \delta_i$ with a tie-line through an equivalent reactance X_T. Define the phase angle deviation $\Delta\delta = \delta_1 - \delta_2$, which is small, implying $sin\,(\Delta\delta) = \Delta\delta$, and actual interchange power on tie-lines can thus be expressed by:

$$P_{1\_2} = T_{12}\Delta\delta = 2\pi\frac{|U_1|\,|U_2|}{X_T}\left(\int f_1 dt - \int f_2 dt\right), \tag{6}$$

where T_{12} stands for synchronous coefficient; f_1 and f_2 are time dependent variables. Based on the simplified equivalent tie-line model, we have the following proposition:

Lemma 1. *TBC can guarantee that system frequency and tie-line interchange power return to nominal values $P_{i\_j,s}$ if and only if:*

$$\frac{P_{1\_2,s}}{T_{12}} + \frac{P_{2\_3,s}}{T_{23}} + \cdots + \frac{P_{m\_1,s}}{T_{m1}} = 0. \tag{7}$$

Proof of Lemma 1.

Sufficient: Define tie-line interchange power deviation by:

$$\Delta P_{i\_j} = P_{i\_j,a} - P_{i\_j,s}. \tag{8}$$

According to Equations (6) and (8), it has:

$$\begin{cases} \int f_1 dt - \int f_2 dt - \frac{P_{1\_2,s}}{T_{12}} = \frac{\Delta P_{1\_2}}{T_{12}} \\ \int f_2 dt - \int f_3 dt - \frac{P_{2\_3,s}}{T_{23}} = \frac{\Delta P_{2\_3}}{T_{23}} \\ \vdots \\ \int f_m dt - \int f_1 dt - \frac{P_{m\_1,s}}{T_{m1}} = \frac{\Delta P_{m\_1}}{T_{m1}}. \end{cases} \tag{9}$$

By summation on both sides of equal signs of Equation (9) with the aid of Condition (7), it follows that:

$$\frac{\Delta P_{1\_2}}{T_{12}} + \frac{\Delta P_{2\_3}}{T_{23}} + \cdots \frac{\Delta P_{m\_1}}{T_{m1}} = 0. \tag{10}$$

Since TBC can guarantee the total interchange power deviation per control area to return to zero:

$$\begin{cases} \Delta P_{1\_2} - \Delta P_{m\_1} = 0 \\ \Delta P_{2\_3} - \Delta P_{1\_2} = 0 \\ \vdots \\ \Delta P_{m\_1} - \Delta P_{m-1\_m} = 0. \end{cases} \tag{11}$$

By combining Equations (10) and (11), it can be proved that $\Delta P_{i\_j} = 0$.

Necessary: Since all tie-line interchange power deviations are zero, then Equation (10) holds. As Equation (10) always holds for the equivalent tie-line model, it can be proved that Condition (7) holds by substituting Equation (10) into Equation (9). □

Remark 5. *LFC based on classical LFC model (systems linearized about the nominal equilibrium points) automatically satisfies Lemma 1 since* $P_{i\_j,s} = 0$.

3.3. Hybrid Load Frequency Control Scheme for a Ring Network

As is discussed above, usually, the sufficient and necessary conditions do not hold for general ring networks. There exists only one ring network of a special structure to which TBC can apply. However, this condition requires specially constructed ring networks which are rare to be found in real-life engineering practice. More usually, a random ring network cannot satisfy this condition. In this section, the hybrid scheme is presented to solve LFC of general ring networks.

From TBC applicability analysis in Section 2, it is learned that if the number of control variables is larger than that of control inputs, the solution of Equation (2) is either non-existential or flexible. For ring networks, the number of control variables outnumbers that of control inputs by one. Hence, an auxiliary controller must be introduced for LFC to balance the number of control variables and control inputs. As for the LFC scheme, the proposed scheme must be designed in a way such that $rank(A) = n+1$. Based on TBC, a hybrid LFC scheme is proposed by alternating TBC of specific areas with pure interchange power control.

Firstly, auxiliary control system should be established. The auxiliary control center is established by dividing one certain area k into two sub-areas. k should contain the transmission line k_q of the maximum transmission capacity. k_q plays the role of new tie-lines of two sub-areas. The diagram of this modified ring network is as shown in Figure 5.

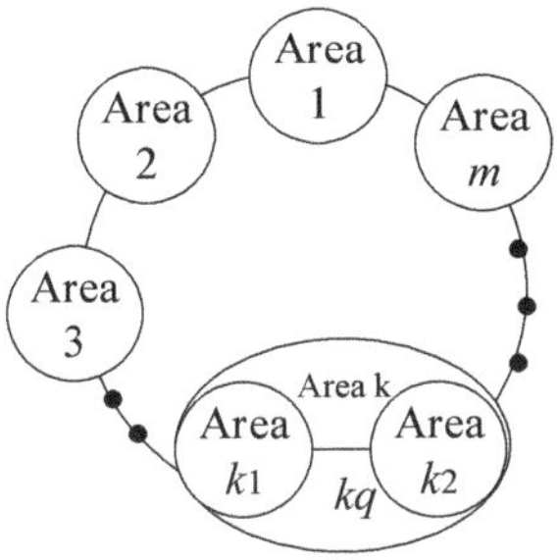

Figure 5. Diagram of a redivided ring network.

After ascertaining auxiliary control system, an LFC scheme should be proposed. In this paper, a hybrid LFC scheme is adopted, the control equations are expressed by:

$$\begin{cases} ACE_1 = \beta_1 \Delta f_1 - \Delta P_{tie(m\_1)} + \Delta P_{tie(1\_2)} \\ ACE_2 = \beta_2 \Delta f_2 - \Delta P_{tie(1\_2)} + \Delta P_{tie(2\_3)} \\ \vdots \\ ACE_{k_1} = \beta_{k1} \Delta f_k - \Delta P_{tie(k-1\_k)} \\ ACE_{k_2} = \Delta P_{tie(k\_k+1)} \\ \vdots \\ ACE_m = \beta_m \Delta f_m - \Delta P_{tie(m-1\_m)} + \Delta P_{tie(m\_1)}, \end{cases} \tag{12}$$

where k_1 and k_2 stands for the new divided sub-areas from Area k. It is obviously known that $\Delta f_i = 0\ (1 \le i \le m)$ when $ACE_i = 0$. Substitute $\Delta f_k = 0$ into the $k+1^{th}$ Equation in (12), and then $\Delta P_{tie(k-1\_k)} = 0$. In addition, substitute it into the k-th equation in (12), and it follows that $\Delta P_{tie(k-2\_k-1)} = 0$. Furthermore, $\Delta P_{tie(j-1\_j)} = 0\ (2 \le j \le k)$ can be obtained. Similarly, $\Delta P_{tie(j-1\_j)} = 0$ $(k+1 \le j \le m)$ can be obtained. Namely, all tie-line interchange power returns to the nominal value.

Remark 6. *The hybrid LFC scheme cannot guarantee the interchange power stability of the newly established tie-line between the two sub-areas. However, since this tie-line of the divided Area k is chosen as Line k_q of a large transmission capacity, it can withstand the power deviations within allowable ranges.*

4. Load Frequency Control of Multi-Area Systems with General Topology

With the scale-up of power girds, interconnection between areas is becoming more and more complex. From the perspective of topology, the structure of interconnection can be classified into ring, tree, mesh, bus and star, etc. It spontaneously results in LFC problems of multi-area systems with these different interconnection manners. In this section, LFC of multi-area systems of general topologies is studied and a corresponding LFC scheme is presented, which offers routines that may guide LFC design of future interconnected multi-area systems.

Assume that there exist m interconnected areas $A_1, A_2, \cdots, A_m$ and A_i is connected by j $(1 \leq i \neq j \leq m)$ (there exists no isolated power system). Then, the control variables are COI frequency deviation Δf_o and and tie-line interchange power deviation $\Delta P_{tie(i\_j)}$ $(1 \leq i \neq j \leq m)$.

Remark 7. *For multi-area systems where there exist isolated areas, by extracting the isolated ones and implementing flat frequency control (FFC), the rest of the system conforms with the topology discussed in this section.*

Problem Definition: The LFC goal of multi-area interconnected systems is to design control schemes such that $\lim_{t\to\infty} \Delta f_i \to 0$, $\lim_{t\to\infty} \Delta P_{tie(i\_j)} \to 0$.

It can be learned that only two types of topologies exist for multi-area systems.

- *Type A*: There exists no ring network in power systems.
- *Type B*: There exist ring networks in power systems.

Definition 1. *The area which has only one tie-line is defined as pseudo-independent area (PIA).*

It is also learned that the Type A network can be characterized by a recursive tree structure through appropriate indexing as is shown in Figure 6.

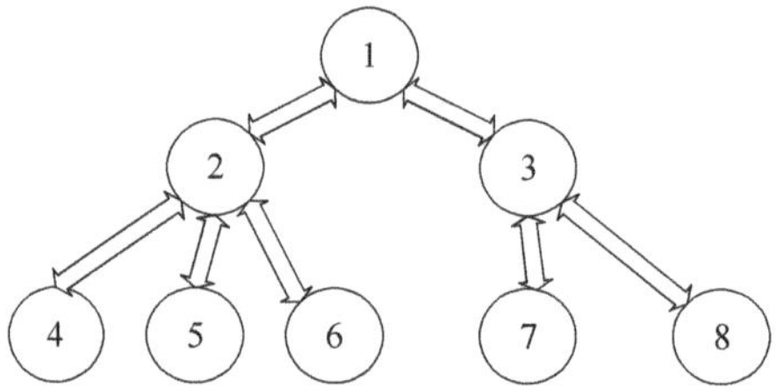

Figure 6. Diagram of the tree structure of the Type A network (without ring).

As for the Type B network, it can be regarded as a composite network by adding tie-lines which can form cycles in the topology of Type A network. Hence, the Type B network can be indexed in the same way as indexing of the Type A one.

4.1. Load Frequency Control Scheme for the Type A Network (without Ring)

Theorem 1. *System frequency deviation Δf_i and tie-line interchange power deviation $\Delta P_{tie(i\_j)}$ $((1 \leq i \neq j \leq m)$ of the Type A network can be guaranteed to return to zero under TBC.*

Proof. Control equations under TBC can be defined by:

$$A_1x_1 = y_1, \tag{13}$$

where x_1 is the control variables vector containing COI frequency deviation Δf_o and tie-line interchange power deviation $\Delta P_{tie(i\_j)}$; $y_1 = \begin{bmatrix} ACE_1 & ACE_2 & \cdots & ACE_m \end{bmatrix}^T$; and A_1 is the coefficient matrix which can be written by:

$$A_1 = \begin{bmatrix} A_{11} & A_{12} \end{bmatrix}, \tag{14}$$

where $A_{11} = \begin{bmatrix} \beta_1 & \beta_2 & \cdots & \beta_m \end{bmatrix}^T$; A_{12} is an $m \times (m-1)$ matrix containing either 0 or ± 1 which satisfies:

$$\sum_{i=1}^{m} A_{12}(i,:) = \underbrace{\begin{bmatrix} 0 & 0 & \cdots & 0 \end{bmatrix}}_{m}. \tag{15}$$

Namely, the row sum of A_{12} is a $1 \times m$ vector containing all zero elements.

To prove that system frequency and tie-line deviations return to zero is to prove that there only exist zero solutions for control variables x_1, which is equivalent to proving that $rank(A_1) = rank(\begin{bmatrix} A_1 & y_1 \end{bmatrix}) = m$. Since control inputs guarantee $ACE_i = 0$, it can then be proved that $rank(A_1) = rank(\begin{bmatrix} A_1 & y_1 \end{bmatrix})$. Therefore, the rest of the proof is to verify $rank(A_1) = m$.

Elementary transformation of the matrix is used here such that the standard form of A_1 is obtained to clearly find the rank. The transformation procedures are as follows:

- Step 1: Let $num = m$. The first row of A_1 is replaced by the sum of all rows of A_1 such that A_{1_1} is obtained. The first row of A_{1_1} is multiplied by $1/\sum \beta_i$. Then, the jth row of A_{1_1} $(2 \leq j \leq m)$ is replaced by the sum of that row and $-\beta_j$ of the first row:

$$A_{1\_2} = \begin{bmatrix} 1 & 0 & 0 & \cdots & 0 \\ 0 & \times & \times & \cdots & \times \\ 0 & \times & \times & \cdots & \times \\ \vdots & \vdots & \ddots & \ddots & \vdots \\ 0 & \times & \cdots & \cdots & \times \end{bmatrix},$$

  where $\times$ stands for an uncertain value which can either be 0, -1 or 1. Notice in A_{1_2} that there still exists the $[\times]$ part, and transformation should be implemented to eliminate it. It can be learned for PIA k that its corresponding row in A_{1_2}, denoted by PIAR (the PIA Row), can be expressed by:

$$\begin{bmatrix} 0 & 0 & \cdots & \underset{s_k}{1} & \cdots & 0 \end{bmatrix}.$$

  For any two different PIA $k \neq l$, $s_k \neq s_l$. In the following steps, the matrix is transformed based on the area reduction (AR) approach.
- Step 2: The AR approach is utilized starting from this step. Firstly, all the areas that are only connected by PIAs are selected. Namely, Area q, which is connected by PIA k $(1 \leq k \leq m-1)$, is selected. Then, the q-th row of A_{1_2} is replaced by the sum of that row and PIAR v_k, such that the new PIAR v_q is obtained and Area q is established as a new PIA. After this transformation, for all these areas that are only connected by PIAs, the AR approach is implemented by establishing new n_o PIAs and eliminating original n_o PIAs, and $num = num - n_o$.
- Step 3: if $num = 1$, go to Step 4, else go to Step 2.
- End

The ultimate matrix A_{1_2} by the transformation above can be defined by:

$$A_{1\_2} = \begin{bmatrix} a_1 \\ A_2 \end{bmatrix},$$

where $a_1 = \begin{bmatrix} 1 & 0 & \cdots & 0 \end{bmatrix}$; for any row j $(1 \leq j \leq m-1)$ of A_2, it satisfies:

$$\begin{bmatrix} 0 & 0 & \cdots & \underset{s_j}{1} & \cdots & 0 \end{bmatrix},$$

where $s_j \neq 1$. In addition, for any $j \neq i$, it follows that $r_j \neq r_i$. Thus, it is apparently known that $rank(A_{1\_2}) = m$ since all the transformation is elementary transformation and then $rank(A_1) = rank(A_{1\_2}) = m$. Namely, the control variables only have zero solutions and the LFC goal is reached. □

4.2. Load Frequency Control Scheme for Type B Network (with Ring)

In this case, there exist ring networks in power grids. The one ring structure has already been addressed in Section 3. Hence, in this section, we emphasize systems containing multi-ring networks and design LFC schemes. Before proposing specific control schemes, a definition of independent ring (IR) is presented.

Definition 2. *Independent ring (IR) is defined by a ring network that has at least one tie-line that is not shared with other ring networks.*

By analysis, it can be learned that the number of IRs is exactly the difference between the number of control variables and that of control inputs. From the discussion of TBC applicability in Section 2, it is known that the necessary conditions for zero unique solutions of control variables are that the number of control inputs and that of control variables are the same. Hence, as with the previous case of a one ring network in Section 3, auxiliary control systems are set up for this Type B network.

The auxiliary controller is introduced by founding an additional control center for the two areas on the endpoints of the unshared tie-line of the IR ring (in the case of multiple unshared tie-lies, choose the one which has the shortest length). The diagram of the auxiliary control center is shown in Figure 7.

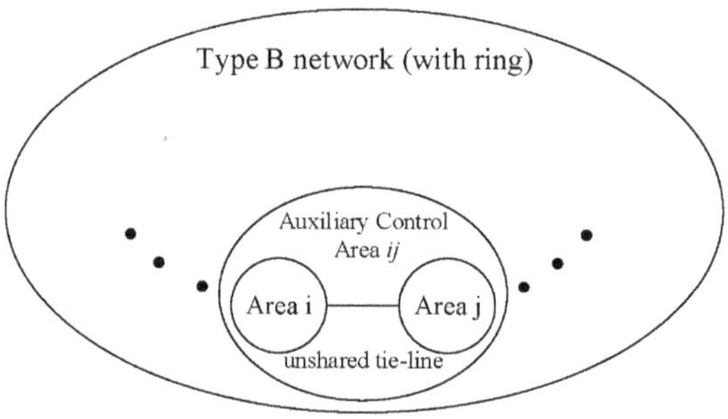

Figure 7. Diagram of the auxiliary control center.

After introducing auxiliary control systems, the LFC scheme is presented as the hybrid control mode below. The hybrid controller is composed of two parts.

- For areas which are not contained in auxiliary control centers, the TBC mode (1) is still used.
- For those pairs of areas (i, j) that are contained in certain auxiliary control centers, the LFC scheme is:

$$\begin{cases} ACE_i = \Delta P_{tie(i\_x)} \\ ACE_j = \Delta P_{tie(j\_y)} \\ ACE_{ij} = \beta_{ij}\Delta f_{ij} + \Delta P_{tie(i\_j)}. \end{cases} \tag{16}$$

Theorem 2. *System frequency deviation Δf_i and tie-line interchange power deviation $\Delta P_{tie(i\_j)}$ ($(1 \le i \le m$, $1 \le j \le m-1)$ of the Type B network can be guaranteed to return to zero under the hybrid law in Equation (1) and Equation (16).*

Proof. As with the proof in Theorem 1, it boils down to verifying that $rank(A_1) = m$. It is easily proved that system frequency deviation Δf_i is regulated back to zero, which is obtained in the same way as through summation of control equations in Equation (3). From the third equation in Equation (16), it is learned that tie-line interchange power deviation of the chosen unshared tie-line of the IR satisfies $\Delta P_{tie(i\_j)} = 0$, which is equivalent to that the $i(j)^{\text{th}}$ row of A_{1_2}, is a PIAR. In addition, Area i and j, which connect the unshared tie-line of the IR in this case, are equivalent to PIAs. Then, the rest of the proof is the same as that in Theorem 1 and is thus omitted due to space limitations. □

5. Simulation and Analysis

In this section, LFC for ring networks is investigated by means of simulations under Matlab/Simulink (R2015b, Mathworks, Natick, Massachusetts, USA). Firstly, TBC is exerted to illustrate inapplicability to LFC for ring networks. Then, the hybrid LFC scheme in Section 3.3 is used to overcome this inapplicability by making both system frequency and interchange power per tie-line return to nominal states.

According to Remark 5, it is known that the condition in Condition (7) holds for the classical LFC model spontaneously. However, the nominal values of the system states (e.g., tie-line interchange power) are definitely not zero. In this paper, the Electricity Market is introduced and the Disco participation matrix (DPM) is used to calculate the nominal/scheduled values of tie-line interchange power [23], which are non-zero values based on contracts between distribution and generation companies of the interconnected areas. The block diagram of a ring network containing three areas is shown in Figure 8. The parameters are given in [22].

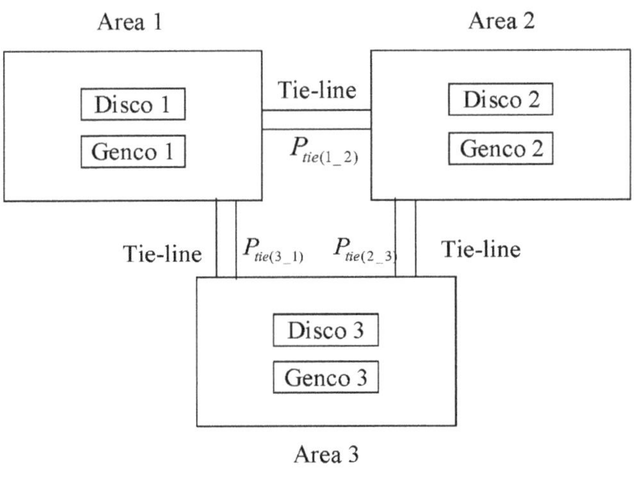

Figure 8. Diagram of a ring network containing three areas.

There is one generation and one distribution company for each area in Figure 8, and the DPM is:

$$\begin{matrix} & Disco_1 & Disco_2 & Disco_3 \\ Genco_1 & 0.3 & 0.3 & 0.3 \\ Genco_2 & 0.5 & 0.6 & 0.2 \\ Genco_3 & 0.2 & 0.1 & 0.5 \end{matrix}, \quad (17)$$

where the element $cpf_{i,j}$ in Equation (17) is called *cpf* (Contract Participation Factor), which is the fraction of the total load contracted by Disco j (Distribution Company j) towards Genco i (Generation Company i). The sum of all entries in row i is one [24]. The nominal tie-line interchange power $P_{i\_j,s}$ is defined by:

$$P_{i\_j,s} = P_{Gi\_j,s} - P_{Gj\_i,s}, \quad (18)$$

where $P_{Gi\_j,s}$ stands for the power contracted by Discos in Area j to Gencos in Area i. The tie-line interchange power deviation is thus expressed as follows:

$$\Delta P_{i\_j} = P_{i\_j,a} - P_{i\_j,s}. \quad (19)$$

Let the power demand of the Disco in Areas 1, 2 and 3 be 1 p.u., 2 p.u. and 2 p.u., respectively. Based on Equations (18) and (17), the nominal interchange power of each tie-line is 0.1 p.u., 0.2 p.u. and 0.4 p.u., respectively.

Next, simulations are executed for the following two circumstances under TBC to illustrate the sufficient and necessary conditions in Lemma 1:

- *Circumstance 1* : The sufficient and necessary condition (7) holds for the ring network,
- *Circumstance 2* : The sufficient and necessary condition (7) does not hold for the ring network.

In the last part of this section, the hybrid LFC scheme in Section 3.3 is tested on an IEEE 39-bus system-based ring network.

Remark 8. *In this paper, all actual control inputs, namely, the ultimate power adjustments for each LFC participating component is obtained by PI controllers:* $P_u = K_p\,(ACE) + K_i \int ACEdt$. *The controller design using advanced control theories is not the emphasis of this paper.*

5.1. Circumstance 1 under Tie-Line Bias Control

In this case, let synchronous coefficient T_{ij} be $T_{12} = 4, T_{23} = 6.28, T_{31} = 7.0364$, such that $\frac{P_{1\_2,s}}{T_{12}} + \frac{P_{2\_3,s}}{T_{23}} + \frac{P_{3\_1,s}}{T_{31}} = 0$ (Condition (7) holds). From the simulation results in Figure 9 (dashed line), it is learned that each tie-line exchange power returns to the nominal value.

5.2. Circumstance 2 under Tie-Line Bias Control

In this case, let synchronous coefficient T_{ij} be $T_{12} = 4, T_{23} = 6.28, T_{31} = 6$, such that Condition (7) does not hold. From the simulation results in Figure 9 (dot line), it is learned that none of the tie-line exchange power returns to the nominal value.

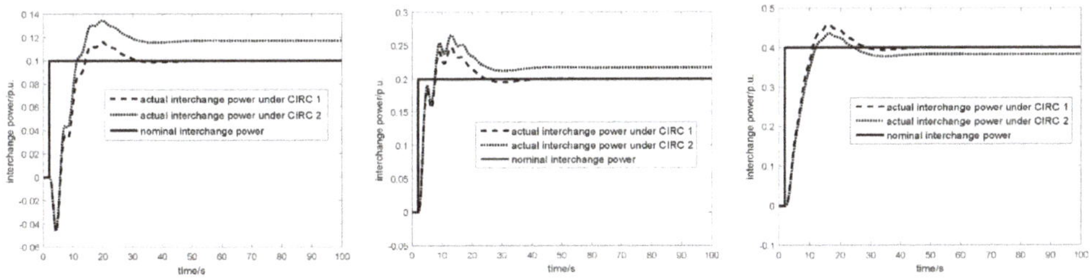

(a) (b) (c)

Figure 9. Simulation results in Sections 5.1 (Condition (7) holds) and 5.2 (Condition (7) does not hold). (**a**) Interchange power between Areas 1 and 2; (**b**) interchange power between Areas 2 and 3; and (**c**) interchange power between Areas 3 and 1.

5.3. Hybrid Load Frequency Control Scheme under Circumstance 2

From simulation results in Sections 5.1 and 5.2, it can be learned that once the ring network does not satisfy Condition (7), TBC can no longer apply. In this section, the hybrid LFC scheme in Section 3.3 is simulated on a general ring network which does not satisfy Condition (7).

The standard IEEE 39-bus system is adopted to more closely reflect the practical power system [25]. In order to construct the ring network under Circumstance 2, the system is divided into three areas as is shown in Figure 10.

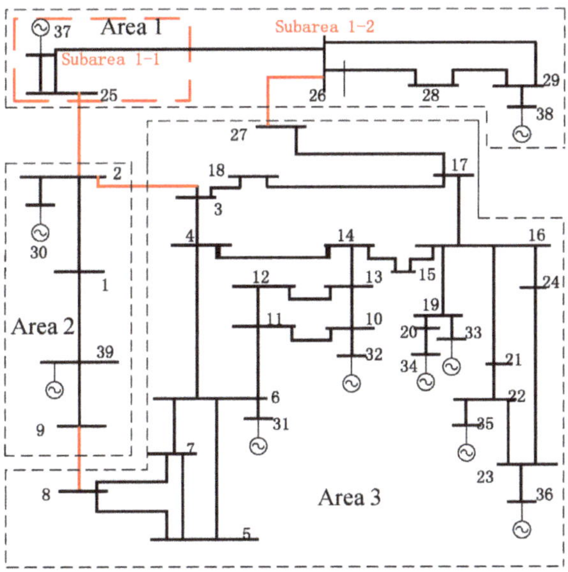

Figure 10. Diagram of an IEEE 39 system based three-area ring network.

Notice that there exist two tie-lines between Areas 2 and 3 (the transmission lines between bus 2 and 3, bus 9 and 8). Based on Remark 1, the two lines are equivalent to one tie-line, and the LFC scheme is executed only to control the interchange power of the equivalent tie-line.

From the simulation results in Figure 11, it can be learned that the proposed LFC scheme effectively solves the TBC inapplicability problem. Interchange power of the tie-line between the sub-areas divided from Area 1, as is discussed in Remark 6, does not return to the nominal state.

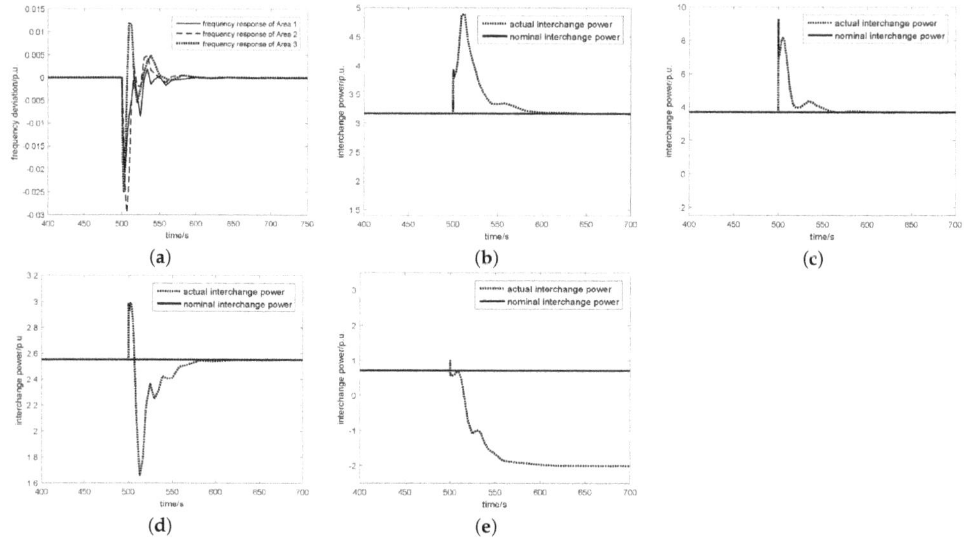

Figure 11. Simulation results in Section 5.3 (Condition (7) does not hold). (**a**) Frequency response of three areas; (**b**) interchange power between Areas 1 and 2; (**c**) interchange power between Areas 2 and 3; (**d**) interchange power between Areas 3 and 1; and (**e**) interchange power between two sub-areas in Area 1.

6. Conclusions

In this paper, applicability of TBC-based LFC schemes toward multi-area interconnected ring networks is analyzed. The sufficient and necessary conditions where TBC can deal with LFC of ring networks is presented. In addition, for general rings that cannot satisfy the condition, a hybrid LFC scheme is designed by introducing an auxiliary controller and modifying TBC in certain areas. In addition, we discuss LFC of multi-area interconnected systems with general interconnection manners/topologies, which offers routines that may guide LFC design of future power grids with more complex topologies. The simulations of an IEEE 39-bus based three-area ring network proves the effectiveness of the proposed scheme.

Acknowledgments: This work is supported by the National Natural Science Foundation of China (No. 51577031 and No. 61473084) and the State Key Laboratory of Smart Grid Protection and Control.

Author Contributions: Chunyu Chen and Kaifeng Zhang conceived and designed the experiments; Kun Yuan performed the experiment; Chunyu Chen wrote the paper; and Kaifeng Zhang and Xianliang Teng reviewed the paper.

Conflicts of Interest: The authors declare no conflict of interest.

References

1. Lee, S.S. Northeast Asia power system interconnection and ESS based balance strategies in South Korea. In Proceedings of the 2014 IEEE Power & Energy Society (PES) General Meeting, National Harbor, MD, USA, 27–31 July 2014; pp. 1–5.
2. Despa, D.; Mitani, Y.; Li, C.; Watanabe, M. PMU based monitoring and estimation of interarea power oscillation for Singapore-Malaysia interconnection power system. In Proceedings of the 9th International Power & Energy Conference (IPEC 2010), Singapore, 27–29 October 2010; pp. 476–480.
3. Jebali, M.; Ben Salah, R.; Kahouli, O.; Bouchoucha, C.; Abdallah, H.H. Stability analysis for large power system interconnections (Tunisia-Libya). In Proceedings of the 4th International Conference on Systems and Control (ICSC), Sousse, Tunisia, 28–30 April 2015; pp. 161–168.

4. Al-Shaalan, A.M. Technical and economical merits of power systems interconnection. *J. Power Energy Eng.* **2013**, *1*, 1–7.
5. Blume, S.W. *Electric Power System Basics for the Nonelectrical Professional*; John Wiley & Sons: New York, NY, USA, 2008; Volume 32.
6. Häger, U.; Rehtanz, C.; Voropai, N. *Monitoring, Control and Protection of Interconnected Power Systems*; Springer: New York, NY, USA, 2014; Volume 36.
7. Liu, Y.; Liu, Q. Introduction of Three-China Grid. *Appl. Energy Technol.* **2011**, *1*, 34–35.
8. Liu, Z.; Zhang, Q. Study on the development mode of national power grid of China. *Proc. Chin. Soc. Electr. Eng.* **2013**, *33*, 1–10.
9. Schweppe, F.C.; Mitter, S.K. Hierarchical system theory and electric power systems. In Proceedings of the Symposium on Real Time Control of Electric Power Systems, Baden, Switzerland, 27–28 September 1972; pp. 259–277.
10. The North American Electric Reliability Corporation (NERC). *Balancing and Frequency Control*; NERC: Atlanta, GA, USA, 2009.
11. Jayawardene, I.; Venayagamoorthy, G.K. Reservoir based learning network for control of two-area power system with variable renewable generation. *Neurocomputing* **2015**, *170*, 428–438.
12. Pandey, S.K.; Mohanty, S.R.; Kishor, N. A literature survey on load–frequency control for conventional and distribution generation power systems. *Renew. Sustain. Energy Rev.* **2013**, *25*, 318–334.
13. Oni, B.; Graham, H.; Walker, L. Investigation of nonlinear tie line bias control of interconnected power systems. *IEEE Trans. Power Appar. Syst.* **1981**, *PAS-100*, 2350–2356.
14. Sudha, K.; Santhi, R.V. Load frequency control of an interconnected reheat thermal system using type-2 fuzzy system including SMES units. *Int. J. Electr. Power Energy Syst.* **2012**, *43*, 1383–1392.
15. Venayagamoorthy, G.K.; Jayawardene, I.; Arunagirinathan, P. Tie-line bias control and oscillations with variable generation in a two-area power system. In Proceedings of the IEEE 7th International Conference on Information and Automation for Sustainability, Colombo, Sri Lanka, 22–24 December 2014; pp. 1–6.
16. Lim, K.; Wang, Y.; Zhou, R. Robust decentralised load-frequency control of multi-area power systems. *IEE Proc. Gener. Transm. Distrib.* **1996**, *143*, 377–386.
17. Dong, L.; Zhang, Y.; Gao, Z. A robust decentralized load frequency controller for interconnected power systems. *ISA Trans.* **2012**, *51*, 410–419.
18. Rahmani, M.; Sadati, N. Hierarchical optimal robust load-frequency control for power systems. *IET Gener. Transm. Distrib.* **2012**, *6*, 303–312.
19. Franze, G.; Tedesco, F. Constrained load/frequency control problems in networked multi-area power systems. *J. Frankl. Inst.* **2011**, *348*, 832–852.
20. Tedesco, F.; Casavola, A. Fault-tolerant distributed load/frequency supervisory strategies for networked multi-area microgrids. *Int. J. Robust Nonlinear Control* **2014**, *24*, 1380–1402.
21. Rerkpreedapong, D.; Hasanović, A.; Feliachi, A. Robust load frequency control using genetic algorithms and linear matrix inequalities. *IEEE Trans. Power Syst.* **2003**, *18*, 855–861.
22. Kundur, P.; Balu, N.J.; Lauby, M.G. *Power System Stability and Control*; McGraw-Hill: New York, NY, USA, 1994; Volume 7.
23. Donde, V.; Pai, M.; Hiskens, I.A. Simulation and optimization in an AGC system after deregulation. *IEEE Trans. Power Syst.* **2001**, *16*, 481–489.
24. Kothari, D.; Nagrath, I. *Power System Engineering*; Tata McGraw-Hill: New Delhi, India, 2008.
25. Pai, A. *Energy Function Analysis for Power System Stability*; Springer Science & Business Media: New York, NY, USA, 2012; Volume 1.

energies

MDPI

Article

Power Control of Low Frequency AC Transmission Systems Using Cycloconverters with Virtual Synchronous Generator Control

Achara Pichetjamroen * and Toshifumi Ise

Graduate School of Engineering, Osaka University, 2-1 Yamada-oka, Suita, Osaka 565-0871, Japan; ise@eei.eng.osaka-u.ac.jp

* Correspondence: achara@pe.eei.eng.osaka-u.ac.jp; Tel.: +81-6-6879-7689

Academic Editor: Paolo Mercorelli

Received: 6 September 2016; Accepted: 19 December 2016; Published: 28 December 2016

Abstract: This paper is focused on the application of a multi-terminal line-commutated converter-type low frequency AC transmission system (MTLF) using a cycloconverter by applying a new power control scheme for multi-terminal operation. With the virtual synchronous generator (VSG) control scheme, the transmitting power among the multi-terminal system can be accomplished without a communication link for frequency synchronization in each terminal. The details of the proposed control scheme are explained in order to understand the advantages of this method. The configuration of a two-phase low frequency AC transmission system (LFAC) is adopted to examine with the proposed control scheme. Simulation results are provided to illustrate the proposed control scheme with respect to the LFAC system's performance.

Keywords: multi-terminal; low frequency AC transmission system (LFAC); power control; cycloconverters; virtual synchronous generator (VSG)

1. Introduction

The low frequency AC transmission system (LFAC) can be considered as promising for a long distance, large scale transmission system using line-commutated converters. One of the promising applications of this system is the operation on a multi-terminal system [1]. The integration of other types of power plants such as offshore wind power, High Voltage Direct Current (HVDC) or multi-terminal HVDC transmission with the main power grid are subjects of ongoing research [2,3]. In the case of LFAC, it could be easier to integrate the existing AC power plant because it can be built with commercially available power system components such as transformers and cables designed for regular frequency [4]. The transformer could be derated by a factor of transmitting frequency, with the same rated current, but partially containing the original rated voltage [5]. Given this information, the installation of LFAC can be easily paralleled with the existing transmission line.

For the application on the multi-terminal system with line-commutated converters, the operation of multi-terminal HVDC is complicated [6]. It is difficult to control power flow directions among terminals without an extra DC power flow controller, which requires greater complexity, a greater number of devices and extra costs.

In terms of the power control scheme for this LFAC transmission system, to date only one reference can be found. The concept of the power control method for the two-terminal system has been briefly mentioned in [7]. However, no details on the power control scheme were presented. In addition, the cycloconverters do not have synchronizing power with respect to the low frequency side. Therefore, absolute phase reference on the low frequency side is required to synchronize power for the low frequency output assisted by the reference time signal from the global positioning system

(GPS) or synchronous digital hierarchy (SDH), which is difficult to operate in the multi-terminal system application [8–10].

The goal of this paper is to present a new power control method for application on the multi-terminal LFAC transmission system. The controller is designed by using the advantages of a visual synchronous generator (VSG) with a governor control [11]. The voltage control is introduced to generate firing angles to the cycloconverters. The current limiter and extinction angle control are included for overcurrent and commutation failure protection of the converters. VSG with a governor control is applied to control the amount of power transmission in the LFAC multi-terminal system application. Furthermore, the proposed control scheme is applied in a two-phase multi-terminal system of LFAC to perform and verify the control operation by PSCAD/EMTDC Software (version 4.2.1, Winnipeg, MB, Canada).

2. Control Strategies for LFAC for the Two-Terminal System

The basic operation of the LFAC transmission system can be described by considering two terminals: sending and receiving ends. A cycloconverter is applied to lower the line frequency from 50/60 Hz to a smaller amount. At the receiving end, another cycloconverter is operated to convert the fraction frequency back to the grid frequency. These two conversions obtain firing angle signals from the designed control scheme, which is achieved by the different phase angles of different terminals. This is one of the common control methods of the LFAC transmitting power control. The active transmitting power over transmission lines can be expressed by Equation (1).

$$P = \frac{V_s V_R}{X} \sin \delta \quad (1)$$

where V_S is the sending end voltage, V_R is the receiving end voltage, X is line reactance, and δ is the transmitting angle.

As indicated by Equation (1), the voltage control scheme concept with the LFAC transmission system can be described in Figure 1.

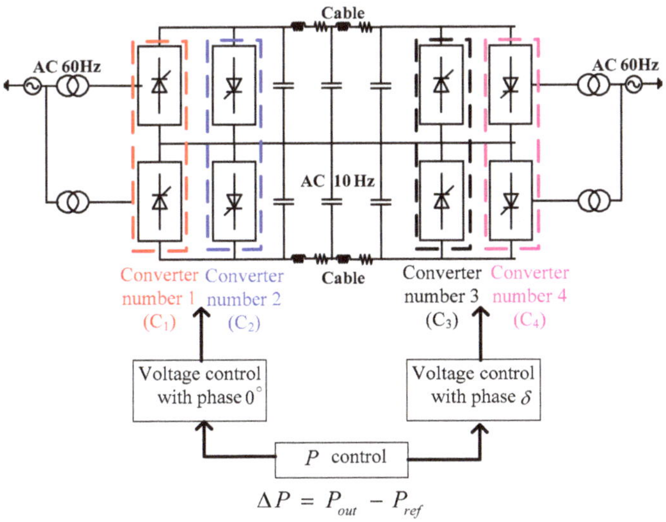

Figure 1. Concept of low frequency AC transmission system (LFAC) control scheme.

As shown in Figure 1, at the sending end, the phase is set to the constant value of 0° and the amount of transmitting power is determined by the phase difference from the receiving end.

A cycloconverter operates by following the modes in Table 1. On the low frequency side, a positive current waveform can be obtained from converter number 1 and number 4 operations. During the zero crossing point, all converters are blocked by the blocking signal from the controller. The negative current waveform can be received from converter number 3 and number 2 by reversing the current direction.

Table 1. Mode operation of the cycloconverter.

Operating Converter	Positive Current	Zero Cross Point	Negative Current
Rectifier No. of operating converter	C_1	Block	C_3
Inverter No. of operating converter	C_4	Block	C_2

2.1. The Proposed Voltage Control Scheme

Considering the mode control of the cycloconverter, this converter works as a two-set of HVDC anti-parallel connections and their inputs are a sinusoidal waveform instead of a constant value. From this point of view, the voltage control scheme can be designed by using HVDC concepts, in which the operation is much easier to understand.

$$\begin{aligned} V_d &= \tfrac{3\sqrt{2}}{\pi} V_{LL} \cos\alpha \\ \tfrac{3\sqrt{2}}{\pi} V_{LL} &= V_{d0} \\ V_d &= V_{d0} \cos\alpha \end{aligned} \tag{2}$$

From Figure 2, the Equation (2) is obtained to design the voltage control scheme of the LFAC by substituting V_d (DC voltage) with V_{dLF} (voltage on low frequency side) in Figure 3, and V_{LL} is the root mean square (*RMS*) value of 60 Hz-side AC line to line voltage. The designed control block diagram is shown in Figure 3, in which the voltage reference value of the low frequency side $v_{LF}^* = E_{max}\sin(\omega t + 0°)$ is for reference angle side, $E_{max}\sin(\omega t + \delta)$ is on another side and v_{LF} is the measured instantaneous voltage on low frequency side.

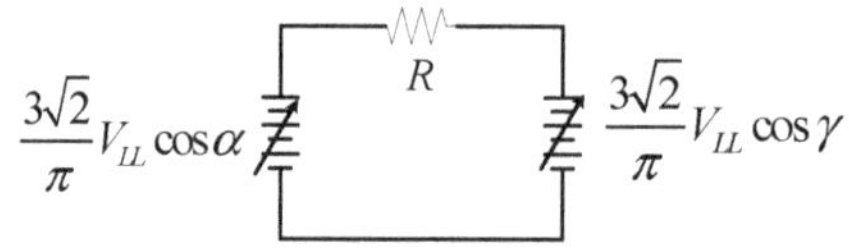

Figure 2. Equivalent circuit of High Voltage Direct Current (HVDC) link.

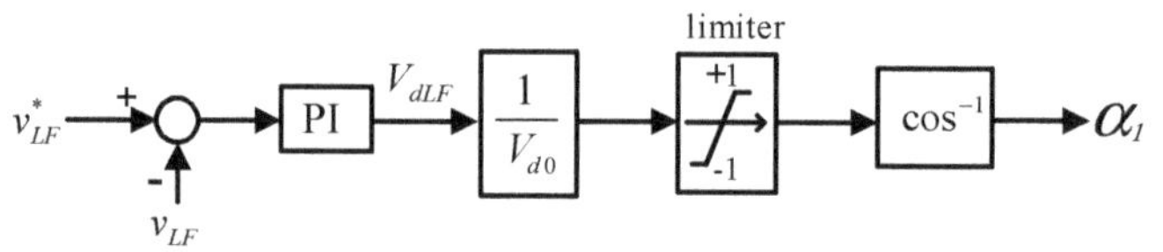

Figure 3. Voltage control of LFAC.

To obtain the phase reference (δ), the power control scheme needs to be considered. Equation (1) shows the amount of transmitting power between two terminals following the reference power. Furthermore, power on the low frequency side is not constant, thus the moving average control method is adopted to average the amount of power on the low frequency side. The power control block diagram can be achieved as shown in Figure 4.

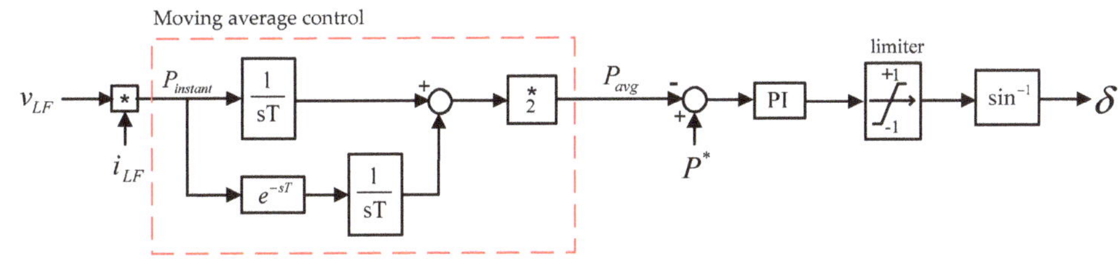

Figure 4. Power control of LFAC.

To control the amount of transmitting power between each terminal, the power control scheme is chosen for application to the cycloconverters in the LFAC transmission system. According to the system circuit mode and transmitting power in (1), power reference (P*) is compared with average power (P_{avg}) to obtain the phase difference (δ) for the voltage reference.

2.2. Proposed Frameworks for Current Limiter and Extinction Angle Control

For protection from overcurrent and commutation failure, a current limiter and extinction angle control are applied in this system. The current limiter is used to limit overcurrent during the transmission of power where i_{LF} = the measured instantaneous current on the low frequency side. The current control scheme is shown in Figure 5.

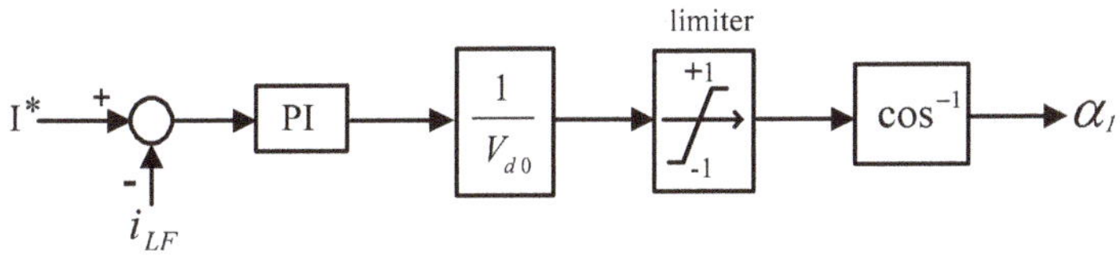

Figure 5. Current limiter of LFAC.

For line-commutated converters, in practice, there are some overlapping angles due to the commutation reactance. The constant extinction angle control (γ) is used to avoid commutation failure. As shown in Figure 6, this extinction angle control works by comparing gamma reference with gamma measure, then the minimum firing angle is chosen to be the gating pulse for cycloconverters. The typical value of extinction angle (γ) is set at 18° for 60 Hz of the AC system.

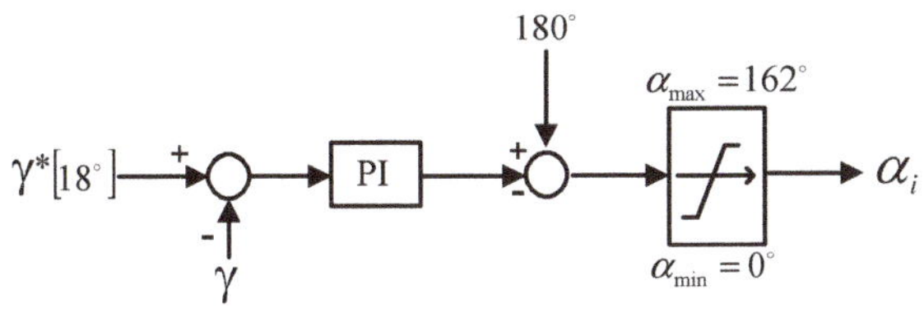

Figure 6. Extinction angle control of LFAC.

During the zero cross point current, a gate block control scheme is applied to every converter as shown in Figure 7.

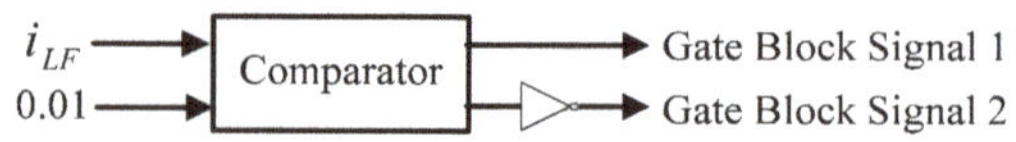

Figure 7. Gate block control scheme of LFAC.

In the complete developed control scheme of LFAC, as shown in Figure 8, the same control system is applied for every converter to control the amount of transmitting power between terminals. Following the converter operation mode, δ is set at 0° for one side of the converter. On another side of the converter, δ depends on the reference of transmitting power to control the amount and direction.

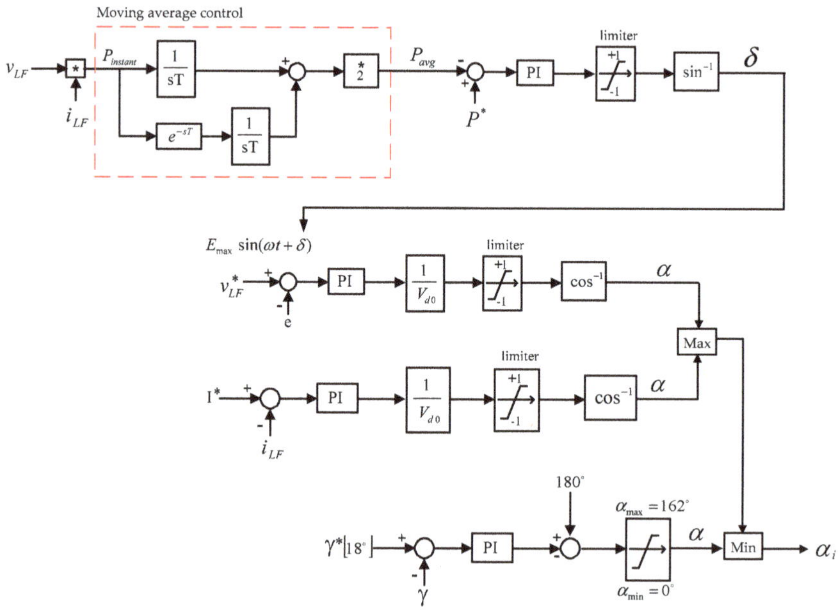

Figure 8. The proposed overall control scheme of LFAC.

3. Control Strategies for Multi-Terminal LFAC

The proposed control scheme to control the amount of transmitting power and direction among several terminals is explained in this section. In a multi-terminal application of LFAC (LCC-MTLF), a VSG is introduced to synchronize the phase and frequency among terminals instead of a communication link or GPS. The VSG model is based on the swing equation of synchronous generators. Based on this equation, the relation between the output power P_{out} and the input power P_{in} is shown in Equation (3).

$$P_{in} - P_{out} = J\omega_m \frac{d}{dt}\omega_m + D(\omega_m - \omega_g) \quad (3)$$

where ω_m is the virtual rotating angular frequency, ω_g is the angular frequency, J is the moment of inertia of rotating mass, and D is the damping factor of the damping power introduced by the damp winding.

The concept of VSG control is shown in Figure 9. For the power balance between P_{in} and P_{out}, according to the concept of VSG control, if the network frequency is lower than the reference frequency (f^*), then the generator increases its power into the network.

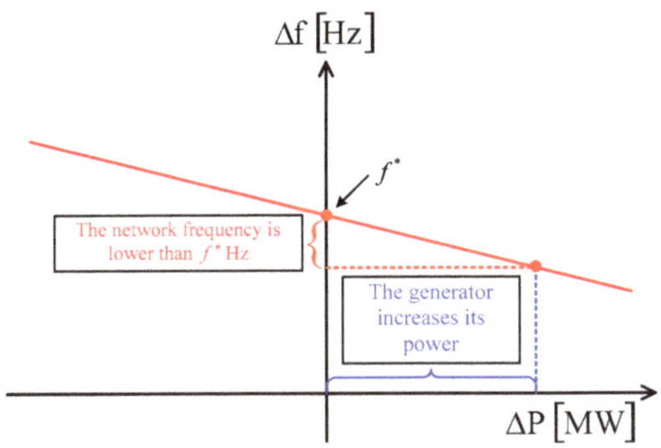

Figure 9. Concept of virtual synchronous generator (VSG) control.

Figure 10 shows the VSG control scheme applied to the LFAC transmission system. The VSG control is emulated to the swing equation in the VSG block.

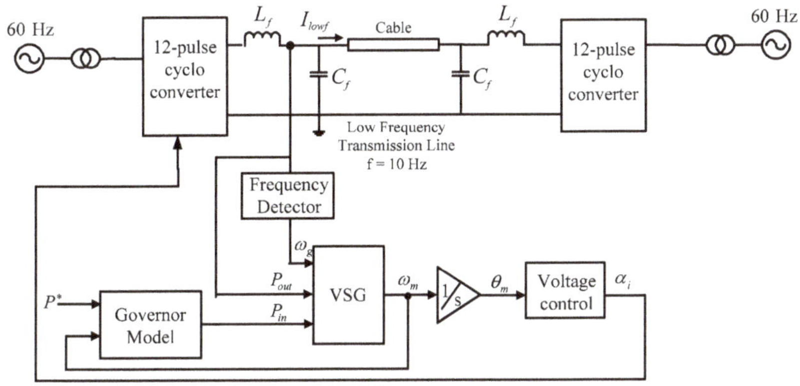

Figure 10. Block diagram of the VSG control scheme for LFAC.

For simplification, the damping factor D is considered as a small constant parameter. As the swing equation is a differential equation, an algorithm based on the Runge–Kutta iterative method is used to solve the angular frequency ω_m following the flow chart as shown in Figure 11. The calculated ω_m provides the frequency reference for the governor model and phase reference θ_m through an integrator. P_{base} is the rated capacity of transmission power.

The block of the governor model is shown in Figure 12. The governor model is implemented to tune the input power command based on the frequency deviation. The transfer function of the governor control can be expressed as Equation (4), where $K_p = 5$, $T_d = 0.1$.

$$P_{in} = P^* + \frac{K_p}{1 + T_d s}(f^* - f_{VSG}) \tag{4}$$

In the VSG control part as shown in Figure 13, P_{in}, P_{out} and low frequency side frequency (f_g) are inputs of the VSG control unit. In each control cycle, the momentary ω_m is calculated via the application of the fourth-order Runge–Kutta iterative method inside the VSG block to generate the virtual phase angle θ_i (i = 1, 2, 3) sent through the voltage control unit.

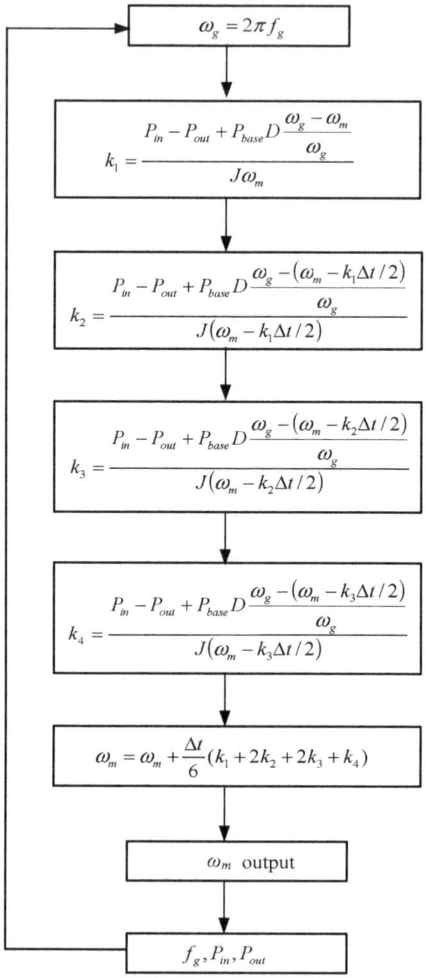

Figure 11. Flow chart of solving swing equation by Runge–Kutta method.

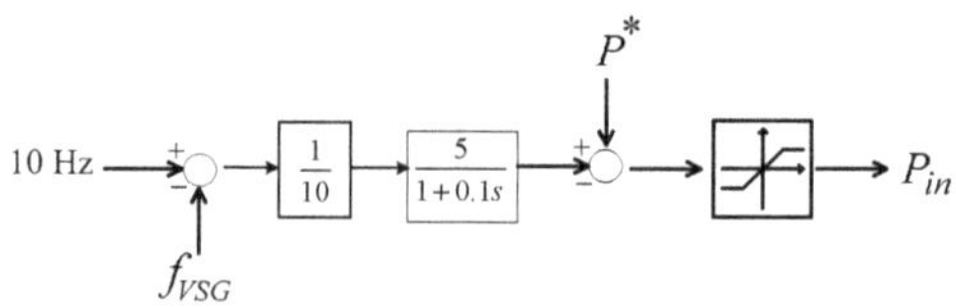

Figure 12. Governor control scheme.

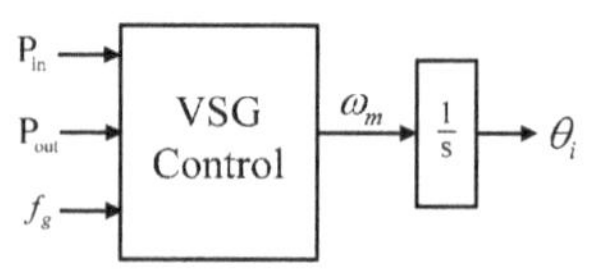

Figure 13. VSG control scheme.

As for the voltage control scheme, the concept of the control operation is the same as the voltage control scheme as described in Section 2.1. When a disturbance occurs, or more load shares are added to the system, the system can stabilize itself by regulating voltage using this control.

The current limiter works by comparing the rating current with measured current from the low frequency side to generate the firing angle to the cycloconverter, which can limit overcurrent by limiting the firing angle to the cycloconverter. Also, extinction angle control (γ) is included to avoid commutation failure in a converter.

By combining each part of the control scheme, the complete control scheme of the multi-terminal LFAC transmission system is shown in Figure 14. The voltage control receives the virtual phase angle θ_i from the VSG block to be the voltage reference, then the firing angle α_i is generated and is sent to the cycloconverters.

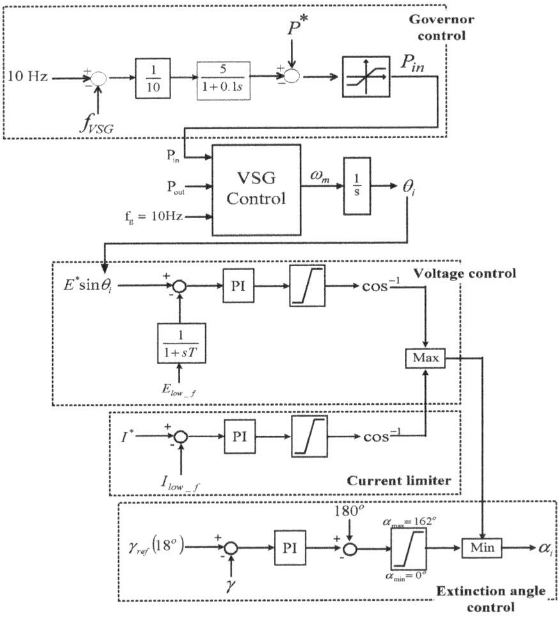

Figure 14. Complete control scheme of multi-terminal LFAC.

4. Simulation Results

To demonstrate the validity of the proposed power control scheme for the multi-terminal LFAC application, the simulations have been carried out using the tool of the PSCAD/EMTDC simulation software. The transmitting power in this system is rated at 1400 MW, and the transmission distance is 500 km. The transmission power cable is modeled by cascading 10 km of cable; its model is shown in Figure 15.

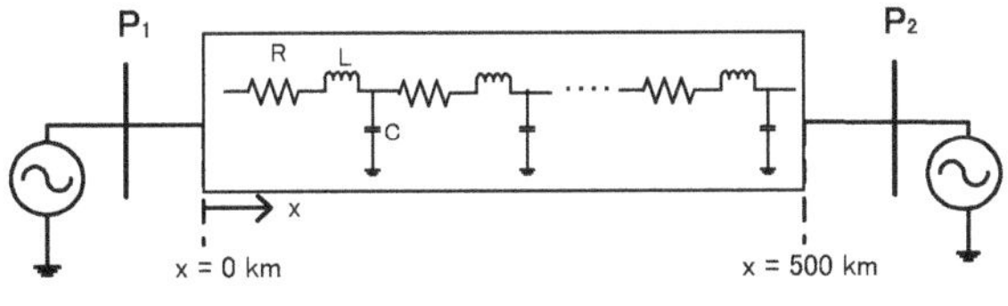

Figure 15. Cable model for the simulation.

For the simulation of the multi-terminal LFAC application, a three-terminal two-phase LFAC configuration [1] is applied with a proposed power control scheme as shown in Figure 16.

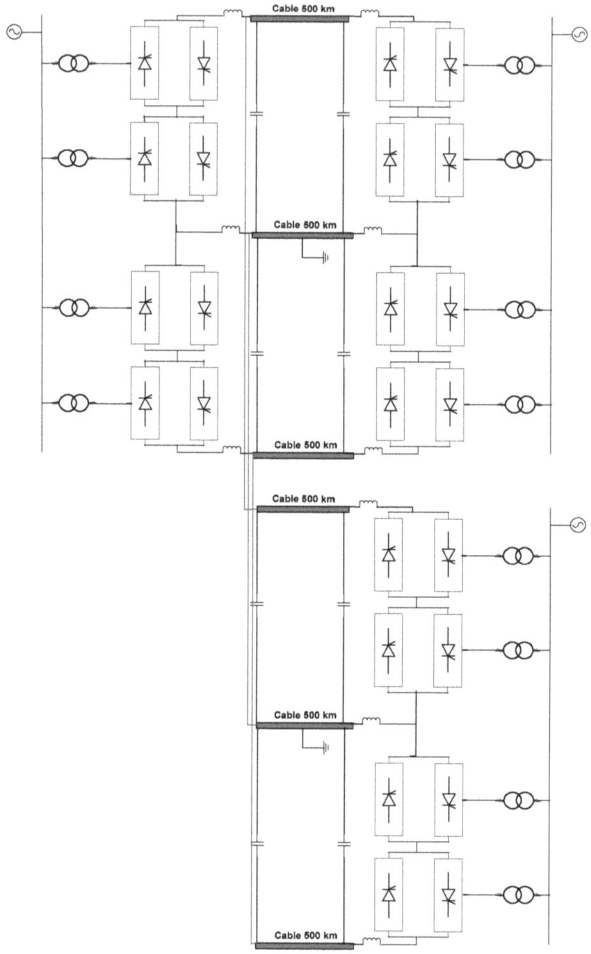

Figure 16. Three-terminal two-phase LFAC.

4.1. Current Limiter and Extinction Angle Control Operation

From the proposed control scheme in Section 2.2, some simulation results are performed to verify the control operation. Figure 17 shows the simulation result of the extinction angle control that cannot exceed the limit at the reference $\gamma = 18°$. An extinction angle γ is calculated from the equation $\gamma = 180° - (\alpha + u)$ where $u = 12°$ as shown in Figure 17b. The overlapping angle is around 12° for 60 Hz and γ is controlled at 18°. This can confirm the extinction angle control scheme works well.

During the period of transition, the mode of the cycloconverter may cause a spike current. This can malfunction with devices in the system so that the current limiter is used for protection. From the control scheme shown in Figure 5, the current limiter works by comparing the rating current with the measure current from the low frequency side to generate the firing angle to the cycloconverter. The current limiter can limit the overcurrent by limiting the firing angle to the cycloconverter. Figure 18 shows the result of the current limiter control that limits current at 20 kA on the low frequency side.

The *E_lowf* waveform is transmitting low frequency AC voltage, following the reference value from the control scheme. The *I_lowf* waveform shows a transmitting current on the low frequency side, which is limited at 20 kA due to the current limiter. From the result, it can be confirmed that the current limiter control can work properly.

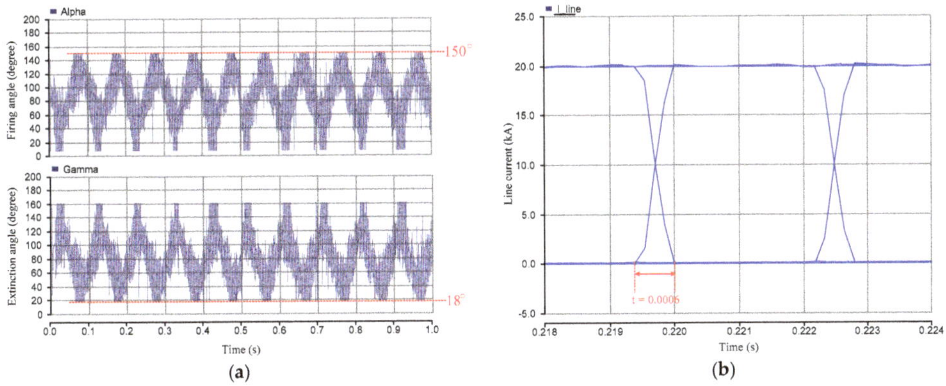

Figure 17. Simulation result of extinction control. (**a**) Firing angle and extinction angle during operation at 20 kA; and (**b**) line current waveforms.

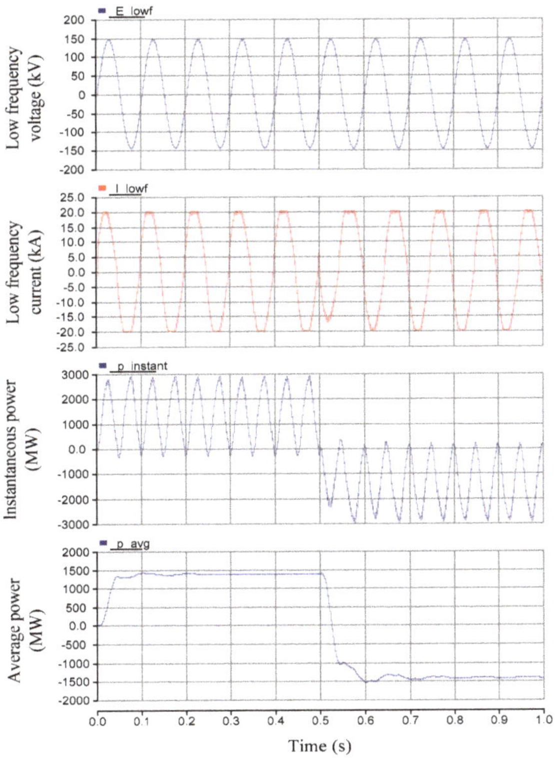

Figure 18. Simulation result for current limiter control.

4.2. Multi-Terminal LFAC Response

The simulation results of the three-terminal two-phase LFAC system are explained in this section. In the performed multi-terminal configuration, as mentioned in Figure 16, the multi-terminal LFAC demonstrates the effectiveness of the proposed system and control method. The total amount of power in this system is 1400 MW. The power flow pattern is shown Figure 19 and the details are shown in Table 2. In the power flow schedule given in Table 2, terminals 1, 2 and 3 are desired to inject a power of 800 MW, 500 MW and −1300 MW into the grid. The minus signs show that the power flow is in the opposite direction. The simulation parameters are shown in Table 3.

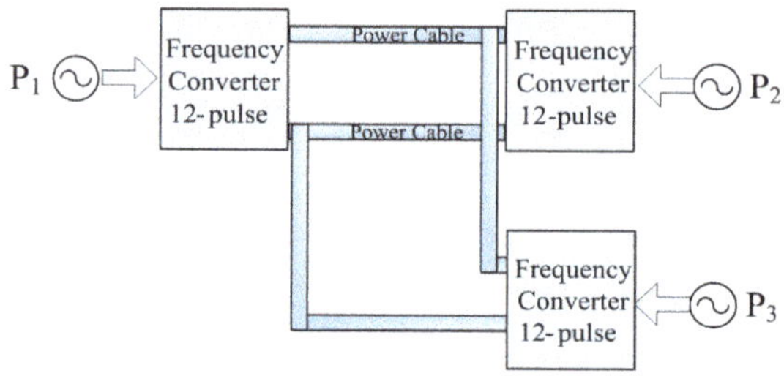

Figure 19. Power flow patterns for simulation.

Table 2. Power flow patterns.

Terminal No.	P^* (Power Reference, MW)	
	Sending	**Receiving**
1	800	−600
2	500	−300
3	−1300	900

Table 3. Parameters of simulated system.

LFAC System	500 kV, 10 Hz
Line frequency	60 Hz
Transmitting frequency	10 Hz
Transformer	500 kV/110 kV
Maximum power transfer	1400 MW
Maximum cable length	500 km
Number of terminals	3
Power reversal	After 0.5 s

Figure 20 depicts the results of a simulation where the transmitting power flows from terminals 1 at 800 MW, and 2 and 3 at 500 MW and 1300 MW, respectively. After 0.5 s, the power direction is reversed following power references (terminal 1 at 600 MW, terminal 2 at 300 MW and terminal 3 at 900 MW). The voltage (*E_lowf*), current (*I_lowf*), and average power (*p_avg*) are the results on the low frequency side. As can be seen from the simulation results, the amount of power in each terminal can be transferred following the power flow patterns given in Table 2. The transmitting power on the low frequency side is 10 Hz following the reference. Figure 19 shows the power that flows from terminal 1 at 800 MW and in the reverse direction at 0.5 s following the power control scheme command.

Figure 20b,c shows the result of the power transmission at terminals 2 and 3. The results show that the amount and direction of power can follow the power flow patterns. It can be confirmed that the proposed power control for the multi-terminal LFAC using the VSG control works properly.

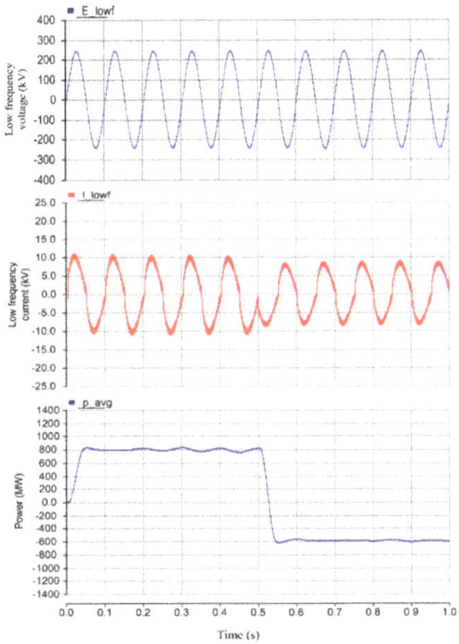

(a)

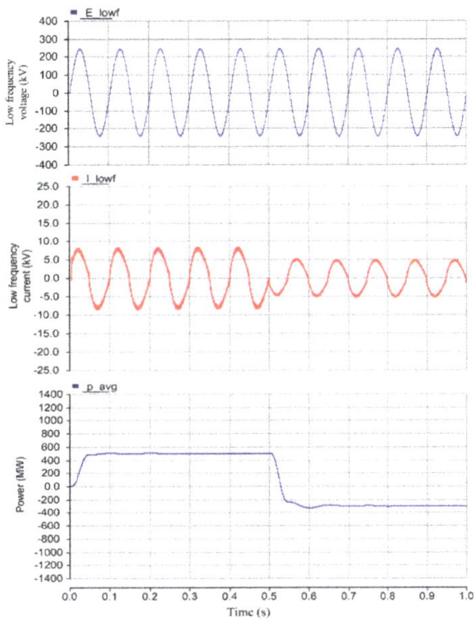

(b)

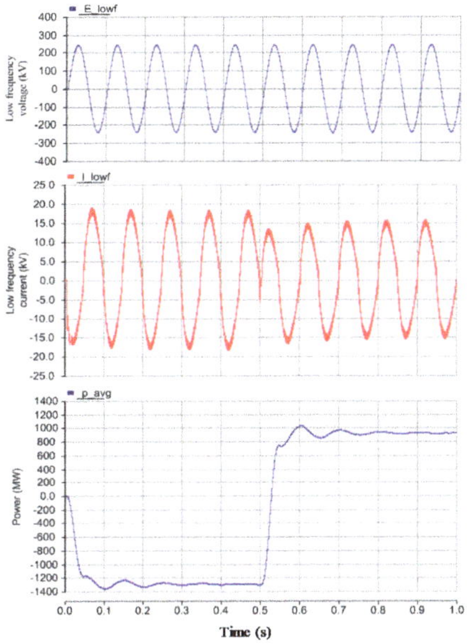

(c)

Figure 20. Low frequency side waveforms (voltage, current and power) on (**a**) terminal 1; (**b**) terminal 2; and (**c**) terminal 3.

5. Conclusions

This paper explains the proposed power control scheme of the multi-terminal LFAC application. A three-terminal two-phase LFAC cycloconverter configuration is applied to operate with the proposed control scheme. There are three main important points to be concluded as follows:

(1) The proposed power control scheme which was explained in detail is feasible for synchronized frequency and regulated voltage by integrating VSG and the governor control to the control scheme;
(2) A two-phase LFAC configuration was adopted to operate as a multi-terminal system LFAC. The simulation results confirm the operation of the current limiter and extinction angle control, which operated using this configuration;
(3) The multi-terminal LFAC application was operated by using the proposed power control scheme to transmit power among three terminals following the reference power patterns. The simulation results verified the operation of the proposed control scheme with the given transmission system without a communication link.

This study can lead us to consider more details such as adding additional terminals to operate with the proposed control scheme and configuration. Comparing merits and drawbacks of the system with other existing multi-terminal transmission systems in the presence of various disturbance scenarios could be considered as the future steps for the present work.

Author Contributions: Achara Pichetjamroen conceived the proposed control strategy, performed calculation, simulation of the system and wrote the paper. Toshifumi Ise as research supervisor provided guidance and key suggestions, participated in proposing the ideas of the system configurations and control scheme, and helped in writing the paper.

Conflicts of Interest: The authors declare no conflict of interest.

References

1. Achara, P.; Ise, T. A proposal on low frequency AC transmission as a multi-terminal transmission system. *Energies* **2016**, *9*, 687.
2. Bozhko, S.; Asher, G.; Li, R.; Clare, J.; Yao, L. Large offshore DFIG based wind farm with line-commutated HVDC connection to the main grid: Engineering studies. *IEEE Trans. Energy Convers.* **2008**, *23*, 119–127. [CrossRef]
3. Mau, C.N.; Rudion, K.; Orths, A.; Eriksen, P.B.; Abildgaard, H.; Styczynski, Z.A. Grid connection of offshore wind farm based DFIG with low frequency AC transmission system. In Proceedings of the Power and Energy Society General Meeting, San Diego, CA, USA, 22–26 July 2012.
4. Tuan, N.; Min, L.; Surya, S. Steady-state analysis and performance of low frequency AC transmission lines. *IEEE Trans. Power Syst.* **2016**, *31*, 3873–3880.
5. Slade, P.G.; Smith, R.K. A comparison of the short circuit interruption performance using transverse magnetic field contacts and axial magnetic field contacts in low frequency circuits with long arcing times. In Proceedings of the Discharges and Electrical Insulation in Vacuum International Symposium, Yalta, Crimea, 27 September–1 October 2004.
6. Etienne, V.; Boon, T.O. Multiterminal HVDC with thyristor power-flow controller. *IEEE Trans. Power Deliv.* **2012**, *27*, 1205–1212.
7. Funaki, T.; Matsuura, K. Feasibility of the low frequency AC transmission. In Proceedings of the Power Engineering Society Winter Meeting, Singapore, 23–27 January 2000.
8. Cho, Y.; Cokkinides, G.J.; Meliopoulos, A.P. Time domain simulation of a three-phase cycloconverter for LFAC transmission systems. In Proceedings of the Transmission and Distribution Conference and Exposition, Orlando, FL, USA, 7–10 May 2012.
9. Nakagawa, R.; Funaki, T.; Matsuura, K. Installation and control of cycloconverter to low frequency AC power cable transmission. In Proceedings of the Power Conversion Conference, Osaka, Japan, 2–5 April 2002.
10. Chen, H.; Johnson, M.H.; Aliprantis, D.C. Low frequency AC transmission for offshore wind power. *IEEE Trans. Power Deliv.* **2013**, *28*, 2236–2244. [CrossRef]
11. Bevrani, H.; Ise, T.; Miura, Y. Virtual synchronous generators: A survey and new perspectives. *Int. J. Electr. Power Energy Syst.* **2014**, *54*, 244–254. [CrossRef]

energies

MDPI

Article

A Causal and Real-Time Capable Power Management Algorithm for Off-Highway Hybrid Propulsion Systems

Johannes Schalk [1,*] and Harald Aschemann [2]

1 MTU Friedrichshafen GmbH, Maybachplatz 1, 88045 Friedrichshafen, Germany
2 Chair of Mechatronics, Rostock University, Justus-von-Liebig Weg 6, 18059 Rostock, Germany; harald.aschemann@uni-rostock.de
* Correspondence: johannes.schalk@mtu-online.com; Tel.: +49-7541-908532

Academic Editor: William Holderbaum
Received: 28 September 2016; Accepted: 16 December 2016; Published: 26 December 2016

Abstract: Hybrid propulsion systems allow for a reduction of fuel consumption and pollutant emissions of future off-highway applications. A challenging aspect of a hybridization is the larger number of system components that further increases both the complexity and the diversification of such systems. Hence, beside a standardization on the hardware side for off-highway systems, a high flexibility and modularity of the control schemes is required to employ them in as many different applications as possible. In this paper, a causal optimization-based power management algorithm is introduced to control the power split between engine and electric machine in a hybrid powertrain. The algorithm optimizes the power split to achieve the maximum power supply efficiency and, thereby, considers the energy cost for maintaining the battery charge. Furthermore, the power management provides an optional function to control the battery state of charge in such a way that a target value is attained. In a simulation case study, the potential and the benefits of the proposed power management for the hybrid powertrain—aiming at a reduction of the fuel consumption of a DMU (diesel multiple unit train) operated on a representative track—will be shown.

Keywords: off-highway propulsion system; hybrid electric vehicle; hybrid electric diesel multiple unit train (DMU); hybridization; power management strategy; energy management strategy

1. Introduction

1.1. Motivation

In the automotive industry, hybrid electric and full electric vehicles play a significant role in efforts to meet future legislated emission targets. Beside stricter limits for pollutant emissions of nitrogen oxides and particulates, European regulations also aim at reducing the average CO_2 emissions of passenger cars to 95 g/km [1]. For off-highway applications—such as diesel multiple unit trains (DMUs) or marine applications—so far no legislative regulation to reduce CO_2 emissions has been implemented. Nevertheless, lowering fuel consumption and thus CO_2 emissions is the major goal in developing future off-highway propulsion systems. This is comprehensible considering that the fuel share of total life cycle costs for off-highway applications can be up to over 90% and, additionally, fuel prices are very likely to increase in the mid- and long-term future [2]. Hybrid powertrain technology can help to significantly lower fuel consumption of rail and marine applications [3–6]. Compared to conventional non-hybrid powertrains, additional operational functionalities of hybrids like recuperation of kinetic energy, engine operation point shifting or electric boosting improve the system performance and reduce pollutant emissions [7]. Figure 1 shows an overview of the functionalities of hybrid powertrains and their impact on emissions, fuel consumption and system performance.

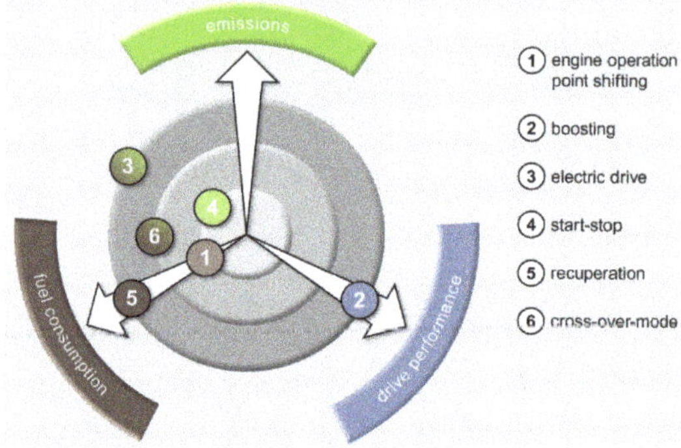

Figure 1. Overview of hybrid powertrain functionalities.

On the other hand, electrification of propulsion systems further increases the number of system components and, therefore, the complexity of already highly sophisticated off-highway architectures consisting of multiple engines, generators, gearboxes, etc. This leads to an even higher diversification of system configurations. To cope with this variety in series production, it is compulsory to standardize components of hybrid propulsion systems and to use them for various applications. Besides the hardware standardization, it is also necessary to harmonize software functions and control algorithms in order to apply them to as many different system configurations and architectures as possible. Therefore, a power management algorithm (PMA) is required to control the power split between combustion engine and electric motor in a hybrid propulsion system. The PMA has to offer the flexibility and modularity to be applicable to a vast variety of systems, including multi-engine system topologies.

1.2. Classification of Power Management Algorithms

In the past, a wide range of different academic approaches of PMAs to optimize the power split in hybrid propulsion systems for on- and non-road applications have been published in the literature. In order to classify multiple PMAs by the basis of their data, one can differentiate between causal and non-causal strategies [8,9]. In causal strategies, the decisions made by the PMAs are based upon present and past state variables. On the contrary, in a non-causal approach, knowledge of the whole drive cycle is processed. Considering data of the complete drive cycle enables an optimization of the fuel consumption for a global optimum [9,10]. For DMUs, where a detailed knowledge of the track and timetable is available, a non-causal PMA in combination with a predictive strategy is feasible. From a functional perspective, heuristic, often called rule-based or sub-optimal, and optimization-based, or also known as optimal, algorithms can be distinguished [9,11–13]. Heuristic strategies include deterministic rule-based and fuzzy logic control algorithms [14–17]. Those approaches are state of the art in most prototype and production hybrid vehicles [9]. They are real-time capable and are processed online. The optimization-based or optimal PMAs are mainly processed offline. Those strategies are normally non-causal and, therefore, process information about the whole drive cycle. The optimal PMAs are divided into numerical methods such as dynamic programming (DP) and analytical approaches, like Pontryagins minimum principle [11,18,19]. DP enables to find a global optimum but is not real-time capable and, therefore, frequently used as benchmark PMA. Other optimization-based PMAs are the so called equivalent consumption minimization strategies (ECMS) [9,12]. The approach associates the stored electric energy to a future increase or decrease of fuel consumption. Therefore, usually an equivalence factor is imposed to convert battery into fuel power, and based on that into an equivalent fuel consumption. In the meantime, many different ECMS approaches, primarily for automotive applications, were introduced. They use rule- or map-based routines as well as online adaption to determine the equivalent factor and optimize power split [20–24].

1.3. Requirements for the Off-Highway Application Power Management

In this paper a new power management algorithm (PMA) which controls the power split between combustion engine and electric motor in a hybrid propulsion system is presented. It is an online optimization-based causal strategy. Only present and past state variables are used, and no knowledge of the whole drive cycle is necessary. The PMA is developed to be applied to various applications and system configurations, including systems with multiple engines and a variety of electrical components. It is suited for use in hardware-in-the-loop (HiL) experiments as well as field testing of systems for off-highway applications. Furthermore it has to fulfill, amongst others, the following requirements:

- Modular structure
- Real-time capability
- Compatibility to series production propulsion system control units
- Suitability for different mission scenarios and velocity/load control modes (including driver-controlled traction torque demand and automatically controlled drive strategies)
- Consideration of variable auxiliary or external load requests in the PMA.

It is important to point out that some of these requirements significantly differ from demands for PMAs used in automotive applications. Especially the necessity of a high modularity and the flexibility of the algorithm for various applications, topologies and system components are challenging concerning algorithm development and code implementation.

The paper comprises a simulation case study where the PMA is exemplarily applied to control a hybrid propulsion system for a DMU application. In Section 2, boundary conditions for the simulation case study are presented, including the specification of the system hardware in Section 2.1, as well as a short description of the drive strategies in Section 2.2, which are used for the simulations. Afterwards, in Section 3, a brief overview over the simulation model approach is given along with examples for a model validation. For the model validation, measured data from a hardware-in-the-loop (HiL) test rig is used. It represents a realistic prototype of a hybrid propulsion system driving a virtual vehicle model of the DMU. In Section 4, the PMA is described comprehensively. At first, the general problem for the control of hybrid systems is defined. Section 4.2 points out the possible operation modes of parallel hybrid powertrains which are relevant for the PMA. Thereafter, Section 4.3 presents the different functions and their interaction with the PMA. The results of the simulation case study are finally discussed in Section 5.

2. Boundaries of the Simulation Case Study

2.1. Hybrid Diesel Multiple Unit Train (DMU) System Specification

The PMA introduced in this paper has been developed for a Siemens Desiro VT642 DMU dedicated for regional passenger transportation. In a simulation case study, the vehicle is equipped with two identical parallel hybrid power units (PU), which form the propulsion system. Each PU consists of a diesel internal combustion engine (ICE) and an electric motor/generator unit (MG), which is placed on the engine output shaft center, see Figure 2. The angular velocities of ICE and MG can be decoupled by a clutch (CL). The gearbox (GR) is a six-speed automatic transmission with a hydrodynamic torque-converter on its input shaft. As energy storage unit, an electrochemical lithium-ion accumulator (BAT) is used in each PU. Furthermore, Figure 2 shows the arrangement of the auxiliary loads, comprising electric auxiliaries (AUX) and mechanical power take-offs (PTO) at the ICE, like the hydraulic fan. Table 1 shows a more detailed specification of the vehicle and system components.

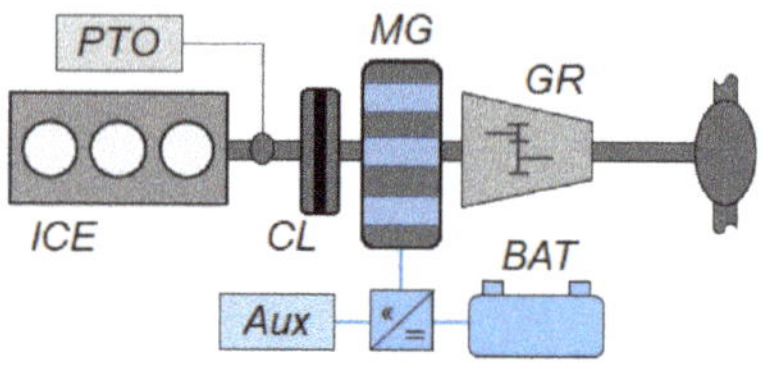

Figure 2. Scheme of a parallel hybrid power unit.

Table 1. System specification of vehicle and components.

Vehicle	type number PUs operating weight	Siemens Desiro VT642 2 83,000 kg
Engine	type number cylinders rated power rated speed	MTU 6H1800R85LP 6 390 kW 1800 rpm
Motor/Generator	type rated power rated speed	p. magnet synchron. motor 200 kW (continuous) 1600 rpm
Gearbox	type number gears	ZF EcoLife 6
Battery	type nom. voltage nom. capacity max./min. current	Lithium-ion battery 670 V 90 Ah +/-300 A

In the simulation case study, a system configuration is analyzed, where two identical PUs operate synchronously. It means, for example, that the power demand for the ICE calculated by the PMA is identical in both PUs. It is possible and, moreover, from an efficiency perspective rather beneficial to consider two PUs that are logically linked but operated individually. Such an operation strategy, however, is not state-of-the-art yet because a high voltage link between the electric circuits of both PUs would be required. Figure 3 outlines a system configuration with two PUs, which are linked on their electric circuit and share a joint battery. The second combustion engine is masked, which implies that such systems enable to restore one ICE. This is possible because—compared to a conventional non-hybrid system—the hybridization increases the system power considerably. Thus, considering track profiles and mission scenarios with a relatively low traction power demand, it is feasible to remove one of the ICEs and compensate the loss of traction power with the MGs. It becomes clear that even quite conventional propulsion system configurations of DMUs with two PUs can differ significantly by their design and operation, not to mention systems layouts with more than two PUs. This shows the necessity of a highly flexible PMA approach for off-highway hybrid systems.

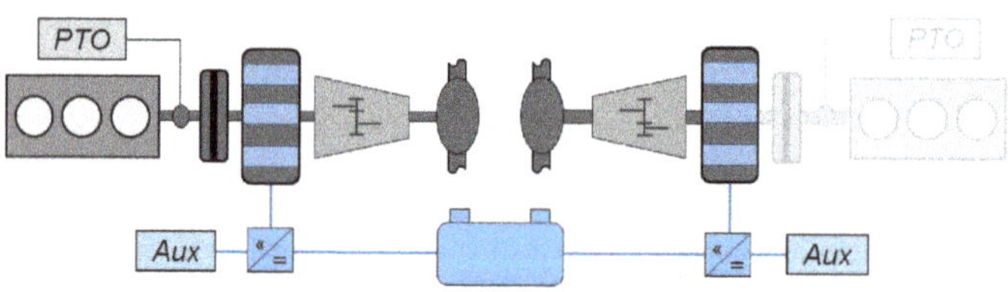

Figure 3. Scheme of a parallel hybrid propulsion system for a diesel multiple unit train (DMU) consisting of two PUs and a combined electric circuit with the potential to replace one ICE.

2.2. Drive Strategy

Optimizations of the drive strategy to operate a DMU on a given track are the object of research in many publications. Often dynamic programming (DP) algorithms are used to optimize the velocity trajectory of hybrid vehicles [19,25]. Due to the fact that DP requires a high computational power, they are not real-time capable. Recently, approaches have been proposed where the DP algorithm is outsourced and computed on a cloud computing platform. The optimized velocity profile is sent back to the vehicle afterwards [26]. Such approaches have significant potential to reduce fuel consumption.

However, for the simulation case study in this paper, a more simple drive strategy is used. Initially, a mission specification for the track, on which the DMU is operated, is defined. It specifies the driving distance, drive and stop durations as well as velocity limits of each track sequence. Based on this information, a driver simulation model, implemented in Simulink and Stateflow, computes the requested velocity profile. The objective of the driving strategy is the reduction of the DMU's energy consumption traveling on the tracks considering all the specified constraints (driving duration, velocity limits etc.). For this purpose, the algorithm aims at minimizing the cruising speed, considering the quadratic influence of velocity on the required traction force, see Equation (1). To operate the vehicle at a cruising speed as low as possible, the driving strategy asks for maximum acceleration and deceleration in the particular parts of each track sequence. Acceleration and deceleration demands are restricted by the performance capabilities of the PU, vehicle limitations or even operational boundaries, like a deceleration limit of $-1\ \mathrm{m/s^2}$ due to comfort considerations. In DMUs with conventional PUs, such a drive strategy, where the energy consumption is minimized, is considerably simple and leads to good results. Due to its simplicity it is fairly realistic that a human DMU operator in real world driving can easily track such a velocity profile.

For a hybrid system with energy storage unit, the possibility of a recovery of kinetic energy arises. For the hybrid drive strategy, hence, the algorithm of the driver model retards the deceleration phase. The strategy demands the vehicle to decelerate with a constant brake power, equivalent to the maximum mechanic brake power of the MG. This leads to slightly higher cruising speeds and, thus, energy consumption, but also significantly increases the amount of recuperated energy. Therefore, this drive strategy is used for hybrid operation mode.

3. System Simulation Model and Validation

The virtual vehicle and propulsion system are built up as a dynamic simulation model in the simulation environment GT-SUITE. GT-SUITE library blocks are used to model the physical components as well as to implement the basic control functions belonging to the components. Physical components and functions are combined to so-called compounds and archived in a compound library. The compounds possess standardized interfaces to enable a high degree of modularity, which allows to build simulation models of various system configurations for diverse applications. Each component provides the actual values of its physical state variables as well as their maximum and minimum values for each of them, as model outputs. More complex control algorithms, like the PMA, are realized in MATLAB/Simulink and Stateflow. The Simulink models can be compiled to dll-code and implemented in the GT-SUITE environment. The simulation models are suited for performance simulations, controls development, system parameter and configuration optimizations as well as hardware-in-the-loop testing.

3.1. Modelling System Components

This section gives a brief overview over the physical modeling theory and the equations, which the component models are based on. Furthermore, the predefined objects of the GT-SUITE library, which are used for the particular component models, are presented. Additionally, the most important model parameters are specified.

3.1.1. Vehicle Model

The dynamic vehicle simulation model uses the "GT VehicleBody", which is based on the equation of motion. The traction force F_{Tr} is equal to the sum of all drive resistance forces:

$$F_{Tr}(t) = c_r\, m\, g \cos(\alpha) + \frac{1}{2} c_D A\, \rho\, v(t)^2 + m\, g \sin(\alpha) + \left(m + \frac{\theta_{red}}{r_{dyn}^2}\right) \dot{v}(t) \tag{1}$$

with track inclination $\propto$, air density ρ, vehicle velocity v and the reduced mass moment of inertia θ_{red}. The required input traction torque M_{Drive} to the vehicle differential is calculated by:

$$M_{Drive}(t) = \frac{F_{Tr}(t) \cdot r_{dyn}}{\eta_{Dif} \cdot i_{Dif}(t)}. \tag{2}$$

Table 2 gives an overview over the most important vehicle model parameters.

Table 2. Vehicle model parameters.

Parameter	Symbol	Unit	Value
Vehicle mass	m	kg	83,000
Track coefficient	c_r	-	0.001
Drag coefficient	c_D	-	0.8
Frontal area	A	m^2	10.8
Dynamic roll radius	r_{dyn}	m	0.38
Differential ratio	i_{Dif}	-	2.59
Differential efficiency	η_{Dif}	-	0.95

3.1.2. Engine Model

The engine simulation model used in this case study is based on the "GT EngineState" object. It computes fuel consumption using a static brake specific fuel consumption ($bsfc$) map, see Figure 4a. The fuel consumption FC_{ICE} is calculated by:

$$FC_{ICE}(t) = \int_0^t M_{ICE}(\tau) \cdot \omega_{ICE}(\tau) \cdot bsfc(\tau)\, d\tau \tag{3}$$

The angular velocity ω_{ICE} of the ICE in $\frac{rad}{sec}$ (or n_{ICE} in rpm) follows from

$$\dot{\omega}_{ICE}(t) = \frac{M_{ICE}(t) + M_{PTO}(t)}{\theta_{ICE}} \tag{4}$$

with the engine torque M_{ICE}, the mass moment of inertia θ_{ICE} of the ICE and the mechanical auxiliary torque M_{PTO}. The mechanical auxiliaries or PTOs (power-take-offs) are directly linked to the engine output shaft. In the real vehicle, M_{PTO} is a function of both angular velocity and PTO load, which depends on coolant or boost temperatures etc. In the simulation model, M_{PTO} or P_{PTO} is a function of the engine angular velocity, whereas the load is set constant, see Figure 4b.

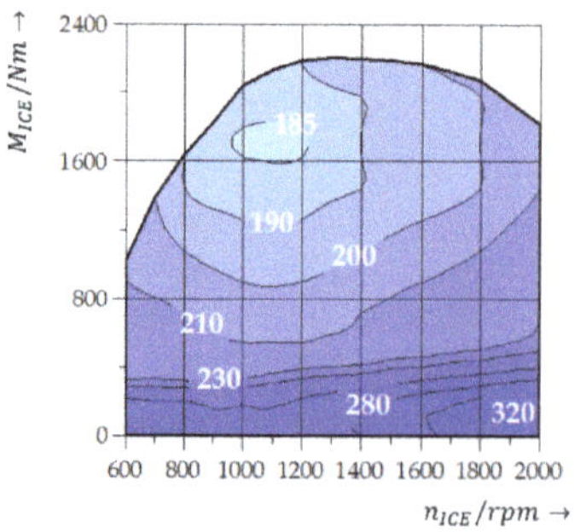

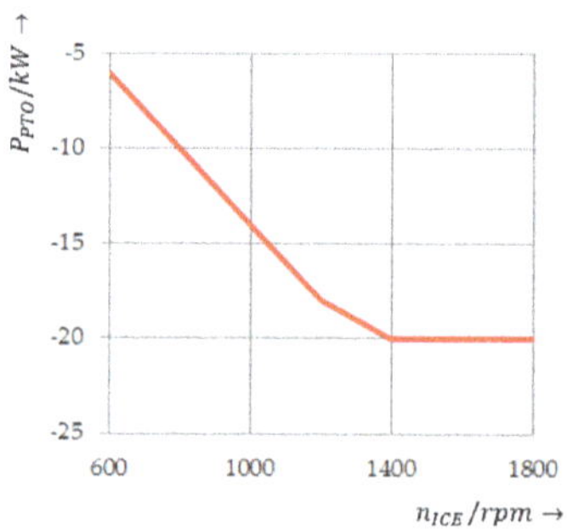

Figure 4. (**a**) Generic bsfc map for a diesel engine; (**b**) Mechanical auxiliary power P_{PTO} over engine speed n_{ICE}.

3.1.3. Gearbox Model

The gearbox is a six-speed transmission with a hydrodynamic torque converter (HTC) and transmits the input torque M_{GR} to the vehicle differential input shaft. The model consists of a continuously-variable gear set, a HTC including a model of the torque converter clutch, and the controls module. It manages the clutch actuation and gear shifts. The controls are implemented in Simulink and integrated into the GT-SUITE model as dll-file. The required gear is computed externally by one module of the power management. The gearbox output torque is calculated by:

$$M_{GR} = M_{ICE} + M_{PTO} + M_{MG}, \tag{5}$$

$$M_{GR\_out} = \eta_{GR} \cdot i_{GR} \cdot M_{GR} \tag{6}$$

(valid at closed torque converter clutch) with the MG effective torque M_{MG} and the gear ratio i_{GR}. For simplicity, the gearbox efficiency η_{GR} is set constant. The HTC enables the railway vehicle to accelerate with high traction forces. In gears with a gear ratio of $z > 1$ it is "deactivated" and locked with a torque converter clutch. To calculate the turbine torque of the HTC, M_{Turb} the impeller torque M_{Imp} is multiplied by the torque ratio i_{HTC}, which is taken from the torque ratio table in Figure 5a. The impeller torque is calculated using the capacity factor K_{HTC}.

$$M_{Imp} = \left(\frac{\omega_{Imp}}{K_{HTC}}\right)^2 \text{ with } K_{HTC} = f_1\left(\frac{\omega_{Turb}}{\omega_{Imp}}\right), \tag{7}$$

$$M_{Turb} = i_{HTC} \cdot M_{Imp} \text{ with } i_{HTC} = f_2\left(\frac{\omega_{Turb}}{\omega_{Imp}}\right). \tag{8}$$

3.1.4. Motor/Generator Model

To determine the electro-mechanic losses in the MG model, the efficiency η_{MG} is interpolated linearly from a map, see Figure 5b. Beside the losses of the MG itself, it also includes the electric efficiency of the electric converter. From the mechanic power P_{MG} and the efficiency, the electric power in the generator mode becomes $Pe_{Gen} = \eta_{MG} \cdot P_{MG}$, whereas the motor mode is characterized by $Pe_{Mot} = \frac{P_{MG}}{\eta_{MG}}$. Generally, electric machines are capable of running on overload for a restricted period of time. The overload duration is thermally restricted. Therefore, the MG involves a simple heat model to simulate heating and cooling sequences:

$$m' \cdot c \cdot \frac{dT}{dt} = \dot{Q}_{Loss}(t) + \dot{Q}_{Cooling}(t). \tag{9}$$

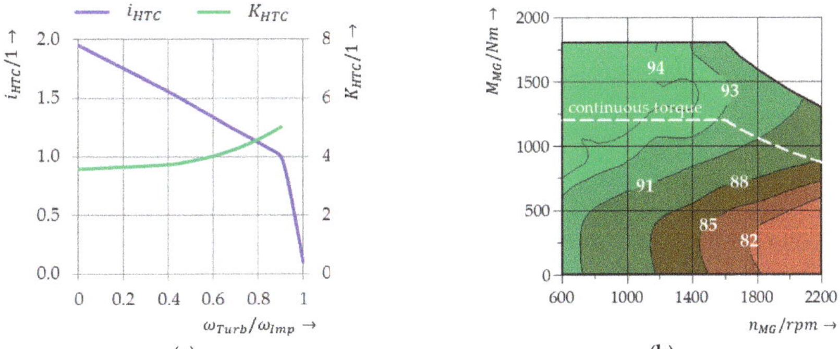

Figure 5. (**a**) Torque converter torque ratio i_{HTC} and capacity factor K_{HTC} over the speed ratio; (**b**) MG efficiency map.

The product of mass m' and specific heat c (here for iron) represents the time constant of the differential Equation (9) and depends on the MG size. The heat flows follow from:

$$\dot{Q}_{Loss}(t) = (1 - \eta_{MG}(t)) \cdot P_{MG}(t), \tag{10}$$

$$\dot{Q}_{Cooling}(t) = \alpha_w \cdot A' \cdot (T_{Coolant} - T(t)). \tag{11}$$

The parameters m', α_w, $T_{Coolant}$ and A' for this simplified approach are set to match pre-defined heating-up and cool-down times as well as to reach thermal steady state at the rated continuous power of the MG. The internal control routine of the MG model deactivates the overload mode and reduces the maximum power as soon as a temperature threshold is exceeded. MG power in overload mode is increased by a factor of 1.5 compared to the continuous operation mode, see Figure 5b.

3.1.5. Battery Model

The battery model is based on the equivalent electric circuit shown in Figure 6a. The voltage of the battery U_{BAT} follows from Kirchhoff's voltage law for the illustrated equivalent circuit:

$$U_{BAT}(t) = U_{OC}(t) - R_i(t) \cdot I_{BAT}(t) \tag{12}$$

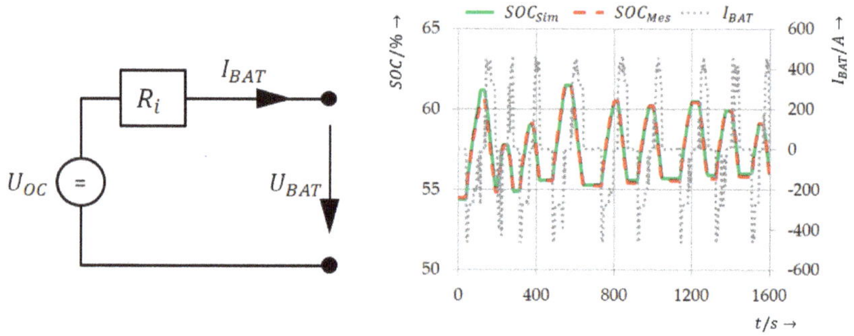

(a) (b)

Figure 6. (**a**) Equivalent electric circuit for the battery; (**b**) Battery validation cycle: simulated SOC_{Sim} and measured SOC_{Mes} over time.

The open-voltage circuit U_{OC} represents the equilibrium potential of the battery, which corresponds to the number n_{ser} of battery cells connected in series (13). The battery stack capacity C_{BAT} follows from the number of parallel linked battery cells n_{par}, with:

$$U_{OC} = n_{ser} \cdot U_{cell}, \tag{13}$$

$$C_{BAT} = n_{par} \cdot C_{cell}. \tag{14}$$

The internal resistance R_i of the battery in (12) is taken from 2-D lookup tables parametrized for each battery cell type. For charging and discharging events, different tables of $R_i(SOC)$ as a function of battery state of charge SOC are employed. From the power sum at the electric circuit, the battery current I_{BAT} can be calculated. Beside the MG, a further power sink of the electric circuit is the electric auxiliary power Pe_{Aux}:

$$I_{BAT}(t) = \frac{Pe_{MG}(t) + Pe_{Aux}(t)}{U_{OC}(t)}. \tag{15}$$

The integration of I_{BAT} according to (15) leads to the battery SOC. It is one of the most important input variables of the PMA of a hybrid propulsion system, because it represents the energetic status of the energy storage. It is often the reference to trigger mode switches between various system operation modes. SOC_0 is the initial state of charge value of the battery:

$$SOC(t) = SOC_0 + \int_0^t \frac{I_{BAT}(\tau)}{C_{BAT}} \, d\tau. \tag{16}$$

3.2. Model Validation

This section gives a brief overview of the model validation. The component models outlined in the previous section are used for open- and closed-loop component tests. The measured data of those tests contributes to a further improvement of the model accuracy.

3.2.1. Battery Model Validation

For model parameterization and validation purposes, measured data from open-loop battery tests was used. In the test procedure, a battery prototype on a battery test bench with a DC/DC power source was stressed with a battery current profile simulated beforehand. The current profile results from a simulation of a DMU with two hybrid PUs operating in regional passenger transportation. Battery state variables as voltage, current and SOC are measured. Afterwards, simulations of the battery were executed to optimize the internal resistance tables in the battery model for a minimum deviation of measured and simulated values for voltage and SOC. The optimization is based on Brent's algorithm [27]. Figure 6b shows a good matching between the calculated and observed SOC.

3.2.2. Powertrain Model Validation

In more extensive test campaigns of the whole propulsion system, validation data for the PU model was recorded. For this purpose, a prototype of a MTU hybrid PU was installed on a system test bench. In hardware-in-the-loop tests (HiL tests), the powertrain on the test bed is coupled with the simulation model of a DMU running on a virtual track. The test bed automation system is linked via a real-time interface to the simulation model of the vehicle and driver. In such a HiL test environment, performance tests can be executed or software functions are validated etc. Figure 7a gives a schematic overview of the experimental set-up. In Figure 7b, the simulated and measured results for the accumulated fuel consumptions over the whole drive cycle—for conventional and hybrid operation—are compared. The stated results are normalized on the fuel consumption for conventional mode. For the measurements campaign and the model validation, a previous version of the PMA, comprising a rule-based approach, was used. Obviously, the results underline and validate the achieved high accuracy of the simulation models.

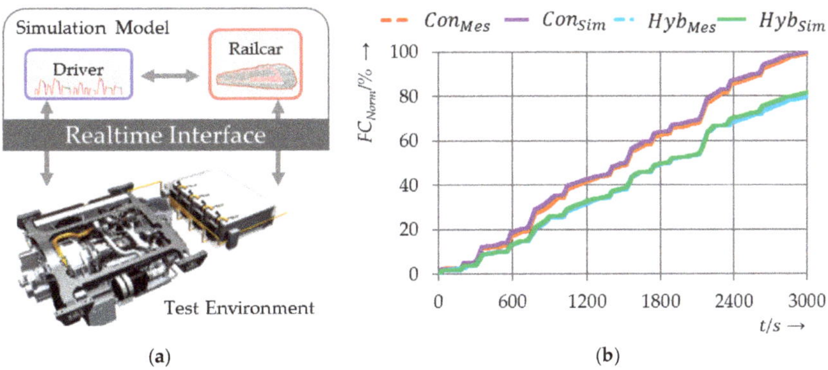

Figure 7. (**a**) Schematic overview over experimental HiL layout; (**b**) Measured and simulated normalized fuel consumption for conventional and hybrid mode.

4. Hybrid Power Management Algorithm (PMA)

4.1. Problem Definition

The PMA controls the power flow in the propulsion system and optimizes the power split between the ICE and the MG. Based on the optimal control problem [26], a suitable cost functional or performance index J is stated in (17). The cost functional J substitutes a corrected amount of fuel consumption FC_{Cor} of the system in a time interval $t \in [t_0, t_{end}]$. The objective of the PMA is to minimize J and FC_{Cor} over the given drive cycle. FC_{Cor} represents the quantity to evaluate the performance of the hybrid system according to its energy consumption and efficiency:

$$J = \phi(SOC(t_{end})) + \int_{t_0}^{t_{end}} \dot{m}_{Fuel}(u(\tau), x(\tau), \tau)\, d\tau \tag{17}$$

J is calculated by the integration of fuel mass flow $\dot{m}_{Fuel}$ in $\frac{kg}{sec}$, as a function of the control inputs $u(t)$ and state variables $x(t)$. The first term in (17) stands for a correction term to penalize the deviation of energy stored in the battery between t_0 and t_{end}:

$$\phi(SOC(t_{end})) = \lambda \cdot [SOC(t_0) - SOC(t_{end})]. \tag{18}$$

Here, λ represents a fuel equivalent factor, which converts the deviation of SOC into a value of fuel consumption in $\frac{kg}{\%}$. In contrast to some ECMS control algorithms, where an online adaption of λ is part of the PMA, λ is set constant in this case. It is determined by the amount of fuel, which is needed to balance battery charge deviation, assuming that charging is performed while the vehicle is at standstill (halt) at the end of each drive cycle:

$$\lambda = \frac{E_{BAT}^{100}}{H_u \cdot \eta_{chrg}^{op}}. \tag{19}$$

E_{BAT}^{100} stands for the maximum amount of energy, which can be stored in the battery at $SOC = 100\%$. H_u is the lower heating value of diesel fuel, and η_{chrg}^{op} denotes the charging efficiency at a predefined ICE operation point. This operating point for charging at standstill (halt) is optimized offline for maximum charging efficiency and set as a parameter.

4.2. Hybrid Operation Modes

The configuration of the parallel hybrid system, see Figure 2, enables different operation modes. Table 3 outlines the various operation modes and assigns them to drive events. The drive events result from the requested traction power P_{Trac}. This table shows the signs of ICE and MG power demands, P_{ICE} and P_{MG}, as well as the time derivative $\dot{SOC}$ of the state of charge for the different drive modes. The signs take the auxiliary powers into account: $P_{PTO} < 0$ at the ICE output shaft and $Pe_{AUX} > 0$ at the electric circuit. In Table 3 the plus sign, +, stands for a power request higher than zero. The minus sign, -, represents a negative power request, whereas, 0, refers to a constant value. The double plus and minus signs indicate very high or low power requests, which for example arise at acceleration or braking events.

During acceleration, two modes are possible. At *Pure ICE* mode, the traction power is exclusively provided by the ICE. The MG runs as generator and supplies electric auxiliaries. In the *Combined Mode*, the traction power is split between ICE and MG. During cruising, *Pure Electric* driving is possible. Furthermore, the battery can be charged by increasing the ICE load. In the case of a negative traction power demand, the deceleration event, the MG serves as generator to recuperate as much energy as possible. The clutch between ICE and MG is opened, to operate the ICE at idle speed. At halt, the vehicle velocity is zero. In *Normal Halt* mode, the ICE powers the PTO and supplies power for the AUX. If necessary, the ICE load can be increased to charges the battery. Additionally, the ICE may be switched off in *Start/Stop* mode, and the AUX power is supplied by the BAT. At deceleration and halt events, the different operation modes are triggered according to pre-defined thresholds of the SOC and other state variables. In those events, the PMA is a heuristic rule-based approach, where no optimization is executed. During acceleration and cruising, the optimization routine for P_{ICE} and P_{MG} is active.

Table 3. Drive Events and Hybrid Operation Modes.

Drive Event	Operation Mode	P_{Trac}	P_{ICE}	P_{MG}	$\dot{SOC}$
Acceleration	*Pure ICE*	++	++	-	0
	Combined Mode	++	+	+	-
Cruising	*Pure ICE*	+	+	-	0
	Pure Electric	+	0	+	-
	Combined Mode	+	+	+	-
	Charge Mode	+	++	-	+
Deceleration	*Recuperation*	−	+	−	+
Halt	*Normal Halt*	0	+	-	0
	Start/Stop	0	0	0	-
	Charge Mode	0	+	-	+

4.3. PMA Optimization Routine

Figure 8 shows a signal flow scheme of the PMA optimization routine including its most important input and output signals. Each block of the scheme represents one function of the routine, which is described in this section. In case of a DMU with more than one PU, P_{ICE} and P_{MG} represent the power demand from the particular power path. This means, for example, that if more than one engine is present, P_{ICE} substitutes the sum of engine power demands. A further function block splits P_{ICE} and distributes the particular share of power to the different engines. In case of PUs running synchronously, as assumed in this paper, power is equally split to the PUs' machines. In this section, for simplicity, the PMA refers to the ICE and MG, neglecting the fact that more than one of each component is involved.

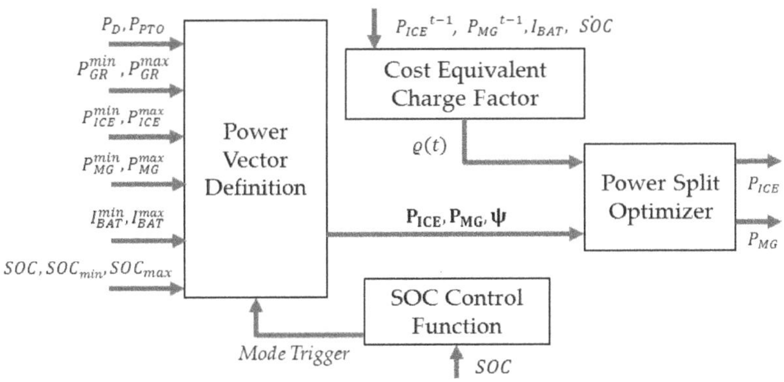

Figure 8. Signal flow scheme of the PMA optimization routine.

4.3.1. Power Vector Definition

From the traction power required and the power request of the mechanical auxiliaries P_{PTO}, this function block calculates the propelling power demand P_D. It always equals the sum of the power demands, P_{ICE} plus P_{MG}. Moreover, the gearbox input power limitations $[P_{GR}^{min}, P_{GR}^{max}]$ as well as further vehicle constraints, have to be considered. In a next functional step, the inappropriate operation modes for the current drive event are rejected. This is necessary because the battery SOC always has to be in a valid state of charge range, $SOC_{min} \leq SOC \leq SOC_{max}$. If, for instance, SOC is above its maximum value, *Charge Mode* is prohibited. Based on the feasible modes and constraints regarding the ICE, MG and BAT the maximum and minimum values for the power split ratio ψ are defined. From those extremes, the operation point vectors $\boldsymbol{\psi}$, $\boldsymbol{P_{ICE}}$ and $\boldsymbol{P_{MG}}$ for the split ratios of the length k are determined. They follow from:

$$\boldsymbol{\psi} = \begin{pmatrix} \psi_1^{max} \\ \vdots \\ \psi_k^{min} \end{pmatrix}, \tag{20}$$

$$\boldsymbol{P_{ICE}} = \boldsymbol{\psi} \cdot P_D \text{ and } \boldsymbol{P_{MG}} = (1 - \boldsymbol{\psi}) \cdot P_D, \tag{21}$$

$$\text{subject to} \left\{ \begin{array}{c} 0 \leq P_{ICE} \leq P_{ICE}^{max} \\ P_{MG}^{min} \leq P_{MG} \leq P_{MG}^{max} \\ SOC_{min} \leq SOC \leq SOC_{max} \\ I_{BAT}^{min} \leq I_{BAT} \leq I_{BAT}^{max} \\ P_{GR}^{min} \leq (P_D + P_{PTO}) \leq P_{GR}^{max} \end{array} \right\}. \tag{22}$$

4.3.2. Power Split Optimization

The optimization function block determines the control output signals for P_{ICE} and P_{MG}. Therefore, the algorithm calculates a power supply efficiency vector of length k, $\boldsymbol{\Omega} = (\Omega_1 \ldots \Omega_k)$, where each element of the vector assigns a power supply efficiency Ω_i to the particular elements of the operation point vectors. The power supply efficiency correlates to the sum of requested powers, P_D, P_{PTO} and Pe_{AUX}, divided by the sum of power consumption, P_{Fuel} plus Pe_{BAT-} (electric discharge power):

$$\Omega^k = \frac{P_D + P_{PTO} + Pe_{AUX} + Pe_{BAT+}^k}{P_{Fuel}^k + \left(\frac{Pe_{BAT-}^k}{\varrho(t)} \right)} \tag{23}$$

with:

$$P_{Fuel}^k = \frac{\boldsymbol{P_{ICE}}(k)}{\eta_{ICE}(\boldsymbol{P_{ICE}}(k), \omega_{ICE})}. \tag{24}$$

It is differentiated between electric battery power for charging Pe_{BAT+} and discharging Pe_{BAT-}. P_{BAT}^k for each vector element k is calculated by:

$$P_{BAT}^k = -\frac{P_{MG}(k)}{\eta_{BAT}(I_{BAT}) \cdot \eta_{MG}(P_{MG}(k), \omega_{MG})}, \tag{25}$$

hence:

$$Pe_{BAT+}^k = max[0, Pe_{BAT}^k] \text{ and } Pe_{BAT-}^k = |min[0, Pe_{BAT}^k]|. \tag{26}$$

The values for the conversion efficiencies η_{ICE}, η_{MG} and η_{BAT} are taken from efficiency maps (e.g., Figure 5b) or, in case of η_{ICE}, are calculated based on lookup tables of $bsfc(\boldsymbol{P_{ICE}}(k), \omega_{ICE})$. The variable $\varrho(t)$ in (23) stands for an energy cost equivalent charge factor and is described more detailed in the next subsection. Finally, the control output signals for P_{ICE} and P_{MG} are determined by the maximum value Ω^{max} of the power supply efficiency vector $\boldsymbol{\Omega}$:

$$arg\ max\{\boldsymbol{\Omega}(\boldsymbol{P_{ICE}}, \boldsymbol{P_{MG}}, P_D, P_{PTO}, Pe_{AUX})\} \rightarrow P_{ICE}, P_{MG}. \tag{27}$$

4.3.3. Energy Cost Equivalent Charge Factor

The energy cost equivalent charge factor $\varrho(t)$ accounts for the amount of fuel energy that was used in the time interval $t \in [t_0, t]$ to charge the battery. Its valid range is $\eta_{chrg}^{min} \leq \varrho(t) \leq 1$. If the battery is charged only by recuperation during braking, without any consumption of fuel, $\varrho(t)$ has the value 1. But if the energy stored in the battery is provided by the ICE, $\varrho(t)$ equals the medium efficiency of the charging procedure. Hence, for $\dot{SOC} > 0$, $\varrho(t)$ is computed by:

$$\varrho(t) = \frac{E^*_{BAT}(t)}{E^*_{ChrgFuel}(t)} \tag{28}$$

with:

$$E^*_{BAT}(t) = E^*_{BAT}(t_0) + \int_{t_0}^{t} U_{OC}(\tau) \cdot I_{BAT}(\tau)\, d\tau \tag{29}$$

and:

$$E^*_{ChrgFuel}(t) = E^*_{ChrgFuel}(t_0) + \int_{t_0}^{t} P_{ChrgFuel}(\tau)\, d\tau. \tag{30}$$

In the case of discharging, $\dot{SOC} \leq 0$, the PMA modifies (28) and holds $\varrho(t)$ constant. It follows from (28) that $\varrho(t)$ substitutes the ratio of energy E^*_{BAT}, stored in the battery, and the fuel energy needed to charge the battery $E^*_{ChrgFuel}$. Please note that $P_{ChrgFuel}$ represents the share of fuel consumed for charging. $P_{ChrgFuel}$ can be calculated from the electric charging power at the battery terminal and the product of conversion efficiencies for the ICE as well as MG, leading to:

$$P_{ChrgFuel}(t) = \frac{U_{BAT}(t) \cdot I_{BAT+}}{\eta_{ICE}(M_{ICE}(t),\, \omega_{ICE}(t)) \cdot \eta_{MG}(M_{MG}(t),\, \omega_{MG}(t))}. \tag{31}$$

The index * implies that E^*_{BAT} represents the amount of energy in between the admissible boundaries of the battery state of charge, $SOC_{min} \leq SOC \leq SOC_{max}$. This makes sense because only this partition of the battery is used during operation.

From a mathematical point of few, $\varrho(t)$ set in (23) is handled like an efficiency value, which increases the energy costs, or in other words, reduces the efficiency of the electrical path. In case of $\varrho(t)$ being close to one, it means that most of the energy in the battery was provided by "free" sources like recuperation or even external sources, for instance, external charging in a depot. As a consequence, a high portion of P_D is provided by the electric path. This results in ψ close to zero due to its higher conversion efficiency as compared to the ICE path. In applications, system configurations or even drive cycles, where not much of the consumed energy can be recuperated during the deceleration and only a relatively small BAT is used, *Charge Mode* is employed quite regularly to maintain *SOC* within its admissible interval. In this case, $\varrho(t)$ falls below one and, thus, increases the electric energy cost. Due to its implementation, $\varrho(t)$ adapts itself dependent on system parameters, the drive cycle and other operational boundaries.

4.3.4. SOC Control Function

There are system configurations or constraints for specific applications, where *SOC* does not only have to be within its bounds but have to match a specific value at the end of the drive cycle or even among single track sequences. This is the case if the additional installed electric power is intended to improve system performance or, for example, to replace one combustion engine, see Figure 3. Therefore, the PMA features a function block that determines an expected amount of energy that is recuperated during braking down from the current vehicle velocity. The computation addresses the driving resistance forces according to (1), estimates a brake trajectory in compliance with the drive strategy, see Section 2.2, and determines the recuperation duration $t_{Rec} = f(v)$. From t_{Rec} and an assumed mean electric recuperation power Pe^{mean}_{Rec}, a recuperated SOC^{calc}_{Rec} is calculated:

$$SOC^{calc}_{Rec} = F^i_{Rec} \cdot \frac{t_{rec} \cdot Pe^{mean}_{Rec}}{E^{100}_{BAT}}. \tag{32}$$

The factor F^i_{Rec} in (32) represents a correction factor and is adapted online for a better match of the predicted SOC^{calc}_{Rec} and the real SOC_{Rec}. It is calculated by a comparably simple approach according to (33) but leads to relatively good results (see Section 5):

$$F_{Rec}^{i} = F_{Rec}^{i-1} + dF_{Rec} \text{ with } dF_{Rec} = \frac{SOC_{Rec}}{SOC_{Rec}^{calc}} - F_{Rec}^{i-1}, \tag{33}$$

$$\text{subject to } -0.15 \leq dF_{Rec} \leq 0.15. \tag{34}$$

The computation of F_{Rec}^{i} is processed at discrete states i, after each recuperation event, where the velocity is reduced from cruising speed to a halt. Based on SOC_{Rec}^{calc} and a target value for SOC_{Trg}, a finite-state machine sets a trigger to initiate the *Charge Mode*. The PMA strives for equality of SOC and SOC_{Trg} at the start of the next cycle sequence. Hence, the *Charge Mode* is activated until $SOC \geq (SOC_{Trg} - SOC_{Rec}^{calc})$ is achieved.

The SOC control function is a straightforward approach to control the state of charge of the battery BAT in order to attain a predefined target value. By a proper choice of the target value, it allows for implementing different operating strategies for hybrid systems. If SOC_{Trg} is close to the upper SOC limit, a performance oriented strategy is applied. Here, it is ensured that at any time a high battery SOC is available for electrically boosted acceleration events. For a low SOC_{Trg}, the PMA allows a wider range of possible operation modes and, moreover, to optimize a power split only from an efficiency point of view. Furthermore, this charge triggering function is intended to be the interface for further optional predictive algorithms, which process the knowledge of the whole drive cycle.

5. Case Study Results

In the simulation case study, the DMU specified in Section 2 with two identical PUs is employed. The vehicle parameters are shown in Table 2. The electric auxiliary power is set to $Pe_{Aux} = 30$ kW for the vehicle. In the reference case, where the DMU runs on *Pure ICE* mode, the auxiliaries are supplied by a generator driven by the ICE. For evaluation purposes, the corrected fuel consumption FC_{Cor}, calculated with (15)–(17), is used. The DMU drives on a track which was generated from an evaluation of various DMUs' field data. This track is representative for the real world operation of such a vehicle.

The fuel consumption of the reference diesel-driven vehicle adds up to 32.8 L. In hybrid operation, the DMU consumes 25.8 L, which denotes a reduction of FC_{Cor} by 21.3%. Figure 9 shows the vehicle velocity profile over time for the diesel and hybrid system as well as the altitude profile. The velocity profiles differ due to the modified drive strategies and their particular brake trajectories (Section 2.2).

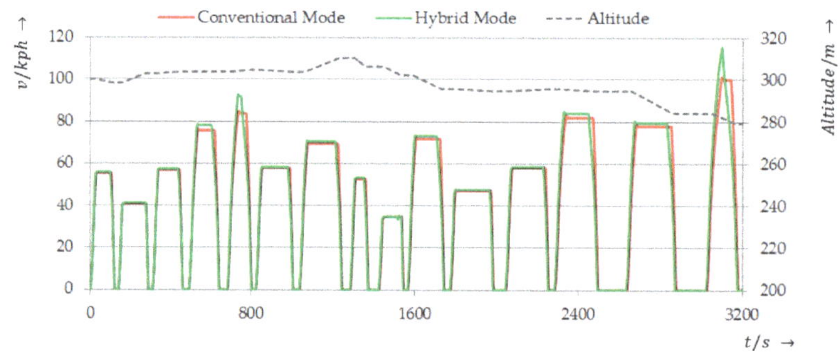

Figure 9. Altitude and velocity profiles for conventional and hybrid mode.

In Figure 10a, the battery SOC is shown. Due to the activated SOC control function, the $SOCs$ at the end of the drive cycle and also at the end of each cycle sequence are fairly close to the target value SOC_{Trg} of 55%. The maximum deviation of the SOC and its target value at halt events occurs after the seventh drive sequence and adds up to 0.59%. Considering the comparatively simple approach from Section 4.3.4, Figure 10b shows the energy cost-equivalent charge factor $\varrho(t)$ over cycle time. It becomes obvious that $\varrho(t)$ falls during cruising and raises throughout recuperation. This is the typical behavior and points out that *Charge Mode* is active quite often. The fact that $\varrho(t)$ is not below one during the whole cycle shows that a relatively high amount of electric charging power for the battery is gained by recuperation. As a result, *Pure Electric* operation is executed rather often during cruising, because electric energy is quite "cheap". Figure 11a shows the percentages of the individual operation modes during acceleration and cruising. The *Combined Mode* extensively occurs during vehicle acceleration. The reason is that a combined use of ICE and MG increases the vehicle performance and, therefore, results in lower cruising speeds. About 56% of the time, the *Pure Electric* mode is activated, whereas the ICE is running on idle speed decoupled from the MG by the clutch. This leads to the question, whether it is feasible to completely switch off the ICE during driving. The employment of *Start/Stop* mode at halt further reduces the fuel consumption by 2.5%, which leads to a FC_{Cor} of 25.0 L. If *Start/Stop* mode is expanded to cruising events as well, the FC_{Cor} is further lowered by 1.2%, see Figure 11b.

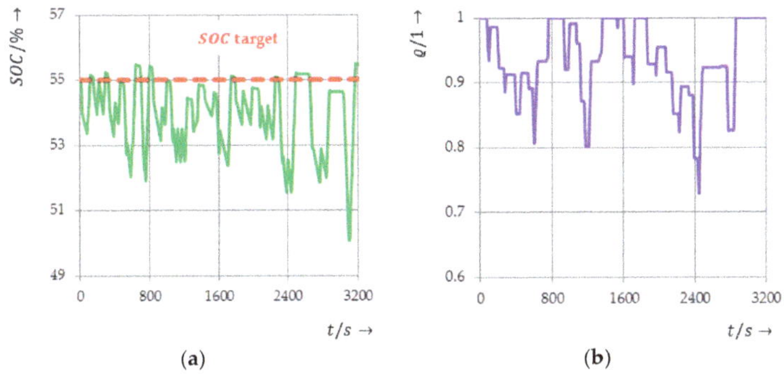

Figure 10. (**a**) *SOC* and its target value over cycle time; (**b**) Energy cost equivalent charge factor over cycle time.

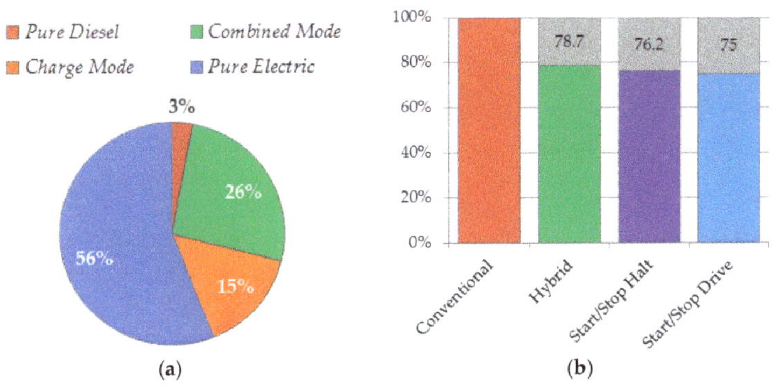

Figure 11. (**a**) Percental proportion of operation modes during acceleration and cruising (hybrid without *Start/Stop*); (**b**) Overview over percental reductions of corrected fuel consumption.

6. Conclusions and Outlook

In this contribution, an optimization-based causal PMA for hybrid propulsion systems is presented. It optimizes the power split between ICE and MG according to vectors of possible operation points. For each vector element, which represents one feasible operational power distribution, the corresponding element of a power supply efficiency vector is calculated. The power split that is related to the highest efficiency determines the power demand for ICE and MG as control output. The computation of the power supply efficiency uses an energy-cost-equivalent charge factor, which accounts for the energy consumption to charge the battery. It differentiates between regenerative charging during braking and "energy expensive" charging via the ICE. Moreover, as an optional feature, the power split can be controlled in such a way that the battery *SOC* matches a pre-defined target value at the end of the drive cycle. In a case study, the presented PMA is employed for the control a hybrid propulsion system of a DMU consisting of two identical PUs. Thereby, the DMU was operated on a track which was generated from logging field data of comparable vehicles. In comparison to the conventional diesel-driven system, a fuel consumption reduction of 21.3% was achieved. Employing *Start/Stop* at halt and, additionally, during *Pure Electric* driving, the fuel consumption was further reduced by 3.7%. It was stated that the PMA leads to high proportion of *Pure Electric* driving during cruising events. This has a significant influence on the engine load profile because especially areas of low engine loads, where engine efficiency is weak, disappear. This is even more true as soon as the engine is switched off during *Pure Electric* driving. Beside the shown fuel consumption benefits, the *Pure Electric* mode leads to shorter engine operation time and extends the time-before overhaul (TBO), which further reduces operational costs for the system operator. Another aspect which has to be considered is the influence of hybridization on exhaust emissions. DMUs' modern diesel engines are compliant to EPA Tier 4i emission standards. Those are commonly equipped with SCR (selective catalytic reduction) exhaust after-treatment systems. Investigations in the HiL test environment show that the changed engine operation in hybrid systems, compared to conventional ones, considerably influence exhaust after-treatment performance, hence real driving emissions [7]. The implementation of *Start/Stop* during *Pure Electric* driving are one possibility to influence exhaust after-treatment temperatures and, for example, prevent the SCR-catalyst from cooling down too quickly during idling.

In future investigations, the introduced PMA will be applied to system configurations with multiple PUs, where the ICEs and MGs are operated individually. Furthermore, the PMA's suitability for other applications will be proven. To demonstrate the "real-world" capability of the PMA, it will be applied to a PU prototype of a DMU propulsion system in hardware-in-the-loop testing campaigns on a system test bench. In those experiments, the impact and the benefits of the proposed PMA and its different operation modes on fuel consumption and emissions will be investigated for different DMU propulsion system configurations.

Author Contributions: Johannes Schalk developed the algorithm and the simulation model. Furthermore, he executed the case study. The HiL-Test procedure had been conducted at a test facility at MTU Friedrichshafen, where Johannes Schalk was part of the team of test engineers. Harald Aschemann supervised the work, contributed his know-how in hybrid Off-Highway applications and co-wrote the paper.

Conflicts of Interest: The authors declare no conflict of interest.

References

1. European Parliament and Council of the European Union. REGULATION (EG) No 443/2009 Setting Emission Performance Standards for New Passenger Cars as Part of the Community's Integrated Approach to Reduce CO_2 Emissions from Light-Duty Vehicles, April 2009. Available online: http://eur-lex.europa.eu/legal-content/EN/TXT/PDF/?uri=CELEX:32009R0443&from=EN (accessed on 20 August 2016).
2. Baffes, J.; Kose, M.A.; Ohnsorge, F.; Stocker, M. *The Great Plunge in Oil Prices: Causes, Consequences and Policy Responses*; Policy Research Note; World Bank Group: Washington, DC, USA, 2015.
3. Dittus, H.; Hülsebusch, D.; Ungethüm, J. Reducing DMU fuel consumption by means of hybrid energy storage. *Eur. Transp. Res. Rev.* **2011**, *3*, 149–159. [CrossRef]
4. Oszfolk, B.; Radke, M.; Ibele, Y. Hybridantrieb stellt Marktreife unter Beweis. *ETR Eisenbahntechnische Rundsch.* **2015**, *9*, 44–49.
5. Yuan, L.C.W.; Tjahjowidodo, T.; Lee, G.S.G.; Chan, R. Equivalant consumption minimization strategy for hybrid all-electric tugboats to optimize fuel savings. In Proceedings of the American Control Conference (ACC), Boston, MA, USA, 6–8 July 2016.
6. Jayaram, V.; Khan, M.Y.; Welch, W.A.; Johnson, K.; Miller, J.W.; Cocker, D.R. A generalized approach for varifying the emission benefits of off-road hybrid mobile sources. *Emiss. Control Sci. Technol.* **2016**, *2*, 89–98. [CrossRef]
7. Hass, C.; Oszfolk, B.; Schalk, J. Emissions of a hybrid propulsion system for regional trains as an example for innovative non-road propulsion systems. In Proceedings of the 8th Emission Control, Dresden, Germany, 2–3 June 2016.

8. Hofman, T.; Steinbuch, M.; Van Druten, R.; Serrarens, A. Rule-based energy management strategies for hybrid vehicles. *Int. J. Electr. Hybrid Veh.* **2007**, *1*, 71–94. [CrossRef]
9. Guzella, L.; Sciarretta, A. *Vehicle Propulsion Systems: Introduction to Modeling and Optimization*; Springer: Berlin/Heidelberg, Germany, 2013.
10. Goerke, D. *Untersuchungen zur Kraftstoffoptimalen Betriebsweise von Parallelhybridfahrzeugen und Darauf Basierende Auslegung Regelbasierter Betriebsstrategien*; Springer: Wiesbaden, Germany, 2016.
11. Guzella, L.; Sciarretta, A. Control of hybrid electric vehicles. *IEEE Control Syst.* **2007**, *27*, 60–70.
12. Helbing, D.I.M.; Uebel, D.I.S.; Tempelhahn, D.I.C.; Bäker, I.B. Bewertender Überblick von Methoden zur Antriebsstrangsteuerung in Hybrid- und Elektrofahrzeugen. *ATZelektronik* **2015**, *10*, 66–71. [CrossRef]
13. Karbaschian, M.A.; Söffker, D. Review and comparison of power management approaches for hybrid vehicles with focus on hydraulic drives. *Energies* **2014**, *7*, 3512–3536. [CrossRef]
14. Lampe, A. Regelbasierte Betriebsstrategien zur Vorauslegung von Hybridantriebssträngen. *ATZ Automobiltech. Z.* **2016**, *116*, 76–82. [CrossRef]
15. Hanho, S.; Hyunsoo, K. Development of near optimal rule-based control for plug-in hybrid electric vehicles taking into account drivetrain component losses. *Energies* **2016**, *9*, 420. [CrossRef]
16. Paschero, M.; Storti, G.L.; Rizzi, A.; Mascioli, F.M.F. Implementation of a fuzzy control system for a parallel hybrid vehicle powertrain on compactrio. *Int. J. Comput. Theory Eng.* **2013**, *5*, 273. [CrossRef]
17. Khoucha, F.; Benbouzid, M.; Kheloui, A. An optimal fuzzy logic power shift strategy for parallel hybrid electric vehicles. In Proceedings of the IEEE Vehicle Power and Propulsion Conference, Lille, France, 1–3 September 2010.
18. Yuan, Z.; Teng, L.; Fengchun, S.; Peng, H. Comparative study of dynamic programming and pontryagin's minimum principle on energy management of a parallel hybrid electric vehicle. *Energies* **2013**, *6*, 2305–2318. [CrossRef]
19. Leska, M.; Aschemann, H. Fuel-optimal combined driving strategy and energy management for a parallel hybrid electric railway vehicle. In Proceedings of the 20th International Conference on Methods and Models in Automation and Robotics, Miedzyzdroje, Poland, 24–27 August 2015.
20. Sivertsson, M.; Sundström, C.; Eriksson, L. *Adaptive Control of a Hybrid Powertrain with Map-Based ECMS*; IFAC World Congress: Prague, Czech Republic, 2011.
21. Mustardo, C.; Rizzoni, G.; Guezennec, Y.; Staccia, B. A-ECMS: An adaptive algorithm for hybrid electric vehicle energy management. *Eur. J. Control* **2005**, *11*, 509–524. [CrossRef]
22. Nüesch, T.; Cerofolini, A.; Mancini, G.; Cavina, N.; Onder, C.; Guzzella, L. Equivalent consumption minimization strategy for control of real driving NOx emissions of a diesel hybrid electric vehicle. *Energies* **2014**, *7*, 3148–3178. [CrossRef]
23. Onori, S.; Serrao, L. On adaptive-ECMS strategies for hybrid electric vehicles. In Proceedings of the International Scientific Conference on Hybrid and Electric Vehicles, Rueil-Malmaison, France, 6–7 December 2011.
24. Winkler, M.; Geulen, S.; Josevski, M.; Tegethoff, M.; Abel, D.; Vöcking, B. Online parameter tuning methods for adaptive ECMS control strategies in hybrid electric vehicles. In Proceedings of the FISTA World Automotive Congress, Maastricht, The Netherlands, 2–6 June 2014.
25. Katsargyri, G.-E.; Kolmanovsky, I.V.; Michelini, J.; Kuang, M.L.; Phillips, A.M.; Rinehart, M.; Dahleh, M.A. Optimally controlling electric hybrid vehicles using path forecasting. In Proceedings of the 2009 American Control Conference, St. Louis, MO, USA, 10–12 June 2009.
26. Geering, H.P. *Optimal Control with Engineering Applications*; Springer: Berlin/Heidelberg, Germany, 2007.
27. Gamma Technologies. *GT-SUITE User Manual. GT-SUITE Optimisation Manual*, version 2016; Gamma Technologies: Westmont, IL, USA, 2015.

energies

MDPI

Article

Simplified Minimum Copper Loss Remedial Control of a Five-Phase Fault-Tolerant Permanent-Magnet Vernier Machine under Short-Circuit Fault

Chenyu Gu, Wenxiang Zhao * and Bufeng Zhang

School of Electrical and Information Engineering, Jiangsu University, Zhenjiang 212013, China; chenyuGu92@163.com (C.G); zhangbufeng1990@163.com (B.Z.)

* Correspondence: zwx@ujs.edu.cn; Tel.: +86-511-8879-1960

Academic Editor: Paolo Mercorelli
Received: 9 July 2016; Accepted: 13 October 2016; Published: 25 October 2016

Abstract: A fault-tolerant permanent-magnet vernier (FT-PMV) machine incorporates the merits of high fault-tolerant capability and high torque density. In this paper, a new remedial control is proposed for a five-phase FT-PMV machine with short-circuit fault of stator windings. Based on the principle of copper loss minimization, the aims of the proposed control strategy are to keep magnetic motive force (MMF) unchanged and minimize torque ripple. The proposed remedial control strategy contains two parts. Firstly, the remedial currents of the healthy phases are used to compensate for the ripple of MMF caused by the short-circuit current. Secondly, an open-circuit fault-tolerant control strategy is used to compensate for the lack of normal torque in the fault phase. Finally, the vector sum of two parts is adopted to derive the remedial currents. The final expression of the proposed remedial current is simpler than that than these previous methods. In addition, the proposed remedial currents are sinusoidal, which can reduce the reactive component in instantaneous power produced by pulsating torque and iron loss of a sine back-EMF machine. A FT-PMV prototype is built. The simulations and the experiments verify the effectiveness of the proposed strategy.

Keywords: remedial control; fault-tolerant permanent-magnet vernier (FT-PMV) machine; short-circuit fault; torque

---

1. Introduction

For some high reliability applications such as aerospace and electric vehicle, continued operation of the drive is essential [1–3]. Using electric vehicles as an example, such capability is related to the ability to continue operation at a modest speed to finish the trip or at least reach a repair shop. In much of the literature, the fault-tolerant operations of conventional three-phase machines under open-circuit fault condition have been reported [4–7]. Compared with traditional three-phase drives, multiphase machine drives have a number of advantages, such as low harmonic currents, smooth torque ripple and high torque capability [8]. Multiphase machines can keep operating with the remaining healthy phases when open-circuit fault occurs in one or two phases. In addition, multiphase machines can operate without additional hardware under faulty conditions [9–12].

On the other hand, fault-tolerant permanent-magnet (PM) machines have attracted much interest in recent years [13–16], but they suffer from the compromise between fault-tolerant capability and torque density. In [17], a five-phase fault-tolerant PM vernier (FT-PMV) machine with improved torque capability was proposed. Also, though many works seek optimal fault-tolerant control strategies, most of them focus on open-circuit fault, rather than short-circuit fault. An optimal control technique for a multiphase PM machine under various open-circuit faults was presented in [18], in which the copper loss and the torque ripple were considered. The proposed fault-tolerant control methods in [19] considered the fundamental and third-harmonic current components together for the excitation of the healthy stator phases. In [20], a mathematical analysis based on the space vector representation of a multiphase system was discussed and assessed under open-circuit fault. In order to generate optimal current references in fault mode, a vector approach was reported in [21], which requires relatively simple computations. The authors of [22] adopted a new control strategy for fault-tolerant operation of the redundant flux-switching PM machine drive by injecting the harmonic currents, in which the capability limitation of the power converter is considered. A generalized optimal fault-tolerant control technique was proposed in [23,24] for a five-phase PM machine drives with open-circuit fault. Few works focus on continued operation of multiphase PM machines under short-circuit fault conditions [25–30]. In [31], a remedial operation was investigated by using the open-end winding drive topology.

In this paper, an offline closed-form solution for short-circuit fault will be proposed, discussed and verified. It should be noted that the existing offline closed form fault-tolerant solutions were usually used for open-circuit fault, rather than for short-circuit fault. This proposed control strategy bases on the copper loss minimization theory and aims to keep magnetic motive force (MMF) unchanged, thus obtaining smooth output torque. The short-circuit fault in one phase can be divided into two parts. Namely, one is caused by the short-circuit current, while the other is due to the lack of normal torque, which can be considered the open-circuit fault condition. As opposed from the existing strategies of multiphase PM machines under short-circuit fault conditions, the proposed strategy is simply divided into two steps. Firstly, currents in healthy phases are used to compensate for the ripple of MMF caused by the short-circuit current. Secondly, the open-circuit fault-tolerant control strategy is used to compensate for the lack of the normal torque in the faulty phase. Finally, a new remedial current can be calculated by utilizing the vector sum of two parts of the strategy.

In Section 2, the FT-PMV machine will be briefly presented. In Section 3, the remedial current control strategy under open-circuit fault condition will be given. The remedial short-circuit current control strategy will be derived in Section 4. Moreover, in Section 5, the finite-element analysis (FEA) and the experimental results will be used to verify the proposed strategy. In Section 6, comparisons between and discussions about the proposed method and these existing methods will be presented. Finally, conclusions will be drawn in Section 7.

2. Topology and Feature

A five-phase FT-PMV machine is shown in Figure 1. The number of stator slots is 20 and the rotor has 31 pole pairs. This machine has each coil wound around a single tooth, which is the so-called single-layer fractional-slot concentrated winding. For conventional direct-drive machine topologies with increased number of poles, slots and windings, magnetic circuits are usually required to generate high torque. Such a machine inevitably increases the machine volume and weight. FT-PMV machines with concentrated windings can offer high torque capability while reducing the number of slots and copper loss. FT-PMV machines can effectively modulate the high-speed rotating field of the armature windings and the low-speed rotating field of the PM outer rotor by adopting flux-modulation poles (FMPs). Hence, the armature poles number should differ from the rotor poles one, which breaks the traditional design rule of PM machines. Also, the volume of the stator with vernier structure will not increase with the large number of PM poles.

Moreover, the predicted FEA-based back electromotive force (back-EMF) of the five-phase FT-PMV machine is shown in Figure 2. Due to the insulation problem, the windings terminal of one phase is short-circuited. Then, its FEA-based short-circuit current is shown in Figure 3. It can be seen that the short-circuit current is limited because fault-tolerant design is adopted [17]. The main design data are given in Table 1.

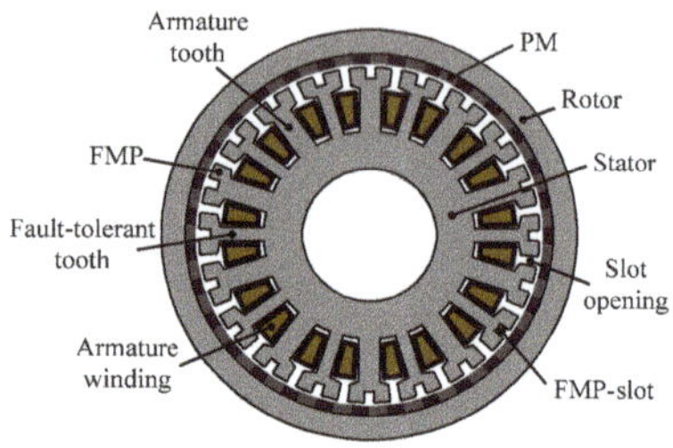

Figure 1. Fault-tolerant permanent-magnet vernier (FT-PMV) machine.

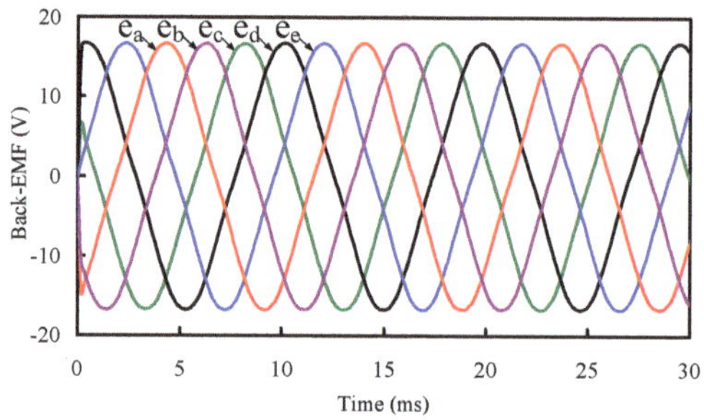

Figure 2. Back-EMF.

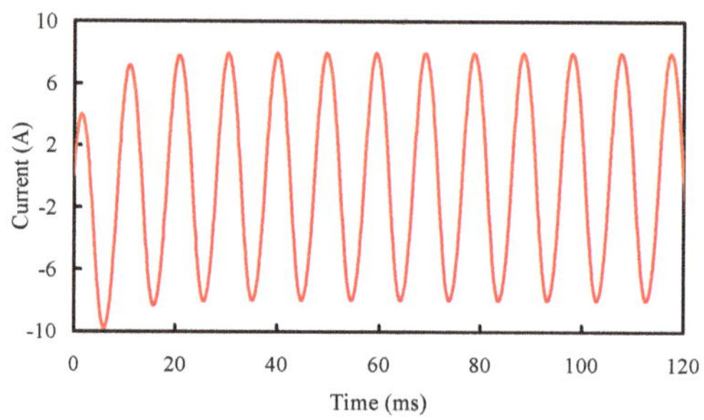

Figure 3. Short-circuit current.

Table 1. Motor parameters.

Rated output power (kW)	1.8
Rated current (A)	10
Rated speed (rpm)	200
Number of phases	5
Number of stator slots	20
Number of rotor pole pairs	31
Number of FMPs	40
Rated voltage (V)	70
Current density (A/mm^2)	6.6
Outside stator radius (mm)	60
Outside rotor radius (mm)	70
Stack length (mm)	60
Air-gap length (mm)	0.5
PM thickness (mm)	3.2
Fault-tolerant tooth length (mm)	1.9
Armature tooth length (mm)	3.3
Number of stator turns per coil	48

The proposed remedial control strategy divides the short-circuit problem into two parts to simplify the remedial problem. One part is the compensation of the open-circuit fault which is used to achieve the required torque. The other part is the compensation of the short-circuit fault which is used to eliminate pulsating torque.

3. Open-Circuit Fault-Tolerant Control

The back-EMFs of the five-phase FT-PMV machine can be described as

$$\begin{cases} e_a = E\cos(\omega t) \\ e_b = E\cos(\omega t - \frac{2\pi}{5}) \\ e_c = E\cos(\omega t - \frac{4\pi}{5}) \\ e_d = E\cos(\omega t + \frac{4\pi}{5}) \\ e_e = E\cos(\omega t + \frac{2\pi}{5}) \end{cases} \quad (1)$$

where E is the amplitude of the back-EMF, and ω is the electrical angular velocity. Due to its sinusoidal back-EMF, this FT-PMV machine is suitable for brushless ac (BLAC) operation. Hence, the currents can be expressed as

$$\begin{cases} i_a = I\cos(\omega t) \\ i_b = I\cos(\omega t - \frac{2\pi}{5}) \\ i_c = I\cos(\omega t - \frac{4\pi}{5}) \\ i_d = I\cos(\omega t + \frac{4\pi}{5}) \\ i_e = I\cos(\omega t + \frac{2\pi}{5}) \end{cases} \quad (2)$$

where I is the amplitude of the phase currents. According to (1) and (2), the stator phases MMF can be written as

$$\begin{cases} MMF_a(\varphi, \omega t) = NI\cos\varphi\cos(\omega t) \\ MMF_b(\varphi, \omega t) = NI\cos(\varphi - \frac{2\pi}{5})\cos(\omega t - \frac{2\pi}{5}) \\ MMF_c(\varphi, \omega t) = NI\cos(\varphi - \frac{4\pi}{5})\cos(\omega t - \frac{4\pi}{5}) \\ MMF_d(\varphi, \omega t) = NI\cos(\varphi + \frac{4\pi}{5})\cos(\omega t + \frac{4\pi}{5}) \\ MMF_e(\varphi, \omega t) = NI\cos(\varphi + \frac{2\pi}{5})\cos(\omega t + \frac{2\pi}{5}) \end{cases} \quad (3)$$

where N is the total number of turn per phase and φ is the spatial angle. Hence, the sum of MMF of all phases can be given as

$$MMF_1(\varphi, \omega t) = MMF_a(\varphi, \omega t) + MMF_b(\varphi, \omega t) + MMF_c(\varphi, \omega t) + MMF_d(\varphi, \omega t) + MMF_e(\varphi, \omega t) \quad (4)$$

After calculation, it can be written as

$$MMF_1(\varphi, \omega t) = \frac{5}{4}NI(\mathrm{e}^{\mathrm{j}\omega t}\mathrm{e}^{-\mathrm{j}\varphi} + \mathrm{e}^{-\mathrm{j}\omega t}\mathrm{e}^{\mathrm{j}\varphi}) \quad (5)$$

Moreover, MMF_1 can also be expressed as

$$MMF_1 = \{N[(i_a + \mathrm{e}^{-\mathrm{j}\frac{2\pi}{5}} i_b + \mathrm{e}^{-\mathrm{j}\frac{4\pi}{5}} i_c + \mathrm{e}^{\mathrm{j}\frac{4\pi}{5}} i_d + \mathrm{e}^{\mathrm{j}\frac{2\pi}{5}} i_e)\mathrm{e}^{\mathrm{j}\varphi} + (i_a + \mathrm{e}^{\mathrm{j}\frac{2\pi}{5}} i_b + \mathrm{e}^{\mathrm{j}\frac{4\pi}{5}} i_c + \mathrm{e}^{-\mathrm{j}\frac{4\pi}{5}} i_d + \mathrm{e}^{-\mathrm{j}\frac{2\pi}{5}} i_e)\mathrm{e}^{-\mathrm{j}\varphi}]\} \quad (6)$$

Equation (7) can be obtained on the basis of (5) and (6).

$$\frac{5}{2}I\mathrm{e}^{\mathrm{j}\omega t} = i_a + \mathrm{e}^{\mathrm{j}\frac{2\pi}{5}} i_b + \mathrm{e}^{\mathrm{j}\frac{4\pi}{5}} i_c + \mathrm{e}^{-\mathrm{j}\frac{4\pi}{5}} i_d + \mathrm{e}^{-\mathrm{j}\frac{2\pi}{5}} i_e \quad (7)$$

Here, the open-circuit fault of phase-a is chosen as an example. MMF can be kept unchanged when i_a is equal to zero in (7). Equation (7) can be written as

$$\frac{5}{2}I\mathrm{e}^{\mathrm{j}\omega t} = \mathrm{e}^{\mathrm{j}\frac{2\pi}{5}} i'_b + \mathrm{e}^{\mathrm{j}\frac{4\pi}{5}} i'_c + \mathrm{e}^{-\mathrm{j}\frac{4\pi}{5}} i'_d + \mathrm{e}^{-\mathrm{j}\frac{2\pi}{5}} i'_e \quad (8)$$

It can be assumed that

$$\begin{cases} i'_b = I\left[x_2\cos(\omega t) + y_2\sin(\omega t)\right] \\ i'_c = I\left[x_3\cos(\omega t) + y_3\sin(\omega t)\right] \\ i'_d = I\left[x_4\cos(\omega t) + y_4\sin(\omega t)\right] \\ i'_e = I\left[x_5\cos(\omega t) + y_5\sin(\omega t)\right] \end{cases} \tag{9}$$

According to (8) and (9), it can be obtained as

$$\begin{cases} 0.309x_2 - 0.809x_3 - 0.809x_4 + 0.309x_5 = \frac{5}{2} \\ 0.309y_2 - 0.809y_3 - 0.809y_4 + 0.309y_5 = 0 \\ 0.951x_2 + 0.588x_3 - 0.588x_4 - 0.951x_5 = 0 \\ 0.951y_2 + 0.588y_3 - 0.588y_4 - 0.951y_5 = \frac{5}{2} \end{cases} \tag{10}$$

An additional reasonable constraint is as follows:

$$i'_b + i'_c + i'_d + i'_e = 0 \tag{11}$$

According to (11), it can be written as

$$\begin{cases} x_2 + x_3 + x_4 + x_5 = 0 \\ y_2 + y_3 + y_4 + y_5 = 0 \end{cases} \tag{12}$$

According to (10) and (12), the results of x_2, y_2, x_3, y_3, x_4, y_4, x_5 and y_5 are not unique, so it is necessary to add a target function to make the final result unique. Then, the target function F aims to minimize copper loss can be given as [32]

$$F(x_2, y_2, x_3, y_3, x_4, y_4, x_5, y_5) = ({x_2}^2 + {y_2}^2) + ({x_3}^2 + {y_3}^2) + ({x_4}^2 + {y_4}^2) + ({x_5}^2 + {y_5}^2) \tag{13}$$

Among all the results, there will be some results of minimizing F, which can be called the optimal results. According to (10), (12) and (13), a built-in function of MATLAB based on SQP algorithm can be used to achieve the optimal results of x_2, y_2, x_3, y_3, x_4, y_4, x_5 and y_5.

$$\begin{cases} i'_b = I\left[1.118\cos\omega t + 0.951\sin\omega t\right] \\ i'_c = I\left[-1.118\cos\omega t + 0.588\sin\omega t\right] \\ i'_d = I\left[-1.118\cos\omega t - 0.588\sin\omega t\right] \\ i'_e = I\left[1.118\cos\omega t - 0.951\sin\omega t\right] \end{cases} \tag{14}$$

Also, it can be given as

$$\begin{cases} i'_b = 1.4678I\cos(\omega t - 0.2244\pi) \\ i'_c = 1.2631I\cos(\omega t - 0.8459\pi) \\ i'_d = 1.2631I\cos(\omega t + 0.8459\pi) \\ i'_e = 1.4678I\cos(\omega t + 0.2244\pi) \end{cases} \tag{15}$$

In this case, when open-circuit fault happens in one phase of stator windings, the currents in remaining healthy phases can be used to keep MMF_1 unchanged based on the principle of copper loss minimization.

4. Short-Circuit Fault-Tolerant Control

A fault-tolerant current control strategy for the five-phase FT-PMV machine under short-circuit condition will be discussed in this section. Compared with that under open-circuit fault condition, the torque ripple of the FT-PMV machine under short-circuit fault condition includes two parts. One is caused by the short-circuit current, while the other owes to the lack of the normal torque which can be considered as the open-circuit fault condition. Corresponding to short-circuit fault, the proposed remedial control strategy also has two aspects. First of all, the currents in healthy phases are used to compensate for the ripple of MMF caused by the short-circuit current. MMF can be kept at zero, so the torque pulsation can be eliminated. Then, open-circuit fault-tolerant control strategy is used to compensate for the lack of the normal torque. Finally, a new remedial control strategy can be derived by utilizing the vector sum of both fault-tolerant control strategies.

When phase-a is with short-circuit fault in stator windings, its short-circuit current can be written as

$$i_{af} = I_f \sin(\omega t - \theta) \tag{16}$$

where I_f is the amplitude of the short-circuit current, and θ is the angle between short-circuit current and back-EMF in faulty phase. The vector sum of the MMF can be written as

$$MMF_2 = Ni_{af} + aNi_b'' + a^2Ni_c'' + a^3Ni_d'' + a^4Ni_e'' \tag{17}$$

where i_{af} is the short-circuit current in phase-a. The healthy currents aim to compensate the torque ripple caused by the short-circuit current and a is $e^{(2/5\pi)j}$.

MMF_2 in (17) is kept to zero, so the total MMF of the FT-PMV machine is the sum of MMF_1 and MMF_2, which can be kept constant. Based on (16) and (17), (18) can be obtained as

$$\begin{cases} 0.309i_b'' - 0.809i_c'' - 0.809i_d'' + 0.309i_e'' + i_{af} = 0 \\ 0.951i_b'' + 0.588i_c'' + 0.588i_d'' + 0.951i_e'' = 0 \end{cases} \tag{18}$$

In (18), there are infinite results for the compensating currents without conditional constraints. In order to compensate for the ripple caused by the short-circuit current in phase-a, the healthy currents can be calculated. It can be assumed that

$$\begin{cases} i_b'' = x_b\cos(\omega t) + y_b\sin(\omega t) \\ i_c'' = x_c\cos(\omega t) + y_c\sin(\omega t) \\ i_d'' = x_d\cos(\omega t) + y_d\sin(\omega t) \\ i_e'' = x_e\cos(\omega t) + y_e\sin(\omega t) \end{cases} \tag{19}$$

where x_b, y_b, x_c, y_c, x_d, y_d, x_e and y_e are unknown quantities. The proposed control strategy aims to minimize the copper loss. So, the target function f can be given as

$$f(x_b, y_b, x_c, y_c, x_d, y_d, x_e, y_e) = ({x_b}^2 + {y_b}^2) + ({x_c}^2 + {y_c}^2) + ({x_d}^2 + {y_d}^2) + ({x_e}^2 + {y_e}^2) \tag{20}$$

According to (18) and (19), (21) can be obtained as

$$\begin{cases} 0.309x_b - 0.809x_c - 0.809x_d + 0.309x_e = -I_f\sin\theta \\ 0.309y_b - 0.809y_c - 0.809y_d + 0.309y_e = I_f\cos\theta \\ 0.951x_b + 0.588x_c - 0.588x_d - 0.951x_e = 0 \\ 0.951y_b + 0.588y_c - 0.588y_d - 0.951y_e = 0 \end{cases} \tag{21}$$

According to (21), the results of x_b, y_b, x_c, y_c, x_d, y_d, x_e and y_e are not unique, so it is necessary to use a target function to make the final result unique. Many results can minimize the result of (20) among all the results and they can be called the optimal results.

The compensation, which owes to the lack of the normal torque, can be considered as the open-circuit fault condition. The strategy has been discussed in Section 3 and described in (15). Based on (15) and (19), the final current control strategy for short-circuit fault can be given as

$$\begin{cases} i_b^* = i_b'' + i_b' \\ i_c^* = i_c'' + i_c' \\ i_d^* = i_d'' + i_d' \\ i_e^* = i_e'' + i_e' \end{cases} \tag{22}$$

The universal current control strategy can also be written as

$$\begin{cases} i_b^* = x_b\cos(\omega t) + y_b\sin(\omega t) + 1.4678I\sin(\omega t - 0.2244\pi) \\ i_c^* = x_c\cos(\omega t) + y_c\sin(\omega t) + 1.2631I\sin(\omega t - 0.8459\pi) \\ i_d^* = x_d\cos(\omega t) + y_d\sin(\omega t) + 1.2631I\sin(\omega t + 0.8459\pi) \\ i_e^* = x_e\cos(\omega t) + y_e\sin(\omega t) + 1.4678I\sin(\omega t + 0.2244\pi) \end{cases} \quad (23)$$

When the windings terminal of one phase is with short-circuit fault, other phases can continue to operate. The rotor speed of 200 rpm will be discussed as example. The short-circuit current is presented in Figure 3. Based on the back-EMF and the short-circuit current, I_f and θ in (16) can be obtained. Then, x_b, y_b, x_c, y_c, x_d, y_d, x_e and y_e in (21) can be calculated based on (20) and (21). I_f and θ will change for different machines and different operating conditions, and then x_b, y_b, x_c, y_c, x_d, y_d, x_e and y_e can be calculated. Based on the FEA results, I_f and θ can be obtained as follows

$$\begin{cases} I_f = 8.04A \\ \theta = 1.42\pi \end{cases} \quad (24)$$

According to (20), (21) and (24), a built-in function of MATLAB based on SQP algorithm can be used to achieve the optimal results of x_b, y_b, x_c, y_c, x_d, y_d, x_e and y_e. Then, the currents compensating for the ripple of MMF caused by the short-circuit current are given as

$$\begin{cases} i_b^{''} = -1.60\cos(\omega t) + 0.43\sin(\omega t) \\ i_c^{''} = 4.19\cos(\omega t) - 1.12\sin(\omega t) \\ i_d^{''} = 4.19\cos(\omega t) - 1.12\sin(\omega t) \\ i_e^{''} = -1.60\cos(\omega t) + 0.43\sin(\omega t) \end{cases} \quad (25)$$

The final remedial currents of the short-circuit current control strategy can be given as

$$\begin{cases} i_b^* = -1.60\cos(\omega t) + 0.43\sin(\omega t) + 1.4678I\sin(\omega t - 0.2244\pi) \\ i_c^* = 4.19\cos(\omega t) - 1.12\sin(\omega t) + 1.2631I\sin(\omega t - 0.8459\pi) \\ i_d^* = 4.19\cos(\omega t) - 1.12\sin(\omega t) + 1.2631I\sin(\omega t + 0.8459\pi) \\ i_e^* = -1.60\cos(\omega t) + 0.43\sin(\omega t) + 1.4678I\sin(\omega t + 0.2244\pi) \end{cases} \quad (26)$$

5. Validation

5.1. Simulation

First, Figure 4 shows the FEA-based currents and output torque waveforms of the normal five-phase FT-PMV machine. It can be seen that the current waveforms are sinusoidal and the corresponding average torque and torque ripple factor of the machine drive are 3.6 Nm and 24.5%, respectively. In [33], the back-EMF harmonics, which interact with the sinusoidal phase current waveforms to generate torque ripple for a five-phase machine, are 9th, 11th, 19th, 21st, etc. Hence, the output torque ripple mainly results from the cogging torque.

Second, the fault currents and torque under short-circuit fault condition are shown in Figure 5. It can be seen from Figure 5b that the torque has high pulsation under fault conditions.

Third, the fault-tolerant performance with the current control strategy in (26) is shown in Figure 6 under single-phase short-circuit fault. The fault-tolerant currents are presented in Figure 6a. The corresponding average torque and torque ripple factor of the machine drive are 3.5 Nm and 37.0%, respectively. By comparing the torque performances at normal and remedial operations, it can be found that torque ripple is minimized and the torque performance is maintained when the strategy is adopted. Since the low-order harmonics in (23) are neglected, the FT-PMV machine at fault-tolerant operation has relatively significant torque ripple.

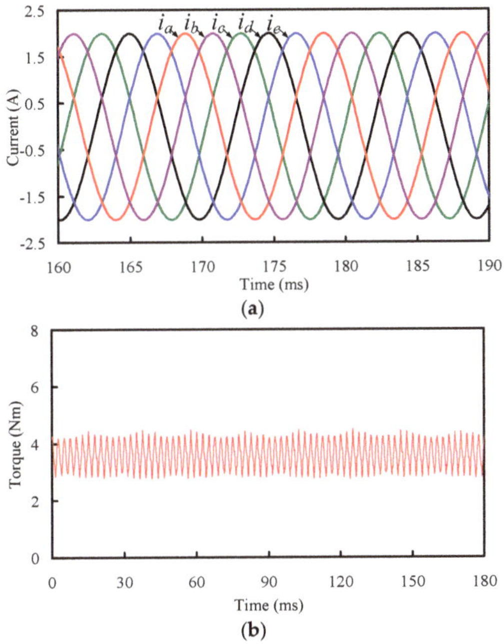

Figure 4. Performance under healthy condition: (**a**) Currents; (**b**) Output torque.

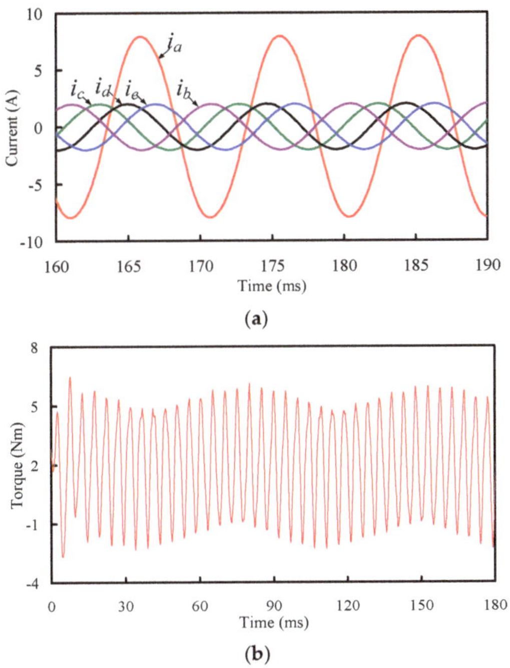

Figure 5. Performance under short-circuit fault condition: (**a**) Currents; (**b**) Output torque.

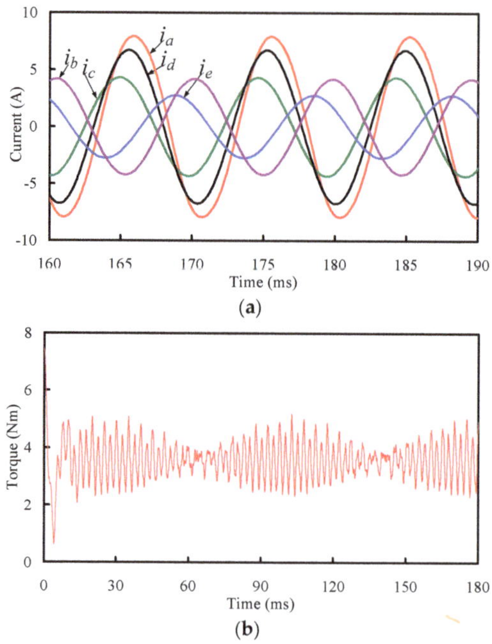

Figure 6. Performance under fault-tolerant condition: (**a**) Currents; (**b**) Output torque.

In addition, output torque and torque ripple for the five-phase FT-PMV machine under healthy, fault and fault-tolerant conditions are compared in Table 2. It can be concluded that torque ripple under fault-tolerant operation can be minimized with the proposed strategy. Table 3 compares the compositions of harmonics for the proposed method and those presented methods in [28,29].

Table 2. Torque performance under various conditions.

Torque	Normal	Fault	Remedial
Maximum torque (Nm)	4.5	6.5	5.0
Minimum torque (Nm)	2.7	−2.7	2.3
Torque ripple (%)	24.5	242.1	37.0

Table 3. Composition of harmonics.

Case	Fundamental	Higher Order
Method in Ref. [28]	√	√
Method in Ref. [29]	√	√
Proposed method	√	×

5.2. Experiment

In order to validate the effectiveness of the theoretical analysis, a five-phase FT-PMV machine was designed and built. For this experiment, a digital signal processing TMS320F2812 is used to implement the control. A separately excited dc generator is used as the variable load. To measure the torque of the proposed machine drive, a transient torque transducer is mounted between the five-phase FT-PMV machine and the dc generator. Moreover, the currents are sensed by the Hall-effect sensors and the position signal is obtained by the optical encoder with an accuracy of 2048 counts per revolution. Experimental setup of the five-phase FT-PMV machine is presented in Figure 7.

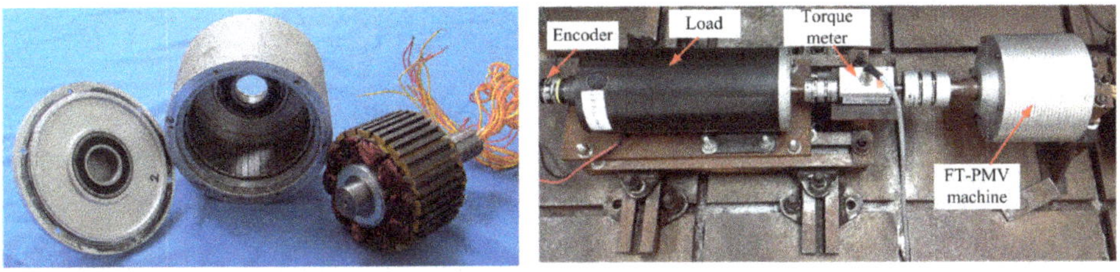

Figure 7. Experimental setup.

The block diagram of the fault-tolerant control scheme is illustrated in Figure 8, in which ω_{ref} is the reference speed, ξ is the position angle feedback from the machine. The currents i^*_b, i^*_c, i^*_d and i^*_e of the short-circuit fault-tolerant control strategy are calculated by using (23). The signals in the gate driver for a five-phase inverter come from a hysteresis current controller.

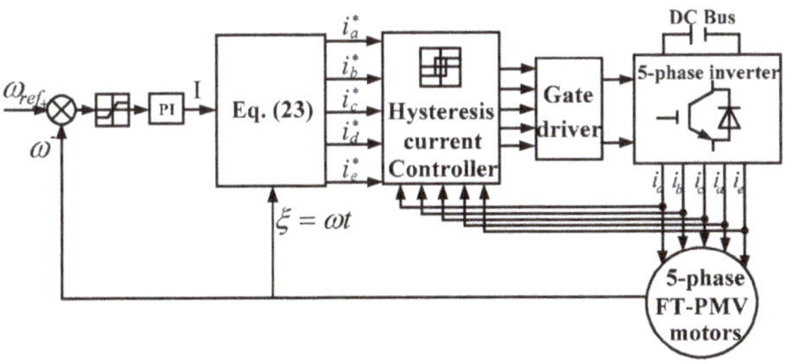

Figure 8. Block diagram of proposed fault-tolerant control scheme.

Moreover, the short-circuit current of the fault phase and the back-EMFs of other healthy phases are shown in Figure 9, verifying the fault-tolerant capability of the FT-PMV machine.

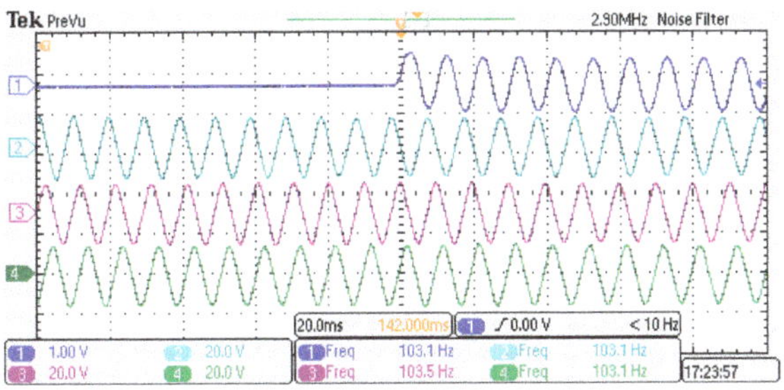

Figure 9. Measured short-circuit phase current (trace 1) and its adjacent phase back-EMFs (traces 2–4) (10 ms/div, 10 A/div, 20 V/div).

Furthermore, Figure 10a shows the output torque and the currents of the FT-PMV machine under healthy conditions, while those under the fault-tolerant condition are shown in Figure 10b (the second trace is the current of the short-circuit fault phase). By comparing the measured torque and current waveforms at normal and fault-tolerant operations, it can be concluded that the proposed fault-tolerant control strategy can maintain the torque capability and minimize the torque ripple under the short-circuit fault condition.

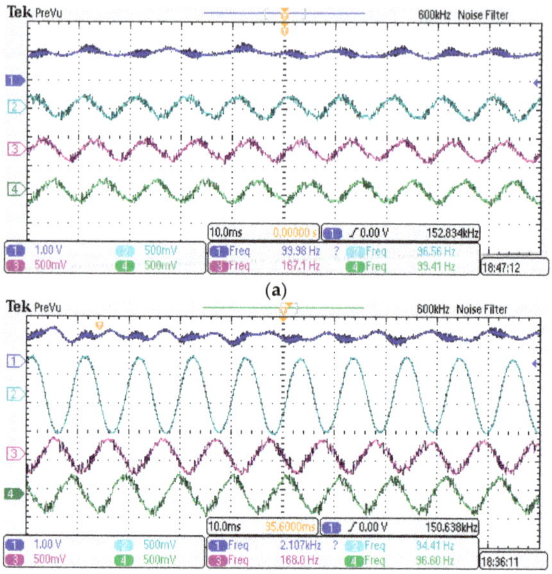

(a)

(b)

Figure 10. Measured torque (trace 1) and currents (traces 2–4). (10 ms/div, 2 Nm/div, 5 A/div): (**a**) Normal condition; (**b**) Fault-tolerant condition.

Figure 11 shows the dynamic responses of the speed and current under sudden changes in load torque when the machine operates in the fault-tolerant mode. It can be seen that the speed regulation is very good.

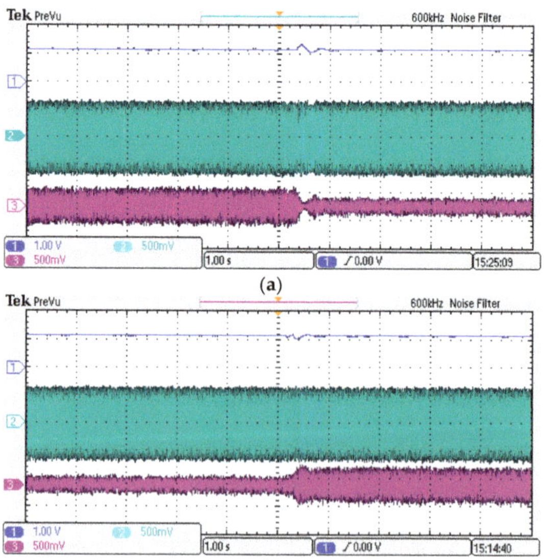

(a)

(b)

Figure 11. Measured speed (trace 1), short-circuit current (trace 2) and normal current (lower trace 3) responses at a sudden change of load under fault-tolerant condition (1 s/div, 160 rpm/div, 5 A/div): (**a**) Load decrease; (**b**) Load increase.

6. Comparisons and Discussion

Compared to these existing methods, the proposed method has obvious difference. First, the expression of the proposed remedial currents is simpler than these previous methods. In fact, the calculation methods in [28,29] are similar to that in [27]. Also, the fault in [31] occurs in the inverter, which is different from the fault discussed in the paper. So, in this section, only the method proposed in [27] is compared as an example.

Applying the method in [27] to the proposed sine back-EMF machine, the final expression can be derived as

$$\begin{bmatrix} i_a \\ i_b \\ i_c \\ i_d \\ i_e \end{bmatrix} = \begin{bmatrix} -i_f \\ \frac{2.5I-8.4\cos(wt)\cos(wt-1.42\pi)}{10-5\cos^2(wt)}[4\cos(wt-\frac{2}{5}\pi)+\cos(wt)] \\ \frac{2.5I-8.4\cos(wt)\cos(wt-1.42\pi)}{10-5\cos^2(wt)}[4\cos(wt-\frac{4}{5}\pi)+\cos(wt)] \\ \frac{2.5I-8.4\cos(wt)\cos(wt-1.42\pi)}{10-5\cos^2(wt)}[4\cos(wt+\frac{4}{5}\pi)+\cos(wt)] \\ \frac{2.5I-8.4\cos(wt)\cos(wt-1.42\pi)}{10-5\cos^2(wt)}[4\cos(wt+\frac{2}{5}\pi)+\cos(wt)] \end{bmatrix} \quad (27)$$

Its waveform plots are shown as Figure 12. The FFT of the currents is shown as Figure 13. Figure 14 shows the torque under fault-tolerant condition. The corresponding average torque and torque ripple factor of the machine drive are 3.6 Nm and 44.8%, respectively. It is obvious that both the proposed method and the method in [27] can maintain the torque performance, whereas the method in [27] leads to higher order harmonics. In addition, for a sinusoidal back-EMF machine, the lower harmonic in currents offers some important advantages. For instance, based on the principle of instantaneous electrical input power and mechanical output balance condition, the reactive component in instantaneous power produced by pulsating torque can be reduced by eliminating higher harmonics in the currents. At the same time, it can reduce iron loss.

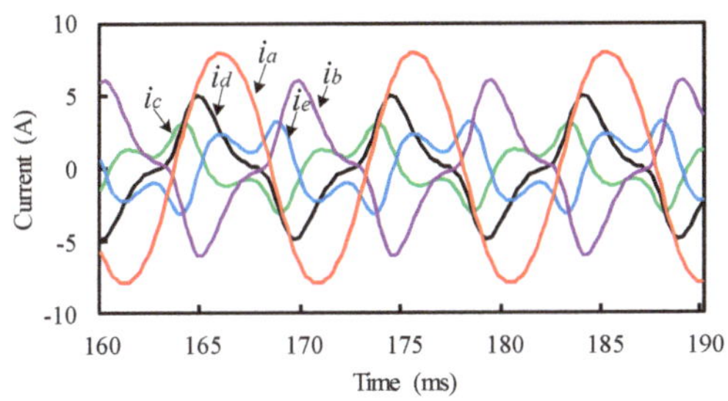

Figure 12. Currents under fault-tolerant conditions.

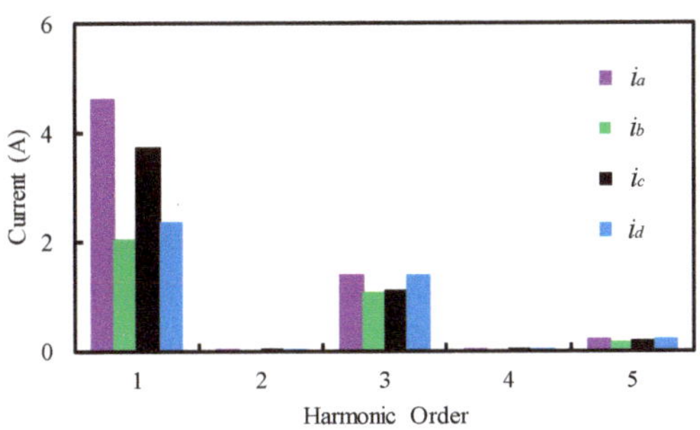

Figure 13. FFT of the currents.

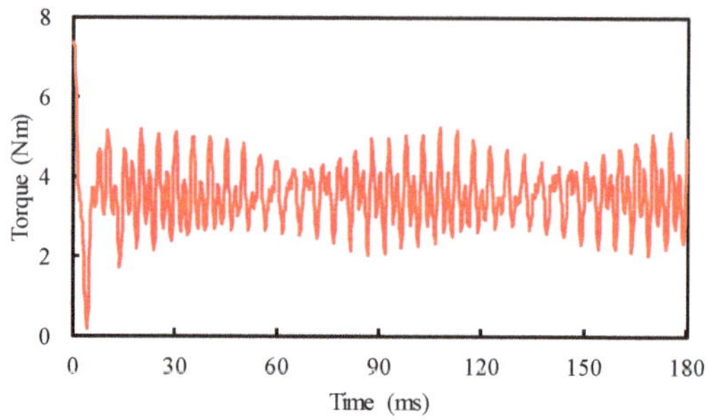

Figure 14. Torque under fault-tolerant conditions.

7. Conclusions

In this paper, a remedial current control strategy for a five-phase FT-PMV machine under short-circuit fault conditions has been proposed, which is based on the cooper loss minimization theory. The remedial currents have been calculated to minimize output torque ripple and machine copper losses. A five-phase FT-PMV machine has been built to verify the proposed current control strategy. It exhibits good torque performance when operating under fault-tolerant conditions. The proposed control strategy has been validated by the simulation and experimental results. The simulation and the experiment show that the proposed control strategy can reduce copper losses, maintain torque performance and minimize torque ripple during a short-circuit fault. Hence, it has a bright future in high-reliability and high torque density applications.

Acknowledgments: This work was supported by the National Natural Science Foundation of China (Project 51422702), by the Qing Lan Project of Jiangsu Province, and by the Priority Academic Program Development of Jiangsu Higher Education Institutions.

Author Contributions: Chenyu Gu and Wenxiang Zhao proposed the remedial control strategy; Chenyu Gu and Bufeng Zhang performed the simulations and the experiments.

Conflicts of Interest: The authors declare no conflict of interest.

References

1. Cheng, M.; Sun, L.; Buja, G.; Song, L. Advanced electrical machines and machine-based systems for electric and hybrid vehicles. *Energies* **2015**, *8*, 9541–9564. [CrossRef]
2. Jiang, X.; Huang, W.; Cao, R.; Hao, Z.; Li, J. Analysis of a dual-winding fault-tolerant permanent magnet machine drive for aerospace applications. *IEEE Trans. Magn.* **2015**, *51*, 8114704. [CrossRef]
3. Zhao, J.; Gao, X.; Li, B.; Liu, X.; Guan, X. Open-phase fault tolerance techniques of five-phase dual-rotor permanent magnet synchronous motor. *Energies* **2015**, *8*, 12810–12838. [CrossRef]
4. Bianchi, N.; Bolognani, S.; Zigliotto, M.; Zordan, M. Innovative remedial strategies for inverter faults in IPM synchronous motor drives. *IEEE Trans. Energy Convers.* **2003**, *18*, 306–314. [CrossRef]
5. Mendes, A.M.S.; Cardoso, A.J.M. Fault-tolerant operating strategies applied to three-phase induction-motor drives. *IEEE Trans. Ind. Electron.* **2006**, *53*, 1807–1817. [CrossRef]
6. Ding, W.; Hu, Y.; Wu, L. Investigation and experimental test of fault-tolerant operation of a mutually coupled dual three-phase SRM drive under faulty conditions. *IEEE Trans. Power Electron.* **2015**, *30*, 6857–6872. [CrossRef]
7. Wang, Z.; Chen, J.; Cheng, M.; Zheng, Y. Fault tolerant control of paralleled voltage-source inverters fed PMSM drives. *IEEE Trans. Ind. Electron.* **2015**, *62*, 4749–4760. [CrossRef]
8. Levi, E.; Bojoi, R.; Profumo, F.; Toliyat, H.; Williamson, S. Multiphase induction motor drives—A technology status review. *IET Electr. Power Appl.* **2007**, *1*, 489–516. [CrossRef]
9. Baek, S.K.; Shin, H.U.; Kang, S.Y.; Park, C.S.; Lee, K.B. Open fault detection and tolerant control for a five phase inverter driving system. *Energies* **2016**, *9*, 355. [CrossRef]

10. Wang, D.; Liu, C.; Li, G. An optimal integrated control scheme for permanent magnet synchronous generator-based wind turbines under asymmetrical grid fault conditions. *Energies* **2016**, *9*, 307. [CrossRef]
11. Duran, M.J.; Barrero, F. Recent advances in the design, modeling, and control of multiphase machines. *IEEE Trans. Ind. Electron.* **2016**, *63*, 459–468. [CrossRef]
12. Mohammadpour, A.; Parsa, L. A unified fault-tolerant current control approach for five-phase PM motors with trapezoidal back EMF under different stator winding connections. *IEEE Trans. Power Electron.* **2013**, *28*, 3517–3527. [CrossRef]
13. Prieto, B.; Martínez-Iturralde, M.; Fontán, L.; Elosegui, I. Fault-tolerant permanent magnet synchronous machine-phase, pole and slot number selection criterion based on inductance calculation. *IET Electr. Power Appl.* **2015**, *9*, 138–145. [CrossRef]
14. Liu, G.; Qu, L.; Zhao, W.; Chen, Q.; Xie, Y. Comparison of two SVPWM control strategies for give-phase fault-tolerant permanent-magnet motor. *IEEE Trans. Power Electron.* **2016**, *31*, 6621–6630. [CrossRef]
15. Zhao, W.; Cheng, M.; Hua, W.; Jia, H.; Cao, R. Back-EMF harmonic analysis and fault-tolerant control of flux-switching permanent-magnet machine with redundancy. *IEEE Trans. Ind. Electron.* **2011**, *58*, 1926–1935. [CrossRef]
16. Chen, Q.; Liu, G.; Zhao, W.; Sun, L.; Shao, M.; Liu, Z. Design and comparison of two fault-tolerant interior-permanent-magnet motors. *IEEE Trans. Ind. Electron.* **2014**, *61*, 6615–6623. [CrossRef]
17. Liu, G.; Yang, J.; Zhao, W.; Ji, J.; Chen, Q.; Gong, W. Design and analysis of a new fault-tolerant permanent-magnet vernier machine for electric vehicles. *IEEE Trans. Magn.* **2012**, *48*, 4176–4179. [CrossRef]
18. Dwari, S.; Parsa, L. An optimal control technique for multiphase PM machines under open-circuit faults. *IEEE Trans. Ind. Electron.* **2008**, *55*, 1988–1995. [CrossRef]
19. Dwari, S.; Parsa, L. Fault-tolerant control of five-phase permanent-magnet motors with trapezoidal back EMF. *IEEE Trans. Ind. Electron.* **2011**, *58*, 476–485. [CrossRef]
20. Tani, A.; Mengoni, M.; Zarri, L.; Serra, G.; Casadei, D. Control of multiphase induction motors with an odd number of phases under open-circuit phase faults. *IEEE Trans. Power Electron.* **2012**, *27*, 565–577. [CrossRef]
21. Kestelyn, X.; Semail, E. A vectorial approach for generation of optimal current references for multiphase permanent magnet synchronous machines in real time. *IEEE Trans. Ind. Electron.* **2011**, *58*, 5057–5065. [CrossRef]
22. Zhao, W.; Cheng, M.; Chau, K.T.; Cao, R.; Ji, J. Remedial injected-harmonic-current operation of redundant flux-switching permanent-magnet motor drives. *IEEE Trans. Ind. Electron.* **2013**, *60*, 151–159. [CrossRef]
23. Mohammadpour, A.; Sadeghi, S.; Parsa, L. A generalized fault-tolerant control strategy for five-phase PM motor drives considering star, pentagon, and pentacle connections of stator windings. *IEEE Trans. Ind. Electron.* **2014**, *61*, 63–75. [CrossRef]
24. Salehifar, M.; Salehi Arashloo, R.; Moreno-Eguilaz, M.; Sala, V.; Romeral, L. Observer-based open transistor fault diagnosis and fault-tolerant control of five-phase permanent magnet motor drive for application in electric vehicles. *IET Power Electron.* **2015**, *8*, 76–87. [CrossRef]
25. Bianchi, N.; Bolognani, S.; Pre, M. Strategies for the fault-tolerant current control of a five-phase permanent-magnet motor. *IEEE Trans. Ind. Appl.* **2007**, *43*, 960–970. [CrossRef]
26. Mohammadpour, A.; Mishra, S.; Parsa, L. Fault-tolerant operation of multiphase permanent-magnet machines using iterative learning control. *IEEE J. Emerg. Sel. Top. Power Electron.* **2014**, *2*, 201–211. [CrossRef]
27. Mohammadpour, A.; Parsa, L. Post-fault control technique for multi-phase PM motor drives under short-circuit faults. In Proceedings of the 2013 Twenty-Eighth Annual IEEE Applied Power Electronics Conference and Exposition (APEC), Long Beach, CA, USA, 17–21 March 2013.
28. Mohammadpour, A.; Parsa, L. Global fault-tolerant control technique for multi-phase permanent-magnet machines. *IEEE Trans. Ind. Appl.* **2015**, *51*, 178–186. [CrossRef]
29. Sen, B.; Wang, J. Stationary frame fault-tolerant current control of polyphase permanent-magnet machines under open-circuit and short-circuit faults. *IEEE Trans. Power Electron.* **2016**, *31*, 4684–4696.
30. Sedrine, E.B.; Ojeda, J.; Gabsi, M.; Slama-Belkhodja, I. Fault-tolerant control using the GA optimization considering the reluctance torque of a five-phase flux switching machine. *IEEE Trans. Energy Convers.* **2015**, *30*, 927–938. [CrossRef]
31. Nguyen, N.; Meinguet, F.; Semail, E.; Kestelyn, X. Fault-tolerant operation of an open-end winding five-phase PMSM drive with short-circuit inverter fault. *IEEE Trans. Ind. Electron.* **2016**, *63*, 595–605. [CrossRef]

32. Jen-Ren, F.; Lipo, T.A. Disturbance-free operation of a multiphase current-regulated motor drive with an opened phase. *IEEE Trans. Ind. Appl.* **1994**, *30*, 1267–1274. [CrossRef]
33. Atallah, K.; Wang, J.; Howe, D. Torque-ripple minimization in modular permanent-magnet brushless machines. *IEEE Trans. Ind. Appl.* **2003**, *39*, 1689–1695. [CrossRef]

energies MDPI

Article

Cost Analysis Method for Estimating Dynamic Reserve Considering Uncertainties in Supply and Demand

Kyung-bin Kwon [1], Hyeongon Park [2,*], Jae-Kun Lyu [3] and Jong-Keun Park [4]

1 Electrical Engineering, Korea Military Academy (KMA), Seoul 01850, Korea; kwon8908kr@gmail.com
2 Department of Statistics Institute of Engineering Research, Seoul National University, Seoul 08826, Korea
3 Korean Electric Power Corporation (KEPCO), Naju 58217, Korea; handyjack@naver.com
4 Department of Electrical and Computer Engineering, Seoul National University, Seoul 08826, Korea; parkjk@snu.ac.kr
* Correspondence: hyeongon@snu.ac.kr; Tel.: +82-2-880-6898

Academic Editor: Paolo Mercorelli
Received: 8 August 2016; Accepted: 12 October 2016; Published: 20 October 2016

Abstract: The use of appropriate hourly reserve margins can maintain power system security by balancing supply and demand in the presence of errors in the forecast demand, generation outages, or errors in the forecast of wind power generation. Because the cost of unit commitment increases with larger reserve margins, cost analysis to determine the most economical reserve margin is an important issue in power system operation. Here, we define the "short-term reliability of balance" and describe a method to determine the reserve margin based on the short-term reliability of balance. We describe a case study, in which we calculate the reserve margin using this method with various standards of short-term reliability of balance. A cost analysis is then performed to determine the most economic standard, and a comparison between our method and a conventional method is carried out. The results show that our method with an economic short-term reliability of balance enables more reliable and efficient operation of the power system. Moreover, with an hourly reserve margin, we show that an increase in wind power generation can result in a significant decrease in the operating cost, which makes wind power generation economically viable.

Keywords: dynamic reserve; unit commitment; reliability; wind power generation; cost analysis; power system operation; uncertainty; generation outage

1. Introduction

One of the most important characteristics of power systems is that demand and supply be balanced at all times. An unexpected change in demand or supply results in fluctuations in the system frequency, and if balance between generation and load is lost, it can lead to frequency deviations, a loss of synchronization between generators, or even a blackout of the entire power system. For this reason, the system operator (SO) should provide operating reserve to account for unexpected fluctuations in demand or supply, or problems that may occur in the electrical grid.

The appropriate operating reserve margin is related to both the reliability and the cost associated with maintaining it. A larger operating reserve margin provides greater ability to accommodate unexpected losses of generators or increases in demand. However, a larger reserve margin also leads to higher generation costs of the reserve generators. Because operating reserve should be available at all times, an excessive operating reserve margin will lead to significant additional costs.

A great deal of discussion has focused on appropriate operating reserve margins, including methods to estimate the frequency regulation reserve [1,2] and consideration of contingencies [3], as well as operating standards for optimal reserve [4,5]. These reports focused on operating reserve margins that apply during time periods greater than one day.

In [1–5], the main uncertainty considered is generation outage, which could result in an insufficient supply of power compared with demand. However, unexpected variations in demand may occur over much shorter timescales, and the required reserve margin will differ for each time period. Therefore, for the SO to manage the power system more economically, the reserve margin should be allowed to vary with time, rather than being fixed for a constant rate of demand all day. Both [6,7] consider the uncertainties in probabilistic outage with expected energy not served (EENS) and estimate an appropriate operating reserve margin that satisfies risk criteria such as the loss of load probability (LOLP). Furthermore, [8,9] also consider the load forecasting error to optimize the spinning reserve requirements while maintaining certain LOLP criteria. In [10], the tradeoff between unit capacity and average production cost is analyzed and the spinning reserve margin is optimized while considering the schedule cost and expected interruption cost. In [11], some indices of credibility assessment are proposed to predict the forecasting error and probabilistic load forecast, which could support decision-making in power system operation. Nonetheless, these papers did not consider instability in supply caused by wind power generation, which could form a significant proportion of the power supply and have a more variable power output than conventional generators.

In addition to the uncertainties caused by generation outage, many investigations have considered uncertainties in wind power forecasting for power systems with a large portion of wind power generation [12–16]. If there is high penetration of wind generation in a power system, the forecasting error cannot be ignored regarding the uncertainties in the supply of power. In [17], wind power uncertainty is modeled explicitly using scenarios, and reserve requirements are enforced on the scenarios to account for the limited number of scenarios represented. A new type of unit commitment, which considers the interruptible load as a reserve to handle the increased uncertainty with wind power, is also suggested in [18]. Both [19,20] also consider wind power generation as an operating risk and propose an algorithm based on non-parametric kernel density estimation and a model for a short-term future forecasting using a conditional probability approach to reduce the risk. In [21,22], wind power uncertainty is modeled as Gaussian distribution and finite-state Markove chain, respectively. Then reserve capacity which could minimize the expected cost is calculated based on each model. However, although these studies considered the uncertainties in wind power generation, they did not fully incorporate other major uncertainties arising from load forecasting and generation outage.

Finally, methods to calculate hourly reserve margins considering the uncertainties of load forecast, generation outage, and wind power generation are researched [23–27]. In [23], the required operating balance reserve is estimated with consideration given to the load forecast error and unavailable capacity due to unit outage in a time horizon of 1–48 h. Both [24,25] generate probability functions of conventional generation outage, load forecasting uncertainty, and wind power forecast uncertainty, and the reserve margin is estimated considering the trade-off between risk and reserve cost. In [26], a new random variable which consists of uncertainties of load shedding and wind curtailment is defined, and conditional value-at-risk (CVaR) is adopted to determine risk reserve requirement. In [27], the spinning reserve margin is determined from a defined term called the short-term reliability of balance, which is calculated from the probabilistic models of uncertainties.

Nevertheless, most papers did not consider the relation between unit commitment and probabilistic generation outage modeling. The most important thing in modeling the uncertainties while considering uncertainties like demand forecasting, generation outage, and wind power forecasting is that they are related and cause differences in the overall demand and supply of power. Specifically, the generation outage should be considered carefully, because the volume of generating output planned to supply the system that cannot be used because of the outage is strongly linked to the result of unit commitment and economic dispatch. The unit commitment and economic dispatch should also be calculated while considering the uncertainties in generation outage when they are dealing with the risk of uncertainties combined with demand forecasting and wind power generation error.

Consequently, algorithms and calculation methods should be developed to consider these uncertainties in the unit commitment and estimating operating reserve margin. Moreover, any estimation of the operating reserve margin should consider not only the risk that an unbalance of supply and demand occurs but also the overall operating cost. Cost analysis should take into account both the generation cost, calculated from the unit commitment and dispatch, and the load-shedding cost, which is related to the risk that the uncertainties would cause, because a greater reserve margin would decrease the risk but increase the generation cost. Therefore, the unit commitment and reserve margin should be estimated to minimize operating costs while maintaining a certain level of risk.

This paper describes a unit commitment method that considers the uncertainties caused by load forecasting, generation outage, and wind power forecasting. An algorithm that considers the generation outage more precisely is included. This algorithm calculates the economic dispatch and estimated reserve margin based on the overall expected cost and amount of risk, which is defined as the short-term reliability of balance. The major contributions of this paper are as follows: (1) precise generation outage modelling is formulated from unit commitment result based on the premise that volume of generation outage occurred equals the volume planned in unit commitment; (2) a new algorithm is proposed that can perform unit commitment with minimum reserve margin constraints in order to maintain the reliability standard in all time period, where the minimum reserve margin constraints are calculated with consideration of demand forecasting and wind power forecasting errors as well as generation outage; and (3) a case study is carried out to identify the usefulness of the proposed method in terms of improved reliability and cost-effectiveness, compared to the conventional method.

The remainder of this paper is organized as follows: Section 2 introduces the probability distribution formulations of uncertainty factors. Demand forecast errors, generation outage, and wind power generation forecast errors are considered in this section along with the short-term reliability of balance to consider these three errors at the same time. In Section 3, the definition of short-term reliability of balance and the algorithm that is used to estimate the dynamic reserve is given. Section 4 describes the formulation of the operating cost, which includes the generation and load-shedding costs. Section 5 describes a case study using the reserve and cost calculation methods, and Section 6 concludes the paper.

2. Probability Distribution Model of Uncertainties in Demand and Supply

2.1. Probability Distribution Model of Demand Forecast Error

The hourly reserve requirement is closely related to the uncertainty in demand, which can be represented by the error associated with the forecast load, and hence by a probability distribution model for demand.

The demand at time t for case s can be calculated as follows [28]:

$$AD_t^s = ED_t^s + L_t \tag{1}$$

where ED_t^s is the forecast demand at time t for case s and L_t is the error in that forecast demand. Here, we assume that the forecast error in the demand follows a normal distribution with a zero mean [29].

Clearly, a larger error corresponds to a larger standard deviation in the demand. The error varied because the demand pattern of customers differed on both an hourly basis and a seasonal basis. These errors can be described using a statistical approach with data from the forecast peak demand and the actual peak demand, i.e.,

$$TE_t = \frac{1}{C_t^s} \sum_s \frac{|AD_t^s - ED_t^s|}{AD_t^s} \times 100\ (\%) \tag{2}$$

Assuming that the probability distribution of the demand is normally distributed [30], the difference between the actual demand and the expected demand $AD_t^s - ED_t^s$, will also follow a normal distribution. Therefore, the absolute value of the difference, $| AD_t^s - ED_t^s |$, can be described by the positive-valued half of a zero-mean normal distribution. Assuming a sufficiently long time period and the values of AD_t^s being similar to each other, the mean of this half-normal distribution can be calculated as follows [31]:

$$E(AD_t^s - ED_t^s) = \sqrt{\frac{2}{\pi}}\sigma_t = \frac{1}{N_t^s}\sum_s |AD_t^s - ED_t^s| = AD_t^s \times \frac{TE_t}{100} \tag{3}$$

Assuming $AD_t^s - ED_t^s$, we find

$$\sqrt{\frac{2}{\pi}}\sigma_t = ED_t^s \times \frac{TE_t}{100} \tag{4}$$

The standard deviation in the demand forecast error during hour t is therefore

$$\sigma_t = \sqrt{\frac{\pi}{2}} \times ED_t^s \times \frac{TE_t}{100} \tag{5}$$

and we can conclude that the probability distribution of the hourly forecast error in the demand is normally distributed with a standard deviation given by (5).

2.2. Probability Distribution Model of Generator Outages

The power output of each generator differs from hour to hour due to unit commitment and variations in the load. The uncertainty during peak times, when a large number of generators are online, is larger than that for off-peak times. Accordingly, a different probability distribution is required during peak times and off-peak times. The probability distribution model of generator outages should be calculated using the forced outage rate (FOR) for each generator, which is defined as the probability that the unit will be unavailable under similar conditions [32].

Because the probability distribution model is calculated hourly, the FOR for each generator is defined as the probability that the generator is not available in that hour. The probability that the generator i is available in that hour is therefore given by $1 - \text{FOR}_i$.

From the results of unit commitment, if a given generator is unavailable, then neither is the power that the generator was planned to provide. It follows that the dispatch schedule and planned power output of generators will change in response to an outage. Because the demand varies as a function of time, the available power will also vary, even though the FOR of generators is constant with the time. Consequently, the probability distribution of outages varied with time.

The time-dependent probability of generator outage G_t can be calculated as follows:

$$P(G_t) = \sum_j P(G_{t,j}) \tag{6}$$

subject to

$$P(G_{t,j}) = \prod_{N \in \mathfrak{F}_t} P_{FOR}^N \tag{7}$$

and

$$\mathfrak{F}_t = \{N \mid \sum_{N \in \mathfrak{F}_t} G_t^N = G_t\} \tag{8}$$

These expressions show that an outage G_t occurs when the sum of the outage generators is G_t and can be calculated as the product of the FOR of each generator.

Unlike the probability distribution for demand, the probability distribution for generator outage is discrete and positive-valued because the lack of capacity is determined from the combination of online generators and the planned generation output.

2.3. Probability Distribution Model of Forecast Error for Wind Power Generation

The expected wind power generation may differ on an hourly basis because of fluctuations in the wind velocity and direction. The resulting probability distribution errors in wind power generation vary depending on the expected wind power generation, which can be approximated by a gamma distribution. In order to build a detailed model of the probability distribution of the error in the wind power generation, the expected wind power can be divided into five classes for separate analysis [23].

The results for Classes 1 and 2 have relatively small wind power generation and tend to have a positive error, whereas Classes 4 and 5 have large expected wind power generation and tend to have negative errors in the probability distribution. It follows that if relatively low wind power generation is expected, it is more likely that the expected value will be an underestimate; i.e., the wind forecast error was typically positive. With higher wind power generation, the expected value tended to overestimate the forecast wind power; i.e., the forecast error in the wind power was typically negative.

To approximate the data using a gamma distribution, some modifications are required because a gamma distribution is defined only in the range greater than zero, and the probability density function has an average value of $k\theta$. Assuming that the average error of these probability distributions in each class is equal to zero, a modified gamma-like distribution was constructed using the basic probability density function with a translation of $-k\theta$ on the x-axis. The probability density functions for Classes 1, 2, and 3 were as follows:

$$f(x;k,\theta) = \frac{(x+k\theta)^{k-1}e^{-(x+k\theta)/\theta}}{\theta^k\Gamma(k)} \quad for\ x > 0,\ k > 0,\ \theta > 0 \tag{9}$$

Classes 4 and 5 exhibit probability distributions that are right-sided. These probability distributions are described by probability density functions that are reflected on the y-axis, so that the probability density functions for Classes 4 and 5 are as follows:

$$f(x;k,\theta) = \frac{(-x-k\theta)^{k-1}e^{(x+k\theta)/\theta}}{\theta^k\Gamma(k)} \quad for\ x > 0,\ k > 0,\ \theta > 0 \tag{10}$$

3. Algorithm for Dynamic Reserve Estimation

To calculate the hourly reserve using the probability distribution models of the demand forecast error, generation outage and wind power generation errors should be defined; i.e., the "standard" that determines the reserve margin should be defined. Here this "standard" is termed the "short-term reliability of balance" and is calculated from the probability that imbalance between supply and demand occurs in a given hour, considering the uncertainties in supply and demand.

If $P_{R,t}$ is the reserve for hour t, the probability that imbalance in supply and demand occurs is as follows:

$$P(P_{\mathrm{Im},t} > P_{R,t}) = P(G_t + L_t + W_t > P_{R,t}) = \sum_{x=0}^{P_{total}} [P(G_t = x) \times P(P_{\mathrm{Im},t} \mid L_t + W_t > P_{\mathrm{Im},t} - x)] \tag{11}$$

and the probability that balance of supply and demand is maintained during an hour, i.e., the short-term reliability of balance is as follows:

$$SR_t = [1 - P(P_{\mathrm{Im},t} > P_{R,t})] \times 100 \quad (\%) \tag{12}$$

This expression describes the probability that the power system maintains balance between supply and demand if the hourly reserve $P_{R,t}$ is allocated: the larger the hourly reserve, the larger the SR_t.

To maintain power balance, the short-term reliability of balance should be maintained above a given threshold every hour. If the short-term reliability of balance is determined, the minimum reserve requirement for each hour can be calculated as follows:

$$P_{MR,t} = \min(P_{R,t} \;:\; P(P_{\text{Im},t} > P_{R,t}) \;<\; 1 - \frac{SR_t}{100}) \tag{13}$$

To calculate the reserve requirements for each hour using (13), probability distributions of the error in the forecast demand, generation outages, and the error in the forecast wind power generation should be determined.

The probability distributions for generator outages are calculated using unit commitment, which provides hourly information on the online/offline status and planned generation for all generators. However, to perform unit commitment, the reserve allocation for each hour should be determined because of the constraints for reserve during each hour, which is part of the unit commitment process. Therefore, to calculate the reserve requirements, a recursive algorithm is used, as shown in Figure 1.

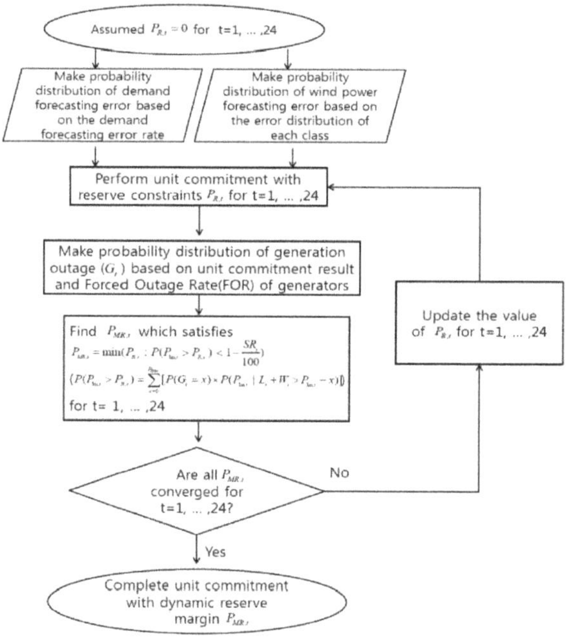

Figure 1. Solution method to determine the hourly spinning reserve margin.

The algorithm in Figure 1 consists of the following four steps:

Step 1. A probability distribution of the demand forecast error and wind generation forecast error is constructed using data for the hourly demand forecast error and the probability distribution for each class of wind farm.

Step 2. In the first iteration, the required hourly reserve is assumed to be zero, and unit commitment is carried out with no reserve. A probability distribution for generator outage is then constructed.

Step 3. The hourly reserve margin is calculated that satisfies (13), and we check for convergence.

Step 4. If convergence is not achieved, re-enter Step 2 using the calculated reserve requirements from Step 3.

During Step 1, unit commitment is carried out without considering outages of generators, or errors in the forecast demand and wind power. The probability distribution for outrages of generators can be calculated using data from unit commitment; however, these errors cannot be considered in the first iteration. During Step 3, the appropriate reserve margin is calculated using the probability distribution for generator outage, and the errors in the forecast demand and the wind power, which were calculated in Step 2. However, because the required reserve margin changes, the algorithm should be repeated until the hourly reserve margin converges and satisfies both unit commitment and the appropriate reserve margin calculated using (13). When the reserve margin converges, short-term reliability of balance should be satisfied.

4. Cost of Reserve

4.1. Operating Cost

Here, the operating cost is the overall daily cost to maintain power system security, which includes the cost of all conventional generators (i.e., the generation cost), the expected cost for imbalance between demand and supply, and the expected cost of outages (i.e., the load-shedding cost).

As the short-term reliability of balance varies, so will the reserve margin, which results in different generation and load-shedding costs. Therefore, to determine the short-term reliability of balance that is most economical, we should calculate the operating cost as a function of the reserve margin and the short-term reliability of balance for each hour.

4.2. Generation Costs

The generation cost is the cost of conventional generators over 24 h, determined using unit commitment. Unit commitment is the process that determines whether generators are online or offline, and is carried out hourly. The objective is to identify the unit combination that has the lowest cost based on cost functions of each generator, while satisfying the predetermined reserve margin.

The cost function of each generator is a quadratic function of the output of the generators, and is determined by the sum of production cost, which has unit cost coefficients a_i, b_i, and c_i, in each generator and start-up/shutdown cost in terms of generation start-up or shutdown, i.e.,

$$\text{(Generation Cost)} = \sum_{t=1}^{24} [\sum_{i=1}^{N_G} \{(a_i + b_i P_{i,t} + c_i P_{i,t}^2) \times MBtu + STC_i(u_{i,t-1}, u_{i,t})\}] \quad (14)$$

Because demand is met by both conventional generators and wind power, the generation cost decreases as the wind power penetration increases. Using the method described here to calculate the reserve, the generation cost increases as the short-term reliability of balance increases because the constraints of short-term reliability of balance in unit commitment increase the operating cost.

4.3. Load-Shedding Cost

When imbalance occurs between demand and supply, the SO should carry out load shedding to maintain the system frequency. Those customers affected by load shedding cannot use electricity, and monetary penalties result. The load-shedding cost is the expected sum of these monetary penalties and increases as more load shedding occurs.

Load shedding is determined by the reserve margin and the error in the forecast demand, generation outage, and wind power. The load shedding in hour t, is given by

$$P_{LS,t} = G_t + L_t + W_t - P_{R,t} \quad (15)$$

and occurs when demand exceeds the sum of the supply and the reserve margin. The probability that load shedding $P_{LS,t}$ occurs during time t can be calculated as follows:

$$P(P_{LS,t}) = P(G_t + L_t + W_t - P_{R,t} = P_{LS,t}) = \sum_{x=0}^{P_{Total}} [P(G_t = x) \times P(P_{\mathrm{Im},t} \mid L_t + W_t = P_{LS,t} - P_{R,t})] \quad (16)$$

Using (15) and (16), the load-shedding cost can be calculated as the product of $P_{LS,t}$, the probability of load shedding, and the outage cost, C_{OC}, i.e.,

$$(\text{Load Shedding Cost}) = \sum_{t=1}^{24} \sum_{P_{LS,t}=0}^{\infty} \{[\sum_{x=0}^{P_{Total}} \{P(G_t = x) \times P(P_{\mathrm{Im},t} \mid L_t + W_t = P_{LS,t} - P_{R,t})\}] \times P_{LS,t}\} \times C_{OC} \quad (17)$$

Because load shedding decreases as the reserve margin increases, the load-shedding cost also decreases. It follows that the load-shedding cost has an inverse relation with the short-term reliability of balance.

5. Case Study

5.1. Test System Description

A case study was carried out based on an IEEE 118 bus system with 54 generators. The demand and the forecast error in the demand for each hour are shown in Figure 2a. These data were from the summer of 2012 in Korea, and modified proportionally considering the generation capacity of the IEEE 118 bus system. The maximum generation capacity of wind farms was set to 500 MW; the hourly forecast wind power generation is shown in Figure 2b. The probability distributions describing the error in the forecast wind power generation of five classes were determined from the models shown in [23].

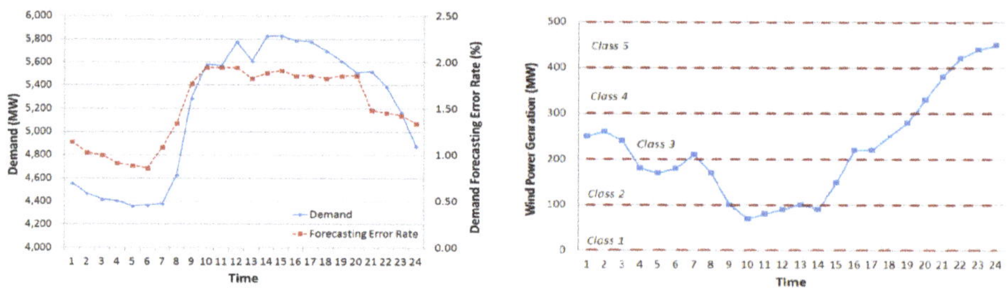

(a) (b)

Figure 2. Forecast demand and wind power generation used in the case study: (**a**) Forecast demand (labeled with left y-axis) and the error in the forecast demand (labeled with right y-axis) based on data for summer 2012 in Korea; (**b**) Hourly forecast wind power generation in a typical day and for Classes 1–5.

The FOR for the 54 generators in the IEEE 118 bus system was divided into 10 (i.e., Type 1 to Type 10), depending on the cost function, and the FOR of each type was set arbitrarily, as listed in Table 1.

The outage cost was assumed to be \$2000/MWh. The unit cost coefficients for each generator were taken from the generator data for the IEEE 118 bus system.

Table 1. Forced Outage Rate (FOR) for the 54 generators in the IEEE 118 bus system.

Type	Unit Number of Generators	Forced Outage Rate (%)
1	4, 5, 10, 29, 36, 40, 43, 44, 45	0.00015
2	7, 14, 16, 19, 22, 23, 26, 34, 35, 37, 47, 48, 51, 52, 53	0.0003
3	33, 41, 46, 49	0.00025
4	20, 21	0.0001
5	1, 2, 3, 6, 8, 9, 12, 13, 15, 17, 18, 31, 32, 38	0.0002
6	11, 39	0.00025
7	59, 61	0.00025
8	42, 50, 54	0.0001
9	27, 28	0.0002
10	30	0.0001

5.2. Costs and Optimal Short-Term Reliability of Balance

To determine the most economical short-term reliability of balance, we should minimize the operating cost. In this case study, the hourly reserve margin was calculated to provide a short-term reliability of balance in the range of 99.1%–99.8%. The operating cost is the sum of the generation cost and the load-shedding cost, which was calculated to determine the optimal short-term reliability of balance.

Figure 3 shows the calculated reserve over 24 h to provide a short-term reliability of balance of 99.2%, 99.4%, 99.6%, and 99.8% using (13). The reserve required during each hour increased as the short-term reliability of balance increased, resulting in an increase in the generation costs and a decrease in the load-shedding cost.

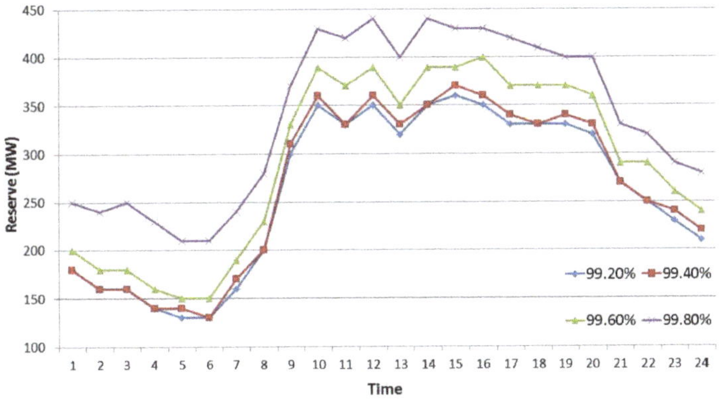

Figure 3. Hourly reserve margin with the short-term reliability of balance varied from 99.2% to 99.8%.

Using the methods described in Section 4, the generation and load-shedding costs can be calculated as a function of the short-term reliability of balance, as listed in Table 2. A minimum in the operating cost corresponded to a short-term reliability of balance of 99.6%, which can therefore be identified as the optimal case.

Figure 4 shows the costs as a function of the short-term reliability of balance. As the short-term reliability of balance increased, so did the generation cost, due to the larger required reserve margin. However, the load-shedding cost decreased as the short-term reliability of balance increased because the increase in the required reserve margin resulted in a smaller probability that demand exceeds supply. The operating cost was minimized for a short-term reliability of balance of 99.6%.

Table 2. The calculated costs for various short-term reliabilities of balance in the range 99.1%–99.8%.

Case	Generation Cost ($)	Load-Shedding Cost ($)	Operating Cost ($)
99.1%	1,738,000	11,907	1,749,907
99.2%	1,738,100	10,894	1,748,994
99.3%	1,739,700	9,954	1,749,654
99.4%	1,740,200	8,797	1,748,997
99.5%	1,740,900	7,836	1,748,736
99.6%	1,741,700	6,445	1,748,145
99.7%	1,744,000	4,550	1,748,550
99.8%	1,746,600	3,189	1,749,789

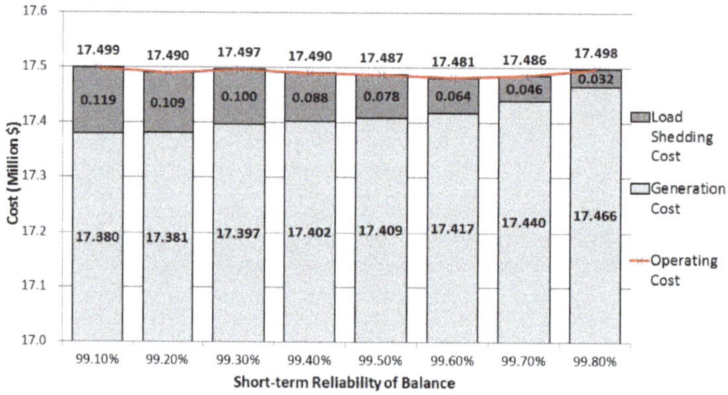

Figure 4. The generation costs and the cost for interruption as a function of the short-term reliability of balance.

5.3. Comparison of Our Method with a Constant Reserve Margin

To compare the method to calculate the reserve described here with the conventional method, a case study was carried out. The conventional method corresponds to a fixed reserve margin, and the proposed method to a reserve margin that varies from hour to hour with a short-term reliability of balance of 99.6% (i.e., our method). The constant reserve margin for the conventional method was determined using the average reserve margin for the proposed method, which gives the same overall reserve for both cases over 24 h.

Figure 5 shows the reserve allocation result for both cases. The reserve margin of the proposed method was varied to maintain a constant short-term reliability of balance. The case of the conventional method had a fixed reserve margin of 292 MW, which is the rounded value of the 24-h average reserve margin in the case of the proposed method.

Table 3 lists the average, maximum, and minimum reserve margin for 24 h using the data shown in Figure 6. Because of uncertainties in the forecast demand, the outages of generators and wind power error vary from hour to hour; the times for the maximum and minimum reserve margin depend on the combination of these three uncertain factors.

The short-term reliability of balance was calculated using (12), as shown in Figure 6. Unlike the proposed method, whereby the short-term reliability of balance was maintained >99.6%, the short-term reliability of balance with the conventional method decreased to 97.69%. Furthermore, the time at which the short-term reliability of balance was a minimum was not the same as that for peak demand, which implies that uncertainties in demand forecast error, generation outage, and wind power generation forecast error should be all considered to properly address imbalance.

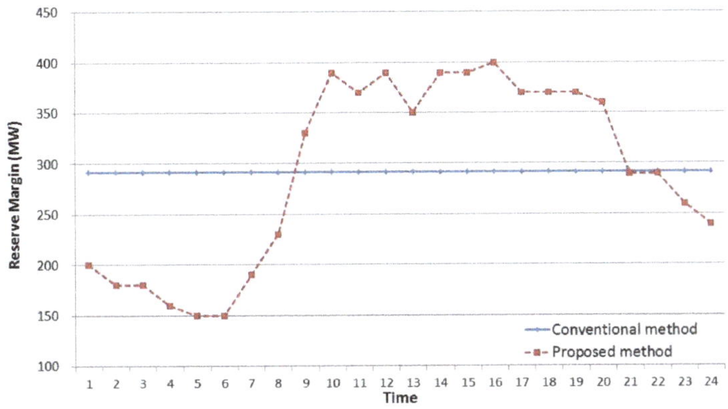

Figure 5. Hourly reserve margins for the conventional method (a fixed reserve margin of 292 MW) and the proposed method (based on a short-term reliability of balance of 99.6%).

Table 3. Maximum, minimum, and mean reserve margins.

Case	Maximum (MW)	Time	Minimum (MW)	Time	Mean (MW)
Conventional method	292	-	292	-	292
Proposed method	400	16	150	5, 6	291.7

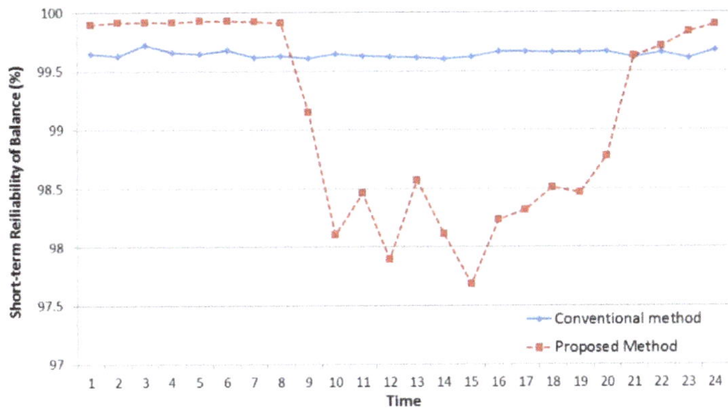

Figure 6. Hourly reliability based on the balance of supply and demand.

Table 4 lists the calculated costs for the conventional and proposed methods. The operating cost for the proposed method was less than that for the conventional method. The conventional method had a lower generation cost because it was better optimized for unit commitment. As shown in Figure 6, the proposed method required more reserve margin during peak hours, resulting in less optimized unit commitment and a larger generation cost; however, because the reserve allocation for the proposed method was based on the constant short-term reliability of balance, the proposed method has a much lower load-shedding cost than the conventional method. It follows that with the conventional method, a larger probability of imbalance existed between supply and demand. Because the operating cost is the sum of the generation and load-shedding costs, the proposed method (which was based on the calculation method described here) proved to be more cost-effective than the conventional method (a constant reserve margin, as with conventional methods).

Table 4. The calculated costs for the conventional method and the proposed method.

Case	Generation Cost ($)	Load-Shedding Cost ($)	Operating Cost ($)
Conventional method	1,737,300	11,681	1,748,981
Proposed method	1,741,700	6,445	1,748,145

5.4. *Costs for Different Wind Farms with the Same Short-Term Reliability of Balance*

As seen from (11), when the penetration of wind power increases, the corresponding decrease in the use of conventional generators leads to fewer generation outages and G_t would decrease, which decreases the required reserve margin. However, an increase in the penetration of wind power would increase the significance of the error in the forecast wind power generation and increase W_t, leading to an increase in reserve requirements. For this reason, it is possible that, as wind power penetration increases, the operating cost would increase because more reserve margin would be required to maintain a constant level of reliability. This assertion opposes claims that wind power generation is more cost-effective with increasing penetration. Therefore, the overall cost should be analyzed to identify if an increase in wind power penetration would be less cost-effective due to an increase in the operating cost. In this subsection, we describe case studies and analyses of the reserve for various different wind power generation scenarios with the same short-term reliability of balance (i.e., 99.6%). In the simulations, the reserve requirements and costs were calculated as a function of the wind power generation. The wind power generation during each hour was as shown in Figure 2b.

Table 5 lists the four different wind power generation scenarios considered in the simulations. The first, Case-WT1, contained one wind farm with a maximum generation capacity of 500 MW. The other scenarios (i.e., Case-WT2, Case-WT3, and Case-WT4) correspond to two, three, and four wind farms, respectively.

Table 5. Wind power scenarios Case-WT1, Case-WT2, Case-WT3, and Case-WT4.

Case	Number of Wind Farm	Maximum Power (MW)	Wind Power Penetration (%)
Case-WT1	1	500	6.48
Case-WT2	2	1000	12.17
Case-WT3	3	1500	17.20
Case-WT4	4	2000	21.69

Figure 7 shows the simulated reserve as a function of time using our proposed method. The reserve margin for each hour increased as the wind power generation capacity increased due to the greater significance of the forecast error in the wind power.

Table 6 lists the simulated costs for the four wind power generation scenarios. Because more wind power was included with Case-WT4 than Case-WT1, the generation costs were lower. In addition, because the error in the forecast wind generation had a greater impact than the outages of conventional generation, the required reserve margin with Case-WT4 was larger than with Case-WT1 (see Figure 7).

Figure 8 plots the data listed in Table 6. Case-WT4 exhibited the lowest operating cost of the four scenarios. The load-shedding cost increased with increasing wind power penetration due to the greater significance of the wind power forecast error. However, the generation cost (which includes the unit commitment cost while satisfying the required reserve margin calculated for a 99.6% short-term reliability of balance) decreased because of the reduced use of conventional generators, leading to a significant drop in the operating cost as the wind power penetration increased. Although the increased wind power penetration resulted in a larger probability of interruptions to the power system, requiring a larger reserve margin to maintain the same short-term reliability of balance, overall, the increase in wind power penetration reduced the operating cost.

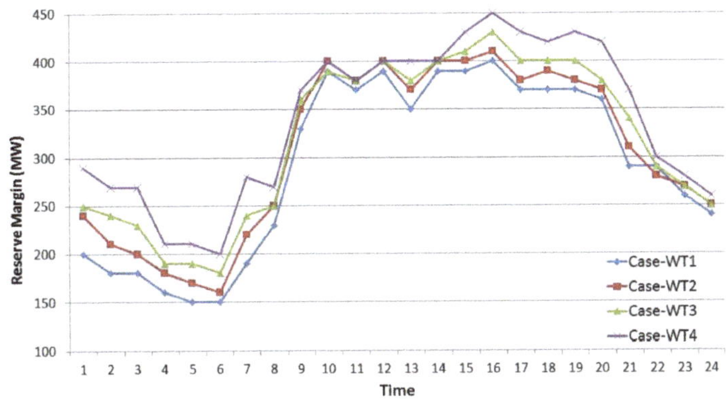

Figure 7. Hourly reserve margin of case-WT1, Case-WT2, Case-WT3, and Case-WT4.

Table 6. Simulated costs for scenarios Case-WT1, Case-WT2, Case-WT3, and Case-WT4.

Case	Generation Cost ($)	Load-Shedding Cost ($)	Operating Cost ($)
Case-WT1	1,741,700	6,445	1,748,145
Case-WT2	1,643,700	6,687	1,650,387
Case-WT3	1,549,100	10,758	1,559,858
Case-WT4	1,459,400	16,413	1,475,813

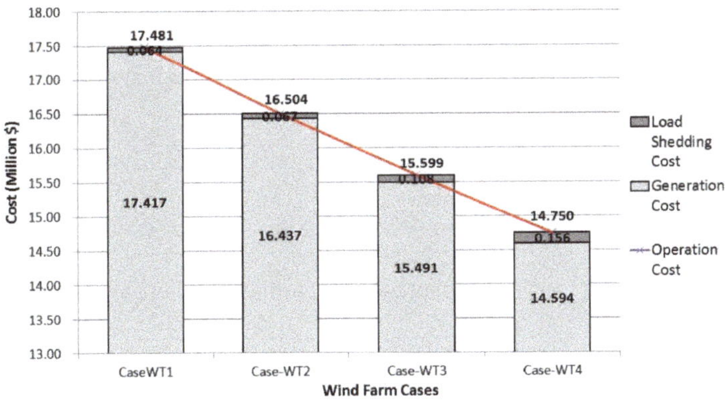

Figure 8. Calculated costs for scenarios Case-WT1, Case-WT2, Case-WT3, and Case-WT4.

6. Conclusions

We have described a method to calculate the hourly reserve margin considering uncertainties in supply and demand—i.e., generation outages and errors in the forecast demand and wind power. These uncertainties were described using probability distribution models, which were generated for every hour. The reserve margin was calculated hourly based on the short-term reliability of balance, which represents the probability of imbalance between supply and demand. To determine an appropriate short-term reliability of balance that minimizes the operating cost, we consider the generation cost and load-shedding cost.

A case study was carried out using the IEEE 118 bus system to calculate the reserve requirements and costs. The short-term reliability of balance was varied, and the hourly reserve margin was calculated; a cost analysis was then performed to identify the most economical short-term reliability of balance. The reliability and costs were calculated using our method and a conventional method with a fixed reserve margin. The results show that the variable hourly reserve margin (i.e., our method) was both more reliable and more economical. In addition, we carried out a cost analysis with various wind power generation scenarios and showed that the operating cost decreased as the wind power penetration increased.

Acknowledgments: This work was supported by the Brain Korea 21 Plus Project in 2016.

Author Contributions: Kyung-bin Kwon designed the study. Kyung-bin Kwon and Hyeongon Park performed the experiments. Jae-Kun Lyu and Jong-Keun Park contributed to the methodology. Hyeongon Park provided professional guidance. Kyung-bin Kwon and Hyeongon Park wrote the paper.

Conflicts of Interest: The authors declare no conflict of interest.

Nomenclature

The following abbreviations are used in this manuscript:

AD_t^s	Actual demand during hour t for case *i* (MW)
ED_t^s	Expected demand during hour t for case *i* (MW)
TE_t	Demand forecast error for hour *t* (%)
C_i^t	Number of cases during hour t of the analysis period
s	Index for case of hour *t* during the analysis period
σ_t	Standard deviation of the probability distribution of the load forecast error during hour *t*
G_t	Random variable for generation outage during hour *t* (MW)
$P(G_t)$	Probability that generation outage G_t occurs
j	Index of the number order in a set
$P(G_{t,j})$	Probability that generation outage G_t occurs by the outages of generators included in $\mathfrak{F}_t$
FOR_i	Forced outage rate (FOR) of generator *i*
$\mathfrak{F}_t$	The set that consists of sets which has generator numbers that satisfy the sum of generation equal to G_t
$G_{t,i}$	Output of generator *i* during hour *t* (MW)
θ	Scale parameter in gamma distribution
k	Shape parameter in gamma distribution
$P_{Im,t}$	Imbalance in demand and supply caused by generation outage, demand forecast error, and wind power forecast error (MW)
L_t	Random variable of demand forecast error during hour *t* (MW)
W_t	Random variable of wind power forecast error during hour *t* (MW)
P_{Total}	Overall generated power/energy during hour *t*, which is the sum of the maximum output of generators dispatched during hour *t* as a result of unit commitment (MW)
$P_{R,t}$	Expected reserve during hour *t* (MW)
SR_t	Short-term reliability of balance during hour *t* (%)
$P_{MR,t}$	Minimum reserve estimation during hour *t* (MW)
a_i, b_i, c_i	Unit cost coefficient of generator *i*
N_i	Number of generators in the power system
STC_i	Startup cost of unit *i* ($)
$u_{i,t}$	Binary variable that is 1 if unit N is online at hour *t* and 0 otherwise
$P_{LS,t}$	Load shedding during hour *t* (MW)
$P(P_{LS,t})$	Probability that load shedding occurs during hour *t*
C_{OC}	Outage cost ($/MWh)

References

1. Chang-Soo, O.; Kang Wan, L. A study on the requirement of frequency regulation reserve in electricity market. In Proceedings of the Korean Institute of Electrical Engineers Summer Conference, Pyeongchang-gun, Korea, 18 July 2005; pp. 298–300.
2. Chang-Soo, O.; Kang Wan, L. A study on an estimate of frequency regulating reserve. In Proceedings of the Korean Institute of Electrical Engineers Fall Conference, Seoul, Korea, 18 November 2005; pp. 264–266.
3. Hyoung-Yong, S.; Jae-Hyung, R.; Jong-Bae, P.; Joong-Rin, S.; Myung-Suk, C. Optimal dispatch for energy and frequency control reserve considering contingency. In Proceedings of the Korean Institute of Electrical Engineers Summer Conference, Busan, Korea, 14 July 2010; pp. 505–506.
4. Ahn, D.-H.; Kwon, S.-K.; Joo, H.-R.; Choi, E.-J. A research of optimum supply reserve levels for stability of power system. *J. Korean Inst. Illum. Electr. Install. Eng.* **2008**, *22*, 55–61.
5. Seong-Ho, R.; Kang-Wan, L.; Gwang-Won, K.; Kab-Ju, H. A study on operation standards for optimal operating reserve in electricity market. *Trans. Korean Inst. Electr. Eng. A* **2003**, *52*, 287–293.
6. Chattopadhyay, D.; Baldick, R. Unit commitment with probabilistic reserve. In Proceedings of the 2002 IEEE Power Engineering Society Winter Meeting, New York, NY, USA, 27–31 January 2002; Volume 281, pp. 280–285.
7. Georgopoulou, C.A.; Giannakoglou, K.C. Metamodel-assisted evolutionary algorithms for the unit commitment problem with probabilistic outages. *Appl. Energy* **2010**, *87*, 1782–1792. [CrossRef]
8. Ortega-Vazquez, M.A.; Kirschen, D.S.; Pudjianto, D. Optimising the scheduling of spinning reserve considering the cost of interruptions. *IEE Proc. Gener. Transm. Distrib.* **2006**, *153*, 570–575. [CrossRef]
9. Ortega-Vazquez, M.A.; Kirschen, D.S. Optimizing the spinning reserve requirements using a cost/benefit analysis. *IEEE Trans. Power Syst.* **2007**, *22*, 24–33. [CrossRef]
10. Wang, M.Q.; Gooi, H.B.; Han, X.S. A tradeoff between unit capacity and average production cost in spinning reserve optimization. *Int. J. Electr. Power Energy Syst.* **2015**, *71*, 215–221. [CrossRef]
11. Dany, G. Power reserve in interconnected systems with high wind power production. In Proceedings of the 2001 IEEE Porto Power Tech Proceedings, Porto, Portugal, 10–13 September 2001; Volume 4, p. 6.
12. Zhang, Z.; Li, C.; Cao, Y.; Tang, L.; Li, J.; Wu, B. Credibility assessment of short-term load forecast in power system. In Proceedings of the IEEE PES Innovative Smart Grid Technologies, Washington, DC, USA, 16–20 January 2012; pp. 1–5.
13. Doherty, R.; O'Malley, M. A new approach to quantify reserve demand in systems with significant installed wind capacity. *IEEE Trans. Power Syst.* **2005**, *20*, 587–595. [CrossRef]
14. Saez-Gallego, J.; Morales, J.M.; Madsen, H.; Jónsson, T. Determining reserve requirements in dk1 area of nord pool using a probabilistic approach. *Energy* **2014**, *74*, 682–693. [CrossRef]
15. Zhou, W.; Sun, H.; Peng, Y. Risk reserve constrained economic dispatch model with wind power penetration. *Energies* **2010**, *3*, 1880–1894. [CrossRef]
16. Chen, J.; Wu, W.; Zhang, B.; Wang, B.; Guo, Q. A spinning reserve allocation method for power generation dispatch accommodating large-scale wind power integration. *Energies* **2013**, *6*, 5357–5381. [CrossRef]
17. Ruiz, P.A.; Philbrick, C.R.; Sauer, P.W. Wind power day-ahead uncertainty management through stochastic unit commitment policies. In Proceedings of the 2009 IEEE/PES Power Systems Conference and Exposition, Seattle, WA, USA, 15–18 March 2009; pp. 1–9.
18. Park, H.-G.; Lyu, J.-K.; Kang, Y.; Park, J.-K. Unit commitment considering interruptible load for power system operation with wind power. *Energies* **2014**, *7*, 4281–4299. [CrossRef]
19. Liao, G.; Ming, J.; Wei, B.; Xiang, H.; Ai, N.J.P.; Dai, C.; Xie, X.; Li, M. Wind power prediction errors model and algorithm based on non-parametric kernel density estimation. In Proceedings of the 2015 5th International Conference on Electric Utility Deregulation and Restructuring and Power Technologies, Changsha, China, 26–29 November 2015; pp. 1864–1868.
20. Thapa, S.; Karki, R.; Billinton, R. Operating risk considerations in wind integrated power systems. In Proceedings of the 2014 International Conference on Probabilistic Methods Applied to Power Systems (PMAPS), Durham, UK, 7–10 July 2014; pp. 1–6.
21. Hedayati-Mehdiabadi, M.; Zhang, J.; Hedman, K.W. Wind Power Dispatch Margin for Flexible Energy and Reserve Scheduling With Increased Wind Generation. *IEEE Trans. Sustain. Energy* **2015**, *6*, 1543–1552. [CrossRef]

22. Wu, J.; Zhang, B.; Deng, W.; Zhang, K. Application of Cost-CVaR model in determining optimal spinning reserve for wind power penetrated system. *Int. J. Electr. Power Energy Syst.* **2015**, *66*, 110–115. [CrossRef]
23. Menemenlis, N.; Huneault, M.; Robitaille, A. Computation of dynamic operating balancing reserve for wind power integration for the time-horizon 1–48 hours. *IEEE Trans. Sustain. Energy* **2012**, *3*, 692–702. [CrossRef]
24. Matos, M.A.; Bessa, R.J. Setting the operating reserve using probabilistic wind power forecasts. *IEEE Trans. Power Syst.* **2011**, *26*, 594–603. [CrossRef]
25. Liu, G.; Tomsovic, K. Qunatifying Spinning Reserve in Systems With Significant Wind Power Penetration. *IEEE Trans. Power Syst.* **2012**, *27*, 2385–2393. [CrossRef]
26. Wang, M.; Bai, H.; Zhou, M.; Wang, J. A generation-reserve co-optimization dispatching model for wind power integrated power system based on risk reserve constraints. In Proceedings of the International Conference on Renewable Power Generation, Beijing, China, 17–18 October 2015; pp. 1–6.
27. Kwon, K.-B.; Park, H.-G.; Lyu, J.-K.; Kim, Y.-C.; Park, J.-K. Dynamic reserve estimating method with consideration of uncertainties in supply and demand. *Trans Korean Inst. Electr. Eng.* **2013**, *62*, 1495–1504. [CrossRef]
28. Ortega-Vazquez, M.A.; Kirschen, D.S. Estimating the spinning reserve requirements in systems with significant wind power generation penetration. *IEEE Trans. Power Syst.* **2009**, *24*, 114–124. [CrossRef]
29. Bo, R.; Li, F. Probabilistic lmp forecasting considering load uncertainty. *IEEE Trans. Power Syst.* **2009**, *24*, 1279–1289.
30. Jong-Uk, L.; Young-Min, W.; Sung-Kwan, J.; Kyung-Bin, S.; Jeong-Do, P.; Burm-Sup, C.; Ki-Jun, S. Analysis of load forecasting error for improving the accuracy of the short-term load forecasting. In Proceedings of the Korean Institute of Electrical Engineers Fall Conference, Cheonan-si, Korea, 11 December 2010; pp. 171–172.
31. The Half-Normal Distribution Method for Measurement Error: Two Case Studies. Available online: https://www-users.york.ac.uk/~mb55/talks/halfnor.pdf (accessed on 1 August 2016).
32. Billiinton, R.; Allen, R.N. Generating capacity—Basic probability methods. In *Reliability Evaluation of Power Systems*; Springer: New York, NY, USA, 1996; pp. 46–47.

energies

Article

A Study on Magnetic Decoupling of Compound-Structure Permanent-Magnet Motor for HEVs Application

Qiwei Xu [1,*], Jing Sun [1], Lingyan Luo [1], Shumei Cui [2] and Qianfan Zhang [2]

1 State Key Laboratory of Power Transmission Equipment & System Security and New Technology, Chongqing University, Chongqing 400044, China; sunjingzzx@163.com (J.S.); hitsnower@163.com (L.L.)
2 Department of Electrical Engineering, Harbin Institute of Technology, Harbin 150080, China; cuism@hit.edu.cn (S.C.); zhang_qianfan@hit.edu.cn (Q.Z.)
* Correspondence: xuqw@cqu.edu.cn; Tel./Fax: +86-23-6510-2434

Academic Editor: Paolo Mercorelli
Received: 6 September 2016; Accepted: 11 October 2016; Published: 14 October 2016

Abstract: The compound-structure permanent-magnet (CSPM) motor is used for an electrical continuously-variable transmission (E-CVT) in a hybrid electric vehicle (HEV). It can make the internal combustion engine (ICE) independent of the road loads and run in the high efficiency area to improve the fuel economy and reduce the emissions. This paper studies the magnetic coupling of a new type of CSPM motor used in HEVs. Firstly, through the analysis of the parameter matching with CSPM in the HEV, we receive the same dynamic properties' design parameters between the CSPM motor and the THS (Toyota Hybrid System) of the Toyota Prius. Next, we establish the equivalent magnetic circuit model of the overall and the secondary model considering the tangential and radial flux distribution in the outer rotor of the CSPM motor. Based on these two models, we explore the internal magnetic coupling rule of the CSPM motor. Finally, finite element method analysis in 2D-ansoft is used to analyze the magnetic field distribution of the CSPM motor in different operation modes. By the result of the finite element method analysis, the internal magnetic decoupling scheme is put forward, laying the theoretical foundation for the further application of the CSPM motor in HEVs.

Keywords: hybrid electric vehicle (HEV); compound-structure permanent-magnet (CSPM) motor; magnetic coupling; parameter matching; decoupling scheme

1. Introduction

With the shortage of fossil fuels, the soaring prices of international energy and the global greenhouse effect, countries have put forward higher requirements for the reduction of fuel consumptions and emissions. In order to solve these problems, many automobile researchers and manufacturers are developing the hybrid electric vehicle (HEV), which combines the engine and the motor as a power plant. Japan has launched many new types of HEVs after the Prius, such as Honda Corporation's Vezel Hybrid, Mercedes-Benz's S400 Hybrid and Ford's Escape, Mercury Mariner series SUV, Ford C-MAX Energi and Fusion HEVs. Germany's Audi also launched the Duo HEVs. Many other countries are also carrying out the research and production of HEVs.

There are many structures of HEVs. The one based on the compound-structure permanent-magnet (CSPM) motor has an electromagnetic power train system, which is one of the forefront techniques in HEV. In the traditional transmission system, we use the settled speed ratios to realize the dynamic performance and improve the efficiency of the HEV. The disadvantages of the fixed speed ratios are complexity of the structure, friction and mechanical loss; at the same time, the expenditure is too high.

Compared with the traditional transmissions system, the electric variable transmission (EVT) has many virtues, which can integrate all of the features in one device, which will be more suitable for application in the HEV. The CSPM motor described in this paper can work as an EVT to realize the continuous change of the transmission ratio, composing a complete hybrid power system. It can serve as: (1) the transmission drive system of the HEV with the function of hybrid power, which simplifies the system structure greatly and realizes the internal combustion engine (ICE) idling stop, power compensation and braking energy regeneration and other purposes; (2) the converters of torque and speed, which can make the ICE always operate at the maximum efficiency district dependent on the working conditions. This will improve the fuel economy and reduce the emissions [1,2]. The theoretical analysis proves that under the suitable control strategies, the CSPM motor based on the electromagnetic energy conversion principle can completely replace the planetary gear in the Prius series HEV.

At present, domestic and foreign scholars mainly study the CSPM in following aspects, such as the structure design, mathematical modeling, magnetic coupling and magnetic system design, parameter matching, control strategy, mode switching, efficiency analysis, temperature rising and cooling system design and dual rotor brushless structure of the CSPM, and have developed different types of prototypes, laying the foundation for further research and practical application. The research mainly exists as follows [3–5]: (1) study of the electromechanical coupling of HEVs based on the CSPM motor; according to the energy transfer mechanisms in different working modes, the CSPM motor can replace the continuously-variable transmission (CVT), starter, generator, clutch and other devices respectively, to achieve CVT mode, starter mode, generator mode, braking feedback mode, dynamic compensation mode and along with the ICE to drive vehicles; (2) research on control the strategy of HEVs based on the CSPM motor. The hybrid power transmission topology of HEVs based on the CSPM motor is changed; thus, the application of the traditional series and parallel HEVs control strategies is analyzed. The output of ICE and the energy storage device can be adjusted by controlling the stator motor (SM) and double rotor motor (DRM) of the CSPM motor, solving the problems, such as dynamic compensation, dynamic mode switching, dynamic control and speed sensor accuracy, etc. Due to the CSPM motor having a complicated structure, this leads to the transmission system being more complex and the control strategy difficult to formulate. Therefore, proposing an effective control strategy can improve the CSPM motors' performances better.

On the outer rotor inside and outside surface of the CSPM motor is installed the permanent magnets, which reduce volume and improve the energy density of the motor.

However, the hybrid excitation source can easily lead to internal magnetic serious coupling, which causes a larger torque ripple, serious heating and is hard to control [6]. There are two reasons for the magnetic coupling: one is the interaction between the magnetic fluxes of the two-layer permanent magnets and changing with the permanent magnets' position; another is the coupling between the magnetic fluxes of the stator armature reaction and permanent magnets. Most of the magnetic fluxes flow along the outer rotor, so the fluxes produced by the two-layer permanent magnets have three relationships in the outer rotor: without interference, mutual overlaps and mutual offset, respectively. The saturation degree of the outer rotor is the lowest when the magnetic fluxes are mutually offset. This paper studies the magnetic coupling in the SM of the CSPM motor. Firstly, the parameter matching of the CSPM motor in the HEV is analyzed, and we receive the dynamic properties' design parameters of the CSPM motor as the THS of the Prius. They both work as the series-parallel power split HEVs to satisfy the same load demand, but the HEV based on the CSPM motor is more compact, intelligent and noiseless. Then, the equivalent magnetic circuit model of the overall and the tangential and radial magnetic flux distribution in the CSPM motor's outer rotor are established; then, we explore the internal magnetic coupling rule of the CSPM motor based on these two models. Finally, by using the finite element method (FEM) electromagnetic simulation to analyze the magnetic field of the CSPM motor in different working conditions, we propose an internal magnetic decoupling scheme, which meets the miniaturization, lightweight and control requirements, laying a foundation for the application of the CSPM motor in the HEV field.

2. Parameters Matching Analysis of the CSPM Motor in the HEV

The CSPM motor is composed of three parts, which are a stator with three-phase windings, an outer rotor with permanent magnets on the inner and outer surfaces and an inner rotor with three-phase windings. The stator and outer rotor with outer surface permanent magnets constitutes the stator motor (SM), which is also called EM2 (electric motor 2), and the outer rotor with the inner surface permanent magnets and inner rotor forms the double rotor motor (DRM), which is also called EM1 [7]. The schematic structure of the HEV based on the CSPM motor is shown in Figure 1. The high degree of integration between DRM and SM makes up the CSPM motor system, which is between the ICE and final gear. The inner rotor is coaxial with the ICE, and the outer rotor is connected to the final gear in the HEV, which has two independent mechanical output shafts. From Figure 1, we can see that there is no mechanical link between the ICE and final gear, which makes the two parts isolated completely to operate better. This new structure integrates the function of the mechanical power distribution devices in the Prius, such as the generator, motor and planetary gear. It can simplify the energy flows to become a pure electric power distribution device, and the entire system structure is more compact and has flexible control, opening up a new direction for the HEV field [8].

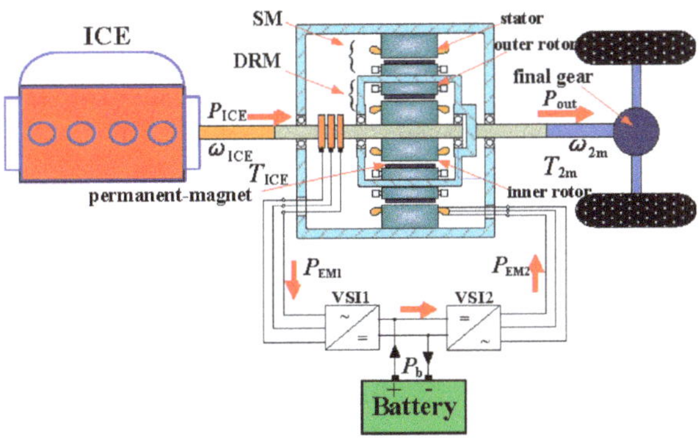

Figure 1. Schematic structure of the HEV based on the CSPM motor. SM, stator motor; DRM, double rotor motor.

The CSPM motor system's function, due to the special structure of the double shafts, is not inferior to the Prius transmission system under the same characteristic parameters [9]. If the output shafts take turns connecting to the ICE or final gear, the different functions of the hybrid power and modes of energy flows are realized. Under the auxiliary adjustment of the CSPM motor, the speed and torque control of ICE can be realized at the same time, and the speed continuous changes of the transmission are simulated. It can be used as an electric variable transmission (EVT), so the operation modes of the system can switch automatically to keep the ICE working at the highest efficiency in different conditions, increasing efficiency, reducing pollution and fuel consumption. The HEVs will become the mainstream car to ease the problems of resources and environment in the future.

2.1. Comparative Analysis of CSPM Motor and THS of Power Split HEVs

THS is the core of the transmission system in the Prius series HEVs, which was launched by Toyota Corporation in 1997 [10]. The speed and torque couplers of THS is a planetary gear, which realizes the functions of electromechanical coupling, power distribution and energy flows in the hybrid power system, solving the difficulty of dynamic coupling in HEVs. The structure diagram of the Prius power transmission system is shown in Figure 2.

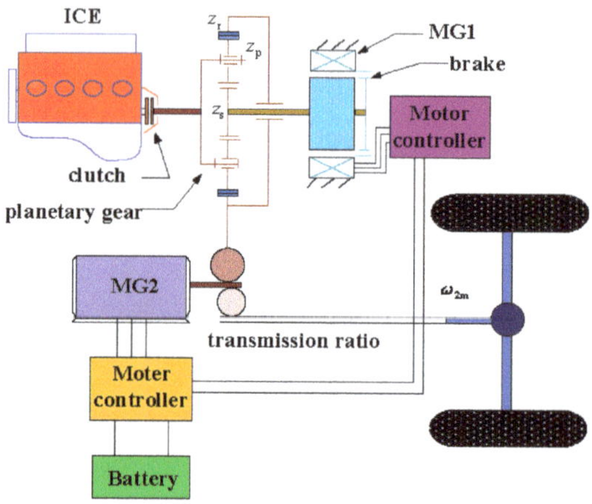

Figure 2. The structure diagram of the Prius power transmission system.

In Figure 2, the output shaft of ICE is connected to the carrier of the planet gear, then the output power is transmitted to the sun and ring gears, respectively. The sun gear is connected to the generator (MG1, (motor generator1)), which can be used as the input or output port. The ring gears act as an output port connected to the motor (MG2) and transmission shaft through a torque coupler. Most output energy of the ICE is engaged with the ring gears through the carrier, which transmits the energy to the driving shaft; while the rest of the energy is transmitted to the MG1 by the mechanical coupling of the carrier and sun gears. The mechanical energy is converted into electrical energy in MG1 to charge the battery or supply MG2, then the driving force is increased. The planetary gear system can be regarded as a torque coupler, which transmits the driving force to the wheel [11]. The speed relationship of the carrier and sun ring is shown in Equation (1):

$$\rho\omega_r - (1+\rho)\,\omega_p + \omega_s = 0 \tag{1}$$

where ω_r is the speed of carrier; ω_p is the speed of the ring gear; ω_s is the speed of the sun gear; $\rho = N_r/N_s$ is the basic ratio of the tooth number (the ratio of the ring and sun gears).

In the HEV, MG1 and MG2 can be used as the control devices of speed and torque, respectively. When the power of the system is allocated, the speed and torque of ICE is adjusted by the control of the central controller. The speed of the three power sources is controlled selectively, i.e., ICE, MG2 and MG1, to simulate the continuous change of the transmission ratio. Therefore, the whole system can be used as an EVT and switch to the best driving mode to suit the road conditions automatically. All parts of this system can work in the optimal area in various conditions, achieving energy optimization matching to reduce emissions and fuel consumption. Most companies use the planetary gears as a hybrid power system. However, the planetary gear is a purely mechanical device, which has serious vibration and noise, high loss and needs regular lubrication, limiting the further application in HEVs.

The diagram of the energy flows in the Prius hybrid power system is shown in Figure 3.

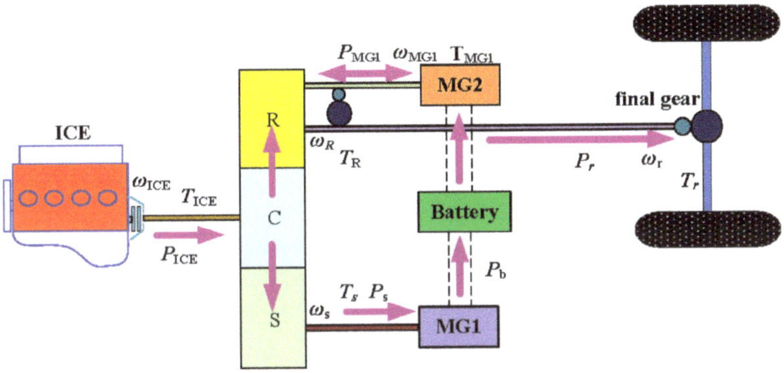

Figure 3. The diagram of the energy flows in the Prius hybrid power system.

The CSPM motor is another power coupling mechanism with two mechanical ports and two electrical ports, which is different from the planetary gear. The unique configuration provides more compactness, thus making the full hybrid function come true. The CSPM can realize the function of the powertrain in the HEV. In addition, the CSPM can also be used as a converter of torque and speed to adjust the ICE to run in the most economical efficiency area. The energy flows are more flexible, and dynamic coupling of the transformation system is more powerful, which has a broad prospect in the application of the HEV. The diagram of the energy flow in the HEV based on the CSPM motor is shown in Figure 4.

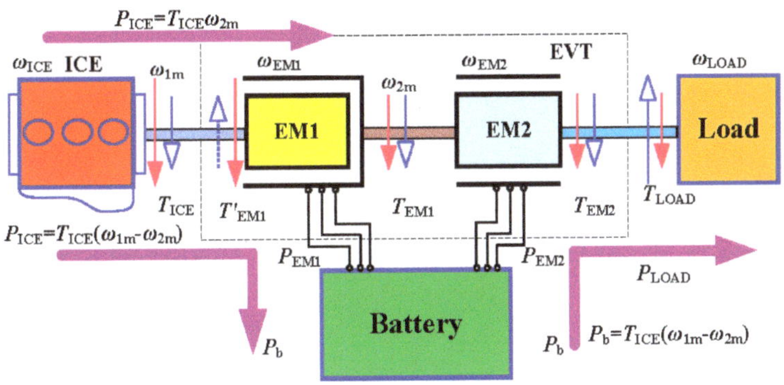

Figure 4. The diagram of the energy flows in the HEV based on the CSPM motor. EVT, electric variable transmission.

The output mechanical energy of ICE (P_{ICE}) is divided into two parts, as shown in Equation (2); a part of the energy is the output power of EM1 (electric motor1) (P_{EM1}) directly passed to the stator of EM2, through the collector ring, brush, rectifier and inverter. Another part of the energy (P_r) is transmitted to the output shaft through the magnetic field, driving the outer rotor of EM1 directly.

$$P_{ICE} = P_{EM1} + P_d \tag{2}$$

The inner rotor of EM1 is mechanically connected to the output shaft of the ICE. In a stable state, the speed of the inner rotor of EM1 is equal to the speed of the ICE (ω_{ICE}), and the electromagnetic torque of EM1 (T_{EM1}) is balanced with the output torque of the ICE (T_{ICE}). Thus, we can get Equation (3):

$$\begin{cases} P_{ICE} = T_{ICE}\omega_{ICE} \\ P_r = P_{ICE} - P_{EM1} = T_{ICE}\omega_{EM2} = T_{EM1}\omega_{EM2} \\ P_{EM1} = T_{EM1}(\omega_{ICE} - \omega_{EM2}) = T_{ICE}(\omega_{ICE} - \omega_{EM2}) \end{cases} \quad (3)$$

The output shaft of the CSPM motor is connected to final gear of the HEV, and the output speed of EM2 (ω_{EM2}) is determined by the speed of the HEV. The output torque of the outer rotor (T_{2m}) is the torque combined from EM1 and EM2, as shown in Equation (4):

$$T_{2m} = T_{EM1} + T_{ICE} = T_{EM1} + T_{EM2} \quad (4)$$

According to the above analysis, both the THS of the Prius series HEVs and the power transmission device of the CSPM motor can realize the output power split of the ICE. We use the ω_{EM1}, ω_{EM2}, ω_{MG1}, ω_{MG2} to express the speed of EM1, EM2, MG1, MG2; T_{EM1}, T_{EM2}, T_{MG1}, T_{MG2} to express the torque of EM1, EM2, MG1, MG2, respectively; T_{ICE}, T_d to express the torque of ICE and driving shaft; ω_{ICE}, ω_d as the speed of ICE and driving shaft and the power transmission system mathematical models of THS and the CSPM motor are compared, as shown in Table 1.

Table 1. The comparison of the power transmission system mathematical models of THS and the CSPM motor.

Relationship	THS (Prius)	CSPM Motor
Speed coupling	$\omega_{ICE1} = \frac{1}{1+\rho}\omega_{MG1} + \frac{\rho}{1+\rho}\omega_{MG2}$	$\omega_{ICE2} = \omega_{EM1} + \omega_{EM2}$
Torque constraint	$T_{ICE1} = (1+\rho)\,T_{MG1}$	$T_{ICE2} = T_{EM1}$
Torque coupling	$T_{d1} = \frac{\rho}{1+\rho}T_{ICE} + T_{MG2}$	$T_{d2} = T_{ICE} + T_{EM2}$
Speed constraint	$\omega_{d1} = \omega_{MG2}$	$\omega_{d2} = \omega_{EM2}$

2.2. *Parameter Matching Between the CSPM Motor and the Prius System*

2.2.1. Parameter Matching of SM in the CSPM Motor

As shown in Table 1, the torque coupling relationships between the Prius system and the driving shaft of the CSPM motor are obtained by Equations (5) and (6):

$$T_{d1} = \frac{\rho}{1+\rho}T_{ICE} + T_{MG2} \quad (5)$$

$$T_{d2} = T_{ICE} + T_{EM2} \quad (6)$$

If we connect a gear with a speed ratio of $(1+\rho)/\rho$ after the ICE of the CSPM motor, then the new torque coupling relationship of the ICE can be computed in Equation (7):

$$T_{d2} = \frac{\rho}{1+\rho}T_{ICE} + T_{EM2} \quad (7)$$

Comparing Equation (5) with Equation (7), we can know that, if the two torques of the ICE are the same, when $T_{d1} = T_{d2}$, then $T_{MG2} = T_{EM2}$, i.e., completing the torque matching of EM2.

From Equations (8) and (9), it can be seen that EM2 and MG2 have the same speed association constraints, which are equal to the speed of the driving shafts, respectively. Under the same speed of driving shaft, the speeds of EM2 and MG2 are equal, as shown in the Equation (10):

$$\omega_{d1} = \omega_{MG2} \quad (8)$$

$$\omega_{d2} = \omega_{EM2} \quad (9)$$

$$\omega_{\text{MG2}} = \omega_{\text{EM2}} \tag{10}$$

According to the above analysis, the parameter matching of torque and speed between the MG2 of the Prius system and the EM2 of the CSPM motor is realized, and the power parameters of the two motors are also equal.

2.2.2. Parameter Matching of DRM in the CSPM Motor

From Table 1, it can be known that the original speed coupling relationship of ICE in the CSPM motor is shown in Equation (11):

$$\omega_{\text{ICE2}} = \omega_{\text{EM1}} + \omega_{\text{EM2}} \tag{11}$$

After the ICE is connected to the deceleration gear, the new speed coupling relationship becomes:

$$\frac{1+\rho}{\rho}\omega_{\text{ICE2}} = \omega_{EM1} + \omega_{\text{EM2}} \tag{12}$$

The speed coupling relationship of the generator (MG1) in the Prius system, as shown in Equation (13):

$$\omega_{\text{ICE1}} = \frac{1}{1+\rho}\omega_{\text{MG1}} + \frac{\rho}{1+\rho}\omega_{\text{MG2}} \tag{13}$$

Then, we multiply both sides of Equation (13) by the proportional coefficient $(1 + \rho)/\rho$, and Equation (14) is obtained:

$$\frac{1+\rho}{\rho}\omega_{\text{ICE1}} = \frac{1}{\rho}\omega_{\text{MG1}} + \omega_{\text{MG2}} \tag{14}$$

Comparing Equation (12) with Equation (14), it is clearly seen that when the speeds of the ICE in the Prius system and the CSPM motor are same, the speed relationship between the MG1 and EM1 is shown in Equation (15):

$$\omega_{\text{MG1}} = \rho \cdot \omega_{\text{EM1}} \tag{15}$$

Because the ICE is connected with a reduction gear, the new torque coupling relationship is:

$$\frac{\rho}{1+\rho}T_{\text{ICE2}} = T_{\text{EM1}} \tag{16}$$

The torque of MG1 is multiplied by the proportional coefficient; the new coupling relationship can be deduced from Equation (17):

$$\frac{\rho}{1+\rho}T_{\text{ICE1}} = \rho \cdot T_{\text{MG1}} \tag{17}$$

Contrast Equations (16) and (17); when the torques of the ICE are the same in the two systems, the torque relationship of EM1 and MG1 can be obtained:

$$T_{\text{EM1}} = \rho \cdot T_{\text{MG1}} \tag{18}$$

It can be seen in Equations (15) and (18), after the parameter matching, that the speed of CSPM motor is reduced to $1/\rho$ of the Prius system, and the torque is ρ times the Prius system. Therefore, the power of the CSPM motor and the Prius system is equal.

2.2.3. Verification of Parameter Matching

The torque and speed curves of MG1 and MG2 in the THS are shown in Figure 5.

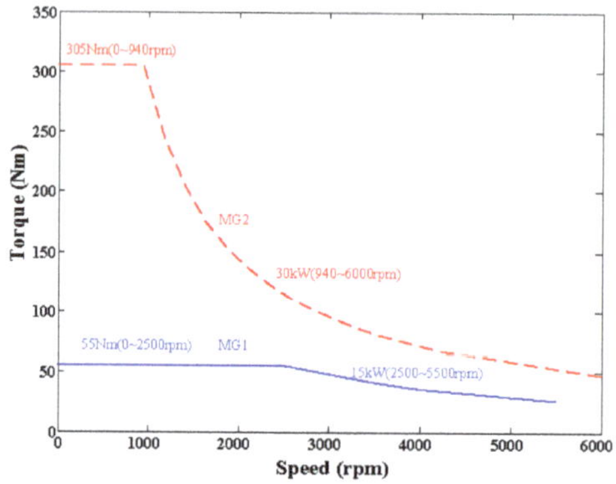

Figure 5. External characteristic curves of MG1 and MG2 in the THS.

According to the parameter matching proposed in this paper, the design parameters of the CSPM motor with the same dynamic performance as the THS (Toyota Hybrid System) are obtained, as shown in Table 2.

Table 2. Design parameters of CSPM and THS.

Power Coupling Device	EM1/MG1		EM2/MG2	
EVT	132 Nm 0~1050 rpm	15 kW 1050~2300 rpm	305 Nm 0~940 rpm	30 kW 940~6000 rpm
THS	55 Nm 0~2500 rpm	15 kW 2500~5500 rpm	305 Nm 0~940 rpm	30 kW 940~6000 rpm

By connecting a gear with a transmission ratio of $(1 + \rho)/\rho$ to the output shaft of the ICE, the parameters between MG2 and EM2 are consistent. The maximum torque of EM1 is ρ times that of MG1, and the maximum speed of EM1 is reduced to ρ times that of MG1.

Then, we put the design parameters of the CSPM motor into the simulation model to calculate the power performance and fuel economy of the HEV. The dynamic performances and fuel economy between the HEVs based on THS and the CSPM motor are compared, as shown in Table 3 and Figure 6.

Table 3. The dynamic performances of the HEVs.

Power Coupling Device	Maximum Speed (km/h)	Acceleration (s/100 km)	Acceleration (s/60 km)
THS	170	15.97	7.3
CSPM	170	15.04	7.1

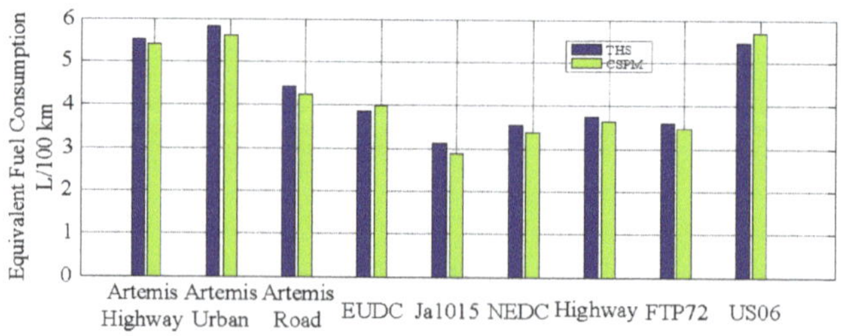

Figure 6. The comparison of fuel economy between THS and the CSPM motor.

The dynamic performances and fuel economy of the HEVs based on the CSPM motor have been improved slightly under the comparison with the Prius system, which shows the reasonableness and effectiveness of the design method.

3. Equivalent Magnetic Circuit Analysis of the CSPM Motor

According to the literature [12], when the two-layer permanent magnets are adopted for a consistent magnetization direction, the no-load equivalent magnetic circuit is as shown in Figure 7.

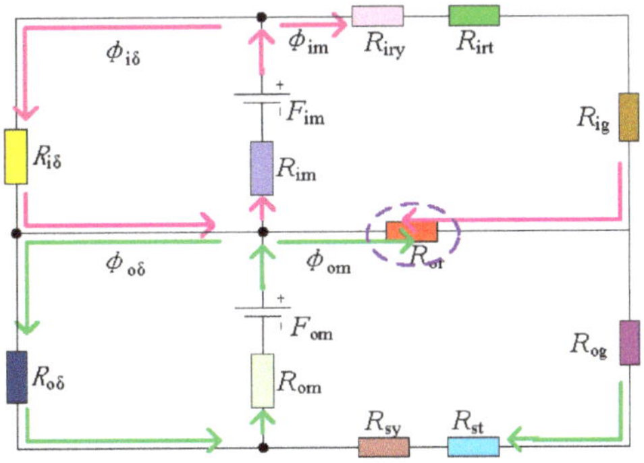

Figure 7. No-load equivalent magnetic circuit in consistent magnetization direction.

F_{im}, F_{om} are the equivalent magnetic motive forces produced by the inner and outer layer permanent magnets; φ_{im}, φ_{om} are the main magnetic fluxes in the DRM and SM; $\varphi_{i\delta}$, $\varphi_{o\delta}$ are the equivalent leakage fluxes in the inner and outer layer permanent magnets; R_{im}, R_{om} are the equivalent reluctances of two-layer permanent magnets; $R_{i\delta}$, $R_{o\delta}$ are the equivalent leakage reluctances of two-layer permanent magnets; R_{iry}, R_{irt} are the yoke, teeth reluctances of the inner rotor; R_{sy}, R_{st} are the yoke, teeth reluctances of the stator; R_{ig}, R_{og} are the reluctances of the inner and outer air-gap; and R_{or} is the reluctance of the outer rotor.

According to Figure 7, the mesh loop equations are analyzed on the basis of the network equation method in circuit analysis, as shown in Equation (19):

$$\begin{cases} F_{im} = \Phi_{im}R_{im} + \Phi_{i\delta}R_{i2} + 0 + 0 \\ F_{im} = \Phi_{i\delta}R_{im} + \Phi_{im}\left(R_{i1} + R_{or}\right) + 0 + \left(-\Phi_{om}R_{or}\right) \\ F_{om} = \Phi_{om}R_{om} + \Phi_{o\delta}R_{o2} + 0 + 0 \\ F_{om} = \Phi_{om}\left(R_{o1} + R_{or}\right) + \Phi_{o\delta}R_{om} - \Phi_{im}R_{or} + 0 \end{cases} \tag{19}$$

where:

$$\begin{cases} R_{i1} = R_{irs} + R_{irt} + R_{ig} + R_{im} \\ R_{i2} = R_{i\delta} + R_{im} \\ R_{o1} = R_{sy} + R_{st} + R_{om} + R_{og} \\ R_{o2} = R_{om} + R_{o\delta} \end{cases} \tag{20}$$

According to Equations (19) and (20), the main magnetic fluxes in the DRM and SM are computed in Equations (21) and (22):

$$\Phi_{im} = \frac{1}{R_{i3}R_{o3} - R_{or}^2}\left(\frac{R_{o\delta}R_{or}}{R_{o2}}F_{om} + \frac{R_{i\delta}R_{o3}}{R_{i2}}F_{im}\right) \tag{21}$$

$$\Phi_{om} = \frac{1}{R_{i3}R_{o3} - R_{or}^2}\left(\frac{R_{o\delta}R_{i3}}{R_{o2}}F_{om} + \frac{R_{i\delta}R_{or}}{R_{i2}}F_{im}\right) \tag{22}$$

As the magnetic fluxes of the outer rotor are in opposite directions, so we can derive the total magnetic flux of the outer rotor from Equations (21) and (22), as shown in Equation (23):

$$\Phi_{or} = \Phi_{om} - \Phi_{im} = \frac{1}{R_{i3}R_{o3} - R_{or}^2}\left[\frac{R_{o\delta}\left(R_{i3} - R_{or}\right)}{R_{o2}}F_{om} - \frac{R_{i\delta}\left(R_{o3} - R_{or}\right)}{R_{i2}}F_{im}\right] \tag{23}$$

where:

$$\begin{cases} R_{i3} = R_{or} + R_{iry} + R_{ig} + R_{irt} + \frac{R_{im}R_{i\delta}}{R_{im}+R_{i\delta}} \\ R_{o3} = R_{st} + R_{sy} + R_{or} + R_{og} + \frac{R_{om}R_{o\delta}}{R_{om}+R_{o\delta}} \end{cases} \tag{24}$$

According to Figure 7, in order to decrease the saturation degree of the outer rotor, we should keep the magnetic motive force unchanged to reduce the magnetic flux in the outer rotor. Therefore, the main magnetic fluxes of DRM and SM flowing through the outer rotor are not limited, i.e., realizing the magnetic decoupling of DRM and SM.

From Equation (24), we can see that R_{i3} includes R_{ig}, $R_{i\delta}$ and R_{o3} includes R_{og}, $R_{o\delta}$. Considering that the magnetic circuit of the air-gap and magnetic leakage circuit of the permanent magnets are closed through the air, which is a non-conducting magnetic material with a much larger reluctance than the magnetic material, so we assume that R_{i3} » R_{or}, R_{o3} » R_{or}.

Thus, the expression of the total magnetic flux in the outer rotor can be approximated as Equation (7):

$$\Phi_{or} \approx \frac{1}{R_{i3}R_{o3} - R_{or}^2}\left[\frac{R_{o\delta}R_{i3}}{R_{o2}}F_{om} - \frac{R_{i\delta}R_{o3}}{R_{i2}}F_{im}\right] \tag{25}$$

The outer rotor of the CSPM motor has a certain thickness, so the magnetic fluxes generated by the two-layer permanent magnets and stator armature reaction flow in the outer rotor will be divided into two paths, i.e., radial and tangential. Tangential magnetic flux is the main magnetic flux, transmitting the electromagnetic energy; the other is the coupling magnetic flux of DRM and SM, which is proportional to the coupling degree. The no-load equivalent magnetic circuit of the tangential and radial magnetic fluxes in the outer rotor is described in Figure 8.

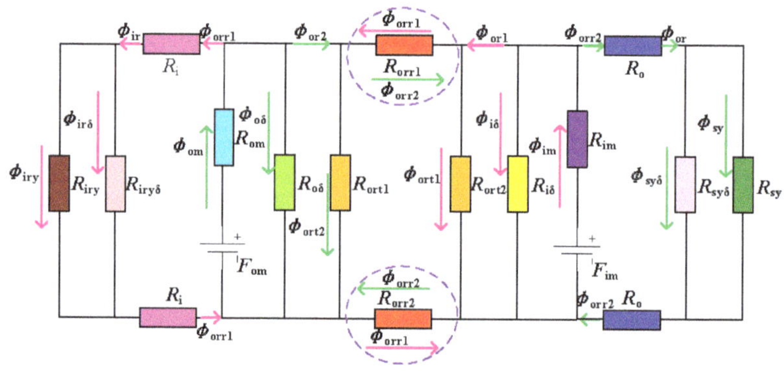

Figure 8. No-load equivalent magnetic circuit of the tangential and radial magnetic fluxes in the outer rotor.

R_{orr1}, R_{orr2} are the equivalent radial reluctances of the outer rotor under the magnetic fluxes of DRM and SM; R_{ort1}, R_{ort2} are the equivalent tangential reluctances of the outer rotor under the magnetic fluxes of DRM and SM; R_i, R_o are the total equivalent reluctances of the inner and outer air-gap; $R_{ir\delta}$, $R_{sy\delta}$ are the equivalent yoke leakage reluctances of the inner rotor and the stator; φ_{or1}, φ_{or2} are the total magnetic fluxes of the outer rotor generated by DRM and SM; φ_{orr1}, φ_{orr2} are the radial magnetic fluxes of the outer rotor generated by DRM and SM; φ_{ort1}, φ_{ort2} are the tangential magnetic fluxes of the outer rotor generated by DRM and SM; φ_{ir} is the magnetic flux of the inner rotor from the inner layer magnets; φ_{or} is the magnetic flux of the stator from the outer layer magnets; φ_{iry}, φ_{sy} are the yoke magnetic fluxes of the inner rotor and the stator.

When the leakage resistances of the inner rotor, stator and two-layer permanent magnets are neglected, the equivalent magnetic motive forces of the two-layer permanent magnets are expressed in Equations (26) and (27):

$$R_{\text{im}} \cdot \Phi_{\text{im}} = F_{\text{im}} = \left(2R_{\text{i}} + R_{\text{iry}}\right)\Phi_{\text{iry}} + \Phi_{\text{ort1}} \cdot R_{\text{ort1}} + \Phi_{\text{orr1}} \cdot R_{\text{orr1}} \tag{26}$$

$$R_{\text{om}} \cdot \Phi_{\text{om}} = F_{\text{om}} = \left(2R_{\text{o}} + R_{\text{sy}}\right)\Phi_{\text{sy}} + \Phi_{\text{ort2}} \cdot R_{\text{ort2}} + \Phi_{\text{orr2}} \cdot R_{\text{orr2}} \tag{27}$$

where:

$$\begin{cases} R_{\text{i}} = R_{\text{irt}} + R_{\text{ig}} \\ R_{\text{o}} = R_{\text{st}} + R_{\text{og}} \\ \Phi_{\text{or1}} = \Phi_{\text{orr1}} + \Phi_{\text{ort1}} \\ \Phi_{\text{or2}} = \Phi_{\text{orr2}} + \Phi_{\text{ort2}} \end{cases} \tag{28}$$

From Equations (26) and (27), the radial and tangential flux ratios of the inner and outer permanent magnets are computed in Equation (29):

$$\begin{cases} \frac{\Phi_{\text{orr1}}}{\Phi_{\text{ort1}}} = \frac{R_{\text{ort1}}}{2R_{\text{i}}+R_{\text{iry}}+2R_{\text{orr1}}} \\ \frac{\Phi_{\text{orr2}}}{\Phi_{\text{ort2}}} = \frac{R_{\text{ort2}}}{2R_{\text{o}}+R_{\text{sy}}+2R_{\text{orr2}}} \end{cases} \tag{29}$$

As $2R_{\text{i}} \gg R_{\text{iry}} + 2R_{\text{orr1}}$ and $2R_{\text{o}} \gg R_{\text{sy}} + 2R_{\text{orr2}}$ simplify Equation (29), we can get the radial coupling magnetic fluxes of the outer rotor generated by DRM and SM, as shown in Equations (30) and (31):

$$\Phi_{\text{orr1}} = \frac{R_{\text{ort1}}}{2R_{\text{i}}}\Phi_{\text{ort1}} \tag{30}$$

$$\Phi_{\text{orr2}} = \frac{R_{\text{ort2}}}{2R_{\text{o}}}\Phi_{\text{ort2}} \tag{31}$$

From the above mathematical analysis, in order to perform the magnetic decoupling of two motors, we should reduce the total magnetic flux in the outer rotor, which is divided into tangential and radial. The radial magnetic flux is the coupling flux between DRM and SM, which should be reduced as far as possible under the required electromagnetic energy. From Equations (30) and (31), we can see that, keeping the main magnetic fluxes unchanged, φ_{orr1} is only associated with R_{ort1}, and φ_{orr2} is only associated with R_{ort2}. Therefore, the magnetic decoupling can be realized if we reduce the tangential reluctance of the outer rotor, i.e., increasing the yoke thickness of the outer rotor properly.

4. Magnetic Coupling Characteristics' Analysis of CSPM

4.1. Analysis of the Magnetization Direction

On both sides of the outer rotor are placed permanent magnets, the magnetization direction of which can be divided into consistent and opposite. In the consistent magnetization direction, the flux line directions in the outer rotor produced by DRM and SM are opposite, which can offset each other, as shown in Figure 9a. When the two magnetic fields in the outer rotor are similarly, the magnetic flux in the outer rotor is less after offsetting. Thus, we can reduce the yoke thickness of the outer rotor properly, without considering the influence of saturation.

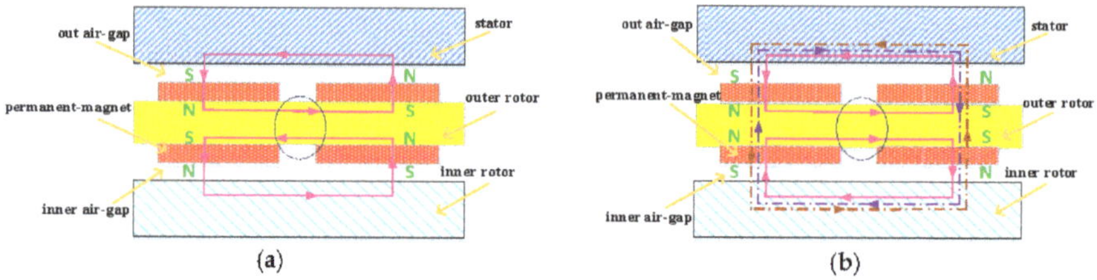

Figure 9. The diagram of the magnetic flux flow in the CSPM motor: (**a**) consistent magnetization direction of the permanent-magnet; (**b**) opposite magnetization direction of the permanent-magnet.

When in the opposite magnetization direction, the magnetic flux directions in the outer rotor are the same, i.e., the magnetic fields overlap each other. This makes the outer rotor easily saturated, and the magnetic flux of DRM will flow into SM and vice versa, as shown in Figure 9b. The original flux lines of the two motors are crippled, and the operations are disrupted. At this point, we must increase the yoke thickness of the outer rotor to reduce the saturation degree and nonlinear magnetic field.

In order to analyze the influence of magnetization directions further, the amplitudes of the air-gap flux density curves of the inner and outer air-gap in different magnetization directions are shown in Figure 10.

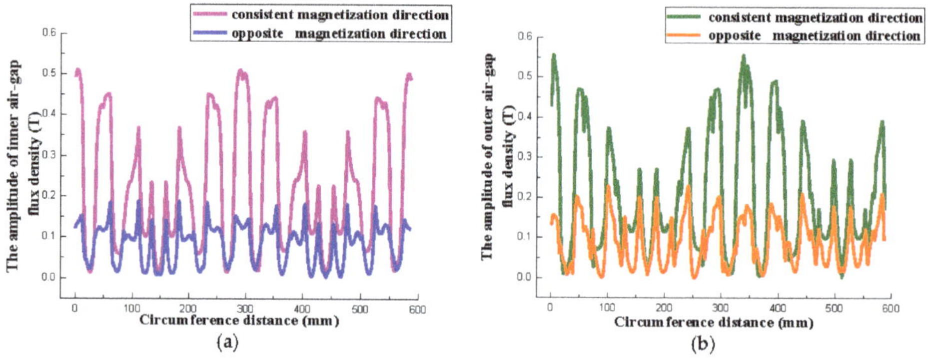

Figure 10. The amplitude of the air-gap flux density curves in different magnetization directions: (**a**) inner air-gap; (**b**) outer air-gap.

We can easily see that the air-gap flux density of SM is higher than DRM in the two kinds of directions, i.e., the magnetic coupling has a greater influence on DRM. The influence of the consistent direction is quite different from the opposite, the former of which is larger. This shows that the two amplitudes of the air-gap flux density are crippled, under the opposite magnetization direction. That is, the original magnetic flux through the outer rotor forms a closed circuit; but, some lines of DRM will now enter into SM, and some lines of SM will enter into DRM. Part of the original work of magnetic flux is offset, and the outer rotor is in a serious saturation state. To realize the magnetic decoupling, we can increase the yoke thickness of the outer rotor or reduce the length of the air-gap, but the power density and magnetic properties of the two motors will be reduced. The reverse lines of the outer rotor can partially offset each other in the consistent magnetization direction, which greatly decreases the saturation degree. The main magnetic fields of the two motors are independent, so the resistances of the whole circuit are small, which increases the flux density of the air-gap.

Therefore, to reduce the saturation degree and yoke thickness of the outer rotor properly, the consistent magnetization direction of the two-layer permanent magnets should be adopted.

4.2. Design the Yoke Thickness of the Outer Rotor

The magnetic fields of DRM and SM are closed through the outer rotor, which provides both the magnetic circuits and magnetic motive forces to the motors. Accordingly, there are mechanical coupling and electromagnetic coupling between the two motors, and the electromagnetic coupling is more serious. Therefore, it is particularly important to opt for a suitable yoke thickness of the outer rotor. Increasing the thickness will gradually reduce the coupling degree between the DRM and SM. The magnetic fluxes are through the outer rotor, then back to the air-gap, respectively, and the interference in the magnetic fields is very small. If the yoke thickness is too thick, the quality and volume of CSPM are increased, leading to inconvenience in the mounting. This will also reduce the power density and work efficiency, leading to the advantages of the CSPM motor not being fully revealed. The outer rotor is easily saturated when the thickness is too thin, which leads to the changes of inductance and resistance, and the magnetic coupling is serious. Moreover, the control difficulty of DRM and SM is increased, i.e., the control goals are hard to achieve, which causes the degradation of the performance.

Therefore, in order to accomplish the magnetic decoupling of the CSPM motor, the key of the design is to select a proper yoke thickness of the outer rotor. This will realize the independent control of the two motors and make full use of the excellent performance of the CSPM motor.

4.2.1. Analysis of the Outer Rotor Based on the Finite Element Method

The finite element method (FEM) model of the CSPM motor is established in Figure 11.

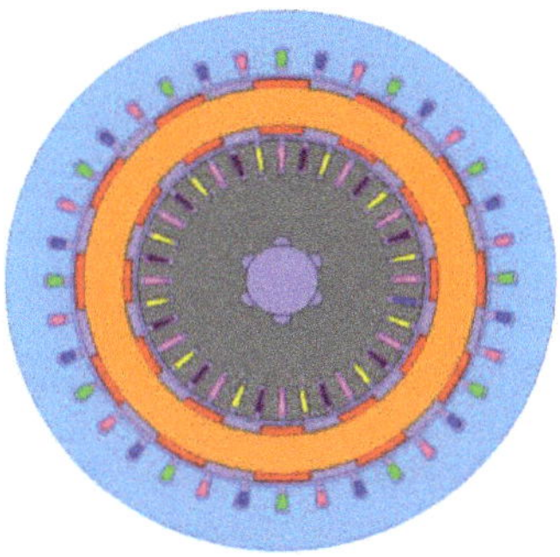

Figure 11. The FEM model of the CSPM motor.

Serious magnetic field saturation is aimed at the outer rotor, with the premise of achieving the same dynamic performance to determine the design parameters of the CSPM motor based on the parameter matching of the Prius series HEVs. The electromagnetic field FEMs of DRM and SM are established, and the electromagnetic simulation calculations are carried out [13], which provide a reference for the electromagnetic design of the CSPM motors. The simulation parameters of the CSPM motor are shown in Table 4.

Table 4. The simulation parameters of DRM and SM.

Parameter	DRM	SM
Rated power (kw)	15	30
Rated voltage (V)	300	300
Number of phase	3	3
Rated speed (rpm)	2000	2000
Number of slots	36	36

Accurate analysis of the distribution of the air-gap magnetic field is the key to study the magnetic properties of CSPM. From the above analysis, we can know that the magnetic flux in the outer rotor is proportional to its resistance, which can be changed by the yoke thickness. Meanwhile, the air-gap flux density can be influenced by the saturation degree of the outer rotor, so we use the air-gap flux density to reflect the magnetic flux in the outer rotor. By changing the yoke thickness of the outer rotor, the curves of the air-gap flux density amplitude in the DRM and SM are shown in Figure 12.

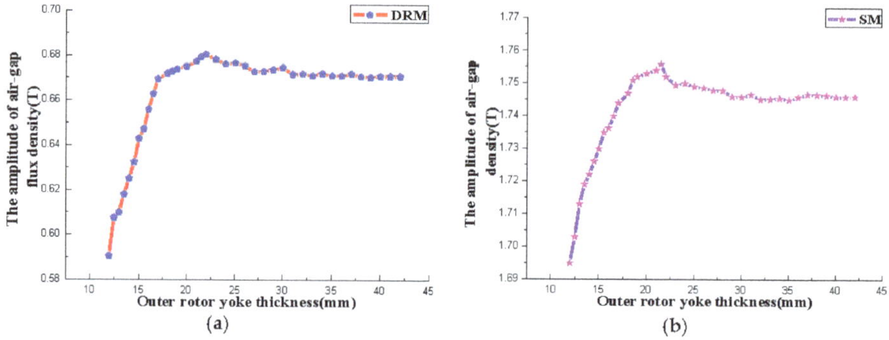

Figure 12. The air-gap flux density amplitude curves in the DRM and SM when the yoke thickness of the outer rotor is changing: (**a**) DRM; (**b**) SM.

It can be seen in Figure 12 that along with the yoke thickness of the outer rotor changing, the variation trends of the air-gap flux density amplitude in the DRM and SM are similarly. The amplitude of SM is higher than DRM, which is caused by the higher rated power. The maximum value of the air-gap magnetic density amplitude is achieved when the yoke thickness of the outer rotor is 22 mm. If the thickness is below 22 mm, the air-gap flux densities of the two motors drop rapidly, and when over 22 mm, the flux densities have little difference.

The flux lines of DRM and SM go through the air-gap into the outer rotor and then return, respectively. Therefore, the smaller the air-gap flux density amplitude is, the more non-linear the outer rotor is. The radial magnetic flux flows into the air-gap more easily, i.e., the radial magnetic flux produced by the inner layer permanent magnets flows into SM, and that produced by the outer layer permanent magnets flows into DRM; the original flux lines are offset. This will cause the great decrease of the air-gap magnetic density amplitude and magnetic properties.

Therefore, if we want to reduce the saturation degree of the outer rotor to realize decoupling, the yoke thickness of the outer rotor with the large air-gap flux density should be selected. We can see that in Figure 8, the air-gap magnetic density is higher when the yoke thickness is greater than 22 mm, that is the DRM and SM are decoupled. However, in order to meet the requirements of being lightweight and the miniaturization of the CSPM motor, we choose 22 mm as the yoke thickness of the outer rotor.

4.2.2. Verification of the Finite Element Method Analysis

According to the selected yoke thickness of the outer rotor and the design parameters of DRM and SM, we establish the electromagnetic simulation model based on 2D-FEM. Then, the magnetic field distribution diagram of the CSPM motor is obtained, as shown in Figure 13.

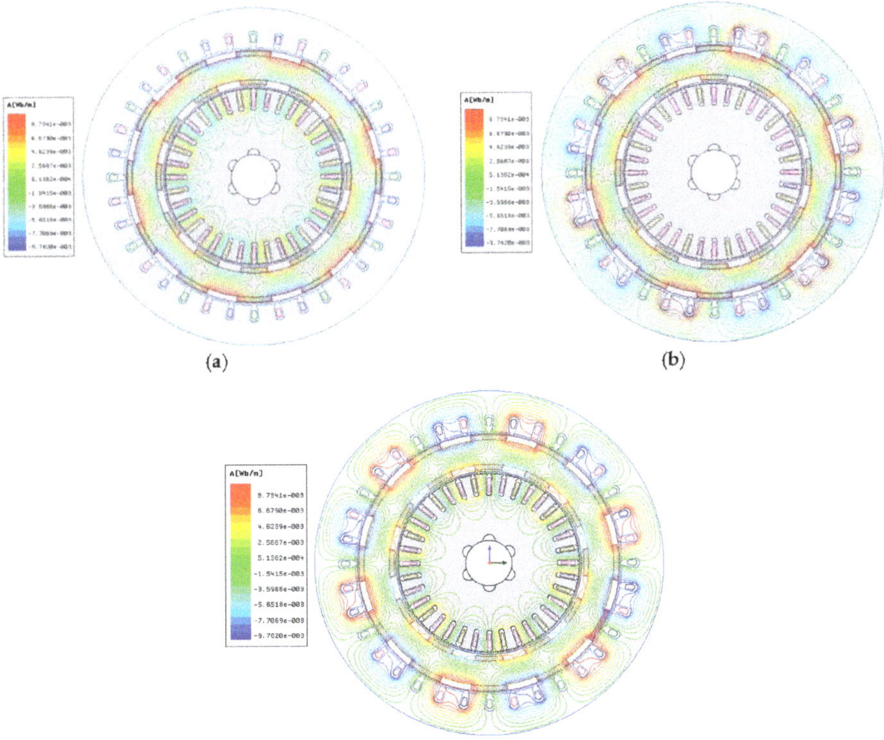

(a) (b) (c)

Figure 13. The magnetic flux line distributions of DRM, SM and the CSPM motor: (**a**) DRM; (**b**) SM; (**c**) CSPM motor.

From Figure 13, it can be clearly seen that when the SM and DRM work alone, the flux lines of the two motors have no cross by choosing the appropriate yoke thickness of the outer rotor. Only a small part of the leakage magnetic flux of SM flows into the DRM, and vice versa. When the two motors operate together, there are few radial coupling magnetic fluxes in the outer rotor. Figure 13c shows the numerical value of the magnetic field intensity of CSPM. We are aware that the values of the stator slots and inner rotor slots are huge, while in the stator and outer rotor yoke, the numerical value of the magnetic field intensity is small. Thus, the two motors can work without interference. At this time, the magnetic coupling degree of the two motors is very small and even can be ignored, i.e., the magnetic decoupling of the CSPM motor is complete. The control difficulty is decreased, and the output torque and power density are increased. Coupled with the appropriate control strategy, the excellent performance of the CSPM motor can be realized. This also avoids a large volume of the device, so further demonstrating the validity of the internal magnetic decoupling scheme, which provides a reference for the optimization design of the CSPM motor.

5. Conclusions

This paper analyzed the magnetic coupling in the CSPM motor resulting from the special structure. Firstly, by the analysis of the parameter matching with CSPM in the HEVs, we determined the design parameters of the CSPM motor, which has the same dynamic performances as the Prius series HEVs. Then, the internal equivalent magnetic circuit model and considering the tangential and radial flux magnetic flux distribution model in the outer rotor of CSPM motor are established. Then, we researched the internal magnetic coupling rule of the CSPM motor based on these two mathematical models. In order to realize the magnetic decoupling of DRM and SM, the magnetic flux passing through the outer rotor should be reduced, and the yoke thickness of the outer rotor should be increased. Finally, through the FEM electromagnetic simulation analysis, we proposed an internal magnetic decoupling scheme of the CSPM motor, which meets the requirements of miniaturization, being lightweight and easily controlled, laying the foundation for the application of the CSPM motor in HEVs.

Acknowledgments: This work was supported by the National Natural Science Foundation of China under Project No. 51507021, the Chongqing Science and Technology Commission of China under Project No. cstc2013jcyjA60001 and The State Key Laboratory of Power Transmission Equipment & System Security and New Technology in Chongqing University of China under Project No. 2007DA10512716303.

Author Contributions: Qiwei Xu and Jing Sun conceived and designed the experiments; Lingyan Luo performed the experiments; Shumei Cui and Qianfan Zhang analyzed the data; Qiwei Xu contributed analysis tools; Jing Sun wrote the paper.

Conflicts of Interest: The authors declare no conflict of interest.

References

1. Wu, L.H.; Wang, Y.N.; Yuan, X.F.; Chen, Z.L. Multiobjective optimization of HEV fuel economy and emissions using the self-adaptive differential evolution algorithm. *IEEE Trans. Veh. Technol.* **2011**, *60*, 2458–2470. [CrossRef]
2. Guo, X.H.; Wen, X.H.; Xu, L.Y.; Zhao, F.; Liu, J. Simulation and Experiment Analysis of Dynamic Process for an EVT Based on Dual Mechanical Port Electric Machine. In Proceedings of the Power Electronics and Motion Control Conference, Wuhan, China, 17–20 May 2009; pp. 2010–2014.
3. Ghayebloo, A.; Radan, A. Superiority of dual-mechanical-port-machine-based structure for series-parallel hybrid electric vehicle applications. *IEEE Trans. Veh. Technol.* **2016**, *65*, 589–602. [CrossRef]
4. Cheng, Y.; Cui, S.M.; Song, L.W.; Chan, C.C. The Study of the operation modes and control strategies of an advanced electromechanical converter for automobiles. *IEEE Trans. Magn.* **2007**, *43*, 430–433. [CrossRef]
5. Pouradbollah, M.; Murgovski, N.; Grauers, A.; Egardt, B. Optimal sizing of a parallel PHEV powertrain. *IEEE Trans. Veh. Technol.* **2013**, *62*, 2469–2479. [CrossRef]
6. Abdelsalam, A.A.; Cui, S.M. A Fuzzy Logic Global Power Management strategy for hybrid electric vehicles based on a permanent magnet electric variable transmission. *Energies* **2012**, *5*, 1175–1198. [CrossRef]
7. Zheng, P.; Thelin, P.; Chen, A.Y.; Nordlund, E. Influence of saturation and saliency on the inductance of a four-quadrant transducer prototype machine. *IEEE Trans. Magn.* **2006**, *42*, 1319–1322. [CrossRef]
8. Cheng, Y.; Espanet, C.; Trigui, R.; Bouscayrol, A.; Cui, S. Design of a Permanent Magnet Electric Variable Transmission for HEV Applications. In Proceedings of the IEEE Vehicle Power and Propulsion Conference, Lille, France, 1–3 September 2010.
9. Xu, Q.W.; Cui, S.M.; Song, L.W.; Zhang, Q.F. Research on the power management strategy of hybrid electric vehicles based on electric variable transmissions. *Energies* **2014**, *7*, 934–960. [CrossRef]
10. Chen, Y.; Trigui, R.; Espanet, C.; Bouscayrol, A.; Cui, S.M. Specifications and design of a PM electric variable transmission for Toyota Prius II. *IEEE Trans. Veh. Technol.* **2011**, *60*, 4106–4114. [CrossRef]
11. Murgovski, N.; Johannesson, L.M.; Egardt, B. Optimal battery dimensioning and control of a CVT PHEV powertrain. *IEEE Trans. Veh. Technol.* **2014**, *63*, 2151–2161. [CrossRef]
12. Zheng, P.; Liu, R.R.; Wu, Q.; Zhao, J.; Yao, Z.Y. Magnetic coupling analysis of four-quadrant transducer used for hybrid electric vehicles. *IEEE Trans. Magn.* **2007**, *43*, 2597–2599. [CrossRef]
13. Chen, L.; Zhu, F.T.; Zhang, M.M.; Huo, Y.; Yin, C.L.; Peng, H.E. Design and analysis of an electrical variable transmission for a series–parallel hybrid electric vehicle. *IEEE Trans. Veh. Technol.* **2011**, *60*, 2354–2363. [CrossRef]

energies

Article

Operation Modeling of Power Systems Integrated with Large-Scale New Energy Power Sources

Hui Li [1], Gengyin Li [1], Yaowu Wu [2], Zhidong Wang [3,*] and Jiaming Wang [3]

1 State Key Laboratory of Alternate Electrical Power System with Renewable Energy Sources, North China Electric Power University, Changping District, Beijing 102206, China; lihui@chinasperi.sgcc.com.cn (H.L.); ligy@ncepu.edu.cn (G.L.)
2 School of Electrical and Electronic Engineering, Huazhong University of Science and Technology, Wuhan 430074, Hubei, China; ywwu@mail.hust.edu.cn
3 State Power Economic Research Institute, Changping District, Beijing 102209, China; wangjiaming@chinasperi.sgcc.com.cn
* Correspondence: wangzhidong802@163.com; Tel.: +86-10-6341-1087

Academic Editor: Paolo Mercorelli
Received: 2 August 2016; Accepted: 30 September 2016; Published: 11 October 2016

Abstract: In the most current methods of probabilistic power system production simulation, the output characteristics of new energy power generation (NEPG) has not been comprehensively considered. In this paper, the power output characteristics of wind power generation and photovoltaic power generation are firstly analyzed based on statistical methods according to their historical operating data. Then the characteristic indexes and the filtering principle of the NEPG historical output scenarios are introduced with the confidence level, and the calculation model of NEPG's credible capacity is proposed. Based on this, taking the minimum production costs or the best energy-saving and emission-reduction effect as the optimization objective, the power system operation model with large-scale integration of new energy power generation (NEPG) is established considering the power balance, the electricity balance and the peak balance. Besides, the constraints of the operating characteristics of different power generation types, the maintenance schedule, the load reservation, the emergency reservation, the water abandonment and the transmitting capacity between different areas are also considered. With the proposed power system operation model, the operation simulations are carried out based on the actual Northwest power grid of China, which resolves the new energy power accommodations considering different system operating conditions. The simulation results well verify the validity of the proposed power system operation model in the accommodation analysis for the power system which is penetrated with large scale NEPG.

Keywords: new energy power generation (NEPG); power output scenario; power system operation model; coordinative optimization

1. Introduction

The technique of power system operation simulation is commonly used to simulate the annual and monthly operating states of all the power generation in a system considering the different boundary conditions of the actual power grid, the different operating characteristics of all kinds of generators, and the utilization efficiency of the new energy power generation (NEPG) and the hydroelectric power generation [1–3]. In recent years, NEPG (which in this paper mainly refers to wind power generation and photovoltaic (PV) power generation) has experienced fast development in China and reached considerable installed capacity, due to the government policy of limiting carbon emissions and the shortage of primary energy resources. Therefore, the operation model of a modern power system which is integrated with large-scale NEPG needs to be carefully reconsidered using correctly established NPEG power output models. An appropriate operation model of a power system considering large-scale integrated NEPG is of great significance for the study of the power system peak regulation ability investigation and the power generation accommodation assessment.

Current literatures on the stochastic production simulation algorithms of power systems mainly include the piecewise linear approximations method [4,5], the slab method [6], the equivalent power function method [7–9] and the half-invariant method [10–12]. The fundamental idea of all these algorithms is to convert the chronological load curve to a sequential load curve during which process the time information is lost and the time-variant characteristics of wind power and PV power cannot be reflected effectively. The multi-state machine modelling method [13] and Monte Carlo simulation method [14,15] are always used in modeling NEPG output for probabilistic production simulations. The multi-state generator modelling method can reflect well the randomness of power output and gauge the impacts of forced outages on the system reliability and production costs [16]. For example, the wind speed Weibull distribution is used for the stochastic production modelling of wind power based on the equivalent multi-state modeling method [17]. In [18,19], the β-distribution is used for the solar radiation intensity modeling for PV power output modeling, and the PV power output is modelled by the radiation intensity prediction based on the ARMA model [20]. The Monte Carlo simulation method based on the sample statistics is clear and simple, and can provide a comprehensive view of the influencing factors in power system operation [21]. The work in [22] introduces a mixed integer linear programming framework to establish the production schedule. In [23], the system efficiency, environmental protection, economic cost and resource availability are considered in the optimization objectives.

This paper aims to propose an effective power system operation model which considers the operation characteristics of large-scale NEPG and is adaptive for the peak optimization analysis and new energy power accommodation assessment. The power output characteristics of wind power and PV power generations are first analyzed according to the practical historical operating data. Then the characteristic indexes and the filtering principle of the NEPG historical output scenarios are introduced with the confidence level, and the calculation model of NEPG's credible capacity is proposed. Based on this, the operation model of power system with large-scale NEPG integrated is established considering the optimization objectives of minimum total costs of the power system and the power balance constraints. A series of comparative operation simulations based on the actual Northwest power grid of China are carried out considering different system operating conditions. With the simulation results, the validity of the proposed power system operation model in NEPG accommodation analysis of the power system with large scale NEPG penetration has been well verified.

2. NEPG Output Characteristics

The output characteristics of wind power generation and PV power generation are largely dependent on the wind source and illumination source, which display strong randomness and volatility. Therefore, the power output prediction of wind power generation and PV power generation requires abundant meteorological information and practical test data, which always have limited accuracy. However, the method based on statistics is proved appropriate to investigate the output characteristics of wind power generation and PV power generation.

2.1. Wind Power Output Characteristics

According to the practical time-series power output data of the test wind farms of Gansu Province in 2007~2008 [24], the indexes of wind power output characteristics, which are outlined below, are extracted and analyzed statistically on an hourly time scale:

(1) The wind power output shows a positive skewed distribution, and reached the top probability with 10% rated output as shown in Figure 1. When the statistical period extends, the probability

distribution curve tends to be smoother (due to space limitations, the following results are indicated with the conclusions only, but no figures).

(2) The cumulative probability of wind power output is a convex quadratic distribution. When the statistical period is extended, the cumulative probability distribution curve tends to be smoother.

(3) The daytime hourly output–guarantee rate of wind power is a concave quadratic distribution, and differs little for different hours.

(4) The hourly output of wind power with different guarantee rates is approximately an even distribution. When the statistical period is extended, the hourly output for different hours tends to be evener.

(5) The wind power output of different hours is highly correlated during a short time period. When the statistical period is extended, the time series correlation distribution tends to be evener.

(6) The wind farm output at the same hour in different days is highly random and volatile during the same season.

(7) The variation probability of wind power output within 10 minutes is approximately a positively skewed distribution. When the wind farm output variation rate is $\leq 1.5\%/\text{min}$, the probability is more than 99% [25].

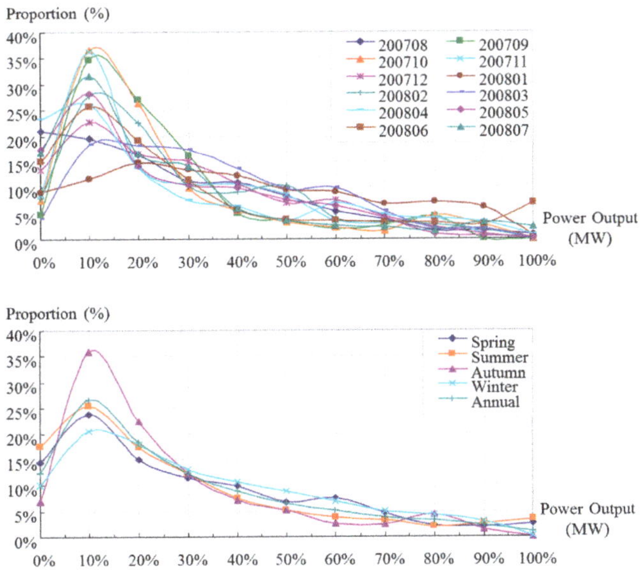

Figure 1. Probability distribution curves of wind power output in 2007~2008 of Gansu Province wind farms.

2.2. Photovoltaic Power Output Characteristics

According to the practical time-series power output data of the test PV farms of Qinghai Province in 2007~2008 [24], the indexes of wind power output characteristic are extracted and analyzed statistically using an hourly time scale, as outlined below:

(1) The PV power output is a positively skewed distribution, which reached the top probability with 70% rated output as shown in Figure 2. When the statistical period is extended, the probability distribution curve tends to smoother.

(2) The cumulative probability of PV power output is a convex quadratic distribution of which the noontime hour in X-axis is generally the symmetry axis. When the statistical period is extended, the cumulative probability distribution curve tends to be smoother.

(3) The PV power daytime hourly output–guarantee rate is a concave quadratic distribution, and differs obviously for different hours, being higher at noon and lower in the morning and evening.

(4) The variation probability of PV power output within 5 min is approximately a positive skewed distribution. When the wind farm output variation rate is $\leq$1%/min, the probability is more than 53.4%.

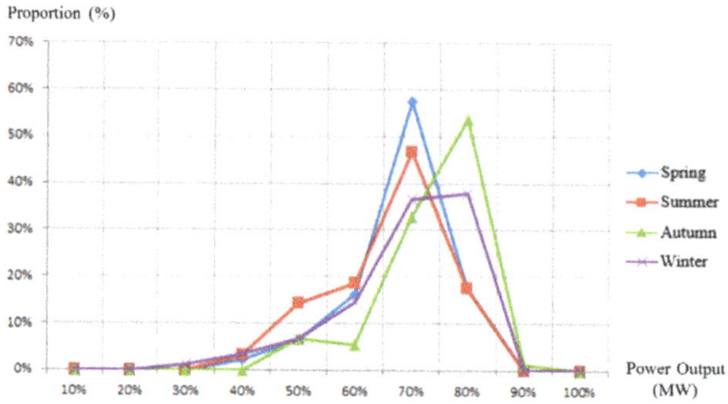

Figure 2. Probability distribution curve of PV power output in 2008 of Qinghai Province PV stations.

3. NEPG Output Scenarios Filtering

When considering the integration of large-scale NEPG into a power system, it's necessary to select the appropriate NEPG output scenarios with a satisfactory confidence level for the power balance analysis, load peaking analysis, and NEPG utilization assessment. To solve this problem, the characteristic indexes of wind power output scenarios and PV power output scenarios are introduced, respectively. The filtering principal of the wind power and photo power output scenarios are proposed consequently.

3.1. Characteristic Indexes of NEPG Output Scenarios

3.1.1. Characteristic Index of Wind Power Output Scenarios

Taking the impact of wind power output on the power system operation simulation into consideration, and considering the daily response characteristics of the load in month m of the area where the wind farm is integrated, the characteristic index X_{im} of wind power output scenarios is defined as follows:

(1) Power output at peak-load period in daytime (WP-PO): in the typical day in month m, the power output of wind farm i at the peak-load period in daytime.
(2) Peaking capacity (WP-PC): in the typical day in month m, the difference of wind farm output between the peak-load period and valley-load period in daytime.
(3) Average power output (WP-AO): the average value of wind farm's daily output in the typical day in month m.

The WP-PO is used for the power balance operation simulation, the WP-PC is used for the peaking balance operation simulation, and the WP-AO is used for the electricity balance operation simulation.

3.1.2. Characteristic index of PV Power Output Scenarios

Considering the inconsistency of the distributions between the daily PV power output and daily load response, the characteristic index X_{im} of PV power output scenarios is defined as follows:

(1) Maximum power output during daytime (PV-MO): in the typical day in month m, the maximum output of PV station i in daytime.

(2) Average power output in daytime (PV-AO): in the typical day in month m, the difference of wind farm output between the peak-load period and valley-load period in daytime.

The PV-MO is used for the power balance and peaking balance operation simulations, and the PV-AO is used for the electricity balance operation simulation. We standardize the characteristic indexes of both the wind power output scenarios and PV power output scenarios from the historical data of month m:

$$I_{im} = \frac{X_{im} - X_{im.\min}}{X_{im.\max} - X_{im.\min}} \quad (1)$$

where $X_{im.\min}$ and $X_{im.\max}$ stand for the minimum and maximum value of all the characteristic indexes in the history, respectively.

3.2. Filtering Principle of NEPG Output Scenarios

The filtering principle of the wind power output scenarios and PV power output scenarios work in the same way. Provided that the characteristic index X_{im} of NEPG station i's output scenarios in the typical day have the same distribution probability compared with the historical data, then it is possible to extract the typical scenario day of NEPG output from the historical scenarios days in the corresponding period with a specified confidence level.

Considering the uncertainty and the safety margin of power system operation, the confidence boundary of NEPG output scenarios is introduced here. We specify a confidence level denoted as α ($0 \leq \alpha \leq 1$), and build up a sequence $\{P_{im}\}$ of the NEPG station i's output scenarios from the historical data, then always exists some scenarios $\{P_{Wim}\}$ among $\{P_{im}\}$ of which the probability that the value of I_{Wim} larger than $\{I_{im}\}$ exceeds α [26]. With a given α, the typical scenario day of NEPG output in the level year is preliminarily filtered, located in the scenario sequences $\{P_{Wim}\}$ which meet the above conditions.

Since wind power and PV power are renewable energy sources so larger characteristic indexes are more favorable for the power system dispatching operation, the confidence boundary of NEPG output scenario is defined as follows: with a identified confidence level α, if there exists some scenarios $\{P_{Wim}\}$ among $\{P_{im}\}$ of which the probability that the value of I_{Wim} larger than $\{I_{im}\}$ exceeds α, the scenario which has the largest value of I_{Wim} among $\{P_{Wim}\}$ is defined as the confidence boundary of NEPG output scenario and the corresponding characteristic index of the can be described by [27]:

$$I_{Wim|\alpha} = \max[I_{im} : p(I_{Wim} > I_{im}) \geq \alpha] \ \forall \ \alpha \in [0,1] \quad (2)$$

Since $[p(I_{Wim} > I_{im}) \geq \alpha] = [p(I_{Wim} \leq I_{im}) < 1 - \alpha]$, the characteristic index of the confidence boundary of NEPG output scenario can also be described by:

$$I_{Wim|\alpha} = \max[I_{im} : p(I_{Wim} \leq I_{im}) < 1 - \alpha] \quad (3)$$

In addition, if known the cumulative probability distribution function $F(I_{Wim})$ or the probability density function $f(I_{Wim})$ of the characteristic index, the probability that $I_{wim} \geq I_{im}$ can be expressed by [28]:

$$p(I_{Wim} > I_{im}) = \int_{I_{im}}^{1} f(I_{Wim}) \mathrm{d}I_{im} = 1 - F(I_{Wim}) \quad (4)$$

With the above definition of the confidence boundary scenario, the filtering process of the typical scenario day of NEPG output in month m in the level year is as follows:

(1) Sequence the historical scenarios

Build up the sequence $\{P_{im}\}$ of the NEPG station i's output scenarios in month m from the historical data, sequence them according to the order of the value of I_{im}, as follows:

$$I_{im}(1) \leq I_{im}(2) \leq \cdots I_{im}(k) \leq \cdots I_{im}(n) \tag{5}$$

where $I_{im}(k)$ is the characteristic index of the kth scenario $P_{im}(k)$; n is the total number of historical scenarios.

(2) Construct the cumulative probability distribution function of the characteristic indexes

Based on (5), the cumulative probability distribution function of the characteristic indexes of NEPG station i in month m in history can be calculated by [26]:

$$F(I_{Wim}) = \begin{cases} 0, & I_{im} < I_{im}(1) \\ \frac{k}{n}, \quad I_{im}(k) \leq I_{im} < I_{im}(k+1), & k = 1, 2, \ldots, n-1 \\ 1, & I_{im} \geq I_{im}(n) \end{cases} \tag{6}$$

(3) Filter the typical scenario day

According to Equations (2)–(4), the characteristic index of the NEPG output confidence boundary $I_{Wim\,|\,\alpha}$ can be obtained by the cumulative probability distribution function from the historical characteristic indexes with a given confidence level. Since the cumulative probability distribution function is a step function but not a continuous function, the characteristic index of the scenario which is closest to the confidence boundary scenario is set as the boundary index $I_{Wim\,|\,\alpha}$, calculated by:

$$\begin{array}{c} [F(I_{im}(j)) - (1-\alpha)][F(I_{im}(j+1)) - (1-\alpha)] \leq 0 \\ \cap (1-\alpha) - F(I_{im}(j)) \leq F(I_{im}(j+1)) - (1-\alpha) \end{array} \tag{7}$$

$$\begin{array}{c} [F(I_{im}(j-1)) - (1-\alpha)][F(I_{im}(j1)) - (1-\alpha)] \leq 0 \\ \cap F(I_{im}(j)) - (1-\alpha) \leq (1-\alpha) - F(I_{im}(j-1)) \end{array} \tag{8}$$

where $I_{im}(j)$ is the characteristic index of wind farm output confidence boundary under the confidence level α. With the determined $I_{Wim\,|\,\alpha}$, the corresponding wind power output scenario $P_{Wim\,|\,\alpha}$ is also assured, set as the wind power output typical day in month m in the lever year.

3.3. Credible Capacity Resolution of NEPG

In the production operation simulation of a power system, a credible NEPG capacity needs to be calculated to make an accurate assessment of the NEPG utilization. The utilized capacity of NEPG station i in area s at hour t in the typical day in month m in the level year is described by:

$$P_{UTsmt} = P_{AVsmt} - P_{Wsmt} \tag{9}$$

where P_{AVsmt} and P_{Wsmt} are the total amount of the available capacity and the wasted capacity of NEPG station i in area s at hour t in the typical day in month m in the level year, respectively.

The total amount of the generated electricity and the available electricity of NEPG station i in area s at hour t in the typical day in month m in the level year are described by:

$$\begin{cases} E_{sm} = \sum\limits_{i \in s} \sum\limits_{t=1}^{24} P_{UTimt} \\ E_{AVsm} = \sum\limits_{i \in s} \sum\limits_{t=1}^{24} P_{AVimt} \end{cases} \tag{10}$$

The total amount of the generated electricity and the available electricity of NEPG station i in area s in month m in the level year are described by:

$$\begin{cases} E_{Msm} = \sum\limits_{i \in s} \sum\limits_{w=1}^{7} \sum\limits_{t=1}^{24} P_{UTimwt}/K_m \\ E_{MAVsm} = \sum\limits_{i \in s} \sum\limits_{w=1}^{7} \sum\limits_{t=1}^{24} P_{AVimwt}/K_m \end{cases} \tag{11}$$

where $P_{UTWimwt}$ and $P_{AVWimwt}$ are the utilized capacity and available capacity of NEPG station i at hour t in the typical day in typical week w in month m in the level year, respectively. K_m is the conversion factor between the weekly generating capacity and the monthly generating capacity of NEPG, described as K_m = day number in month $m/7$.

The daily utilization hour of wind power in a typical day in month m in the level year:

$$T_{sm} = E_{sm}/\sum_{i \in s} C_i \tag{12}$$

where C_i is the rated capacity of NEPG station i.

4. Power System Operation Modeling Considering Large-Scale NEPG Integration

4.1. Objective Function

In this work, the main optimization objective of the power system operation simulation considering large-scale integrated NEPG is set as the lowest power generation cost or the best energy saving effect. The sub-series optimization objectives also include the optimization of areas power exchange, the optimization of thermal power plants' boot capacity, and the optimization of the unit commitment scheme of different generation types. In the optimization process, the technical and economic characteristics of different power generation types are not identical, and their contributions to the power system operation optimization also differs. Besides this, several other technical and economic indicators are also considered, such as: (1) the power plant's maintenance plan; (2) the system load reserve and emergency reserve; (3) the electricity consumption and peak load balance in each area; (4) the annual electricity production, utilization hours, fuel demand and variable operating costs of different generations; (5) the wasted electricity of hydropower stations; and (6) the exchanged power flow between different areas. The objective function is expressed by:

$$\begin{aligned} \min F(P) &= \min E_{WQ} \\ &\cap \min \left\{ E_{HQ} \middle| \max \left(\overline{P}_{Hm} + R_{Hm} \right), m = 1,2,...,12 \right\} \\ &\cap \max \left\{ \overline{P}_{Pm}(T_E, E_Q) + R_{Pm}(T_E, E_Q), m = 1,2,...,12 \right\} \\ &\cap \max \left\{ P_{Am}(T_E) + R_{Am}(T_E), m = 1,2,...,12 \right\} \\ &\cap \max \left\{ \min(\Delta P_{sm}(n_{Mm}), m = 1,2,...,12) \right\} \\ &\cap \min\{ F(P_T) \mid \max \left(\min \left(\Delta P_{sm}, m = 1,2,...,12 \right) \right) \cap \min \left(\Delta E_s \right) \} \end{aligned} \tag{13}$$

where: $F(P)$ represents the total generation cost of the power system; E_{WQ} and E_{HQ} are the wasted electricity of NEPG and hydropower generation, respectively; $\overline{P}_{Hm}$ and $\overline{P}_{Pm}$ are the maximum power output of hydropower plants and pumped storage plants in month m in the level year, respectively; P_{Am} is the power output of nuclear generation in month m in the level year; R_{Hm}, R_{Pm} *and* R_{Am} are the arranged reserve capacity of hydropower plants, pumped storage plants and nuclear plants in month m, respectively; T_E is the expected annual utilization hours; ΔP_{sm} is the electricity surplus in area s in month m in the level year; n_{Mm} is the number of maintenance generators in month m in the level year; $F(P_T)$ is thermal power generations(or the primary energy consumption, the pollution emissions and the thermal power purchase cost), respectively; ΔE_s is the electricity shortage in area s in the level year.

4.2. Restraint Conditions

The power/electricity balance restraints and the generation operating restraints are the two main aspect mainly considered in power system operation simulation with the integration of large-scale NEPG.

4.2.1. Power/Electricity Balance Restraint

(1) Capacity balance restraints:

$$\sum_{i\in s} P_{imt} + \sum_{l\in s} P_{Llmt} = L_{smt} \tag{14}$$

$$\sum_{s\neq 0} L_{smt} = L_{0mt} \tag{15}$$

where L_{smt} is the power load of area s at moment t in month m in the level year. P_{imt} is power output of the hydropower plant i at moment t in month m in the level year. P_{Llmt} is the receiving power of area s through the transmission lines from other areas at moment t in month m in the level year.

(2) Restraint of load reserve and emergency reserve:

$$\begin{cases} \sum\limits_{i\in s} R_{Rim} + \sum\limits_{l\in s} R_{LRlm} = R_{Rsm} \geq R_{RNsm} \\ \sum\limits_{i\in s} R_{Sim} + \sum\limits_{l\in s} R_{LSlm} = R_{Ssm} \end{cases} \tag{16}$$

$$\begin{cases} \sum\limits_{s\neq 0} R_{Rsm} = R_{R0m} \\ \sum\limits_{s\neq 0} R_{Ssm} = R_{S0m} \end{cases} \tag{17}$$

where R_{Rsm} and R_{Ssm} are the spinning reserve capacity and the cold reserve capacity of area s in month m in the level year, respectively. R_{RNsm} is the lower limit of spinning reserve capacity of area s in month m in the level year. R_{Rim} and R_{Sim} are the spinning reserve capacity and the cold reserve capacity generated by generator i in month m in the level year, respectively. R_{LRlm} and R_{LSlm} are the receiving spinning reserve capacity and cold reserve capacity of area s through the transmission lines from other areas in month m in the level year.

(3) Peak load balance restraint

$$\sum_{i\in s} \Delta P_{im} + \sum_{l\in s} \Delta P_{Llm} \geq \Delta L_{sm} + R_{Rsm} \tag{18}$$

where ΔP_{im} and ΔP_{Llm} are the peaking capacity of the power plant i and transmission line l in the typical day in month m in the level year, respectively. ΔL_{sm} is the load peak-valley difference of area s in month m in the level year.

(4) Restraint of maximum transmission line capability:

$$C_{L-lm} \leq P_{Llmt} + R_{LRlm} + R_{LSlm} \leq C_{L+lm} \tag{19}$$

where C_{L+lm} and $C_{L\text{-}lm}$ are the maximum receiving capacity and maximum exporting capacity in transmission line l in month m in the level year.

(5) Electricity balance restraint:

$$\sum_{i\in s} E_{im} + \sum_{l\in s} E_{Llm} = E_{sm} \cap \sum_{s\neq 0} E_{sm} = E_{0m} \tag{20}$$

where E_{sm} is the forecasting load of area s in month m in the level year. E_{im} is the generation capacity of power plant i in month m in the level year. E_{Llm} is the receiving electricity of area s through transmission line l in month m in the level year.

(6) Security boot constraint:

$$\sum_{i \in s} n_{Tim} C_i + \sum_{i \in s} n_{Aim} C_i \geq C_{s\min} \tag{21}$$

where n_{Tim} and n_{Aim} are the boot numbers of thermal plants and nuclear plants in the level year, respectively. C_i is the capacity of power plant i. $C_{s\min}$ is the security boot capacity of area s.

(7) Restraint of lower limit of thermal plant's spinning reserve capacity:

$$\sum_{i \in s} R_{TRim} \geq R_{Ts\min} R_{Rsm} \tag{22}$$

where R_{TRim} is the spinning reserve capacity of thermal plant i in month m in the level year, and $R_{Ts\min}$ is the proportion of thermal plant i's spinning reserve capacity in the total amount of area s.

(8) Restraint of reserve capacity upper limit of hydropower plants and pumped storage power plants:

$$\sum_{i \in s} R_{Him} + \sum_{i \in s} R_{Pim} \leq R_{Hs\max} (R_{Rsm} + R_{Ssm}) \tag{23}$$

where R_{Him} and R_{Pim} are the reserve capacity of hydropower plant and pumped storage plant i in month m in the level year, respectively, and $R_{Hs\max}$ is the corresponding proportion of plant i's reserve capacity in the total amount of area s.

(9) Maintenance ability restraint of thermal plants

$$R_{Ms\min} C_{Ts} \leq \sum_{i \in s} n_{TMim} C_i \leq R_{Ms\max} C_{Ts} \tag{24}$$

where $R_{Ms\max}$ and $R_{Ms\min}$ are the upper limit and lower limit of thermal plant's maintenance ability, respectively. C_{Ts} is the total installed capacity of the thermal plants in area s. n_{TMim} is the number of the planned maintained generator in month m in the level year.

(10) Restraint of the emission of CO_2, SO_2, NO_x and soot

$$\begin{cases} \sum\limits_{i \in s} CO_{Ei} \leq CO_{Es\max} \\ \sum\limits_{i \in s} SO_{Ei} \leq SO_{Es\max} \\ \sum\limits_{i \in s} NO_{Ei} \leq NO_{Es\max} \\ \sum\limits_{i \in s} DS_{Ei} \leq DS_{Es\max} \end{cases} \tag{25}$$

where CO_{Ei}, SO_{Ei}, NO_{Ei} and DS_{Ei} are the CO_2, SO_2, NO_x and soot emissions of power plant i in the level year, respectively. $CO_{Es\max}$, $SO_{Es\max}$, $NO_{Es\max}$ and $DS_{Es\max}$ are the upper limit of CO_2, SO_2, NO_x and soot emissions of area s in the level year.

4.2.2. Generation Operating Restraints

(1) Output restraint:

$$\underline{P}_{im} \leq P_{imt} \leq \overline{P}_{im} \tag{26}$$

where $\overline{P}_m$ and $\underline{P}_{im}$ are the upper limit and lower limit of the power plant i's output in month m in the level year, respectively.

(2) Reserve capacity restraint:

$$0 \leq R_{im} \leq R_{i\max} \tag{27}$$

where R_i is the power plant i's reserve capacity in month m in the level year, and $R_{i\max}$ is the upper limit of R_i.

(3) Energy consumption restraint (represented by utilization hours):

$$\underline{T}_i \leq T_{im} \leq \overline{T}_i \tag{28}$$

where T_{im} is the power plant i's annual utilization hour in the level year, and $\overline{T}_i$ and $\underline{T}_i$ are the upper limit and lower limit of T_{im}.

(4) Maintenance restraint:

$$n_{Mim} \leq \overline{n}_{Mi} \tag{29}$$

where n_{Mim} is the actual maintained number of power plant and $\overline{n}_{Mi}$ is the upper limit.

(5) Electricity balance restraint of hydropower plants:

$$\sum_{t=1}^{24\times D_m} (P_{Himt} + P_{Qimt}) = 24D_m P_{HAVim} \tag{30}$$

where P_{HAVim} is the average output of hydropower plant i in month m, P_{Himt} is the output at hour t, and P_{Qimt} is the abandoned electricity in load peaking. D_m is the number of the days in month m.

(6) Daily electricity balance restraint of pumped storage plants:

$$\sum_{t=1}^{24} P_{Pimt} = \eta_i \sum_{t=1}^{24} L_{Pimt} \tag{31}$$

$$E_{PVmt} = \sum_{t=1}^{24} L_{Pimt} \leq E_{PVi} \tag{32}$$

where E_{PVi} and η_i are the max pumping capacity and the pumping-power conversion efficiency of the pumped storage plant i. E_{PVm}, P_{Pimt} and L_{Pimt} are the pumping power in the typical day, power output at hour t, and pumping load of the pumped storage plant i in month m in the level year, respectively.

(7) Operating quantity restraint of thermal plants:

$$\underline{n}_{im} \leq n_{Tim} \leq \overline{n}_{im} \tag{33}$$

where $\overline{n}_{im}$ and $\underline{n}_{im}$ are the upper limit and lower limit of the operating quantity of thermal plant i in month m in the level year, respectively.

(8) Start-off time restraint of thermal plants in load peaking:

$$t_{Rim} \geq \underline{t}_{Ri} \cap t_{Sim} \geq \underline{t}_{Si} \tag{34}$$

where t_{Rim} and t_{Sim} are the continuous boot hour and the continuous shutdown hour of the thermal plant i when during peaking period in the typical day in month m in the level year, respectively. $\underline{t}_{Ri}$ and $\underline{t}_{Si}$ are the lower limit of the continuous boot hour and the continuous shutdown hour, respectively.

4.3. Power System Operation Simulation Process

In this work, taking the technical and economic characteristics of different power generation types, and the specific power output scenarios of wind power generation and PV power generation into consideration, a power system operation simulation is conducted:

(1) Firstly, considering the limits of transmission capacity between different areas, the technical and economic characteristics of different power generations are evaluated. Applied with the sub-optimization processes on unit generation prioritizing, load distributing, load peaking and unit maintenance/reservation arrangement, the best area power exchange results are calculated to achieve the optimal technical and economic effect of all generation types.

(2) Then, according to the power balance of each area, the boot capacity of thermal power plants and unit combinations are optimized. When the peaking capacity of an area is lacking, the output of NEPG in other areas which transmit power into this area are previously reduced, and then we get the NEPG in the local area.

(3) Finally, with the optimized shutdown schedule of all the power plants, the detailed operating point and generation capacity of each power plant are calculated, according to the priority sequence of "New energy power → Hydro power → Pumped storage power → Nuclear power → Thermal power".

The operation simulation process of a multi-provinces interconnected power system considering the characteristics of large-scale integrated NEPG is shown in Figure 3.

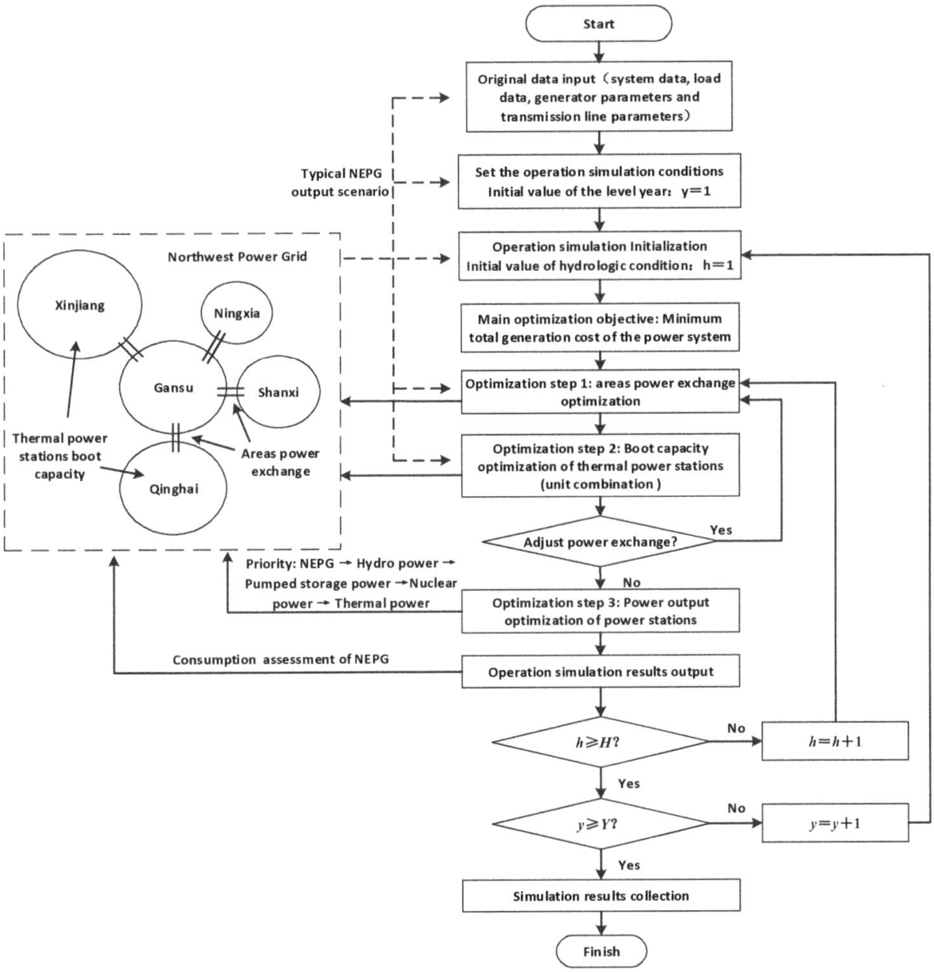

Figure 3. The operation flow diagram of a power system integrated with large-scale new energy power.

5. Simulation Analysis

5.1. Simulation Models

The power system operation simulation is carried out on the power system operation simulation software SPER-ProS, developed by the State Grid Corporation of China. The simulation model is established based on the Northwest power grid of China, which contains diverse energy resources and is integrated with large-scale thermal power, hydropower, wind power and PV power generations. Based on the power system operation model proposed above, the accommodation capability of NEPG in the five northwestern provinces are investigated through the power system operation simulation, in which the practical data of power loads, power exchanges between provinces, installed capacity of power plants, and power output characteristics of NEPG are utilized.

The 750 kV-level power grid structure of the Northwest power grid of China in 2014 is shown in Figure 4. It contains 41,750 kV transformer substations and five high-voltage direct current (HVDC) converter stations. As shown in Figure 4, the grey area represents the planning coal base in Zhundong, Xinjiang Province, the blue area represents the planning GW-level wind power base in Jiuquan, Gansu Province, and the yellow area represents the planning PV power plants centrally integrated in Qinghai Province.

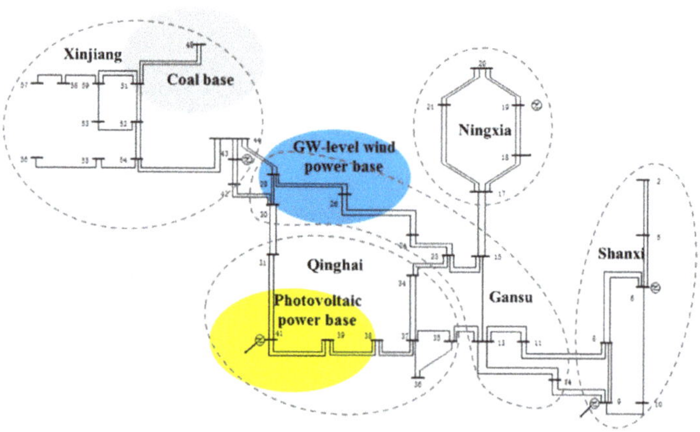

Figure 4. 750 kV power grid diagram of the Northwest power grid in 2014.

5.2. Boundary Conditions

Taking the year 2020 as the level year, the total power load of the Northwest power grid is expected to reach 147,010 MW, the installed hydropower capacity is 39,363 MW, the installed thermal power capacity is 222,642 MW, the installed wind power capacity is 75,480 MW, and the installed PV power capacity is 35,000 MW, as shown in Table 1.

Table 1. The planning power generation capacity of the Northwest power grid.

Generation Type	Installed Capacity (MW)
Hydropower station	39,363
Thermal power station	222,642
Wind power station	75,480
Photovoltaic power station	35,000

Considering the influence of the Hexi Corridor on the NEPG accommodation of the Northwest power grid, the Gansu power grid is modelled divided into three districts: the west, the central and the east, among which the west part is mainly referred to as the new energy centralized-integrated area which includes Jiuquan and Dunhuang, the central part is also referred to as the new energy centralized-integrated area which includes Jinchang and Wuwei, and the east part consists of the load center which includes East Lanzhou and Baiyin, and the coal base which includes Pingliang and Qingyang. For the same reason, the Qinghai power grid is modelled as divided into two districts:

the west and the east, among which the west part mainly refers to the large-scale PV integrated area which includes Haixi and Ge'ermu, and the east part consists of the load center which includes Xining, Riyueshan and Tala, the hydropower integrated area of the upreaches of the Huanghe River, and the PV power integrated area of Hainan.

The power flows of the key transmission sections in province or between provinces are calculated according to the transmission capacity shown as below:

- The first and second transmission sections between Xinjiang and the Northwest power grid: the maximum forward transmission capacity from Xinjiang to the Northwest power grid is up to 4000 MW and the minimum forward transmission power flow is 800 MW.
- The transmission section from Xinjiang to Qinghai: the maximum forward and backward transmission capacity is up to 2000 MW and the minimum transmission power flow is 400 MW.
- The transmission sections between Shanxi and Gansu: the maximum forward and backward transmission capacity are both 6000 MW.
- The transmission sections between Gansu and Ningxia: the maximum forward and backward transmission capacity are both 6000 MW.
- The transmission sections between Gansu and Hexi Corridor: the maximum transmission capacity of the 750 kV transmission line from Jiuquan to Hexi is 3000 MW.

In the power system operation simulation, the load peaking by hydropower is considered first, and the minimum output of the thermoelectric plants is 0.8 p.u. with an annual utilization of more than 4500 h.

5.3. NEPG Accommodation Ability Analysis

Based on the power system operation model proposed in this paper, the power output characteristic of NEPG has been taken into full consideration in the optimization process of our power system operation simulation. The NEPG's accommodation capacity in each province and the transmitted capacity between provinces can be both obtained clearly from the operation simulation results, which is really important for the modern power system operation simulation study with large-scale NEPG integration.

With the boundary conditions described above in the last section, a power system operation simulation of the Northwest power grid of China was conducted and the NEPG accommodation in the five provinces is clearly resolved, as shown in Tables 2 and 3. The accommodated electricity of NEPG in the global Northwest power grid reaches 1558.5 billion kWh and the wasted electricity is about 458 billion kWh, with the wasted electricity proportion being 22.7%. If the installed capacity of NEPG in the provinces where new energy power is heavily wasted is cut down the wasted electricity proportion of the global Northwest power grid declines to 5%, then the wind power accommodation capacity of the Northwest power grid is 40,910 MW and the PV power reaches 28,760 MW. The rest of the NEPG capacity transmitted to interconnected provinces is 40,810 MW, of which the wind power represents 34,570 MW and the PV power is 6240 MW.

In the situation that if 26,750 MW NEPG is transmitted from the Northwest power grid to other district power grids through the DC transmission lines, including the Hami-Zhengzhou ±800 kV DC transmission line, the Jiuquan-Hunan ±800 kV DC transmission line, the Zhundong-Huainan ±1100 kV DC transmission line, there will be 14,060 MW NEPG left that cannot being accommodated.

Table 2. Wind power accommodation in the Northwest power gird.

Power Grid	Installed Capacity (MW)	Local Accommodated Power (MW)	Outside Transmitted Power (MW)
Northwest power grid	75,480	40,910	34,570
Shanxi Province	4050	4050	0
Gansu Province	24,350	8780	15,570
Qinghai Province	1960	1960	0
Ningxia Province	14,080	14,080	0
Xinjiang Province	31,040	12,040	19,000

Table 3. Photovoltaic power accommodation in the Northwest power gird.

Power Grid	Installed Capacity (MW)	Local Accommodated Power (MW)	Outside Transmitted Power (MW)
Northwest power grid	35,000	28,760	6240
Shanxi Province	4000	4000	0
Gansu Province	8500	4260	4240
Qinghai Province	10,000	10,000	0
Ningxia Province	4000	4000	0
Xinjiang Province	8500	6500	2000

5.4. *Sensitivity Analysis*

With the sensitivity analysis of the thermoelectric plants' output, the system peak capacity and the transmission lines' capacity, the NEPG accommodation conditions of the Northwest power gird are further investigated in this section.

5.4.1. Sensitivity Analysis of the Thermoelectric Plants' Output

In an actual power system, during the early period and the final period of the heating season, the minimum technical output of the thermoelectric plants is generally more than 0.55 p.u. and during the middle period of the heating season, the minimum technical output of the thermoelectric plants is generally more than 0.75 p.u. According to practical surveys that the minimum technical output of the thermoelectric plants can even be adjusted to 0.5 p.u. during the early period and the final period of the heating season, and 0.6 p.u. during the middle period of the heating season. Therefore, compared to the present operation status, there is generally 0.1 to 0.15 p.u. space in the thermoelectric plant's capacity to peak load. We sdjust the thermoelectric plants' output according to this margin, and conduct the power system operation simulation again. The power system operation simulation results gives the new accommodation conditions of NEPG showing that the wind power and PV power in the Northwest power grid can be well accommodated.

5.4.2. Sensitivity Analysis of System Peaking Capacity

Based on the boundary conditions described in Section 4.2, in this comparative case, the power system operation simulation is conducted with the minimum power output level of thermal power plant adjusted to 0.7 p.u. from 0.5 p.u., while the minimum technical output of the thermoelectric plants remains the same. The power system operation simulation gives the new NEPG accommodation conditions, and from the simulation results presented in Tables 4 and 5, the wind power accommodation capacity of the global Northwest power grid declines to 26,160 MW and the PV power accommodation capacity declines to 17,890 MW. It's suggested that the NEPG accommodation abilities in the five provinces have all been depressed due to the decline of system peaking capacity which is caused by higher minimum technical output of the thermoelectric plants.

Table 4. Local accommodated power of wind power considering sensitivity analysis (MW).

Power Grid	Basic Scheme	Change the Thermal Power Plants' Minimum Output	Change the Transmission Lines' Capacity
Northwest power grid	40,910	26,160	37,750
Shanxi Province	4050	3850	4050
Gansu Province	8780	3230	7620
Qinghai Province	1960	1960	1960
Ningxia Province	14,080	11,080	14,080
Xinjiang Province	12,040	6040	10,040

Table 5. Local accommodated power of PV power considering sensitivity analysis (MW).

Power Grid	Basic Scheme	Change the Thermal Power Plants' Minimum Output	Change the Transmission Lines' Capacity
Northwest power grid	28,760	17,890	26,940
Shanxi Province	4000	3160	4000
Gansu Province	4260	1000	2440
Qinghai Province	10,000	7500	10,000
Ningxia Province	4000	2730	4000
Xinjiang Province	6500	3500	6500

5.4.3. Sensitivity Analysis of Transmission Lines' Capacity

Still based on the above boundary conditions, but in this sensitivity study case the capacity of the transmission lines from Gansu province to other areas is cut down by 2000 MW. The power system operation simulation results are also presented in Tables 4 and 5. Since the transmission lines from Gansu Province to other areas are the main channel which transmits the electricity generated in Xinjiang and Gansu to the load center in the Northwest power grid, the cutting down in these transmission lines' capacity directly results in the delineation of the NEPG accommodation in Xinjiang and Gansu provinces. However, the NEPG accommodation in the other three provinces has not been affected.

6. Conclusions

This paper aimed to propose an effective power system operation model which considers the operation characteristics of large-scale NEPG and is adaptive for the peak optimization analysis and new energy power accommodation assessment. The power output characteristics of wind power and PV power generations are firstly analyzed according to the practical operating data. Then the characteristic indexes and the filtering principle of the NEPG historical output scenarios are introduced with the confidence level, and s calculation model of NEPG's credible capacity is proposed. Based on this, the operation model of power systems with large-scale NEPG integrated is established considering the optimization objectives of minimum total costs of the power system and the power balance constraints. The operation simulation results based on the actual Northwest power grid of China under different system operating conditions shows that the new energy power accommodation can be resolved effectively based on the proposed power system operation model.

Though three kinds of scenarios of wind power output has been taken into consideration in this paper, the objective function of power system operation simulation and the wind power output scenarios are much more complex in actual power systems, which may result in different power system operation models and this needs to be further analyzed in future.

Acknowledgments: This research work was supported by the National High Technology Research and Development Program of China: "Planning Research and Operation Basic Theory of Power Systems with Integration of High Penetration of Renewable Energies". The authors are grateful for this support.

Author Contributions: Hui Li and Gengyin Li conceived and designed the experiments; Hui Li performed the experiments; Hui Li, Yaowu Wu and Zhidong Wang analyzed the data; Jiaming Wang contributed analysis tools; Hui Li wrote the paper.

Conflicts of Interest: The authors declare no conflict of interest.

References

1. Lou, S.H.; Lu, S.Y.; Wu, Y.W.; Yin, X.G. An Overview on Low-Carbon Power System Planning and Operation Optimization. *Power Syst. Technol.* **2013**, *6*, 1483–1490.
2. Ma, J.; Shi, J.L.; Li, W.Q.; Wang, Z.P. A day-ahead dispatching strategy for power pool composed of wind farms, photovoltaic generations, pumped-storage power stations, gas turbine power plants and energy storage systems based on multi frequency scale analysis. *Power Syst. Technol.* **2013**, *6*, 1491–1498.
3. Wu, H.L.; Du, E.; Men, K.; Yang, L.; Su, L.; Lu, S.Y.; Luo, X. A Low-Carbon Oriented Energy-Saving and Economic Operation Evaluation System. *Power Syst. Technol.* **2015**, *5*, 1179–1185.
4. Baleriaux, E.J.; Guertechin, L.E. Simulation de I'exploitation d'un Parc de Machines Thermiques de Production d'Electricite Couple a des Stations de Pompage. *Revue E* **1967**, *5*, 3–24.
5. Booth, R.R. Power system simulation model based on probability analysis. *IEEE Trans. Power Syst.* **1972**, *91*, 62–69. [CrossRef]
6. Schenk, K.F.; Misra, R.B.; Vassos, S. A new method for the evaluating of expected energy generation and loss of probability. *IEEE Trans. Power Syst.* **1984**, *103*, 294–303. [CrossRef]
7. Wang, X. Equivalent energy function approach to power system probabilistic modeling. *IEEE Trans. Power Syst.* **1988**, *3*, 823–829. [CrossRef]
8. Wang, X. EEF approach to power system probabilistic modeling. *J. Xi'an Jiao Tong Univ.* **1984**, *18*, 13–26.
9. Li, L.; Wang, X.; Wang, X. Probabilistic modeling for inter connected power systems based on the equivalent energy function approach. *Proc. CSEE* **1996**, *16*, 180–184.
10. Rau, N.S.; Toy, P.; Schenk, K.F. Expected energy production costs by the method moments. *IEEE Trans. Power Apparatus Syst.* **1980**, *99*, 1908–1917. [CrossRef]
11. Stremel, J.P.; Jenkins, R.T.; Babb, R.A. Production costing using the cumulant method of representing the equivalent load curve. *IEEE Trans. Power Apparatus Syst.* **1980**, *99*, 1947–1956. [CrossRef]
12. Chen, G.; Xiang, N.; Chen, X. A new method for probabilistic production simulation based on load decomposition. *Proc. CSEE* **1992**, *12*, 47–52.
13. Liao, Y. Research on power system probabilistic production simulation including wind farms. Master's Thesis, North China Electric Power University, Beijing, China, 2012.
14. Wang, X.; Dai, H.; Thomas, R.J. Reliability modeling of large wind farms and associated electric utility interface systems. *IEEE Trans. Power Apparatus Syst.* **1984**, *103*, 569–575. [CrossRef]
15. Zhang, J.T.; Cheng, H.Z.; Hu, Z.C.; Ma, Z.L.; Zhang, J.P.; Yao, L.Z. Power system probabilistic production simulation including wind farms. *Proc. CSEE* **2009**, *29*, 34–39.
16. Billinton, R.; Yi, G.; Karki, R. Application of a joint deterministic-probabilistic criterion to wind integrated bulk power system planning. *IEEE Trans. Power Syst.* **2010**, *25*, 1384–1392. [CrossRef]
17. Zou, X. Research of probabilistic production simulation of power system with wind power and efficiency evaluation of wind farm. Master's Thesis, Xi'an Jiaotong University, Xi'an, China, 2008.
18. Li, Y.; Zio, E. Uncertainty analysis of the adequacy assessment model of a distributed generation system. *Renew. Energy* **2012**, *41*, 235–244. [CrossRef]
19. Atwa, Y.M.; ElSaadany, E.F.; Salama, M.M.A. Optimal renewable resource mix for distribution system energy loss minimization. *IEEE Trans. Power Syst.* **2010**, *25*, 360–370. [CrossRef]
20. Guo, X.Y.; Xie, K.; Hu, B. A time-interval based probabilistic production simulation of power system with grid-connected photovoltaic generation. *Power Syst. Technol.* **2013**, *37*, 1500–1505.
21. Dobakhshari, A.S.; Fotuhi-Firuzabad, M. A reliability model of large wind farms for power system adequacy studies. *IEEE Trans. Energy Convers.* **2009**, *24*, 792–801. [CrossRef]
22. Ben-Awuah, E.; Elkington, T.; Askari-Nasab, H.; Blanchfield, F. Simultaneous production scheduling and waste management optimization for an oil sands Application. *J. Environ. Inform.* **2015**, *26*, 80–90.
23. Tan, Q.; Huang, G.; Cai, Y. A fuzzy evacuation management model oriented toward the mitigation of emissions. *J. Environ. Inform.* **2015**, *25*, 117–125. [CrossRef]
24. Li, H.; Wang, Z.; Guo, F. *Research on the Theory and Method of Large Power Network Planning with Large-Scale Centralized Integration of Multi Form Power Generations*; State Power Economic Research Institute: Beijing, China, 2013.

25. Li, H.; Wang, Z. *Transmission Planning and Designing for the Seven 10 Million kW Level Wind Power Bases*; State Power Economic Research Institute: Beijing, China, 2008.
26. Degroot, M.H.; Schervish, M.J. *Probability and Statistics*; Addison-Wesley: Boston, MA, USA, 2002.
27. Siegmund, D. *Sequential Analysis: Tests and Confidence Intervals*; Springer Science & Business Media: New York, NY, USA, 2013.
28. Papoulis, A.; Pillai, S.U. *Probability, Random Variables, and Stochastic Processes*; Tata McGraw-Hill Education: Boston, MA, USA, 2002.

energies

MDPI

Article

Feedback Linearization Controller for a Wind Energy Power System

Muthana Alrifai *, Mohamed Zribi and Mohamed Rayan

Department of Electrical Engineering, Kuwait University, P. O. Box 5969, Safat 13060, Kuwait; mohamed.zribi@ku.edu.kw (M.Z.); mmsrayan@gmail.com (M.R.)
* Correspondence: m.alrifai@ku.edu.kw; Tel.: +965-2498-7381

Academic Editor: Paolo Mercorelli
Received: 14 June 2016; Accepted: 19 September 2016; Published: 23 September 2016

Abstract: This paper deals with the control of a doubly-fed induction generator (DFIG)-based variable speed wind turbine power system. A system of eight ordinary differential equations is used to model the wind energy conversion system. The generator has a wound rotor type with back-to-back three-phase power converter bridges between its rotor and the grid; it is modeled using the direct-quadrature rotating reference frame with aligned stator flux. An input-state feedback linearization controller is proposed for the wind energy power system. The controller guarantees that the states of the system track the desired states. Simulation results are presented to validate the proposed control scheme. Moreover, further simulation results are shown to investigate the robustness of the proposed control scheme to changes in some of the parameters of the system.

Keywords: wind energy; doubly-fed induction generator (DFIG); feedback linearization

1. Introduction

There has been a global interest in renewable energy resources due to the worldwide increase in power demand and the limitation of fossil fuels and their harmful impact on the environment. Renewable energy resources, such as wind energy, are naturally available, clean and have a much less harmful impact on the environment than fossil fuels.

Wind energy is the fastest growing renewable energy resource. Globally, the annual cumulative installed wind energy capacity has increased rapidly during the period from 1997–2014 [1]. In addition, the advancement on the design of the components of wind energy power systems that include power electronics inverters, electric generators and drive train systems has also contributed to the fast growth and high demand of wind energy conversion systems (WECs). The increasing market share of variable speed wind energy conversion systems has led to further investigations of wind turbines' control technology. Furthermore, researchers have recognized that appropriate control algorithms can greatly improve the efficiency of wind power conversion systems.

Among the various types of available wind turbines, the doubly-fed induction generator-based wind turbines are widely used for variable speed wind turbine systems because of their simple structures, their reliable operations, their high power densities and their energy efficiency. Moreover, doubly-fed induction generators-based energy systems have other advantages, such as the reduction of mechanical stress, the flexibility in controlling active and reactive powers and the ability to track maximum power using different control techniques.

Many research articles dealing with the design of different types of control schemes for DFIG-based variable speed wind turbine systems were published in the last few years; for example, the reader can refer to the works in [2–16] for a detailed review of wind turbine systems and their control. Some of these control techniques are highlighted below.

Researchers have used different control techniques to control DFIG-based wind turbine energy systems. For example, PID controllers and several modified PID control schemes were used to control wind energy conversion systems [17–19]. These types of schemes are believed to be the easiest and simplest design methods. However, DFIG-based wind energy systems are highly nonlinear systems, which are characterized by strong couplings between the different variables of the systems. Thus, nonlinear control schemes need to be designed to control wind energy power systems so that the desired performances are attained.

The work in [20] proposed a method for obtaining the maximum power output of a DFIG-based wind energy system. The proposed control technique was based on the usage of an improved maximum power point tracking (MPPT) curve method and a Lyapunov function. In [21], an artificial neural network technique based on Markov scheme approaches was developed to optimize the output of a wind turbine system. An artificial neural network, which is based on the Jordan re-current concept, was used to estimate the reference tracking speed of the rotor in [22]. A direct and an indirect control structure using a quantum neural network to enhance the efficiency of the system and to optimize the output of a wind energy power system was proposed in [23]. Furthermore, the combination of fuzzy logic control and sliding mode control was investigated in the literature; for example, refer to the works in [24–28].

Investigations to determine the best nonlinear control technique that can be used to improve the performance of DFIG-based WEC systems are still ongoing [29]. The control of both the grid side converter (GSC) and the rotor side converter (RCS) in one control loop was reported in [30]. In [31], a direct active power and reactive power controller based on the estimation of the stator flux was proposed, and a basic hysteresis controller was used. Vector control and direct power control were proposed in [32] to regulate the active power and reactive power generated by a DFIG-based wind energy system. In [33], two distinct control strategies were used to control a DFIG-based WEC system; backstepping and sliding mode control strategies were first used to control the rotor side converter (RCS); then, the same strategies were used to control the grid side converter (GSC). A perturbation observer-based control scheme for the control of DFIG in a multi-machine power system was introduced in [34]. The controller is achieved with a four-loop perturbation observer-based control configuration, and the controlled dynamics were investigated through simulation studies.

Several researchers investigated the usage of the feedback linearization technique for the control of DFIG-based WEC systems. Some of these works are highlighted below. An input-output feedback linearization controller for a DFIG-based system connected to an infinite bus was discussed in [35]; the authors differentiated the electromagnetic torque, the stator power and the grid side converter power outputs of the DFIG to obtain a linear relationship with the rotor voltages, which serve as control inputs. Then, decoupled torque control and decoupled power control were developed. In [36], an input-output linearizing and decoupling control strategy was proposed for a doubly-fed induction generator system. The developed controller was verified by using a 7.5 kW wind power test rig. A mathematical model of the DFIG based on the so-called stator magnetizing current field orientation was given in [37]; then, an input-output linearizing controller is proposed. The controller results in decoupled control and the tracking of the generated active and reactive powers of the system. Direct control of the torque and direct control of the power outputs of the DFIG were established based on the input-output feedback linearization technique in [38,39]. In [40], an adaptive nonlinear control strategy based on a feedback linearization scheme was proposed for a DFIG-based WEC system; a disturbance observer was added to estimate the uncertainties in the parameters of the system. The performance of the system is checked in the presence of nearby faults. Moreover, an exact feedback linearization scheme was proposed in [41] for a DFIG-based WEC system. The wind turbine with DFIG is represented by a third-order model. This model is exactly linearized, and a linear quadratic regulator is used to design an optimal controller for the linearized system. Simulations were performed on a single machine infinite bus system and on a four-machine system; the simulation results indicate that the proposed control scheme improves the transient stability of the power system, and it enhances the damping of the system.

The contribution of this paper involves the design of an input-state feedback linearization controller for a wind energy conversion system. Previous works on feedback linearization of WEC systems, generally, used fifth order models or sixth order models. Our work uses an eighth order model of the WEC system; the state space representation of the eighth order model under consideration uses four electrical states and four mechanical states. The proposed controller controls the rotor side converter (RCS) of a DFIG. It guarantees the stability of the closed loop system and enables the tracking of the desired states of the system. In addition, the proposed controller maximizes the power extracted from the wind. The proposed theoretical work is verified through MATLAB simulations; moreover, the validity of the proposed controller was tested when some of the parameters of the system change.

The rest of the paper is organized as follows. Section 2 presents the model of a DFIG-based wind energy power system; the system consists of eight nonlinear ordinary differential equations. In Section 3, the computations of the values of the desired states of the system, as well as the reference inputs are discussed in detail. Section 4 presents the design of an input-state feedback linearization controller for the wind energy conversion system. To show the effectiveness of the proposed controller, simulation studies are presented in Section 5. In addition, some simulation results are shown to investigate the robustness of the controller to changes in some of the parameters of the system. Finally, the conclusion is drawn in Section 6.

2. Model of the Wind Energy Conversion System

The model of a DFIG-based wind energy conversion power system is presented in this section. A block diagram depiction of the system is shown in Figure 1. The system consists of a wind turbine connected through a gear box to a doubly-fed induction generator. The stator windings of the standard wound induction generator are directly connected to the grid. The rotor windings are connected to the grid through a voltage source and power electronics converters (AC/DC/AC). The wind energy is captured by the motor blades and transferred to the motor hub. The hub has a gearbox, which attaches the low-speed shaft to the high-speed shaft that drives the DFIG, which converts the mechanical energy to electrical energy. This electrical energy is delivered to the grid.

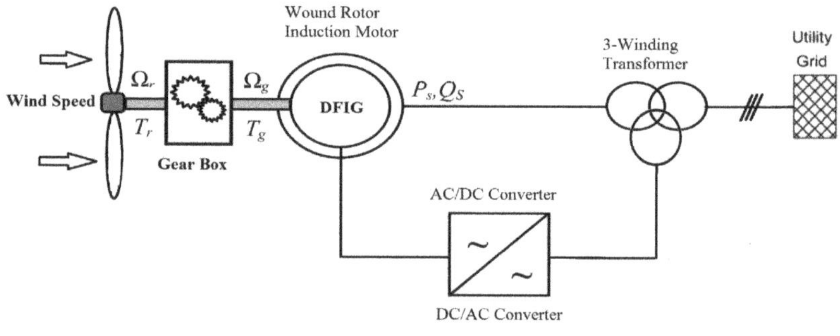

Figure 1. A block diagram representation of the variable speed DFIG-based wind energy conversion system.

Note that Ω_r, Ω_g, T_h and T_g depicted in Figure 1 represent the rotational speed of the turbine on the low-speed side of the gearbox, the mechanical speed of the generator, the high-speed shaft torque and the generator torque, respectively. Furthermore P_s and Q_s are the stator active power and the stator reactive power.

The mechanical power extracted from the wind can be expressed as [42],

$$P_m = \frac{1}{2}C_p(\lambda, \beta)\rho\pi R^2 V^3 \tag{1}$$

where:

P_m: the captured wind power (the mechanical power),
$C_p(\lambda, \beta)$: the power coefficient,
λ: the tip speed ratio (TSR),
β: the pitch angle,
ρ: the air density,
R: the rotor-plane radius,
V: the speed of the wind.

The tip speed ratio can be expressed as,

$$\lambda = \Omega_r R/V \tag{2}$$

where Ω_r is the rotational speed of the turbine on the low-speed side of the gearbox.

The power coefficient C_p versus the tip speed ratio λ for different values of the pitch angle β is depicted in Figure 2. It can be seen from the figure that when $\beta = 0$, the maximum value of C_p is achieved when $C_{p(\max)} = 0.48$ at an optimal value of the tip speed ratio $\lambda_{opt} = 8$. To capture the maximum power from the wind, a variable speed wind turbine normally follows the $C_{p(\max)}$ up to the rated speed by varying the rotor speed to keep the tip speed ratio at its optimum value λ_{opt}.

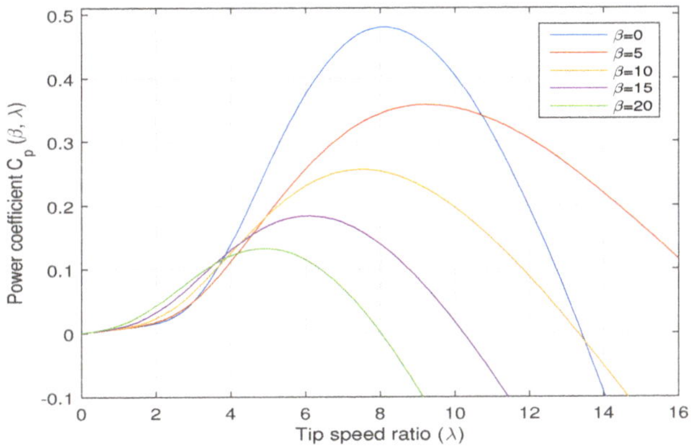

Figure 2. The power coefficient C_p versus the tip speed ratio λ.

Using Equation (1), the maximum captured wind power, $P_{m(\max)}$, can be written as,

$$P_{m(\max)} = \frac{1}{2} C_{p(\max)}(\lambda_{opt}, \beta)\rho\pi R^2 V^3. \tag{3}$$

The aerodynamic torque, T_r, can be expressed as follows,

$$T_r = P_m/\Omega_r = \frac{1}{2} C_p(\lambda, \beta)\rho\pi R^2 V^3/\Omega_r. \tag{4}$$

The wind turbine varies its speed by following the maximum value of the power coefficient $C_{p(\max)}$ so that maximum power is captured. This is done by changing the speed of the rotor to maintain the tip speed ratio at its optimum value λ_{opt} [42]. Using Equations (2)–(4), the aerodynamic torque, T_r, can be written as follows,

$$T_r = K_{opt}\Omega_r^2 \tag{5}$$

where $K_{opt} = 0.5\rho\pi R^5 C_{p(\max)}/\lambda_{opt}^3$.

To model the wind energy power system, we first need to define the state variables of the system. The system consists of an electrical part and a mechanical part. The state variables of the electrical part of the system are the stator currents in the d and q axes (i_{sd} and i_{sq}) and the rotor currents in the d and q axes (i_{rd} and i_{rq}). The state variables of the mechanical part of the system are the rotational speed of the turbine on the low-speed side of the gearbox (Ω_r), the mechanical speed of the generator (Ω_g), the high-speed shaft torque (T_h) and the generator torque (T_g). The inputs for the system are the stator and rotor voltages and the required generator torque ($T_{g,r}$). Note that the voltage u_{sd} is the d-axis stator voltage, and u_{sq} is the q-axis stator voltage. The voltage u_{rd} is the d-axis rotor voltage, and u_{rq} is the q-axis rotor voltage.

The rotational speed of the generator Ω_g is related to the rotor angular frequency ω_r, such that,

$$\Omega_g = \frac{1}{n_p}\omega_r \tag{6}$$

where n_p is the number of pole pairs of the generator.

The stator angular frequency ω_s is related to the stator frequency, f_s, such that,

$$\omega_s = 2\pi f_s. \tag{7}$$

Furthermore, the leakage coefficient can be written, such that,

$$\xi = \frac{1 - L_m^2}{L_s L_r} \tag{8}$$

where L_m is the magnetizing inductance, L_r is the rotor leakage inductance and L_s is the stator leakage inductance.

The model of the wind energy power system consists of eight first order ordinary differential equations (odes). The first four odes are related to the electrical part of the system; the second four odes are related to the mechanical part of the system. Hence, the wind turbine energy conversion system can be described using the following set of nonlinear ordinary differential equations, such that [42,43],

$$\begin{aligned}
\frac{di_{sd}}{dt} &= \frac{-R_s}{\xi L_s}i_{sd} + \omega_s i_{sq} + \frac{1-\xi}{\xi}n_p\Omega_g i_{sq} + \frac{R_r L_m}{\xi L_r L_s}i_{rd} + \frac{L_m}{\xi L_s}n_p i_{rq}\Omega_g - \frac{1}{\xi L_s}u_{sd} + \frac{L_m}{\xi L_r L_s}u_{rd} \\
\frac{di_{sq}}{dt} &= -\omega_s i_{sd} - \frac{1-\xi}{\xi}n_p\Omega_g i_{sd} + \frac{-R_s}{\xi L_s}i_{sq} - \frac{L_m}{\xi L_s}n_p\Omega_g i_{rd} + \frac{R_r L_m}{\xi L_r L_s}i_{rq} - \frac{1}{\xi L_s}u_{sq} + \frac{L_m}{\xi L_r L_s}u_{rq} \\
\frac{di_{rd}}{dt} &= \frac{R_s L_m}{\xi L_r L_s}i_{sd} - \frac{L_m}{\xi L_r}n_p\Omega_g i_{sq} - \frac{R_r}{\xi L_r}i_{rd} + \omega_s i_{rq} - \frac{1}{\xi}n_p\Omega_g i_{rq} + \frac{L_m}{\xi L_r L_s}u_{sd} - \frac{1}{\xi L_r}u_{rd} \\
\frac{di_{rq}}{dt} &= \frac{L_m}{\xi L_r}n_p\Omega_g i_{sd} + \frac{R_s L_m}{\xi L_r L_s}i_{sq} - \omega_s i_{rd} + \frac{1}{\xi}n_p\Omega_g i_{rd} - \frac{R_r}{\xi L_r}i_{rq} + \frac{L_m}{\xi L_r L_s}u_{sq} - \frac{1}{\xi L_r}u_{rq} \\
\frac{d\Omega_r}{dt} &= -\frac{D_r}{J_r}\Omega_r + \frac{K_{opt}}{J_r}\Omega_r^2 - \frac{n_b}{J_r}T_h \\
\frac{d\Omega_g}{dt} &= -\frac{D_g}{J_g}\Omega_g + \frac{1}{J_g}T_h - \frac{1}{J_g}T_g \\
\frac{dT_h}{dt} &= \frac{1}{n_b}(K_{ls} - \frac{D_r D_{ls}}{J_r})\Omega_r + \frac{1}{n_b}(\frac{D_{ls}K_{opt}}{J_r})\Omega_r^2 - \frac{1}{n_b^2}(K_{ls} - \frac{D_g D_{ls}}{J_g})\Omega_g \\
&\quad - D_{ls}(\frac{1}{J_r} + \frac{1}{n_b^2 J_g})T_h + \frac{D_{ls}}{n_b^2 J_g}T_g \\
\frac{dT_g}{dt} &= -\frac{1}{\tau_g}T_g + \frac{1}{\tau_g}T_{g,r}
\end{aligned} \tag{9}$$

where:

D_g: the damping constants for the generator,
D_{ls}: the damping constants for the the equivalent low-speed shaft,
D_r: the damping constants for the rotor,
i_m: the magnetizing current,
J_g: the moment of inertia of the generator,
J_r: the moment of inertia of the rotor,
K_{ls}: the equivalent torsional stiffness of the low-speed shaft,
n_b: the gearbox ratio (the gearbox is considered as a lossless device for this model),
R_r: the rotor resistance,
R_s: the stator resistance,
τ_g: the time constant of the model,
V_s: the stator voltage magnitude.

For the ease of presentation, we define the following parameters for the wind energy conversion system, $p_1 = \frac{-R_s}{\xi L_s}$, $p_2 = \omega_s$, $p_3 = \frac{1-\xi}{\xi} n_p$, $p_4 = \frac{R_r L_m}{\xi L_r L_s}$, $p_5 = \frac{L_m}{\xi L_s} n_p$, $p_6 = -\frac{1}{\xi L_s}$, $p_7 = \frac{L_m}{\xi L_r L_s}$, $p_8 = -\frac{L_m}{\xi L_r} n_p$, $p_9 = -\frac{R_r}{\xi L_r}$, $p_{10} = -\frac{1}{\xi} n_p$, $p_{11} = -\frac{1}{\xi L_r}$, $p_{12} = \frac{R_s L_m}{\xi L_r L_s}$, $p_{13} = -\frac{D_r}{J_r}$, $p_{14} = \frac{K_{opt}}{J_r}$, $p_{15} = -\frac{n_b}{J_r}$, $p_{16} = -\frac{D_g}{J_g}$, $p_{17} = \frac{1}{J_g}$, $p_{18} = \frac{1}{n_b}(K_{ls} - \frac{D_r D_{ls}}{J_r})$, $p_{19} = \frac{1}{n_b}(\frac{D_{ls} K_{opt}}{J_r})$, $p_{20} = -\frac{1}{n_b^2}(K_{ls} - \frac{D_g D_{ls}}{J_g})$, $p_{21} = -D_{ls}(\frac{1}{J_r} + \frac{1}{n_b^2 J_g})$, $p_{22} = \frac{D_{ls}}{n_b^2 J_g}$, $p_{23} = -\frac{1}{\tau_g}$.
Moreover, we define the state variable vector x, which contains the states x_1–x_8, such that,

$$x = [x_1 \;\; x_2 \;\; x_3 \;\; x_4 \;\; x_5 \;\; x_6 \;\; x_7 \;\; x_8]^T = [i_{sd} \;\; i_{sq} \;\; i_{rd} \;\; i_{rq} \;\; \Omega_r \;\; \Omega_g \;\; T_h \;\; T_g]^T. \tag{10}$$

Therefore, the model of the wind energy power system in (10) can be written in compact form as follows,

$$\begin{aligned}
\dot{x}_1 &= p_1 x_1 + p_2 x_2 + p_3 x_6 x_2 + p_4 x_3 + p_5 x_4 x_6 + p_6 u_{sd} + p_7 u_{rd} \\
\dot{x}_2 &= -p_2 x_1 - p_3 x_6 x_1 + p_1 x_2 - p_5 x_6 x_3 + p_4 x_4 + p_6 u_{sq} + p_7 u_{rq} \\
\dot{x}_3 &= p_{12} x_1 + p_8 x_6 x_2 + p_9 x_3 + p_2 x_4 + p_{10} x_6 x_4 + p_7 u_{sd} + p_{11} u_{rd} \\
\dot{x}_4 &= -p_8 x_6 x_1 + p_{12} x_2 - p_2 x_3 - p_{10} x_6 x_3 + p_9 x_4 + p_7 u_{sq} + p_{11} u_{rq} \\
\dot{x}_5 &= p_{13} x_5 + p_{14} x_5^2 + p_{15} x_7 \\
\dot{x}_6 &= p_{16} x_6 + p_{17} x_7 - p_{17} x_8 \\
\dot{x}_7 &= p_{18} x_5 + p_{19} x_5^2 + p_{20} x_6 + p_{21} x_7 + p_{22} x_8 \\
\dot{x}_8 &= p_{23} x_8 - p_{23} T_{g,r}
\end{aligned} \tag{11}$$

Let the desired state vector x_d be such that $x_d = [x_{1d} \;\; x_{2d} \;\; x_{3d} \;\; x_{4d} \;\; x_{5d} \;\; x_{6d} \;\; x_{7d} \;\; x_{8d}]^T$, where x_{id} $(i = 1, 2, ..., 8)$ are the desired values of the states x_i $(i = 1, 2, ..., 8)$ of the power system. Since the desired states have to be an operating point of the power system, then they must satisfy the equations of the model of the wind energy power system in (12). Therefore, the desired states of the power system are governed by the following set of differential equations:

$$\dot{x}_{1d} = p_1 x_{1d} + p_2 x_{2d} + p_3 x_{6d} x_{2d} + p_4 x_{3d} + p_5 x_{4d} x_{6d} + p_6 \bar{u}_{sd} + p_7 \bar{u}_{rd} \tag{12}$$

$$\dot{x}_{2d} = -p_2 x_{1d} - p_3 x_{6d} x_{1d} + p_1 x_{2d} - p_5 x_{6d} x_{3d} + p_4 x_{4d} + p_6 \bar{u}_{sq} + p_7 \bar{u}_{rq} \tag{13}$$

$$\dot{x}_{3d} = p_{12} x_{1d} + p_8 x_{6d} x_{2d} + p_9 x_{3d} + p_2 x_{4d} + p_{10} x_{6d} x_{4d} + p_7 \bar{u}_{sd} + p_{11} \bar{u}_{rd} \tag{14}$$

$$\dot{x}_{4d} = -p_8 x_{6d} x_{1d} + p_{12} x_{2d} - p_2 x_{3d} - p_{10} x_{6d} x_{3d} + p_9 x_{4d} + p_7 \bar{u}_{sq} + p_{11} \bar{u}_{rq} \tag{15}$$

$$\dot{x}_{5d} = p_{13} x_{5d} + p_{14} x_{5d}^2 + p_{15} x_{7d} \tag{16}$$

$$\dot{x}_{6d} = p_{16} x_{6d} + p_{17} x_{7d} - p_{17} x_{8d} \tag{17}$$

$$\dot{x}_{7d} = p_{18} x_{5d} + p_{19} x_{5d}^2 + p_{20} x_{6d} + p_{21} x_{7d} + p_{22} x_{8d} \tag{18}$$

$$\dot{x}_{8d} = p_{23} x_{8d} - p_{23} \bar{T}_{g,r} \tag{19}$$

Note that $\bar{u}_{sd}$, $\bar{u}_{sq}$, $\bar{u}_{rd}$, $\bar{u}_{rq}$ and $\bar{T}_{g,r}$ in (12)–(19) represent the reference inputs of the wind energy conversion system. The computation of these reference inputs will be discussed in the next section.

Define the errors e_i $(i = 1, 2, ..., 8)$, such that:

$$\begin{aligned} e_1 &= x_1 - x_{1d} \\ e_2 &= x_2 - x_{2d} \\ e_3 &= x_3 - x_{3d} \\ e_4 &= x_4 - x_{4d} \\ e_5 &= x_5 - x_{5d} \\ e_6 &= x_6 - x_{6d} \\ e_7 &= x_7 - x_{7d} \\ e_8 &= x_8 - x_{8d} \end{aligned} \quad (20)$$

Using Equations (12)–(21), the dynamic model of the error system can be written, such that:

$$\begin{aligned} \dot{e}_1 &= p_1 e_1 + p_2 e_2 + p_3(x_6 x_2 - x_{6d} x_{2d}) + p_4 e_3 + p_5(x_4 x_6 - x_{4d} x_{6d}) + v_1 \\ \dot{e}_2 &= -p_2 e_1 - p_3(x_6 x_1 - x_{6d} x_{1d}) + p_1 e_2 - p_5(x_6 x_3 - x_{6d} x_{3d}) + p_4 e_4 + v_2 \\ \dot{e}_3 &= p_{12} e_1 + p_8(x_6 x_2 - x_{6d} x_{2d}) + p_9 e_3 + p_2 e_4 + p_{10}(x_6 x_4 - x_{6d} x_{4d}) + v_3 \\ \dot{e}_4 &= -p_8(x_6 x_1 - x_{6d} x_{1d}) + p_{12} e_2 - p_2 e_3 - p_{10}(x_6 x_3 - x_{6d} x_{3d}) + p_9 e_4 + v_4 \\ \dot{e}_5 &= p_{13} e_5 + p_{14}(x_5^2 - x_{5d}^2) + p_{15} e_7 \\ \dot{e}_6 &= p_{16} e_6 + p_{17} e_7 - p_{17} e_8 \\ \dot{e}_7 &= p_{18} e_5 + p_{19}(x_5^2 - x_{5d}^2) + p_{20} e_6 + p_{21} e_7 + p_{22} e_8 \\ \dot{e}_8 &= p_{23} e_8 + v_g \end{aligned} \quad (21)$$

where the inputs v_1, v_2, v_3, v_4 and v_g are defined, such that,

$$\begin{aligned} v_1 &= p_6(u_{sd} - \bar{u}_{sd}) + p_7(u_{rd} - \bar{u}_{rd}) \\ v_2 &= p_6(u_{sq} - \bar{u}_{sq}) + p_7(u_{rq} - \bar{u}_{rq}) \\ v_3 &= p_7(u_{sd} - \bar{u}_{sd}) + p_{11}(u_{rd} - \bar{u}_{rd}) \\ v_4 &= p_7(u_{sq} - \bar{u}_{sq}) + p_{11}(u_{rq} - \bar{u}_{rq}) \\ v_g &= -p_{23}(T_{g,r} - \bar{T}_{g,r}). \end{aligned} \quad (22)$$

After some manipulations, the error dynamics in (22) can be written as follows,

$$\dot{e}_1 = p_1 e_1 + p_2 e_2 + p_3(e_6 e_2 + x_{6d} e_2 + x_{2d} e_6) + p_4 e_3 + p_5(e_4 e_6 + x_{4d} e_6 + x_{6d} e_4) + v_1 \quad (23)$$

$$\dot{e}_2 = -p_2 e_1 - p_3(e_6 e_1 + x_{6d} e_1 + x_{1d} e_6) + p_1 e_2 - p_5(e_6 e_3 + x_{6d} e_3 + x_{3d} e_6) + p_4 e_4 + v_2 \quad (24)$$

$$\dot{e}_3 = p_{12} e_1 + p_8(e_6 e_2 + x_{6d} e_2 + x_{2d} e_6) + p_9 e_3 + p_2 e_4 + p_{10}(e_4 e_6 + x_{4d} e_6 + x_{6d} e_4) + v_3 \quad (25)$$

$$\dot{e}_4 = -p_8(e_6 e_1 + x_{6d} e_1 + x_{1d} e_6) + p_{12} e_2 - p_2 e_3 - p_{10}(e_6 e_3 + x_{6d} e_3 + x_{3d} e_6) + v_4 \quad (26)$$

$$\dot{e}_5 = p_{13} e_5 + p_{14}(e_5^2 + 2 x_{5d} e_5) + p_{15} e_7 \quad (27)$$

$$\dot{e}_6 = p_{16} e_6 + p_{17} e_7 - p_{17} e_8 \quad (28)$$

$$\dot{e}_7 = p_{18} e_5 + p_{19}(e_5^2 + 2 x_{5d} e_5) + p_{20} e_6 + p_{21} e_7 + p_{22} e_8 \quad (29)$$

$$\dot{e}_8 = p_{23} e_8 + v_g \quad (30)$$

Define the matrix M_u and the vectors u, v and $\bar{u}_r$, such that,

$$M_u = \begin{bmatrix} p_6 & p_7 & 0 & 0 & 0 \\ 0 & 0 & p_6 & p_7 & 0 \\ p_7 & p_{11} & 0 & 0 & 0 \\ 0 & 0 & p_7 & p_{11} & 0 \\ 0 & 0 & 0 & 0 & -p_{23} \end{bmatrix}, \quad u = \begin{bmatrix} u_{sd} \\ u_{rd} \\ u_{sq} \\ u_{rq} \\ T_{g,r} \end{bmatrix}, \quad v = \begin{bmatrix} v_1 \\ v_2 \\ v_3 \\ v_4 \\ v_g \end{bmatrix}, \quad \bar{u}_r = \begin{bmatrix} \bar{u}_{sd} \\ \bar{u}_{rd} \\ \bar{u}_{sq} \\ \bar{u}_{rq} \\ \bar{T}_{g,r} \end{bmatrix}. \tag{31}$$

Remark 1. *The controller* $u = \begin{bmatrix} u_{sd} & u_{rd} & u_{sq} & u_{rq} & T_{g,r} \end{bmatrix}^T$ *of the wind energy conversion system model in (12) is computed using the controller* $v = \begin{bmatrix} v_1 & v_2 & v_3 & v_4 & v_g \end{bmatrix}^T$ *and the reference inputs* $\bar{u}_{sd}$, $\bar{u}_{sq}$, $\bar{u}_{rd}$, $\bar{u}_{rq}$ *and* $\bar{T}_{g,r}$, *such that,*

$$\begin{bmatrix} u_{sd} \\ u_{rd} \\ u_{sq} \\ u_{rq} \\ T_{g,r} \end{bmatrix} = \begin{bmatrix} p_6 & p_7 & 0 & 0 & 0 \\ 0 & 0 & p_6 & p_7 & 0 \\ p_7 & p_{11} & 0 & 0 & 0 \\ 0 & 0 & p_7 & p_{11} & 0 \\ 0 & 0 & 0 & 0 & -p_{23} \end{bmatrix}^{-1} \begin{bmatrix} v_1 \\ v_2 \\ v_3 \\ v_4 \\ v_g \end{bmatrix} + \begin{bmatrix} \bar{u}_{sd} \\ \bar{u}_{rd} \\ \bar{u}_{sq} \\ \bar{u}_{rq} \\ \bar{T}_{g,r} \end{bmatrix} \tag{32}$$

Equation (32) can be written in compact form, such that,

$$u = M_u^{-1} v + \bar{u}_r. \tag{33}$$

Note that the determinant of the matrix M_u *equals* $p_{23}(p_7^2 - p_6 p_{11})^2$, *which is different from zero. Hence, the matrix* M_u *is nonsingular, and its inverse always exists.*

3. Computations of the Desired States and the Reference Inputs of the WEC System

This section deals with the computations of the values of the desired states of the wind energy conversion system, as well as the values of the reference inputs of the system.

3.1. Computation of the Desired States of the WEC System

The flux linkages of the stator and the rotor of the generator can be expressed as,

$$\psi_{sd} = L_s i_{sd} + L_m i_{rd} \tag{34}$$

$$\psi_{sq} = L_s i_{sq} + L_m i_{rq} \tag{35}$$

$$\psi_{rd} = L_r i_{rd} + L_m i_{sd} \tag{36}$$

$$\psi_{rq} = L_r i_{rq} + L_m i_{sq} \tag{37}$$

where ψ_{sd} is the stator d-axis flux linkage and ψ_{sq} is the stator q-axis flux linkage. The flux ψ_{rd} is the rotor d-axis flux linkage, and ψ_{rq} is the rotor q-axis flux linkage.

The stator active power, P_s, and the stator reactive power, Q_s, can be expressed as follows,

$$P_s = \frac{3}{2}(u_{sd} i_{sd} + u_{sq} i_{sq}) \tag{38}$$

$$Q_s = \frac{3}{2}(u_{sq} i_{sd} - u_{sd} i_{sq}) \tag{39}$$

where i_{sd} and i_{sq} are the stator currents; u_{sd} and u_{sq} are the stator voltages.

In the stator voltage oriented reference frame, the q-axis is aligned with the supply voltage V_s; then, the reference inputs $\bar{u}_{sd}$ and $\bar{u}_{sq}$ are such that $\bar{u}_{sd} = 0$ and $\bar{u}_{sq} = V_s$.

Using Equations (38)–(39), the desired stator active and reactive powers can be written as follows,

$$P_{s_d} = \frac{3}{2}(\bar{u}_{sd}x_{1d} + \bar{u}_{sq}x_{2d}) = \frac{3}{2}x_{2d}V_s \quad (40)$$
$$Q_{s_d} = \frac{3}{2}(\bar{u}_{sq}x_{1d} - \bar{u}_{sd}x_{2d}) = \frac{3}{2}x_{1d}V_s \quad (41)$$

where x_{1d} is the desired value of i_{sd} and x_{2d} is the desired value of i_{sq}.

Moreover, the stator flux ψ_s is set to be aligned with the d-axis; hence, the stator d-axis and q-axis flux linkages are such that $\psi_{sd} = \psi_s = V_s/\omega_s$ and $\psi_{sq} = 0$. In addition, we choose to set the desired stator active power P_{s_d} to be equal to the maximum power that can be captured from the wind and the desired stator reactive power Q_{s_d} to be zero, then $P_{s_d} = P_{m(\max)}$ and $Q_{s_d} = 0$.

Remark 2. *Note that the desired reactive power is selected to ensure a unity power factor.*

Hence, the desired states x_{1d}, x_{2d}, x_{3d}, x_{4d}, x_{5d}, x_{6d}, x_{7d} *and* x_{8d} *are computed as follows.*

At first, using Equations (40) and (41), we can compute x_{1d} *and* x_{2d} *as follows,*

$$x_{1d} = \frac{2}{3}\frac{Q_{sd}}{V_s} = 0 \quad (42)$$
$$x_{2d} = \frac{2}{3}\frac{P_{sd}}{V_s} = \frac{2}{3}\frac{P_{m(max)}}{V_s}. \quad (43)$$

Then, using Equations (34) and (35) with $\psi_{sd} = \psi_s = V_s/\omega_s$ *and* $\psi_{sq} = 0$, *we can compute* x_{3d} *and* x_{4d}, *which are the desired values of* i_{rd} *and* i_{rq}, *as follows,*

$$x_{3d} = \frac{1}{L_m}\frac{V_s}{\omega_s} \quad (44)$$
$$x_{4d} = -\frac{2}{3}\frac{L_s}{L_m}\frac{P_{m(max)}}{V_s}. \quad (45)$$

Furthermore, the desired value of generator torque x_{8d} *is chosen to provide an appropriate opposing torque. Using Equations (2) and (4),* x_{8d} *can be written as follows,*

$$x_{8d} = -\frac{1}{2}n_b\rho\pi R^3 C_{p(max)}V^2/\lambda_{opt}. \quad (46)$$

Finally, the remaining desired states x_{5d}, x_{6d} *and* x_{7d}, *which are the desired values of* Ω_r, Ω_g, T_h, *are determined by solving the ordinary differential Equations (16)–(18). Recall that these equations are as follows,*

$$\dot{x}_{5d} = p_{13}x_{5d} + p_{14}x_{5d}^2 + p_{15}x_{7d} \quad (47)$$
$$\dot{x}_{6d} = p_{16}x_{6d} + p_{17}x_{7d} - p_{17}x_{8d} \quad (48)$$
$$\dot{x}_{7d} = p_{18}x_{5d} + p_{19}x_{5d}^2 + p_{20}x_{6d} + p_{21}x_{7d} + p_{22}x_{8d} \quad (49)$$

3.2. Computation of the Reference Inputs of the WEC System

Recall that in the stator voltage-oriented reference frame, it is assumed that the q-axis is aligned with the supply voltage V_s. Hence, the reference inputs $\bar{u}_{sd}$ and $\bar{u}_{sq}$ are such that $\bar{u}_{sd} = 0$ and $\bar{u}_{sq} = V_s$. On the other hand, the reference inputs $\bar{u}_{rd}$, $\bar{u}_{rq}$ and $\bar{T}_{g,r}$ are computed as follows. First, we define the terms c_1, c_2, c_3, c_4 and c_5, such that,

$$c_1 = p_1x_{1d} + p_2x_{2d} + p_3x_{6d}x_{2d} + p_4x_{3d} + p_5x_{4d}x_{6d} \quad (50)$$
$$c_2 = -p_2x_{1d} - p_3x_{6d}x_{1d} + p_1x_{2d} - p_5x_{6d}x_{3d} + p_4x_{4d} \quad (51)$$
$$c_3 = p_{23}x_{8d} \quad (52)$$
$$c_4 = \dot{x}_{2d} = C_{p(\max)}(\lambda_{opt}, \beta)\rho\pi R^2V^2\dot{V} \quad (53)$$
$$c_5 = \dot{x}_{8d} = -\frac{n_b}{\lambda_{opt}}C_{p(\max)}(\lambda_{opt}, \beta)\rho\pi R^2V^2\dot{V} \quad (54)$$

where $\dot{V}$ represents the derivative with respect to time of the wind speed model.

Then, using Equations (12), (13) and (19), the reference inputs are obtained, such that,

$$\begin{aligned} \bar{u}_{sd} &= 0 && (55)\\ \bar{u}_{rd} &= -\frac{1}{p_7}c_1 && (56)\\ \bar{u}_{sq} &= V_s && (57)\\ \bar{u}_{rq} &= \frac{1}{p_7}(c_4 - c_2 - p_6V_s) && (58)\\ \bar{T}_{g,r} &= \frac{1}{p_{23}}(p_{23}x_{8d} - c_5) && (59) \end{aligned}$$

4. A Feedback Linearization Controller for the Wind Energy Conversion System

This section deals with the design of a feedback linearization control law for the wind energy conversion system represented by the nonlinear system of odes given by (12).

We start with defining the variables z_1, z_2 and z_3, such that,

$$\begin{aligned} z_1 &= e_5 && (60)\\ z_2 &= p_{13}e_5 + p_{14}(e_5^2 + 2x_{5d}e_5) + p_{15}e_7 && (61)\\ z_3 &= (p_{13} + 2p_{14}e_5 + 2p_{14}x_{5d})[p_{13}e_5 + p_{14}(e_5^2 + 2x_{5d}e_5) + p_{15}e_7] \\ &\quad +2p_{14}\dot{x}_{5d}e_5 + p_{15}[p_{18}e_5 + p_{19}(e_5^2 + 2x_{5d}e_5) + p_{20}e_6 + p_{21}e_7 + p_{22}e_8]. && (62) \end{aligned}$$

Furthermore, define the functions f_1 and f_2, such that:

$$\begin{aligned} f_1 &= (p_{13} + 2p_{14}e_5 + 2p_{14}x_{5d})[p_{13}e_5 + p_{14}(e_5^2 + 2x_{5d}e_5) + p_{15}e_7] \\ &\quad +2p_{14}\dot{x}_{5d}e_5 + p_{15}[p_{18}e_5 + p_{19}(e_5^2 + 2x_{5d}e_5) + p_{20}e_6 + p_{21}e_7] && (63)\\ f_2 &= 2p_{14}(p_{13}e_5 + p_{14}(e_5^2 + 2x_{5d}e_5) + p_{15}e_7 + 2\dot{x}_{5d})z_2 \\ &\quad +(p_{13} + 2p_{14}e_5 + 2p_{14}x_{5d})z_3 + 2p_{14}\ddot{x}_{5d}e_5 \\ &\quad +p_{15}[p_{18}z_2 + 2p_{19}(e_5 + x_{5d})z_2 + 2p_{19}\dot{x}_{5d}e_5] + p_{15}p_{20}(p_{16}e_6 + p_{17}e_7 - p_{17}e_8) \\ &\quad +p_{15}p_{21}(p_{18}e_5 + p_{19}(e_5^2 + 2x_{5d}e_5) + p_{20}e_6 + p_{21}e_7 + p_{22}e_8) && (64) \end{aligned}$$

Let the control parameters α_1, α_2, α_3 and α_4 be positive scalars and choose the parameters β_1, β_2 and β_3 to be positive scalars, such that the polynomial $P_1(s) = s^3 + \beta_3 s^2 + \beta_2 s + \beta_1$ is Hurwitz.

The following proposition gives the feedback linearization controller.

Proposition 1. *The feedback linearization controller,*

$$u = M_u^{-1}v + \bar{u}_r \qquad (65)$$

with M_u and $\bar{u}_r$ given by (31) and (55)–(59) and $v = [v_1 \;\; v_1 \;\; v_1 \;\; v_1 \;\; v_g]^T$, such that,

$$\begin{aligned} v_1 &= -p_2e_2 - p_3(e_6e_2 + x_{6d}e_2 + x_{2d}e_6) - p_4e_3 - p_5(e_4e_6 + x_{4d}e_6 + x_{6d}e_4) - \alpha_1e_1 && (66)\\ v_2 &= p_2e_1 + p_3(e_6e_1 + x_{6d}e_1 + x_{1d}e_6) + p_5(e_6e_3 + x_{6d}e_3 + x_{3d}e_6) - p_4e_4 - \alpha_2e_2 && (67)\\ v_3 &= -p_{12}e_1 - p_8(e_6e_2 + x_{6d}e_2 + x_{2d}e_6) - p_2e_4 - p_{10}(e_4e_6 + x_{4d}e_6 + x_{6d}e_4) - \alpha_3e_3 && (68)\\ v_4 &= p_8(e_6e_1 + x_{6d}e_1 + x_{1d}e_6) - p_{12}e_2 + p_2e_3 + p_{10}(e_6e_3 + x_{6d}e_3 + x_{3d}e_6) - \alpha_4e_4 && (69)\\ v_g &= \frac{1}{p_{15}p_{22}}(-f_2 - p_{15}p_{22}p_{23}e_8 - \beta_1z_1 - \beta_2z_2 - \beta_3z_3) && (70) \end{aligned}$$

when applied to the model of the wind energy power system given by (12) guarantees the asymptotic convergence of the states of the system to their desired values.

Note that the block diagram representation of the DFIG-based WEC system with the proposed controller is depicted in Figure 3.

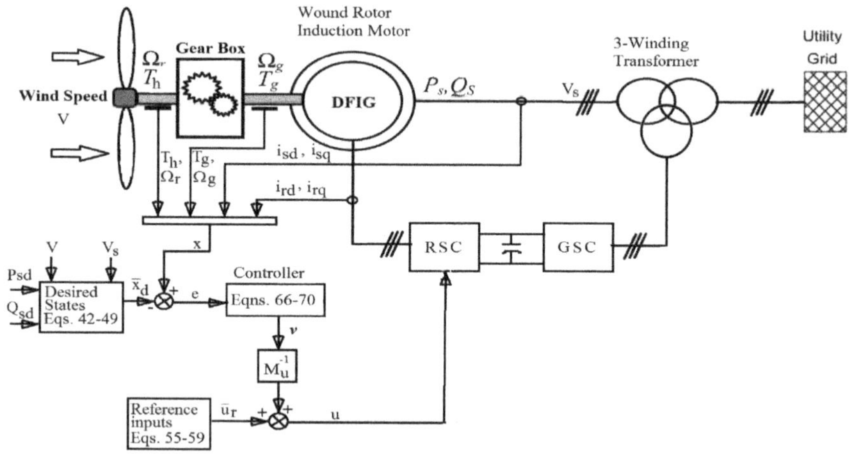

Figure 3. A block diagram representation of the variable speed DFIG-based wind energy conversion system with the proposed controller. GSC, grid side converter.

Proof. Applying the controllers given by (66)–(69) to the first four differential equations of the error system given by (23)–(26), we obtain:

$$\begin{aligned} \dot{e}_1 &= -(\alpha_1 - p_1)e_1 \\ \dot{e}_2 &= -(\alpha_2 - p_1)e_2 \\ \dot{e}_3 &= -(\alpha_3 - p_9)e_3 \\ \dot{e}_4 &= -\alpha_4 e_4. \end{aligned} \tag{71}$$

Because the α_i's ($i = 1, ..., 4$) are chosen to be positive scalars and since the parameters $p_1 = -\frac{R_s}{\xi L_s}$ and $p_9 = -\frac{R_r}{\xi L_r}$ are negative, then it can be concluded from the system of odes in (72) that the errors e_1, e_2, e_3 and e_4 asymptotically converge to zero as t tends to infinity.

The second part of the proof involves proving the asymptotic convergence to zero of the errors e_5, e_6, e_7 and e_8. Taking the time derivative of the variables z_1, z_2 and z_3 defined in Equations (60)–(62) and using the equations of the error system given by (23)–(30), we obtain:

$$\begin{aligned} \dot{z}_1 &= z_2 \\ \dot{z}_2 &= z_3 \\ \dot{z}_3 &= f_2 + p_{15}p_{22}(p_{23}e_8 + v_g) \end{aligned} \tag{72}$$

The application of the controller v_g in (70) to the system of odes in (73), yields,

$$\begin{aligned} \dot{z}_1 &= z_2 \\ \dot{z}_2 &= z_3 \\ \dot{z}_3 &= -\beta_1 z_1 - \beta_2 z_2 - \beta_3 z_3 \end{aligned} \tag{73}$$

The above system can be written in compact form, such that:

$$\dot{z} = M_z z \tag{74}$$

with the matrix M_z and the vector z being such that,

$$M_z = \begin{bmatrix} 0 & 1 & 0 \\ 0 & 0 & 1 \\ -\beta_1 & -\beta_2 & -\beta_3 \end{bmatrix}, \quad z = \begin{bmatrix} z_1 \\ z_2 \\ z_3 \end{bmatrix}.$$

Since β_1, β_2 and β_3 are positive scalars, such that the polynomial $P_1(s) = s^3 + \beta_3 s^2 + \beta_2 s + \beta_1$ is Hurwitz, then the matrix M_z is a stable matrix (i.e., its eigenvalues are located in the left half of the s-plane). The solution of the Equation (74) is $z(t) = exp(M_z t)z(0)$, where $z(0)$ is the value of $z(t)$ when $t = 0$. Therefore, the asymptotic convergence of z_1, z_2 and z_3 to zero as t tends to infinity is guaranteed because M_z is a stable matrix.

The zero dynamics [44] is defined as the internal dynamics of the system when the output is kept identically zero by a suitable input function. For the third order system given by (73), the zero dynamics is analyzed by studying Equations (27)–(30) when $z_1 = 0$, $z_2 = 0$ and $z_3 = 0$.

Using Equations (60) and (61), it is clear that the asymptotic convergence of z_1 and z_2 to zero implies the asymptotic convergence of e_5 and e_7 to zero as t tends to infinity.

Moreover, since z_1, z_2, z_3, e_5 and e_7 converge to zero as t tends to infinity, then the equation given by (62) yields,

$$e_8 = -\frac{p_{20}}{p_{22}} e_6. \tag{75}$$

As t tends to infinity, the differential Equation (28) of the error system given by (23)–(30), which represents the zero dynamics of the system, reduces to,

$$\begin{aligned} \dot{e}_6 &= p_{16}e_6 - p_{17}e_8 \\ &= (p_{16} + \frac{p_{17}p_{20}}{p_{22}})e_6 = a_e e_6. \end{aligned} \tag{76}$$

The constant $a_e = p_{16} + \frac{p_{17}p_{20}}{p_{22}}$ in (76) is such that,

$$\begin{aligned} a_e &= p_{16} + \frac{p_{17}p_{20}}{p_{22}} \\ &= -\frac{D_g}{J_g} - \frac{1}{J_g}\frac{1}{n_b^2}(K_{ls} - \frac{D_g D_{ls}}{J_g})\frac{n_b^2 J_g}{D_{ls}} \\ &= -\frac{D_g}{J_g} - (\frac{K_{ls}}{D_{ls}} - \frac{D_g}{J_g}) = -\frac{K_{ls}}{D_{ls}}. \end{aligned} \tag{77}$$

Therefore, since $a_e = -\frac{K_{ls}}{D_{ls}}$ is always negative, Equation (76) guarantees the asymptotic convergence of the error e_6 to zero as t tends to infinity. Moreover, Equation (75) implies the asymptotic convergence of the error e_8 to zero as t tends to infinity.

The asymptotic convergence to zero of the errors e_i $(i = 1, ..., 8)$ as t tends to infinity implies the asymptotic convergence of the states of the wind energy power system given by (12) to their desired values as t tends to infinity. □

5. Simulation Results of the Controlled Wind Energy Power System

The performance of the wind energy conversion system controlled using the proposed feedback linearization controller was simulated using the MATLAB software. The parameters of the system used for the simulation studies are given in Table 1.

Table 1. The parameters of the DFIG-based wind turbine power system.

Parameter	Value
Rated power	1.5 MW
Rated apparent power	1.5 MVA
Rated voltage (line to line)	575 v
Frequency/angular speed	$2\pi 60$ rad/s
Nominal system frequency	60 Hz
Stator resistance	0.0014 Ω
Stator leakage inductance	89.98 mH
Rotor resistance	0.99187 mΩ
Rotor leakage inductance	82.088 mH
Magnetizing inductance	1.526 mH
Inertia of the generator	53.036 kg·m^2
Pole pairs	3
Wind turbine with a rotor diameter	70 m
Air density	0.55 kg/m^3
Cut-in wind speed	4 m/s
Cut-out wind speed	25 m/s
Rated wind speed	12 m/s
Rated rotor speed	19.7 rpm
Drive-train torsion damper	1.0×10^7 Nm/s
Drive-train torsion spring	5.6×10^9 Nm/rad
Gearbox ratio	75.7098
Inertia of the rotor	34.6×10^3 kg·m^2

The wind speed model used for simulation purposes is given by the following formula:

$$\begin{aligned} V &= 12 + 0.55[\sin(0.0625w) - 0.875\sin(0.1875w) + 0.75\sin(0.3125w) \\ &- 0.625\sin(0.625w) + 0.5\sin(1.875w) + 0.25\sin(3.125w) + 0.125\sin(6.25w)] \end{aligned} \quad (78)$$

with $w = \frac{2\pi}{10}t$. The wind speed profile versus time generated using Equation (78) is depicted in Figure 4.

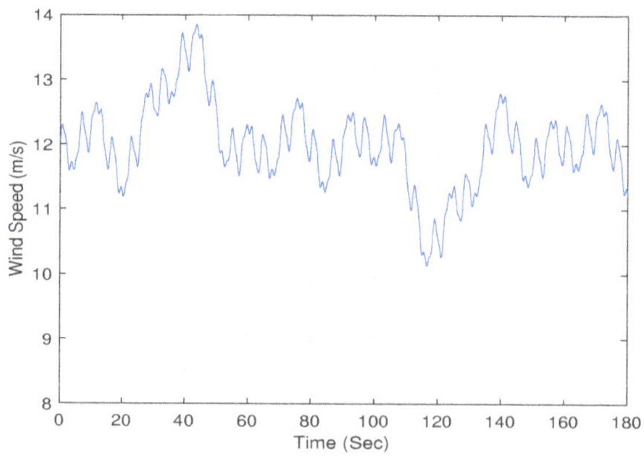

Figure 4. The wind speed profile versus time.

The wind energy conversion system was modeled using Equation (12) and controlled using the proposed feedback linearization control scheme given by Equations (65)–(70). The desired values of the states of the wind energy conversion system are obtained using Equations (42)–(49). The gains of the controller are taken to be: $\alpha_1 = 1e4$, $\alpha_2 = 3e5$, $\alpha_3 = 2e6$, $\alpha_4 = 3.5e8$, $\beta_1 = 3.15e10$, $\beta_2 = 6.35e8$ and $\beta_3 = 9.7e4$.

5.1. Simulation Studies of the WEC System When Using the Nominal Parameters of the System

The simulation results of the controlled system when using the nominal parameters are presented in Figures 5–13. Figure 5 shows the trajectory of the rotor speed Ω_r versus time, and Figure 6 depicts the trajectory of the generator speed Ω_g versus time. The high speed shaft torque and the generator torque versus time are shown in Figures 7 and 8. The errors between the actual and the desired values of the eight states of the system are shown in in Figures 9 and 10. Figure 9 shows the errors in the electrical state variables of the system versus time. It is clear from this figure that the errors e_1, e_3 and e_4 converge to zero. However, the error e_2 shows some fluctuations around zero; these fluctuations are due to the fact that e_2, which is equal to $x_2 - x_{2d}$, where x_{2d} is proportional to V^3 (V is the speed of the wind). Figure 10 shows the errors of the mechanical states variables of the system. It is clear from this figure that the errors e_5 and e_6 converge to zero. However, the errors e_7 and e_8 show some fluctuations around zero; these fluctuations are due to the fact that their desired values are dependent on the wind speed. Note that the average value of the generated torque $T_g = x_8$ is about $-2.5e7$ Nm. Therefore, it can be concluded that the states of the wind energy power system track the desired states. Figures 11 and 12 present the trajectories of the stator active and reactive powers P_s and Q_s versus time. The average of the stator active power P_s is about -0.6 MW. The average of the stator reactive power Q_s is about -0.002 MVAR. Figure 13 depicts the power coefficient C_p versus time; it is clear that the maximum value of the power coefficient is achieved when $C_{p(\max)} \approx 0.48$. It is clear from these figures that all of the closed loop system signals are bounded and that all of the states of the power system converge to their desired values.

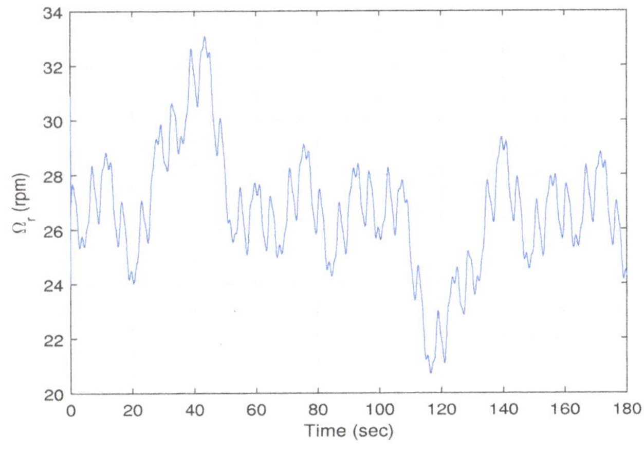

Figure 5. The turbine rotational speed on the low-speed side of the gearbox, Ω_r, versus time.

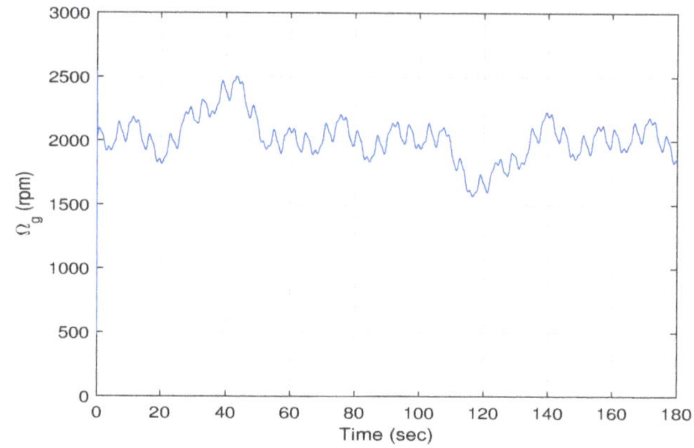

Figure 6. The mechanical generator speed Ω_g versus time.

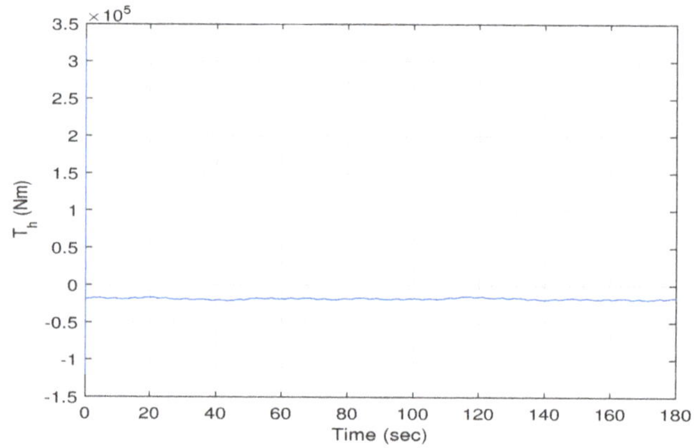

Figure 7. The high-speed shaft torque T_h versus time.

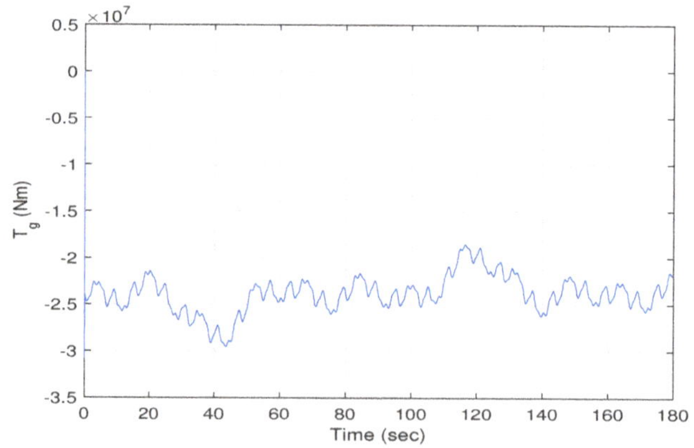

Figure 8. The generator torque T_g versus time.

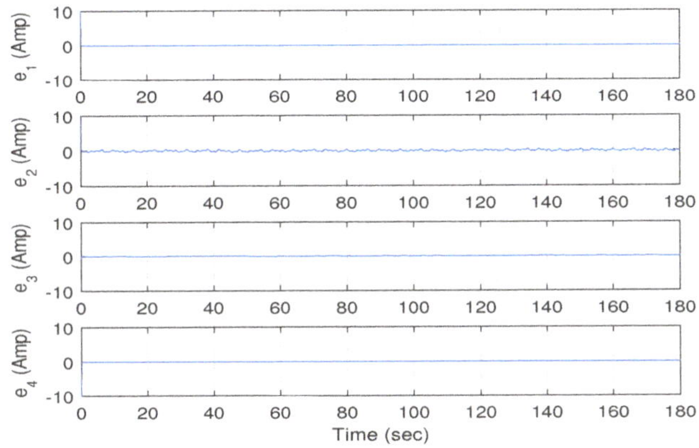

Figure 9. The errors e_1, e_2, e_3, e_4 of the electrical state variables of the WEC system versus time.

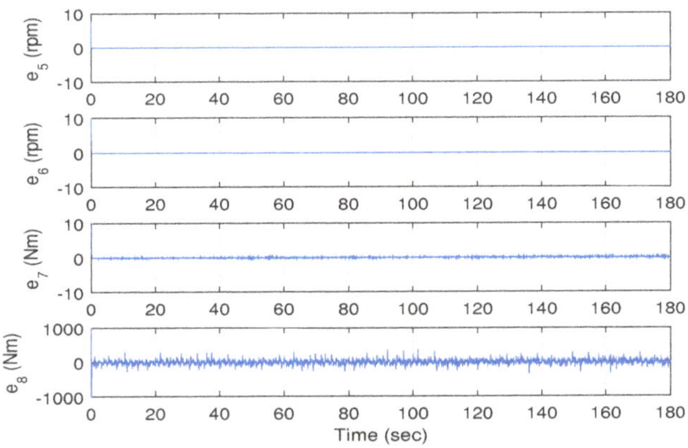

Figure 10. The errors e_5, e_6, e_7, e_8 of the mechanical state variables of the WEC system versus time.

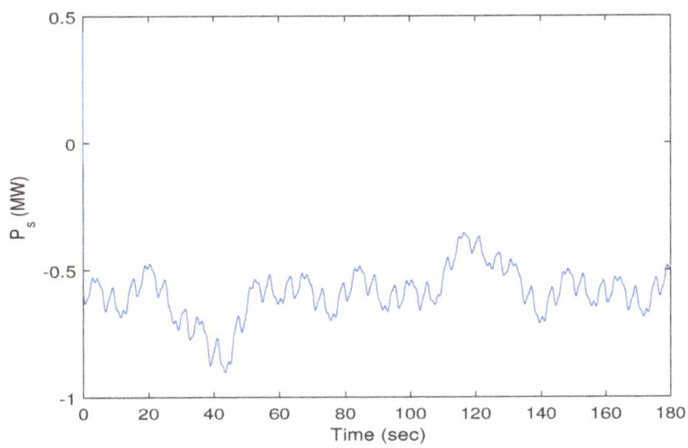

Figure 11. The stator active power P_s versus time.

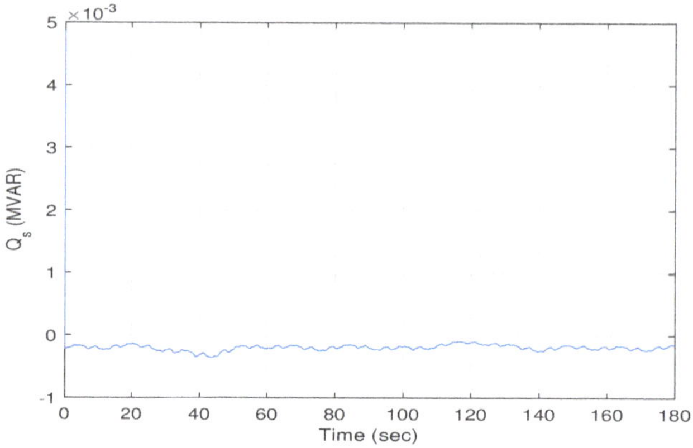

Figure 12. The stator reactive power Q_s versus time.

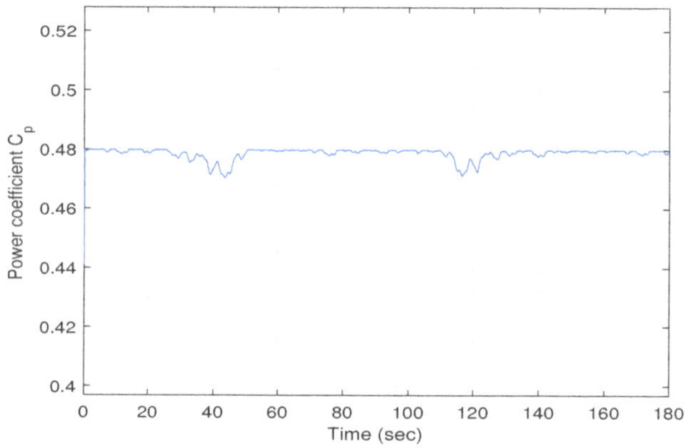

Figure 13. The power Coefficient C_p versus time.

Therefore, it can be concluded that the simulation results indicate that the wind energy conversion system controlled using the proposed feedback linearization controller shows good performance.

5.2. Robustness of the Proposed Control Scheme

Simulation studies were carried out to investigate the robustness of the proposed control scheme to changes in some of the parameters of the wind energy conversion system. At first, the effects of the changes in the stator resistance R_s, the rotor resistance R_r and the rotor inductance L_r are investigated. We simulated the performance of the closed loop system when the stator resistance R_s, the rotor resistance R_r and the rotor inductance L_r are increased by 30% of their nominal values. The responses of the system are not shown because of space limitations. However, the steady state performance of the system with the changed parameters is very similar to the steady state performance of the system with the nominal parameters. Hence, the change in some of the electrical parameters of the system did not affect the steady state performance of the system.

In addition, simulation studies were carried out to investigate the robustness of the proposed control scheme to changes in some of the electrical and some of the mechanical parameters of the wind energy conversion system. The effects of the changes in the stator resistance R_s, the rotor resistance R_r, the rotor inductance L_r, the moment of inertia of the rotor J_r and the moment of inertia of the generator J_g are investigated. Figures 14–22 show the system responses when the stator resistance

R_s, the rotor resistance R_r and the rotor inductance L_r are increased by 30%; and when the moment of inertia of the rotor J_r and the moment of inertia of the generator J_g are increased by 15% of their nominal values. Figures 14 and 15 show the rotor and the generator speed responses versus time. The high speed shaft torque and the generator torque versus time are shown in Figures 16 and 17. The errors between the actual and the desired values of the eight states of the system are shown in in Figures 18 and 19. Figure 18 shows the errors of the electrical state variables of the system versus time. It is clear from this figure that the errors e_1, e_3 and e_4 converge to zero. However, the error e_2 shows some fluctuations around zero. Figure 19 shows the errors of the mechanical states variables of the system. It is clear from this figure that the error e_5 converges to zero. However, the errors e_6, e_7 and e_8 show some fluctuations around zero. Note that in this case, the error e_7 fluctuates between -10 and 10 Nm. The error e_8 fluctuates between $-5e3$ and $5e3$ Nm. The fluctuations in the error e_8 seems to be high. However, it should be kept in mind that the average value of T_g is a bout $-2.5e7$ Nm. Even though the error is high in absolute value, it represents less than one percent of the value of T_g. Figures 20 and 21 depict the stator active and reactive powers versus time. Figure 22 depicts the power coefficient C_p. In this case, the reader can see the differences in the steady state performances of the system with the nominal parameters and the system with the changed parameters.

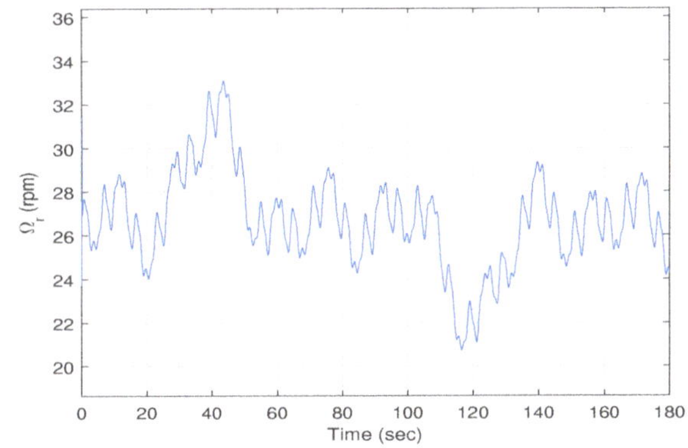

Figure 14. The turbine rotational speed on the low-speed side of the gearbox, Ω_r, versus time when R_s, R_L, L_r, J_r and J_g are different from their nominal values.

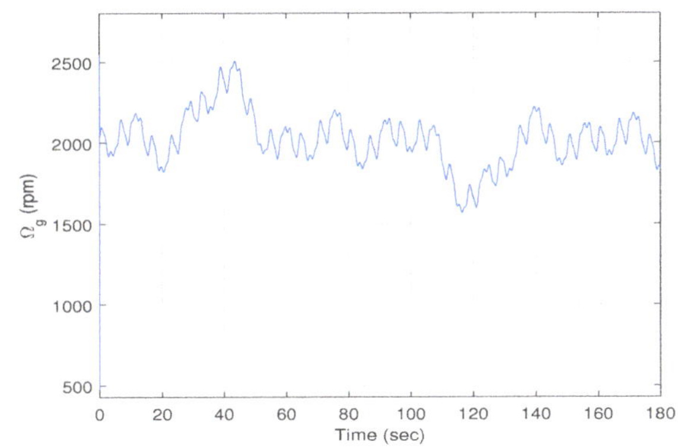

Figure 15. The mechanical generator speed Ω_g versus time when R_s, R_L, L_r, J_r and J_g are different from their nominal values.

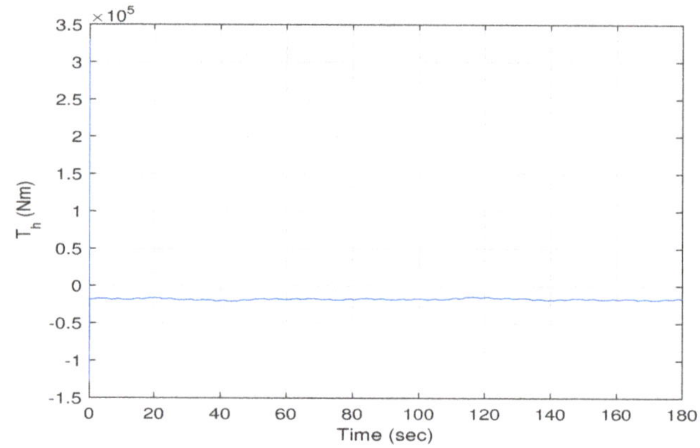

Figure 16. The high-speed shaft torque T_h versus time when R_s, R_L, L_r, J_r and J_g are different from their nominal values.

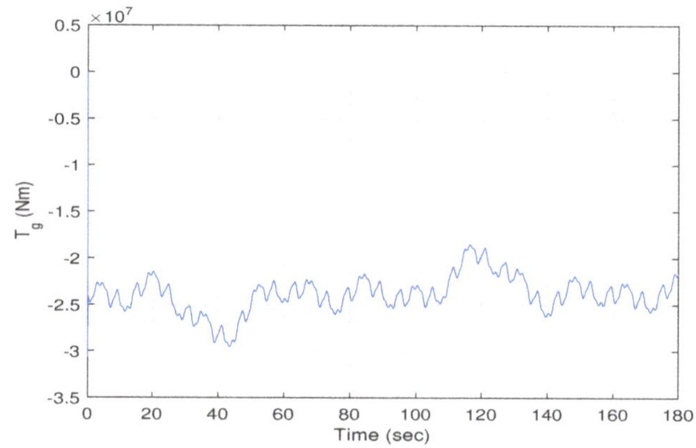

Figure 17. The generator torque T_g versus time when R_s, R_L, L_r, J_r and J_g are different from their nominal values.

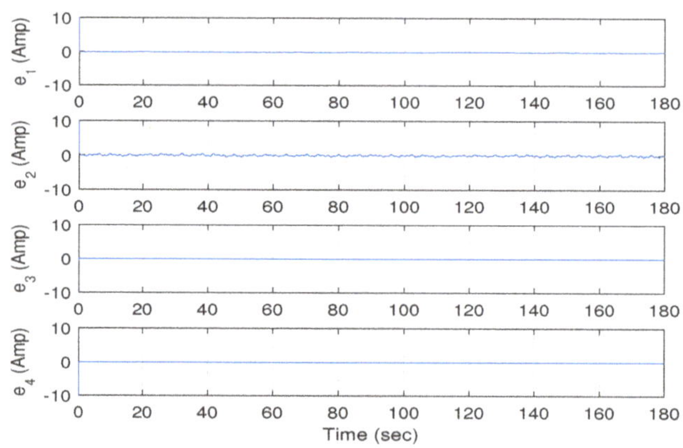

Figure 18. The errors e_1, e_2, e_3, e_4 of the electrical state variables of the WEC system versus time when R_s, R_L, L_r, J_r and J_g are different from their nominal values.

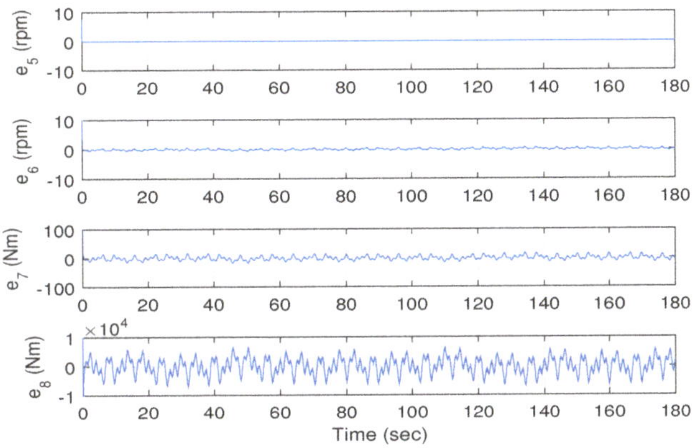

Figure 19. The errors e_5, e_6, e_7, e_8 of the mechanical state variables of the WEC system versus time when R_s, R_L, L_r, J_r and J_g are different from their nominal values.

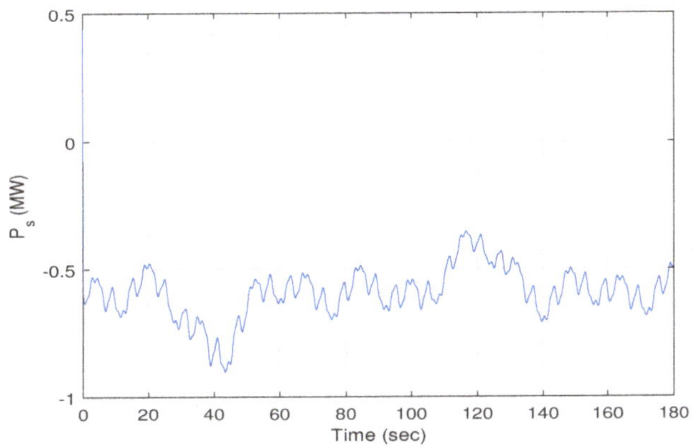

Figure 20. The stator active power P_s versus time when R_s, R_L, L_r, J_r and J_g are different from their nominal values.

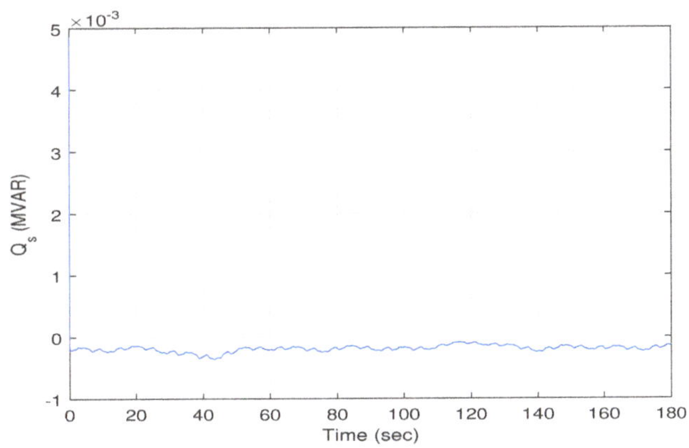

Figure 21. The stator reactive power Q_s versus time when R_s, R_L, L_r, J_r and J_g are different from their nominal values.

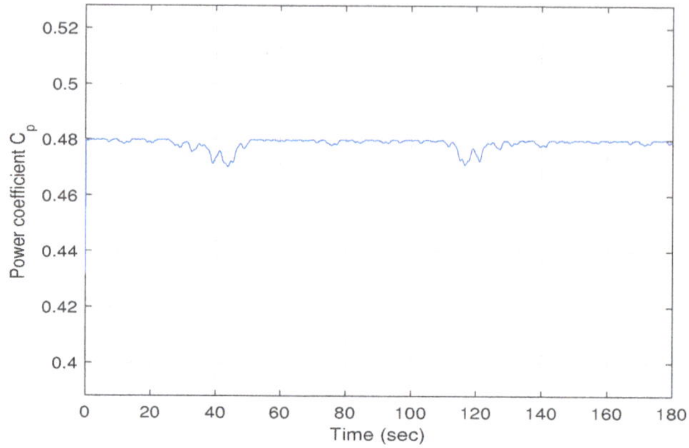

Figure 22. The power Coefficient C_p versus time when R_s, R_L, L_r, J_r and J_g are different from their nominal values.

Hence, it can be concluded that the simulation results show that the proposed controller is robust to changes in some of the electrical parameters of the wind turbine conversion system. However, the simulation results indicate that the proposed controller is sensitive to changes in some parameters of the mechanical system.

6. Conclusions

In this paper, the control of a DFIG-based wind energy power system is investigated. The power system is modeled using a system of eight nonlinear ordinary differential equations. The proposed controller is based on the feedback linearization technique. It is proven that the control scheme guarantees the asymptotic convergence of the states of the power system to their desired values. Simulation results are presented and discussed to show the effectiveness of the proposed scheme. In addition, the effects of the change of some of the parameters of the power system are studied through simulations.

Future work will address the design of other types of controllers, such as observer-based controllers and sliding mode controllers for DFIG-based wind energy power systems.

Acknowledgments: The authors would like to thank Kuwait University for covering the costs to publish this paper in open access.

Author Contributions: The first two authors contributed equally to this work. The third author contributed to the simulations of the performance of the system. All authors approved the final manuscript.

Conflicts of Interest: The authors declare no conflict of interest.

References

1. Global Wind Energy Council. 2014. Available online: http://www.gwec.net/publications/global-wind-report-2 (accessed on 14 June 2016).
2. Ling, Y.; Wu, G.; Cai, X. Comparison of wind turbine efficiency in maximum power extraction of wind turbines with doubly fed induction generator. *Prz. Elektrotech. Electr. Rev.* **2012**, *5*, 157–163.
3. Arturo Soriano, L.; Yu, W.; Rubio, J.D. Modeling and control of wind turbine. *Math. Probl. Eng.* **2013**, *2013*, 982597.
4. Haris, D.A.; Sedighizadeh, M. Nonlinear Model Identification and Adaptive Control Of Variable Speed Wind Turbine Using Recurrent Neural Network. *Int. J. Technol. Phys. Probl. Eng.* **2014**, *6*, 29–37.
5. Ghasemi, S.; Tabesh, A.; Askari-Marnani, J. Application of fractional calculus theory to robust controller design for wind turbine generators. *IEEE Trans. Energy Convers.* **2014**, *29*, 780–787.

6. Beltran, B.; Ahmed-Ali, T.; Benbouzid, M.E. Sliding mode power control of variable-speed wind energy conversion systems. *IEEE Trans. Energy Convers.* **2008**, *23*, 551–558.
7. Kairous, D.; Wamkeue, R.; Belmadani, B. Advanced control of variable speed wind energy conversion system with DFIG. In Proceedings of the 2010 IEEE 9th International Conference on the Environment and Electrical Engineering (EEEIC), Prague, Czech Republic, 16–19 May 2010; pp. 41–44.
8. Zhang, X.; Wang, Y. Robust Fuzzy Control for Doubly Fed Wind Power Systems with Variable Speed Based on Variable Structure Control Technique. *Math. Probl. Eng.* **2014**, 2014, 750101.
9. Tremblay, E.; Atayde, S.; Chandra, A. Comparative study of control strategies for the doubly fed induction generator in wind energy conversion systems: A DSP-based implementation approach. *IEEE Trans. Sustain. Energy* **2011**, *2*, 288–299.
10. Chen, Z.; Guerrero, J.M.; Blaabjerg, F. A review of the state of the art of power electronics for wind turbines. *IEEE Trans. Power Electron.* **2009**, *24*, 1859–1875.
11. Kayikci, M.; Milanovic, JV. Reactive power control strategies for DFIG-based plants. *IEEE Trans. Energy Convers.* **2007**, *22*, 389–396.
12. Xiang, D.; Ran, L.; Tavner, P.J.; Yang, S. Control of a doubly fed induction generator in a wind turbine during grid fault ride-through. *IEEE Trans. Energy Convers.* **2006**, *21*, 652–662.
13. Carrasco, J.M.; Franquelo, L.G.; Bialasiewicz, J.T.; Galván, E.; Guisado, R.C.; Prats, M.Á.; León, J.I.; Moreno-Alfonso, N. Power-electronic systems for the grid integration of renewable energy sources: A survey. *IEEE Trans. Ind. Electron.* **2006**, *53*, 1002–1016.
14. Cheng, M.; Zhu, Y. The state of the art of wind energy conversion systems and technologies: A review. *Energy Convers. Manag.* **2014**, *88*, 332–347.
15. Barambones, O. Sliding Mode Control Strategy for Wind Turbine Power Maximization. *Energies* **2012**, *5*, 2310–2330.
16. Chen, J.; Jiang, L.; Yao, W.; Wu, Q.H. Perturbation estimation based nonlinear adaptive control of a full-rated converter wind turbine for fault ride-through capability enhancement. *IEEE Trans. Power Syst.* **2014**, *29*, 2733–2743.
17. Ren, M.; Zhang, J.; Tian, Y.; Hou, G. A neural network controller for variable-speed variable-pitch wind energy conversion systems using generalized minimum entropy criterion. *Math. Probl. Eng.* **2014**, *2014*, 412027.
18. Yao, X.; Su, X.; Tian, L. Wind turbine control strategy at lower wind velocity based on neural network PID control. In Proceedings of the International Workshop on Intelligent Systems and Applications (ISA '09), Wuhan, China, 23–24 May 2009; pp. 1–5.
19. Perng, J.W.; Chen, G.Y.; Hsieh, S.C. Optimal pid controller design based on PSO-RBFNN for wind turbine systems. *Energies* **2014**, *7*, 191–209.
20. Phan, D.C.; Yamamoto, S. Maximum Energy Output of a DFIG Wind Turbine Using an Improved MPPT-Curve Method. *Energies* **2015**, *8*, 11718–11736.
21. Li, T.; Feng, A.J.; Zhao, L. Neural network compensation control for output power optimization of wind energy conversion system based on data-driven control. *J. Control Sci. Eng.* **2012**, *2012*, 736586.
22. Thongam, J.S.; Bouchard, P.; Ezzaidi, H.; Ouhrouche, M. Artificial neural network-based maximum power point tracking control for variable speed wind energy conversion systems. In Proceedings of the 18th IEEE International Conference on Control Applications, Part of 2009 IEEE Multi-conference on Systems and Control, Saint Petersburg, Russia, 8–10 July 2009; pp. 1667–1671.
23. Ganjefar, S.; Ghassemi, A.A.; Ahmadi, M.M. Improving efficiency of two-type maximum power point tracking methods of tip-speed ratio and optimum torque in wind turbine system using a quantum neural network. *Energy* **2014**, *67*, 444–453.
24. Aissaoui, A.G.; Tahour, A.; Abid, M.; Essounbouli, N.; Nollet, F. A fuzzy sliding mode-based power control design for wind turbine. In Proceedings of the 2012 2nd International Symposium on Environment-Friendly Energies and Applications (EFEA), Newcastle upon Tyne, UK, 25–27 June 2012; pp. 329–335.
25. Yao, X.; Liu, Y.; Guo, C. Adaptive fuzzy sliding-mode control in variable speed adjustable pitch wind turbine. In Proceedings of the IEEE International Conference on Automation and Logistics (ICAL '07), Jinan, China, 18–21 August 2007; pp. 313–318.
26. Abdeddaim, S.; Betka, A. Optimal tracking and robust power control of the DFIG wind turbine. *Int. J. Electr. Power Energy Syst.* **2013**, *49*, 234–242.

27. Kairous, D.; Wamkeue, R. DFIG-based fuzzy sliding-mode control of WECs with a flywheel energy storage. *Electr. Power Syst. Res.* **2012**, *93*, 16–23.
28. Boulâam, K.; Boukhelifa, A.A. Fuzzy sliding mode control for DFIG-based wind turbine power maximisation. In Proceedings of the 7th IET International Conference on Power Electronics, Machines and Drives (PEMD '14), Manchester, UK, 8–10 April 2014; pp. 1–6.
29. Tanvir, A.A.; Merabet, A.; Beguenane, R. Real-Time Control of Active and Reactive Power for Doubly Fed Induction Generator (DFIG)-Based Wind Energy Conversion System. *Energies* **2015**, *8*, 10389–10408.
30. Xu, L. Coordinated control of DFIG's rotor and grid side converters during network unbalance. *IEEE Trans. Power Electron.* **2008**, *23*, 1041–1049.
31. Singh, B.; Naidu, S.; Krishna, N. Direct power control of single VSC-based DFIG without rotor position sensor. *IEEE Trans. Ind. Appl.* **2014**, *50*, 4152–4163.
32. Zarei, M.E.; Asaei, B. Combined vector control and direct power control methods for DFIG under normal and unbalanced and distorted grid voltage conditions. In Proceedings of the 4th Power Electronics, Drive Systems and Technologies Conference (PEDSTC), Tehran, Iran, 13–14 February 2013; pp. 107–112.
33. Khedher, A.; Khemiri, N.; Mimouni, M.F. Wind energy conversion system using DFIG controlled by backstepping and sliding mode strategies. *Int. J. Renew. Energy Res.* **2012**, *2*, 421–430.
34. Liu, Y; Wu, Q.H.; Zhou, X.X.; Jiang, L. Perturbation observer based multiloop control for the DFIG-WT in multi-machine power system. *IEEE Trans. Power Syst.* **2014**, *29*, 2905–2915.
35. Balogun, A.; Ojo, O.; Okafor, F. Decoupled direct control of natural and power variables of doubly fed induction generator for extended wind speed range using feedback linearization. *IEEE J. Emerg. Sel. Top. Power Electron.* **2013**, *1*, 226–237.
36. Chen, G.; Zhang, L.; Cai, X. Optimized Control of the Doubly Fed Induction Generator System Based on input-output Linearizing Scheme. *Wind Eng.* **2014**, *38*, 101–108.
37. Nemmour, A.; Abdessemed, R. The input-output linearizing control scheme of the doubly-fed induction machine as a wind power generation. *Wind Eng.* **2008**, *32*, 285–297.
38. Balogun, A.; Ojo, O.; Okafor, F.; Karugaba, S. Determination of steady state anddynamic control laws of doubly-fed induction generators using natural and power variables. *IEEE Trans. Ind. Appl.* **2013**, *49*, 1343–1357.
39. Balogun, A.; Ojo, O.; Okafor, F. Decoupledcontrol of natural and power variables of doubly-fed induction generator using feedback linearization. In Proceedings of the 2012 IEEE Energy Conversion Congress and Exposition (ECCE), Raleigh, NC, USA, 15–20 September 2012; pp. 801–808.
40. Mauricio, J.; Leon, A.; Gomez-Exposito A.; Solsona, J. An adaptive nonlinear controller for DFIM-based wind energy conversion systems. *IEEE Trans. Energy Convers.* **2008**, *23*, 1025–1035.
41. Wu, F.; Zhang, X.; Ju, P.; Sterling, M. Decentralized nonlinear control of wind turbine with doubly fed induction generator. *IEEE Trans. Power Syst.* **2008**, *23*, 613–621.
42. Kamal, E.; Aitouche, A.; Ghorbani, R.; Bayart, M. Robust nonlinear control of wind energy conversion systems. *Int. J. Electr. Power Energy Syst.* **2013**, *44*, 202–209.
43. Kamal, E.; Oueidat, M.; Aitouche, A.; Ghorbani, R. Robust Scheduler Fuzzy Controller of DFIG Wind Energy Systems. *IEEE Trans. Sustain. Energy* **2013**, *4*, 706–715.
44. Isidori A. The zero dynamics of a nonlinear system: From the origin to the latest progresses of a long successful story. *Eur. J. Control* **2013**, *19*, 369–378.

energies

MDPI

Article

Curtailment in a Highly Renewable Power System and Its Effect on Capacity Factors

Alexander Kies [1,2,*], Bruno U. Schyska [1,2] and Lueder von Bremen [1]

[1] ForWind, Center for Wind Energy Research, Ammerlaender Heerstreet 136, 26129 Oldenburg, Germany; bruno.schyska@forwind.de (B.U.S.); lueder.von.bremen@forwind.de (L.V.B.)
[2] Institute of Physics, University of Oldenburg, Ammerlaender Heerstreet 114, 26129 Oldenburg, Germany
* Correspondence: alexander.kies@uni-oldenburg.de; Tel.: +49-441-798-5078

Academic Editor: Paolo Mercorelli
Received: 26 April 2016; Accepted: 9 June 2016; Published: 30 June 2016

Abstract: The capacity factor of a power plant is the ratio of generation over its potential generation. It is an important measure to describe wind and solar resources. However, the fluctuating nature of renewable power generation makes it difficult to integrate all generation at times. Whenever generation exceeds the load, curtailment or storage of energy is required. With increasing renewable shares in the power system, the level of curtailment will further increase. In this work, the influence of the curtailment on the capacity factors for a highly renewable German power system is studied. An effective capacity factor is introduced, and the implications for the distribution of renewable power plants are discussed. Three years of highly-resolved weather data were used to model wind and solar power generation. Together with historical load data and a transmission model, a possible future German power system was simulated. It is shown that effective capacity factors for unlimited transmission are strongly reduced by up to 60% (wind) and 70% (photovoltaics) and therefore of limited value in a highly renewable power system. Furthermore, the results demonstrate that wind power benefits more strongly from a reinforced transmission grid than photovoltaics (PV) does.

Keywords: renewable energy systems; capacity factor; effective capacity factor; energy system modeling; German power system; renewable site assessment

1. Introduction

The German Energiewende is the process that dispatchable conventional generation units in Germany are replaced by generation from renewable sources. The share of renewable generation in the electricity mix has increased in Germany from 6.3% in 2000 to 29.7% for the first eleven months of 2014 [1]. This 29.7% is divided into 9% wind power, 6.9 % photovoltaics (PV), 3.6% hydro and 10.2% biomass. Concerning the net installed capacities, wind and PV are the leading technologies with 35.6 GW wind (including 0.6 GW offshore) and 38.1 GW PV. Wind and PV are usually considered the most promising renewable technologies for energy generation in the world [2,3]. However, the strongly fluctuating and non-controllable nature of renewable power sources makes the system integration difficult (despite the spatial smoothing of wind [4] and PV [5] feed-in).

The capacity factor of a power plant is the ratio between its actual output in a period of time over the maximal theoretical output in the same period [6]. It is an important measure to decide on new sites for further deployment of renewable energy sources and to derive an optimal wind turbine layout for a specific site [7]. Furthermore, it can also be computed for an ensemble of power plants or an entire country and has direct influence on the financial performance of renewable power plants [8]. For wind turbines, the capacity factor is determined mainly by the geographic location of the wind turbine, as well as the turbine design itself. This renders the absolute value of the capacity factor for a site somewhat arbitrary. One might imagine a small generator combined with a very large rotor leading for most locations to an extremely high capacity factor. Consequently, the technical assumptions are fixed throughout the paper. Details are given in Section 3. For renewable sources, the capacity factor highly relies on meteorological conditions and is therefore also referred to as the meteorological capacity factor. However, for very high shares of renewables and therefore increased curtailment, this capacity factor seems not sufficient to describe the resource of a site adequately. This leads to the introduction of the effective capacity factor. This modified capacity factor considers only generation that is really used and not curtailed. This means that the interplay of generation at different locations and the load patterns need to be taken into account. The effect of curtailment on capacity factors was already discussed for the U.S. without transmission constraints in [9]. Curtailment of wind power induced by congestion was discussed in [10]. The work in [11] introduced the idle capacity as a measure for ineffectively deployed wind power capacity and connected it to several socio-economic factors. However, this work focuses on socio-economic aspects.

Germany is still far from a dominance of fluctuating renewables on the generation side. Nevertheless, a renewable share in the electricity mix of 100% is possible by 2050 [12]. In addition, recent numbers for Germany published by the Bundesnetzagentur (Federal Network Agency) [13] show that curtailment is an issue already: although the share from renewable sources in the electricity mix was below 30% in 2014, 1.6 TWh of wind/PV and biomass generation (ca. 1.2% of total generation from these sources) were curtailed. This was explained to be mainly due to wind power generated in the north not being able to reach the south due to congestion. In comparison to 2013, this value has tripled and demonstrates that curtailment can even be of some importance if renewable shares of generation are rather low.

Furthermore, several events of curtailment of renewable generation have been reported across Europe in the past few years: The economic impact of curtailment in Spain was described as a loss of approximately 70 million Euro from January to March 2013, and the amount of curtailment is expected to increase to yearly 2.3 TWh (3.1% of renewable generation) in 2020 [14]. In Italy, the amount of curtailed wind energy was reduced from 10.7% in 2009 to 1.24% in 2013 due to a significant reinforcement of the transmission grid capacities. In Ireland, 110 GWh wind energy were curtailed in the year 2012 [15]. This is approximately 2% of the total wind generation. Congestion in the power transmission system was responsible for about 80% of the losses, while 20% was due to local network constraints.

The main objective of this paper is to demonstrate the large reduction of the capacity factor due to curtailment in a highly renewable power system and the therefore questionable relevance of the meteorological capacity for site assessment. For this purpose, the following steps are taken: (1) the introduction of an effective capacity factor as an alternative to the meteorological capacity factor; (2) the computation of the effective capacity factors for two highly renewable scenarios of the future German power system; (3) investigation of the dependency of the effective capacity factor on the capacity distribution and the transmission grid strength; (4) drawing of conclusions from the investigation: what does curtailment imply for the preferable design of the future German power system with a very high share of renewables?

2. Methodology

A possible future German power system is simulated. Counties (and independent cities, also referred to as counties) are considered as generators and consumers and represent the nodes in the system, which are connected by links. To derive the grid topology, the simple assumption is made that every county is connected to its direct neighbors via a link of length d_l (average length: 33.5 km), where the length is given by the distance between the geographical centers. Hence, the power grid consists of 402 nodes and 1049 links (Figure 1).

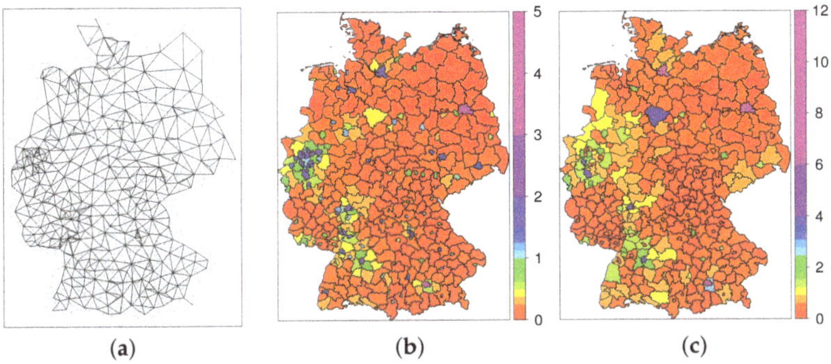

Figure 1. (**a**) Grid topology of the investigated system (for details, see the text); (**b**) population density (1000 inhabitants per km^2) [16]; (**c**) gross domestic product (10^{10}€) of German counties in 2012 [17].

The counties themselves are assumed to be ideal nodes. In existing systems, transmission losses are usually small (1%–2% per 100 km at 380 kV [18]) and neglected. This model setup has been chosen for the following reasons: Firstly, the transmission system should impose as little restrictions as possible on the exchange of energy within the system. An even larger system with more nodes and links, however, would become increasingly computationally expensive and numerically unstable. Secondly, counties are the political-administrative entity responsible for transmission grid planning, supervision and approval.

Two major restrictions apply to the model: first, offshore wind is not considered, and second, Germany is investigated as an isolated power system. These restrictions are briefly discussed in Section 5. For the following investigation, time series of generation for wind $W_n(t)$ and PV $S_n(t)$, based on highly resolved weather data, were computed for each node. Together, they compose the generation of the node n,

$$G_n(t) = W_n(t) + S_n(t) \tag{1}$$

The corresponding time series of the mismatch between generation and load is:

$$\Delta_n(t) = G_n(t) - L_n(t) \tag{2}$$

or, if transmission is included,

$$\Delta_n(t) = G_n(t) - L_n(t) + I_n(t) - E_n(t) \tag{3}$$

where $L_n(t)$ is the time series of the load, $E_n(t)$ are the exports and $I_n(t)$ are the imports of node n. At each node and at all times, generation must equal load to ensure the stability of the system. This leads to the nodal balancing equation:

$$G_n(t) - L_n(t) = P_n(t) - B_n(t) + C_n(t) \tag{4}$$

where $P_n(t)$ is the injection pattern (imports-exports), $B_n(t)$ is the backup (i.e., dispatchable generation) and $C_n(t)$ is curtailed excess energy. Throughout the paper, it is assumed that excess energy is curtailed, and remaining residual loads are covered by backup energy. It follows directly that backup and curtailment do not occur at the same node for the same time. Backup power is not explicitly investigated in the following, since this investigation focuses on the aspect of curtailment. The curtailment time series is calculated as:

$$C_n(t) = \max\left(\{0, \Delta_n(t)\}\right) \tag{5}$$

where $\Delta_n(t)$ is the mismatch after transmission. The left part of the balancing Equation (4) is the active part that is determined by the given data, while the right side is the reactive part, i.e., the response of the system. More generally, this equation could be extended by additional terms to account for storage, demand side management, etc. The transmission model, which determines the injection pattern, is described in Section 2.2. After generation, load and transmission, the remaining residual mismatch is handled by either the backup (if $\Delta_n < 0$, i.e., energy deficit) or curtailment (if $\Delta_n > 0$, i.e., energy excess).

2.1. Definitions

In a given county, the ratio of average renewable generation to average load is described by the share of renewables α_n,

$$\alpha_n = \frac{\langle G_n(t)\rangle}{\langle L_n(t)\rangle} \tag{6}$$

Equivalently, the share of renewables of the whole system is given by:

$$\alpha = \frac{\langle G(t)\rangle}{\langle L(t)\rangle} = \frac{\sum_n \langle G_n(t)\rangle}{\sum_n \langle L_n(t)\rangle} \tag{7}$$

For $\alpha = 1.0$, the generation in all counties combined is on average equal to the average load combined. In this paper, only $\alpha = 1.0$ is considered. Besides α, the parameter β describing the solar share in the renewable generation is defined as:

$$\beta_n = \frac{\langle G_n^S(t)\rangle}{\langle G_n^S(t) + G_n^W(t)\rangle} \tag{8}$$

and similarly for the whole system. For a highly renewable European power system, the optimal β was found to be around 0.4 with respect to the standard deviation of the mismatch [19] or around 0.2 with respect to the need for balancing energy [20].

The capacity factor is the major terminology of this study. For a given county, it is defined as:

$$\nu_n^{w/s} = \frac{\left\langle G_n^{w/s}(t)\right\rangle}{\left\langle G_{n,nom}^{w/s}\right\rangle} \tag{9}$$

where $\left\langle G_{n,nom}^{w/s}\right\rangle$ is the rated power. Consequently, the effective capacity factor for wind $\tilde{\nu}^w$ and photovoltaics $\tilde{\nu}^s$ is defined by:

$$\tilde{\nu}_n^{w/s} = \nu_n^{w/s}\left(1 - \frac{\left\langle C_n^{w/s}(t)\right\rangle}{\left\langle G_n^{w/s}(t)\right\rangle}\right) \tag{10}$$

This definition of an effective capacity is similar to the definition of idle capacity in [11]. Note that the effective capacity factor is the capacity factor that only counts the energy consumed either locally or after transmission.

2.2. Transmission

Transmission is modeled as a two-step optimization problem for every time step. The equations of the full electric power-flow in an AC electricity network are used in a common DC approximation.

For a full derivation of this DC approximation, see, for example, [21,22]. In the first step, the total backup energy is minimized:

$$\underset{F_l}{\text{minimize}} \quad \sum_n B_n(t) =: B^{\min}(t) \tag{11}$$

$$\text{subject to} \quad F_l^- \leq F_l \leq F_l^+ \tag{12}$$

where $F_l^{\pm}$ are the transmission capacities of link l in both directions. This optimization does not necessarily result in a unique solution for the flows F_l. Therefore, the uniqueness of the solution is ensured by minimizing the dissipation $\sum_l d_l F_l^2$, while keeping the overall backup energy need $B^{\min}(t)$;

$$\underset{F_l}{\text{minimize}} \quad \sum_l d_l F_l^2 \tag{13}$$

$$\text{subject to} \quad F_l^- \leq F_l \leq F_l^+ \tag{14}$$

$$\sum_n B_n(t) = B^{\min}(t) \tag{15}$$

where d_l is the length of link l. The injection pattern $P(t)$ is afterwards calculated via the unique solution for the flows F_l of the optimization,

$$P_n = \sum_l K_{nl} F_l \tag{16}$$

K is the incidence matrix, whose elements are defined by:

$$K_{nl} = \begin{cases} 1 & \text{if link l begins at node n} \\ -1 & \text{if link l ends at node n} \\ 0 & \text{otherwise} \end{cases} \tag{17}$$

An equivalent transmission model is used in [23–29]. The effectively-used transmission capacity κ_l^T of a link l connecting two nodes is defined by the 99th percentile,

$$0.99 = \int_0^{\kappa_l^T} p(|F_l|)dp \tag{18}$$

where $p(|F_l|)$ is the time sampled distribution of the absolute values of the unlimited flow over the link l. The transmission capacity of every single link is thus sufficient 99% of the time. This means a transmission grid where every link has the capacity computed by Equation (18) is assumed to be the grid required for unlimited transmission (under the minor restriction that it uses the 99th percentile). Using the 99th percentile for the definition of the transfer capacities aims at reducing the sensitivity towards the weather database.

2.3. Scenario Definition

The following two 100% renewable scenarios (i.e., $\alpha = 1$) were defined for this study:

1. Wind and PV capacities are distributed in such a way that $\alpha_n = \alpha = 1$. This means that every county produces on average as much renewable energy as it consumes.
2. For every county, the currently-installed capacity of wind and PV (2013) is multiplied with the same factor. In this case, today's spatial distribution of capacities is kept, but every installed wind turbine or PV module is upgraded by a certain factor. This process is usually referred to as repowering.

The solar share β_n (see Equation (8)) is fixed for all German counties as it was at the end of 2013. β_n, and the capacity densities for wind and PV can be seen in Figure 2. In northern Germany, β_n is marginal and approaches unity in southern Germany. In the following, the first scenario is referred to as the *self-sufficient* scenario, while the second is called the *upscaled* scenario. The key characteristics of both scenarios can be seen in Table 1 (Section 4). One-hundred percent of generated electricity by 2050 from renewable sources in Germany is in line with a statement of the German Advisory Council on the Environment (SRU) [12]. Thus, the aforementioned scenarios of capacity distribution might be realized at that point. Obviously, they represent the most extreme cases (repowering of all existing generation facilities vs. a homogenous distribution). Thus, the deployment will likely be in between these scenarios by 2050. Figure 3 depicts the general proceeding of this study. The optimization steps were performed using the Python API of Gurobi.

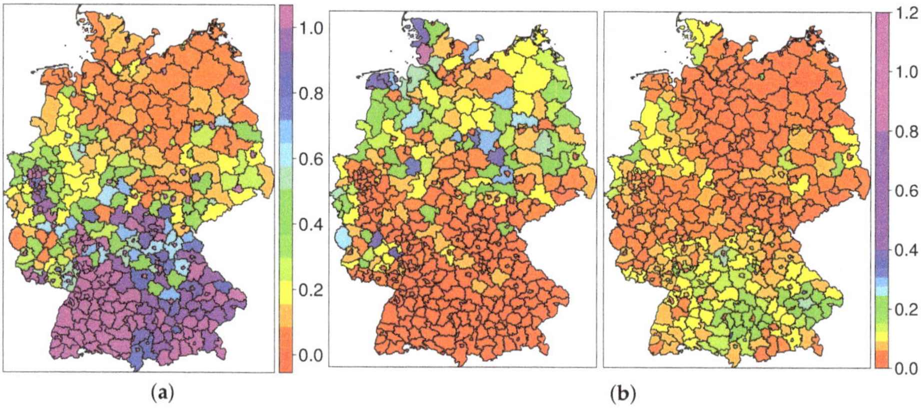

Figure 2. (**a**) Solar share β_n as defined by Equation (8); (**b**,**c**) installed capacity densities (MW/km^2) in all German counties in 2013 for (**b**) wind and (**c**) PV.

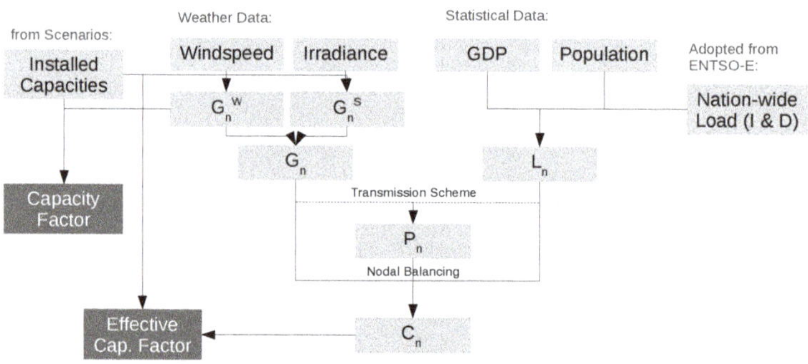

Figure 3. Flowchart describing the model.

3. Data

Curtailment is determined by the interplay of generation and load, as well as by the given possibilities to balance mismatches by importing and exporting energy, i.e., using the grid. To model renewable generation from wind and PV, three years of highly resolved weather data were used. The used data period ranges from January 2010 to December 2012. Wind speeds and irradiance are converted to generation by state-of-the-art methods that are described in more detail in this section.

3.1. Weather Data

Computations of generation are based on data from the following sources:

- Wind speed data at hub height of 77 m and 2-m temperature data from the analysis product of the numerical weather prediction model (NWP) COSMO-DE [30], provided by the German Weather Service (Deutscher Wetterdienst, DWD) and
- Solar irradiation data from Meteosat Second Generation (MSG).

COSMO-DE data have a spatial resolution of 0.025° (ca. 3 km) and hourly temporal resolution. The model domain is centered over Germany and consists of 461 × 421 grid cells. The satellite data were used with a spatial resolution of 0.0625° (ca. 7 km).

3.2. Generation Data

The weather data described in the previous section are used to compute generation time series from wind and PV for each of the 402 German counties. For wind power, the hub height is assumed to be 77 m, the average hub height of wind power plants in Germany at the end of 2013. In the past few years, a trend towards increasing hub heights was observed. Thus, it is likely that the average hub height will increase further leading to higher capacity factors for better wind regimes at these heights. However, this mainly affects the absolute value of the capacity factor. The effect on relative changes due to curtailment are assumed to be negligible.

Wind speeds are converted to power via the power curve model described in [31]. This power curve was fitted to obtain the best match of simulated and observed wind power in the largest German transmission grid zone. We use data for the geographical distribution of installed wind and PV capacities from 2013. For PV power generation, the Klucher model [32] is used to model irradiation on the inclined surface. The PV modules are assumed to have a tilt of 30° and a south-facing orientation (azimuth = 0°). Module temperatures were computed as:

$$T_m = T_a + \sigma I_t \tag{19}$$

where T_a is the ambient temperature, T_m the module temperature and I_t the irradiance on the inclined surface. σ is a factor that was chosen to be $0.036\frac{\text{km}^2}{\text{W}}$. The module efficiency $\eta(T_m)$ is calculated as a function of the module temperature,

$$\eta(T_m) = \eta(25^\circ)\,(1 + aT_m) \tag{20}$$

where $a \in R$ is device-specific. DC power can then be calculated by using the installed capacity,

$$P = \eta(T_m) I_t \text{Cap}^s \tag{21}$$

for any grid point. DC power was converted into AC power using the parameters of a Sunny Mini Central 8000 TL converter [33]. Generation for wind and PV is calculated on the grid point level and aggregated to counties afterwards.

3.3. Load Data

Historical load data for Germany provided by the *European Network of Transmission System Operators for Electricity* (ENTSO-E) was taken. These data were modified within the RESTORE2050 project (Frank Merten, Wuppertal Institute, private communication via e-mail, 2014). Modifications include modeled load profiles from e-mobility and heat pumps.Since no information about the energy demand on a sub-national level is publicly available in Germany, load time series of the counties are estimated based on the county's gross domestic product GDP_n and on the county's number of inhabitants D_n; $L(t) = L_D(t) + L_I(t)$ is the modified ENTSO-E load time series for the entirety Germany, consisting of a domestic and an industrial part. The total load of county n, l_n, is expressed as:

$$l_n(t) = l_{I,n}(t) + l_{D,n}(t) \tag{22}$$

where $l_{I,n}$ and $l_{D,n}$ are the industrial and domestic load time series, respectively. The partial loads are computed as:

$$l_{I,n}(t) = \text{GDP}_n \cdot \frac{L_I(t)}{\sum_n \text{GDP}_n} \tag{23}$$

$$l_{D,n}(t) = D_n \cdot \frac{L_D(t)}{\sum_n D_n} \tag{24}$$

The population density and the gross domestic product of the German counties can be seen in Figure 1 (center and right) for 2012. GDP per capita tends to increase from north to south and from east to west.

4. Results

The results Section is structured as follows: First, capacity factors are calculated from the meteorological data alone. Second, the effective capacity factors are calculated. The dependency on the share of renewables, the solar share and the transmission grid capacities are discussed in Section 5.

Table 1 shows the shares of wind and PV power in the electricity mix, as well as the resulting major quantities, like curtailment and transmission capacity needs for both investigated scenarios. The transmission capacities are determined according to Equation (18) for every link and multiplied with the corresponding length d_l. Thus, the displayed values show the need for unlimited transmission as defined by the 99th percentile in this work. In the *self-sufficient* scenario, every county produces on average as much as it consumes, leading to a very homogenous distribution of generation facilities in Germany. This is reflected by the fact that the *self-sufficient* scenario has a much lower demand for transmission grid capacities, which is about half the need of the *upscaled* scenario. In turn, needed generation capacities are around 10% less, and only 28% of the generation is curtailed in the *upscaled* scenario (39% in the *self-sufficient* scenario).

Table 1. Summary of the main differences between the scenarios for unlimited transmission.

Scenario	β	$\left\langle \frac{C^w}{G^w} \right\rangle$	$\left\langle \frac{C^s}{G^s} \right\rangle$	$\left\langle \frac{C}{G} \right\rangle$	Cap^w (GW)	Cap^s (GW)	$\sum_l d_l \kappa_l^T$ (GWkm)
Self-Sufficient	0.68	0.14	0.52	0.39	160	389	13,800
Upscaled	0.35	0.27	0.29	0.28	289	206	27,065

4.1. Meteorological Capacity Factors

The meteorological capacity factor of a renewable power plant (often simply referred to as the capacity factor) is the actual output in a certain period of time over its theoretical output. The theoretical output is defined by the rated power of the generation unit, for example the wind turbine or PV module. These meteorological capacity factors were calculated under the technical specifications described in Section 3 for every county and are shown in Figure 4. For wind, they are strongly increasing up north, while the opposite is true for PV. Capacity factors for wind power plants are around 20% in the northern part of Germany, but only around 10% in southern Germany. Capacity factors for PV vary less, reaching around 14% in southern Germany and 10% in northern Germany. Total capacity factors for wind and PV in Germany are found to be 16.1% and 12.3% for the *upscaled* scenario, respectively. This compares to the capacity factors published by the German Federal Ministry of Economic Affairs and Energy for the years 2010 to 2012 of 17.9% for onshore wind and of 8.6% for PV [34]. The small apparent discrepancy is likely caused by the technological assumptions made in this study (note: the capacity distribution used is the actual capacity distribution of today and can therefore not be responsible for the differences observed).

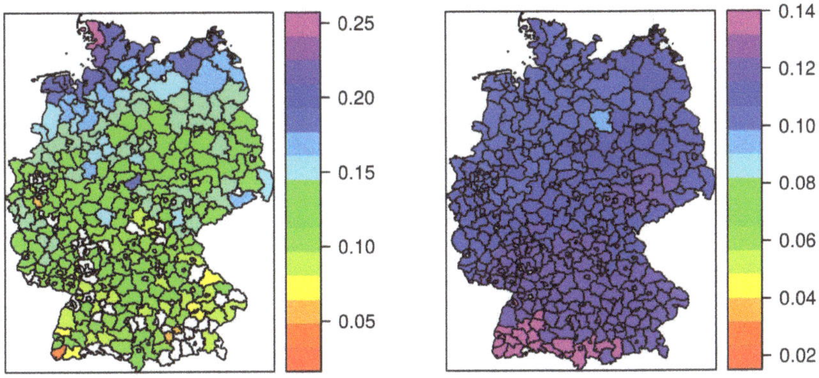

(a) (b)

Figure 4. Meteorological capacity factors without considering curtailment for (**a**) wind and (**b**) PV.

4.2. Effective Capacity Factors

In contrast to the meteorological capacity factors, the effective capacity factor measures only the power output that is effectively used. As they were introduced in Section 2.3, effective capacity factors $\tilde{\nu}^{w/s}$ (see Equation (10)) are investigated for two scenarios of capacity distribution. In the *self-sufficient scenario*, the effective capacity factors are reduced to 6% to 20% for wind and 5% to 12% for PV (Figure 5). While the general qualitative distribution of the effective capacity factor remains for wind (high values in the north and low values in the south), it changes considerably for PV. A strong relationship between the change from ν^s to $\tilde{\nu}^s$ and the share of PV β exists. The highest values for $\tilde{\nu}^s$ can be found towards the northwest and in the central German regions, i.e., in regions with relatively low β (see Figure 2, left). On the contrary, in urban areas of western Germany and in the south, i.e., the regions where the highest meteorological capacity factors and the highest shares of PV occur, the effective capacity factors decrease the most compared to the meteorological capacity factors.

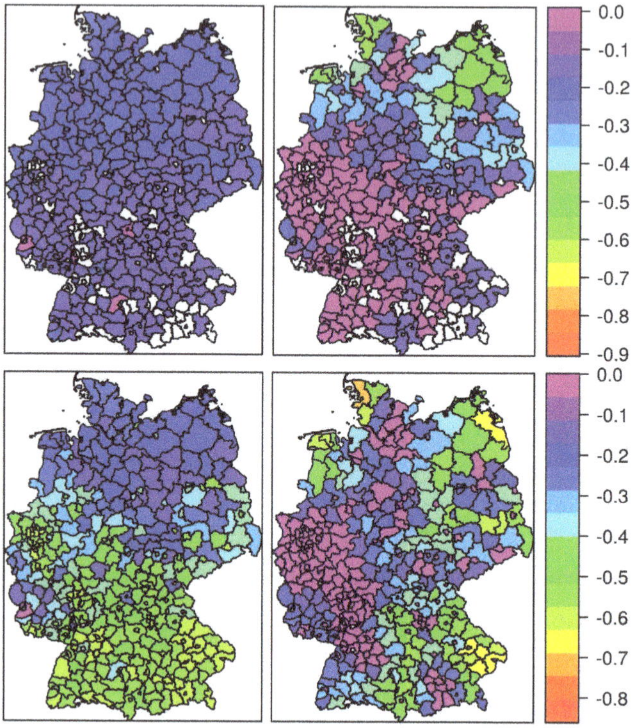

Figure 5. Relative change in capacity factors for wind (**top**) and PV (**bottom**), as well as for the two scenarios *self-sufficient* (**left**) and *upscaled* (**right**). α is set to one, and transmission capacities are unlimited.

In the *upscaled* scenario, effective capacity factors range from 10% to 15% for wind and from 3% to 13% for PV. The strongest reduction for both wind and PV occurs in the rural areas in northeastern and southeastern Germany, as well as the coastal regions. In urban areas and their direct surroundings, the differences between capacity factors and effective capacity factors for PV are very small. For wind the north-south gradient disappears in almost all regions. The high amount of produced power in northern regions leads to strong curtailment and therefore largely reduced capacity factors.

5. Discussion

Where does curtailment occur and why? This paper revolves around the idea of the meteorological capacity factor being no appropriate measure to describe the quality of a site, if shares of renewables are high and thus events of curtailment common. This is closely linked to the problem of decoupled generation and transmission in the power market leading to suboptimal investment choices [35–38].

At what locations does curtailment occur? This depends on different aspects, like the electricity market and the transmission grid. Today, transmission system operators decide on curtailment, and affected owners are almost fully compensated in Germany [39]. The cost of curtailment events are therefore borne by the consumers as a lump sum via network access charges (Netznutzungsentgelte). However, this might change in the future, rendering the question of curtailment more important for the operational side. The coalition agreement of the present German Federal Government aims at modifying the compensation for curtailment so as to provide more incentive to consider the grid situation for new renewable power plants [40].

The markets plays a role mainly via the question of how the cost of transmission is distributed among the market participants. Consider the simple three-node example shown in Figure 6. The three nodes {A,B,C} are connected via two links from A to B and B to C. A and B have similar marginal generation cost and place offers. C bids. If transmission cost does not add to the marginal cost by being provided independently and offers are pooled (shared among market participants), Scenario (II) is realized: A and B both transfer half of their excess energy to C, thus curtailing the other half. If instead transmission cost is above zero and included in the marginal cost, Scenario (III) is realized. B transfers its excess energy to C, and A curtails its power. To generalize this idea: for generation at topologically-favorable positions in the network, marginal costs are comparably low, and curtailment is less likely than at nodes at less favorable positions.

How does the market model connect to the abstract grid topology of this paper? It was assumed that every county is connected to its neighbors. Obviously, this does not resemble today's historically grown power grid. Today, it connects regions and their distribution grids via the high voltage transmission grid. However, the authors of this paper believe that the detailed topology of the grid becomes less important, if the transmission grid is strongly reinforced. In addition, the optimal topology for a highly renewable grid still needs to be determined and will likely differ to some extent from the grid Germany has today. Therefore, it is likely that the beneficial positions in the grid with respect to curtailment will be less dependent on today's grid. Instead, it will heavily depend on the geographical location (distance to load centers). In that way, the chosen abstract topology can be justified.

Two major restrictions apply to the model: offshore wind is not considered, and Germany is investigated as an isolated power system. The first point is probably the less problematic one. The largest differences between meteorological and effective capacity factors, as described in the following sections, are observed in the northernmost regions. These regions border the German offshore regions, which implies that the results would be even more prominent, if offshore wind power generation were included. Concerning imports and exports: under the transmission scheme applied in this paper, it is shown in [23] that Germany imports and exports (for a share of renewables in Germany and Europe: 70%) ca. 7% in units of its consumption. Under the methodology of this investigation, imports can increase curtailment while exports can reduce curtailment. Therefore, it might be an interesting extension of this work to embed Germany into a fully renewable European power system and to study the effect of this on the effective capacity factors.

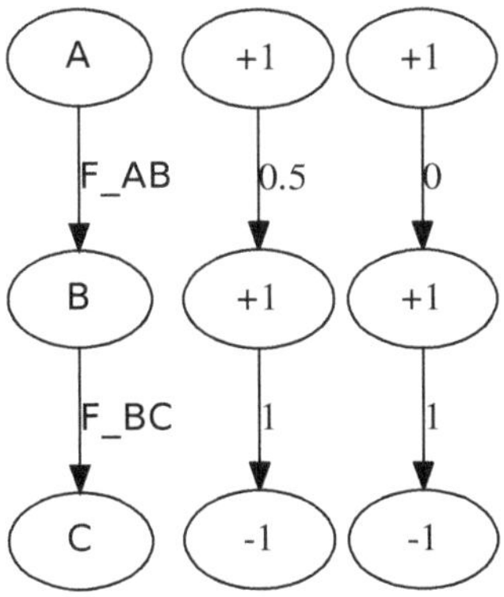

Figure 6. (**a**) Illustration of the example system consisting of three nodes {A,B,C} connected via transmission links; (**b**) B transfers its excess energy to C; (**c**) A and B transfer parts of their excess energy to C. Remaining excess energy is curtailed in both cases.

5.1. Wind/PV Ratio

The results indicate that the wind/PV mix of a node has a large effect on the effective capacity factor. This is easily understandable: The wind/PV ratio determines the temporal correlation of generation and load, which, in turn, has a strong impact on the curtailment. Besides, PV power has a strong diurnal pattern and therefore resembles to some extent the typical diurnal load pattern (peaks during the day and minima at night). On the contrary, wind power generation only exhibits a very weak diurnal cycle. Figure 7 summarizes the reduction of the capacity factor,

$$\Delta\nu^{w/s} = \frac{\tilde{\nu}^{w/s} - \nu^{w/s}}{\nu^{w/s}} \tag{25}$$

for the four quartiles $[0, Q1], ..., (Q3, 1]$ of the solar share β. Focusing on wind, several things can be observed: the largest reduction of capacity factors occurs in the *upscaled* scenario and for small β, i.e., in counties with high shares of wind power generation. In the case of limited transmission (definition in Section 5.3), shown as yellow boxes, this reduction amounts up to 80%. In the *self-sufficient* scenario, the difference between the different intervals of β is minor, i.e., the ratio between PV power and wind power generation is not the driving factor here. The difference between the two scenarios can be explained by the fact that wind power generation accounts for ca. two thirds of total power generation in the *upscaled* scenario, while it contributes only one third in the *self-sufficient* case. Furthermore, the correlation between the mismatch of the county and the overall mismatch is relatively high in counties with high shares of wind leading to much curtailment in these counties.

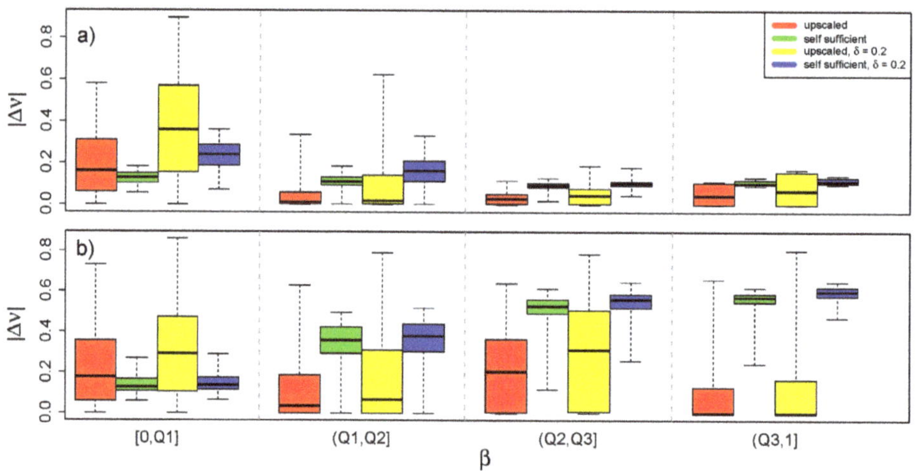

Figure 7. Relative reduction of capacity factors for (**a**) wind and (**b**) PV caused by curtailment with dependency on the solar share β_n for the 402 German counties investigated. Black horizontal lines indicate the medians; boxes embrace values between the first and the third quartiles; and whiskers extend to the maximum and minimum of the range. The curtailment is shown for two different scenarios and for unlimited and limited (parameter $\delta = 0.2$) transmission.

For the curtailment of PV (Figure 7b), things are the reverse: In the *self-sufficient* scenario, PV curtailment is low on average for small β, but increases sharply with increasing β. For β in (Q2,1], around 60% of PV generation is curtailed. In the *upscaled* case, overall β is relatively low ($\beta \approx 0.35$). Hence, almost no PV power is curtailed. In the *self-sufficient* scenario, PV generation contributes the most. The correlation of the generation time series between nodes is larger in this case, which leads to relatively much curtailment, even in counties with a somewhat balanced mix of PV and wind; and to more curtailment in total.

5.2. Share of Renewables

In the *self-sufficient* scenario, the average generation from wind and PV equals the average load in all nodes ($\alpha_n = \alpha = 1.0$). In the *upscaled* scenario, some nodes produce more than they consume, while others produce less. The strong dependency of the curtailment on the share of renewables α_n is shown in Figure 8. High excess generation (i.e., high α_n) leads to the need for exports. In case of common excess generation, energy cannot be exported and is consequently curtailed. For wind and PV, curtailment of up to 60% (unlimited transmission) and 80% (limited transmission) of generated energy can be observed for counties with very high α_n. In counties with low α_n, no curtailment is observed.

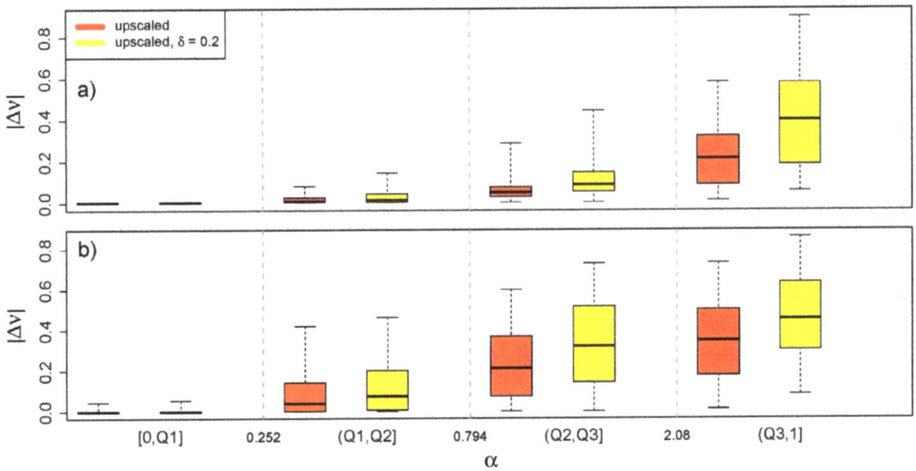

Figure 8. Relative reduction of capacity factors for (**a**) wind and (**b**) PV caused by curtailment with dependency on the share of renewables α_n for the 402 German counties investigated. Black horizontal lines indicate the medians; boxes embrace values between the first and the third quartiles; and whiskers extend to the maximum and minimum of the range. The curtailment is shown for unlimited and limited (parameter $\delta = 0.2$) transmission.

5.3. Limited Transmission

A major obstacle and the subject of intense political controversy is the reinforcement of the transmission grid. All previous results were calculated for unlimited transmission. In this section, it is investigated how curtailment changes if transmission capacities are reduced in an abstract way. First, transmission capacities for every link κ_l are computed by the definitions in Equation (18). These transmission capacities are then all varied with a grid strength parameter $\delta \in [0, 1]$ according to:

$$\kappa_l' = \delta \kappa_l^C \tag{26}$$

where $\delta = 0$ means no transmission between nodes and $\delta = 1$ means transmission capacities are sufficient to transfer all energy for every link 99% of the time. To investigate the impact of limited transmission on the curtailment, δ was varied between zero and one (Figure 9). For low transmission grid strength ($\delta < 0.25$), curtailment is higher in the *upscaled* scenario compared to the *self-sufficient* scenario. For $\delta > 0.25$, overall curtailment in the *self-sufficient* scenario exceeds the overall curtailment in the *upscaled* scenario. In general, transmission is more important in the *upscaled* scenario, i.e., total curtailment is reduced from 67% (no transmission) to 28% for $\delta = 1.0$. For wind and PV considered separately, the reduction is similar: from 73% to 27% and from 57% to 29%, respectively. It also can be observed that for low transmission capacities, the curtailment of PV power is less than the curtailment of wind power, but this changes, if transmission capacities are increased. Therefore, it can be concluded that wind power benefits more from a strong transmission grid than PV does. For the *self-sufficient* scenario, almost no benefit from transmission for PV power, being the dominant source of generation, is noticeable. Wind contributes a smaller share of generation in this scenario, and the curtailment of wind energy can be reduced by strengthening transmission from 32% to 14%. The overall curtailment can be reduced in the *self-sufficient* scenario from 49% to 39%.

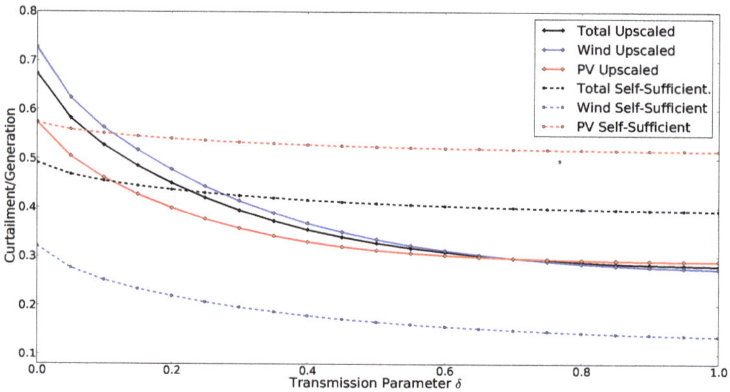

Figure 9. Curtailment with the dependency of the transmission grid strength δ for the *self-sufficient* and the *upscaled* scenario.

Figure 10 shows the effective capacity factors for $\delta = 0.2$. Limiting the transmission capacities to 20% of the maximum transmission capacities decreases effective capacity factors for PV in counties far from the load centers (coastal regions, in Lower Bavaria and southern Brandenburg) by up to 50% in the *upscaled* scenario. In heavily-populated regions, like North Rhine-Westphalia or southern Lower Saxony, no effect is observed. In the *self-sufficient* case, almost no differences occur for most German counties. However, some counties see a further reduction of their effective capacity factors by ca. 15%. Limiting transmission hampers the export of the high surpluses and leads to high curtailment rates. The effect of limited transmission is stronger on wind than on PV (Figure 10, top). For the *self-sufficient* scenario, effective capacity factors in northern Germany decrease significantly, while in most other regions, no significant reduction occurs. In this case, most counties have very similar effective capacity factors for wind in the range of 10% to 14%, and the strong north-south gradient disappears. For the *upscaled* scenario, the transmission limitation has the most prominent (negative) effects, especially on the northern and eastern German regions. In fact, limiting transmission renders them the least favorable for wind power installations in Germany. This is a crucial result, as today, the wind power deployment is still strongest in these regions [41]. In a 100% renewable German power system with an inappropriate transmission grid, these wind power plants would not be economical any longer, because their capacity factor would be partly only 2%, because their surpluses cannot be exported in times of overall generation surpluses. In fact, the surpluses in the north are also the major cause for increased transmission capacity needs in the *upscaled* scenario (see Table 1).

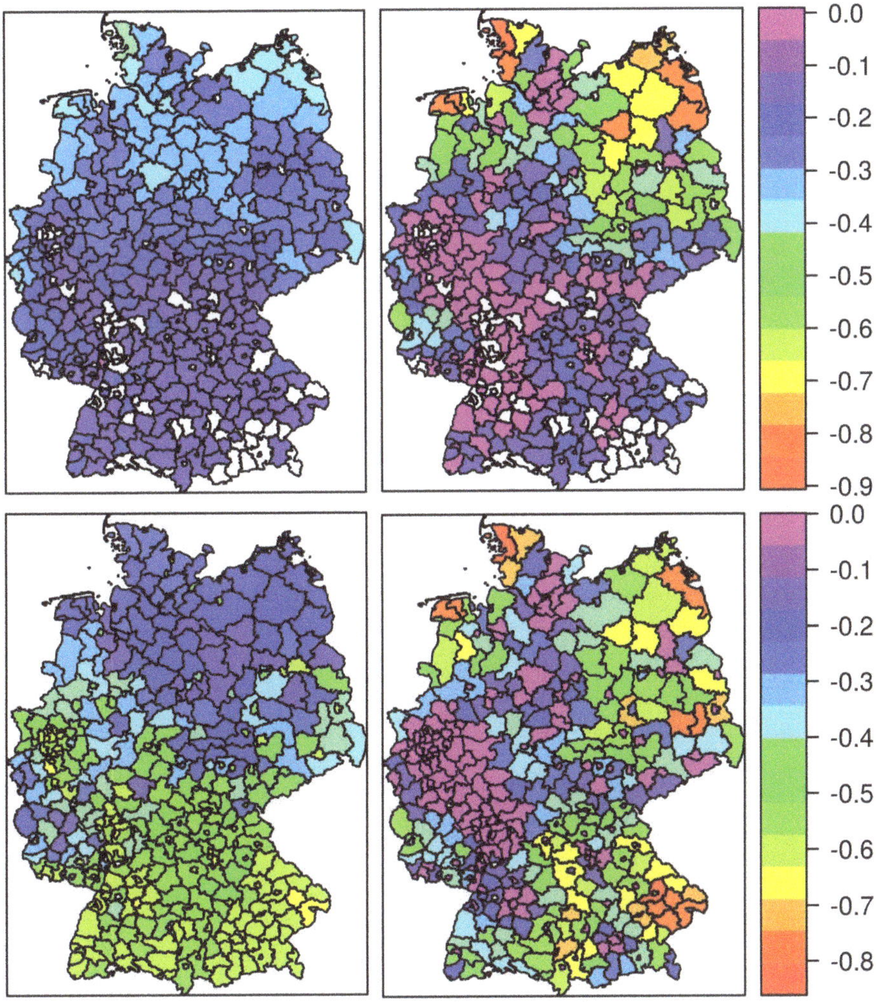

Figure 10. Relative change in capacity factors for wind (**top**) and PV (**bottom**), as well as for the two scenarios *self-sufficient* (**left**) and upscaled (**right**). α is set to one, and transmission capacities are limited ($\delta = 0.2$).

6. Summary and Conclusions

Curtailment is the reduction in the output of a generator not induced by the non-availability of resources. With increasing renewable shares, the curtailed amount of energy consequently grows. Therefore, considering curtailment will be of increasing importance in the future. In this paper, the curtailment for 402 German counties that are all fully connected to their neighboring counties was investigated. Today, the common capacity factor is primarily considered for the site assessment of renewable power plants, because revenue generated by a power plant is linearly linked to its generated energy. In power systems with high shares of renewables, an effective capacity factor seems to be more appropriate to compare different sites with respect to their quality for renewable power deployment. For this investigation, an effective capacity factor was defined. It differs from the common capacity factor by considering only generation that is not curtailed. It was shown that several aspects influence curtailment and hence effective capacity factors. Two renewable scenarios have been defined for the distribution of capacities. In the *upscaled* scenario, today's spatial capacity distribution within Germany is preserved, and the capacity of each generation facility was multiplied by a constant factor to meet demand. This scenario requires strong grid reinforcement to dampen curtailment as much as possible. In the *self-sufficient* scenario, every county was autarkic in the way that its average production meets its average consumption. This scenario is advantageous if the transmission grid is not very well developed. It only requires approximately 50% of the transmission capacity of the *upscaled* scenario to distribute all useable energy surpluses.

The share between wind and photovoltaics in a county and the whole country indirectly influences the curtailment and the overall need for transmission. This is caused by the strong correlation of wind and PV generation between different counties. If a county has a high wind share while the overall share of wind is low, this county experiences several export opportunities, when PV generation is low.

The effect of transmission limitations on effective capacity factors is complex. For PV transmission capacity reductions lead only to a small decrease in most regions. For wind, the effect is much stronger. For some regions, especially in eastern Germany and the coastal areas, effective capacity factors are lower than in central German regions, although meteorological capacity factors behave vice versa, if the transmission grid strength is significantly reduced. The importance of the transmission grid was investigated by reducing the transmission capacity between every pair of nodes to a fraction of the maximum required to transmit all surplus energies. In both scenarios, it was demonstrated that transmission has the potential to reduce curtailment of wind power by two thirds. The benefit for PV by increased transmission is smaller. In the *self-sufficient* scenario, no significant reduction of curtailment for PV power by increased transmission capacities was observed. In the *upscaled* scenario, the maximum reduction of curtailment for PV by transmission is ca. 50%. Future work should investigate the interplay of storage and curtailment, power exports/imports to/from neighboring countries and the role of German offshore wind power.

Acknowledgments: Acknowledgments: We thank Stephan Späth and Ontje Lünsdorf (both University of Oldenburg) for help with the data and discussions. We thank Martin Greiner (Aarhus University) for helpful discussions about transmission models. We thank the Wuppertal Institute for providing load data within RESTORE 2050 (Bundesministerium für Bildung und Forschung, BMBF). This work was supported by the Federal Ministry for Science and Culture of Lower Saxony and by the national R&D Project RESTORE 2050 (FKZ 03SFF0439A, Ministry for Education and Research), as well as the EU project "Integrated Research Programme on Wind Energy" funded by the European Union's Seventh Programme for research, technological development and demonstration under Grant Agreement No. 609795. We thank three anonymous reviewers for their helpful comments and suggestions, which helped to improve this manuscript.

Author Contributions: Author Contributions: The first two authors (Alexander Kies and Bruno U. Schyska) have contributed equally to this work.All authors contributed to the writing of this article and approved the final manuscript.

Conflicts of Interest: Conflicts of Interest: The authors declare no conflict of interest.

References

1. Burger, B. *Electricity Production from Solar and Wind in Germany in 2014*; Fraunhofer ISE: Freiburg, Germany, 2013.
2. Jacobson, M.Z.; Delucchi, M.A. Providing all global energy with wind, water, and solar power, Part I: Technologies, energy resources, quantities and areas of infrastructure, and materials. *Energy Policy* **2011**, *3*, 1154–1169.
3. De Vries, B.J.; van Vuuren, D.P.; Hoogwijk, M.M. Renewable energy sources: Their global potential for the first-half of the 21st century at a global level: An integrated approach. *Energy Policy* **2007**, *35*, 2590–2610.
4. Hasche, B. General statistics of geographically dispersed wind power. *Wind Energy* **2010**, *13*, 773–784.
5. Wiemken, E.; Beyer, H.; Heydenreich, W.; Kiefer, K. Power characteristics of PV ensembles: Experiences from the combined power production of 100 grid connected PV systems distributed over the area of Germany. *Sol. Energy* **2001**, *70*, 513–518.

6. Boccard, N. Capacity factor of wind power realized values *vs.* estimates. *Energy Policy* **2009**, *37*, 2679–2688.
7. Abed, K.A.; El-Mallah, A.A. Capacity factor of wind turbines. *Energy* **1997**, *22*, 487–491.
8. Justus, C.; Hargraves, W.; Yalcin, A. Nationwide assessment of potential output from wind-powered generators. *J. Appl. Meteorol.* **1976**, *15*, 673–678.
9. Diakov, V. Wind resource quality affected by high levels of renewables. *Resources* **2015**, *4*, 378–383.
10. Gu, Y.; Xie, L.; Rollow, B.; Hesselbaek, B. Congestion-induced wind curtailment: Sensitivity analysis and case studies. In Proceedings of the North American Power Symposium (NAPS), Boston, MA, USA, 4–6 August 2011; IEEE: Boston, MA, USA, 2011; pp. 1–7.
11. Flora, R.; Marques, A.C.; Fuinhas, J.A. Wind power idle capacity in a panel of European countries. *Energy* **2014**, *66*, 823–830.
12. Faulstich, M.; Foth, H.; Calliess, C.; Hohmeyer, O.; Holm-Mueller, K.; Niekisch, M.; Schreurs, M. *Climate-friendly, Reliable, Affordable: 100% Renewable Electricity Supply by 2050*; Technical Report; German Advisory Council on the Environment (SRU): Berlin, Germany, 2010.
13. Bundesnetzagentur. EEG in Zahlen. 2015. Available online: http://www.bundesnetzagentur.de/cln_1422/DE/Sachgebiete/ElektrizitaetundGas/Unternehmen_Institutionen/ErneuerbareEnergien/ZahlenDatenInformationen/zahlenunddaten-node.html (accessed on 3 October 2015).
14. Lew, D.; Bird, L.; Milligan, M.; Speer, B.; Wang, X.; Carlini, E.M.; Estanqueiro, A.; Flynn, D.; Gomez-Lazaro, E.; Menemenlis, N.; *et al.* Wind and solar curtailment. In Proceedings of the 12th International Workshop on Large-Scale Integration of Wind Power into Power Systems as well as on Transmission Networks for Offshore Wind Power Plants, London, UK, 22–24 October 2013.
15. EirGrid; SONI. *Delivering a Secure Sustainable Electricity System*; Project Presentation; EirGrid: Dublin, Ireland, 2013.
16. Bundesamt für Kartographie und Geodäsie. 2015. Available online: http://www.bkg.bund.de (accessed on 31 July 2015).
17. Statistische Ämter des Bundes und der Länder. 2015. Available online: http://www.regionalstatistik.de (accessed on 31 July 2015).
18. Negra, N.B.; Todorovic, J.; Ackermann, T. Loss evaluation of HVAC and HVDC transmission solutions for large offshore wind farms. *Electr. Power Syst. Res.* **2006**, *76*, 916–927.
19. Heide, D.; von Bremen, L.; Greiner, M.; Hoffmann, C.; Speckmann, M.; Bofinger, S. Seasonal optimal mix of wind and solar power in a future, highly renewable Europe. *Renew. Energy* **2010**, *35*, 2483–2489.
20. Kies, A.; Nag, K.; von Bremen, L.; Lorenz, E.; Heinemann, D. Investigation of balancing effects in long term renewable energy feed-in with respect to the transmission grid. *Adv. Sci. Res.* **2015**, *12*, 91–95.
21. Oswald, B.; Oeding, D. *Elektrische Kraftwerke und Netze*, 6th ed.; Springer-Verlag: Berlin/Heidelberg, Germany; New York, NY, USA, 2004.
22. Heide, D. Statistical Physics of Power Flows on Networks with a High Share of Fluctuating Renewable Generation. Ph.D. Thesis, Johann Wolfgang Goethe-Universitaet Frankfurt am Main, Frankfurt, Germany, 2010.
23. Rodriguez, R.A.; Dahl, M.; Becker, S.; Greiner, M. Localized *vs.* synchronized exports across a highly renewable pan-European transmission network. *Energy Sustain. Soc.* **2015**, *5*, 1–9.
24. Heide, D.; Greiner, M.; von Bremen, L.; Hoffmann, C. Reduced storage and balancing needs in a fully renewable European power system with excess wind and solar power generation. *Renew. Energy* **2011**, *36*, 2515–2523.
25. Rodriguez, R.A.; Becker, S.; Greiner, M. Cost-optimal design of a simplified, highly renewable pan-European electricity system. *Energy* **2015**, *83*, 658–668.
26. Kies, A.; von Bremen, L.; Chattopadhyay, K.; Lorenz, E.; Heinemann, D. Backup, storage and transmission estimates of a supra-European electricity grid with high shares of renewables. In Proceedings of the 14th Wind Integration Workshop, Brussels, Belgium, 20–22 October 2015.
27. Becker, S.; Rodriguez, R.; Andresen, G.; Schramm, S.; Greiner, M. Transmission grid extensions during the build-up of a fully renewable pan-european electricity supply. *Energy* **2014**, *64*, 404–418.
28. Becker, S.; Frew, B.A.; Andresen, G.B.; Zeyer, T.; Schramm, S.; Greiner, M.; Jacobson, M.Z. Features of a fully renewable US electricity system: Optimized mixes of wind and solar PV and transmission grid extensions. *Energy* **2014**, *72*, 443–458.
29. Rodriguez, R.A.; Becker, S.; Andresen, G.B.; Heide, D.; Greiner, M. Transmission needs across a fully renewable European power system. *Renew. Energy* **2014**, *63*, 467–476.

30. Doms, G.; Schättler, U.; Baldauf, M. *A Description of the Nonhydrostatic Regional COSMO Model, Part I: Dynamics and Numerics*; Technical Report; Consortium for Small-Scale Modelling: Offenbach, Germany, 2011.
31. Späth, S.; von Bremen, L.; Junk, C.; Heinemann, D. Time-consistent calibration of short-term regional wind power ensemble forecasts. *Meteorol. Z.* **2015**, *24*, 381–392.
32. Klucher, T. Evaluation of models to predict insolation on tilted surfaces. *Sol. Energy* **1979**, *23*, 111–114.
33. SMA Solar Technology AG. *Sunny Mini Central 6000TL/7000TL/8000TL*; SMA Solar Technology AG: Niestetal, Germany, 2010.
34. Bundesministerium für Wirtschaft und Energie. *Erneuerbare Energien in Zahlen—Nationale und Internationale Entwicklung im Jahr 2013*; Bundesministerium für Wirtschaft und Energie: Berlin, Germany, 2014.
35. Grimm, V.; Martin, A.; Weibenzahl, M.; Zoettl, G. Transmission and generation investment in electricity markets: The effects of market splitting and network fee regimes. *Eur. J. Oper. Res.* **2016**, *254*, 493–509.
36. Kemfert, C.; Kunz, F.; Rosellón, J. A Welfare Analysis of the Electricity Transmission Regulatory Regime in Germany; DIW: Berlin, Germany, 2015.
37. Baringo, L.; Conejo, A.J. Transmission and wind power investment. *IEEE Trans. Power Syst.* **2012**, *27*, 885–893.
38. Orfanos, G.A.; Georgilakis, P.S.; Hatziargyriou, N.D. Transmission expansion planning of systems with increasing wind power integration. *IEEE Trans. Power Syst.* **2013**, *28*, 1355–1362.
39. Erneuerbare-Energien-Gesetz, 2014. Available online: http://dejure.org/gesetze/EEG/12.html (accessed on 15 March 2016).
40. Deutschlands Zukunft Gestalten—Koalitionsvertrag Zwischen CDU, CSU und SPD, 2013. Available online: http://www.bundesregierung.de/Content/DE/_Anlagen/2013/2013-12-17-koalitionsvertrag.pdf (accessed on 18 March 2016).
41. Betreiberdatenbasis: Betriebsdaten von Windanlagen. Available online: http://www.btrdb.de/ (accessed on 10 January 2015).

energies

MDPI

Article

Sliding Mode Variable Structure Control of a Bearingless Induction Motor Based on a Novel Reaching Law

Zebin Yang [1], Ling Wan [1], Xiaodong Sun [2,*], Fangli Li [1] and Lin Chen [1]

[1] School of Electrical and Information Engineering, Jiangsu University, Zhenjiang 212013, China; zbyang@ujs.edu.cn (Z.Y.); wanlinjsdx@163.com (L.W.); lifanglijsdx@163.com (F.L.); chenlinjsdx@163.com (L.C.)

[2] Automotive Engineering Research Institute, Jiangsu University, Zhenjiang 212013, China

* Correspondence: xdsun@ujs.edu.cn; Tel.: +86-511-8878-2845

Academic Editor: Paolo Mercorelli

Received: 12 April 2016; Accepted: 27 May 2016; Published: 13 June 2016

Abstract: In order to improve the performance of the Bearingless Induction Motor (BIM) under large disturbances (such as parameter variations and load disturbances), an adaptive variable-rated sliding mode controller (ASMC) is designed to obtain better performance of the speed regulation system. Firstly, the L_1 norm of state variables is applied to the conventional exponential reaching law and an adaptive variable-rated exponential reaching law is proposed to reduce system chattering and improve bad convergence performance of the sliding mode variable structure. Secondly, an integral sliding-mode hyper plane is produced according to the speed error in speed regulation system of BIM. Current signal is extracted by the combination of the sliding-mode hyper plane, the electromagnetic torque and the equation of motion. Finally, the feedback speed can adjust operating state adaptively according to speed error and make system chattering-free moving. The simulation and experiment results show that the proposed ASMC can not only enhance the robustness of the system's uncertainties, but also improve the dynamic performance and suppress system chattering.

Keywords: Bearingless Induction Motor (BIM); adaptive variable-rated sliding mode controller (ASMC); adaptive variable-rated reaching law; chattering; robustness

1. Introduction

The Bearingless Induction Motor (BIM) integrates motor rotation and suspended function. It has a series of better characteristics than the conventional motor, such as no friction, no loss, no lubrication, longer service life, high speed and high precision running. BIM holds an irreplaceable status in some special fields including high speed electric spindles, high speed gyros, fly wheel energy storage and aerospace, *etc.* [1–5]. However, BIM is a multivariable, nonlinear, strongly coupled and complex system. When it experiences motor parameter variations or external large disturbances, BIM can not achieve high precision control performance under the conventional Proportional-Integral (PI) controller [6]. As a consequence, it is extremely important to design a fast, accurate and stable high-performance controller.

At present, many relative scholars around the world have done a great deal of research, and some advanced control strategies have been gradually applied to the motor system, such as neural network control [7,8], fuzzy control [9], *etc.* In fact, the control strategies above are difficult to be used in practical engineering due to their complicated design and harsh application conditions. Nevertheless, the sliding mode control strategy belonging to the category of modern control strategy appears greatly superior due to the fact that it does not require a high precision system mathematical model and has strong adaptability to uncertain factors (such as parameter variations, speed mutations, and load disturbances). Most of all, this strategy can be easily realized in the engineering. Therefore, it presents a good application prospect in the field of the alternating current servo control system. In [10], the sliding mode control strategy was applied to the electromagnetic actuator, which avoided the saturation of actuator. In [11,12], the sliding mode observer was used in a permanent magnet synchronous motor (PMSM) and brushless direct current (DC) motor. As a result, it reduced the speed error and improved the accuracy and stability of the motor system without a position sensor control. In [13], a switching estimator combined with a two-stage observer structure was proposed. It reduced the uncertainties and disturbance effects to the sensorless control in automotive applications. In [14], an improved terminal-sliding-mode control was used to implement the clutch control for automotive transmissions. This experiment proved that the control further improved the convergence rate and tracking precision of the closed-loop clutch system. In [15], the use of a single sliding mode controller (SMC) successfully generated a stable DC bus voltage for the charger/discharger DC/DC converter in any operation condition. It offered the necessary conditions for energy storage devices. In [16], a robust fuzzy neural network sliding-mode control for a two-axis motion control system was proposed, which was realized with a TMS320C32 DSP. The results demonstrated that the dynamic behaviors of the proposed control systems were robust to the uncertainties. In [17], an improved sliding mode control strategy was constructed for the model reference adaptive system by adding the speed section switch method. It was shown that the proposed method reduced the system chattering and achieved good effects. In [18], the speed controller and the load torque observer were designed on the basis of a sliding mode control strategy, which reduced the adverse effects on the control performance caused by the load mutation in the PMSM and enhanced the robustness of the system. In [19], based on inverse hyperbolic function, a novel integral SMC was formulated to estimate disturbances and suppress system chattering. Experimental results showed that it had low sensitivity with respect to disturbance and decreased system chattering. In [20], an SMC for the anti-lock braking system was used to maintain the optimal slip value according to optimizing the conventional reaching law. The results proved that the braking stability and the conversion energy was enhanced. In [21], with a novel exponential reaching law, an adaptive nonsingular terminal SMC was designed to gain the fast tracking of speed and position. To a certain extent, the novel reaching law diminished the system chattering; however, it ignored the essential reason for the chattering existing in the exponential reaching law and enlarged the amount of calculations, so it was unable to acquire the best control performance. In [22], a novel high-order sliding mode observer constructed by imposing a super twisting algorithm was proposed to obtain the counter electromotive force of PMSM. The results presented that the method followed the speed quickly, whereas the high-order super twisting algorithm added the complexity of system design and was not easily able to be applied in engineering.

In this paper, an adaptive variable-rated sliding mode controller (ASMC) is proposed to regulate the speed in the external loop of BIM. In order to obtain the optimal speed regulation performance, an adaptive variable-rated exponential reaching law is developed by adding the L_1 norm of state variables to the conventional exponential reaching law. It makes the system able to adjust the approach speed adaptively according to the distance between the state variables and expected point. At the same time, it can diminish the system chattering. Moreover, smoothing the sign function can further suppress chattering. The simulation and experiment results indicate that the proposed controller can effectively improve the response speed and precision of the system, eliminate system chattering, and enhance robustness to uncertain disturbance.

2. Design of the Adaptive Variable-Rated Exponential Reaching Law and ASMC

A nonlinear control system can be described as:

$$\begin{cases} \dot{x} = f(t,x) + g(t,x)u + b(t) \\ s = s(t,x) \end{cases} \tag{1}$$

where $x = [x_1, x_2]^T$ is the system state variable; $g(t,x) \neq 0$; $b(t)$ is the uncertain disturbances of system, $|b(t)| \leqslant \varsigma$ and ς is the upper bound of disturbance; $u \in R^n$ can be expressed by the following equation:

$$u_i(x,t) = \begin{cases} u_i^+(x,t), s_i(x) > 0 \\ u_i^-(x,t), s_i(x) < 0 \end{cases} \tag{2}$$

where $s_i(x)$ is a switching function and can be given as:

$$s_i(x) = \{x \in R^n \; ; s_i(x) = 0\}, i = 1, 2, \cdots k \tag{3}$$

Under the control of $u_i(x,t)$, the system runs along the sliding mode surface and obtains higher control performance. However, as a kind of discrete switch, the sliding mode variable structure has chattering. Thus, Gao Weibing, a control theory expert in China, designed a exponential reaching law to eliminate the chattering [23]. The function is written as:

$$\dot{s} = -\varepsilon \text{sgn}(s) - \lambda s \tag{4}$$

The reaching law improved the quality of the sliding mode to some degree, but taking its banded switching strip into consideration, the system just moved back and forth around the origin and chattered near the origin. The chattering affects the performance of the controller. In addition, ε and λ didn't have the automatic adjustment function according to the position of the state variables, so it can't reach the best convergence property. Devoting to solving the problems above, this paper proposes a new exponential reaching law—the adaptive variable-rated exponential reaching law:

$$\begin{cases} \dot{s} = -\varepsilon \,||\, x \,||_1 \text{sgn}(s) - \lambda \frac{1}{1+\alpha ||x||_1} s \\ \lim\limits_{t \to \infty} ||\, x \,||_1 = 0 \end{cases} \tag{5}$$

where $||\, x \,||_1 = \sum\limits_{i=1}^{n} |x|$ is the L_1 norm of state variables and $\lambda > 0$, $\varepsilon > 0$, $c > 0$, $n > 0$.

The approach speed of the system is associated with the distance between the state variable and the stable point under control of the adaptive variable-rated exponential reaching law. When the system state variables are far away from the sliding mode surface, the system state variables are mainly close to the sliding mode surface through $\lambda s/(1 + \alpha ||\, x \,||_1)$. At the same time, it can accelerate the system's approach speed by reducing the value of α. When the system state variables are already near the sliding mode surface, $-\varepsilon ||\, x \,||_1 \text{sgn}(s)$ plays a key role. As the state variables are close to the stable point, the $||\, x \,||_1$ is decreasing and gets close to zero. This makes the value of $-\varepsilon ||\, x \,||_1 \text{sgn}(s)$ zero. Hence, it weakens the chattering caused by the constant speed $\varepsilon \text{sgn}(s)$, and the sliding mode dynamics will stabilize at the origin. In addition, the sign function can be processed as shown in Formula (6) to achieve the purpose of further weakening chattering [24]:

$$\text{sgn}(s) = \frac{s}{|s| + \delta} \tag{6}$$

where δ is a positive constant.

Considering the high-frequency noise in the traditional sliding mode surface, this paper selects the first-order integral sliding mode surface. It can smooth the speed, reduce the system overshoot and suppress the chattering. The design of ASMC is defined as Theorem 1.

Theorem 1. *For the nonlinear control system shown in (1), the integral sliding mode surface is chosen as:*

$$s = x_1 + c_0 x_2 \tag{7}$$

The controller is designed as:

$$u = g^{-1}(t,x)[-f(x) - b(t) - C^{-1}\varepsilon\,||x||_1 \mathrm{sgn}(s) - C^{-1}\lambda\frac{1}{1+\alpha||x||_1}s] \tag{8}$$

where $C = \begin{bmatrix} 1 & c_0 \end{bmatrix}$. From above, it can be seen that the system can converge to the stable point in a limited time.

Stability Proof of Theorem 1. According to the Lyapunov stability theory, the existence and reachability of sliding mode are expressed as:

$$\dot{V} = s\dot{s} < 0 \tag{9}$$

Formula (9) is combined with Formula (5). Then, Formula (9) turns into:

$$\begin{aligned} s\dot{s} &= s(-\varepsilon||x||_1\mathrm{sgn}(s) - \lambda\tfrac{1}{1+\alpha||x||_1}s) \\ &= -\varepsilon||x||_1|s| - \lambda\tfrac{1}{1+\alpha||x||_1}s^2 \\ &< 0 \end{aligned} \tag{10}$$

Hence, the adaptive variable-rated exponential reaching law meets the demand of the Lyapunov stability theory. The system can arrive at the sliding mode surface within a limited time from any state.

In addition, by a typical nonlinear system $\begin{cases} \dot{x} = \mathbf{A}x + Bu + D \\ s = Cx \end{cases}$, an SMC and an ASMC are designed, respectively, based on the index reaching law and adaptive variable-rated exponential reaching law to verify the superiority of the proposed method in this paper, where $\mathbf{A} = \begin{bmatrix} 0 & 12 \\ 0 & 0 \end{bmatrix}$, $B = \begin{bmatrix} 0 \\ -150 \end{bmatrix}$, $C = \begin{bmatrix} 1 & 10 \end{bmatrix}$, and $D = [1]$. The initial state variables are set as $X(0) = \begin{bmatrix} 6 & 6 \end{bmatrix}$. The simulation results are shown in Figure 1. It can be seen from the diagram that the ASMC has faster convergence speed and smaller chattering than the SMC.

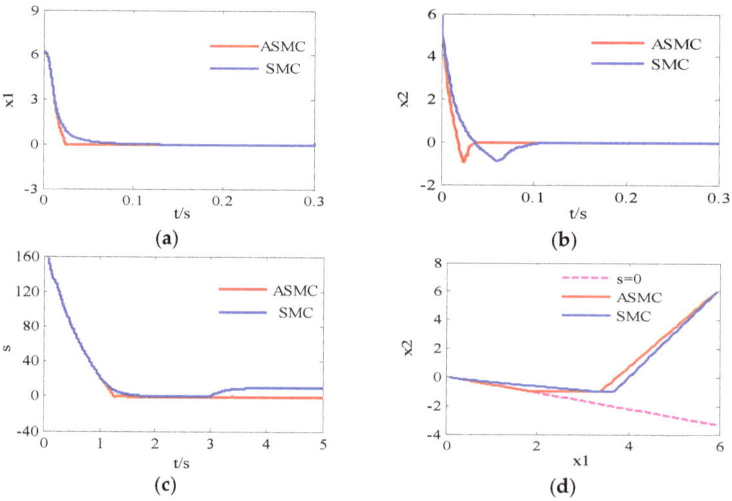

Figure 1. The performance contrast of adaptive variable-rated sliding mode controller (ASMC) and sliding mode controller (SMC): (**a**) The response of x_1; (**b**) The response of x_2; (**c**) The time of approach process; and (**d**) The phase track of system.

3. Design of the BIM Speed ASMC

3.1. The Dynamics Model of BIM

BIM is formed by superimposing other radial levitation force windings in the stator. The rapid rotation and stable suspension of the rotor are realized by controlling the winding current to change the distribution of the air-gap magnetic field. In the d-q coordinate system, the rotor voltage equation in torque windings of BIM is established as:

$$\begin{cases} U_{1rd} = R_{1r}i_{1rd} + p\psi_{1rd} - (\psi_{1q} + L_{1rl}i_{1rq})(\omega_1 - \omega_r) = 0 \\ U_{1rq} = R_{1r}i_{1rq} + p\psi_{1rq} + (\psi_{1d} + L_{1rl}i_{1rd})(\omega_1 - \omega_r) = 0 \end{cases} \tag{11}$$

Air-gap flux linkage equation is given as:

$$\begin{cases} \psi_{1d} = L_{1m}(i_{1sd} + i_{1rd}) \\ \psi_{1q} = L_{1m}(i_{1sq} + i_{1rq}) \end{cases} \tag{12}$$

The electromagnetic torque equation is set up as:

$$T_e = p_1(i_{1sq}\psi_{1d} - i_{1sd}\psi_{1q}) \tag{13}$$

The equation of motion is written as:

$$T_e - T_l = \frac{J}{P_1}\frac{d\omega_r}{dt} \tag{14}$$

where the subscript "1" represents the torque windings, the subscript "2" represents radial levitation force windings, the subscript "s" represents stator and the subscript "r" represents rotor; i_{2sd} and i_{2sq} are current components of the stator in levitation force winding under the d-q axis; ω_1 is the air-gap field speed; ω_r is the rotor speed; J is the rotational inertia of the rotor; T_l is the load torque; and p is the differential operator.

The radial levitation force is described as

$$\begin{cases} F_x = K(i_{2sd}\psi_{1d} + i_{2sq}\psi_{1q}) \\ F_y = K(i_{2sq}\psi_{1d} - i_{2sd}\psi_{1q}) \end{cases} \tag{15}$$

where $K = K_m + K_l$, $K_m = \frac{\pi p_1 p_2 L_{2m}}{18 l r \mu_0 N_1 N_2}$, and $K_l = \frac{p_1 N_2}{2 r N_1}$; p_1 and p_2 are pole pairs of the torque windings and suspension windings, respectively; L_{2m} is mutual inductance of the levitation force windings; l is the effective length of the rotor; r is the stator inner diameter; μ_0 is the permeability of vacuum; N_1 and N_2, respectively, show the effective number of turns of the torque windings and the levitation force windings; ψ_{1d} and ψ_{1q}, respectively, mean the components of flux linkage for the torque winding in the d-q axis.

Because of the magnetic field coupling [25] existing in the radial levitation force and the electromagnetic torque, the air-gap field-oriented control is used to decouple:

$$\psi_{1d} = \psi_1, \psi_{1q} = 0 \tag{16}$$

After plunging Formula (16) into Formula (12) and combining with Formula (11), the slip ω_s and i_{1sd} can be calculated as:

$$\omega_s = \frac{(1 + T_{1rl}p)\, i_{1sq}}{\frac{T_{1r}}{L_{1m}}\Psi_1 - T_{1rl}i_{1sd}} \tag{17}$$

$$i_{1sd} = \frac{1}{1 + T_{1rl}p}\left[(1 + T_{1r}p)\frac{\psi_1}{L_{1m}} + \omega_s T_{1rl} i_{1sq}\right] \tag{18}$$

where $\omega_s = \omega_1 - \omega_r$; i_{1sd} is an excitation component of the torque winding current; $T_{1rl} = L_{1rl}/R_{1r}$; T_{1r} is a time constant of the rotor, $T_{1r} = L_{1r}/R_{1r}$.

Then, the electromagnetic torque can be simplified as:

$$T_e = P_1\psi_1 i_{1sq} \tag{19}$$

At the same time, the radial levitation force can be simplified as:

$$\begin{cases} F_x = Ki_{2sd}\psi_1 \\ F_y = Ki_{2sq}\psi_1 \end{cases} \tag{20}$$

From Formulas (19) and (20), it can be seen that the radial levitation force and torque, respectively, are relevant to their own winding current. Hence, the radial levitation force and electromagnetic torque have been decoupled.

3.2. The Design of the Speed of the ASMC

The speed regulation system of BIM is shown in Figure 2. The input of ASMC is the difference between the given speed ω^* and the actual speed ω. The output of ASMC is the current i_{qs}^*.

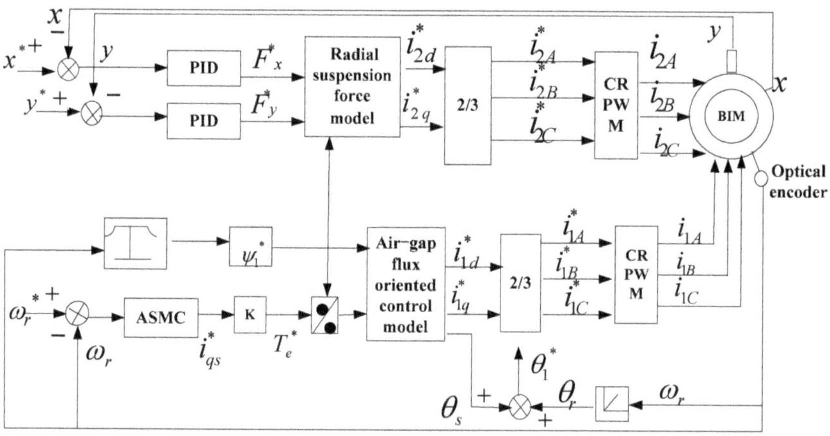

Figure 2. Control diagram of Bearingless Induction Motor (BIM) speed regulation system.

The system state variables are defined as:

$$\begin{cases} e_{\omega 1} = \omega^* - \omega \\ e_{\omega 2} = \int_{-\infty}^{t} e_{\omega 1} dt \end{cases} \tag{21}$$

Differentiating (21) with respect to time and by substituting (14), $e'_{\omega 1}$ can be obtained as:

$$e'_{\omega 1} = -\omega' = -\frac{p_1^2\psi_1}{J} i_{qs} + \frac{p_1}{J} T_l \tag{22}$$

Because of a lot of uncertain disturbances in motor system, $e'_{\omega 1}$ can be given as:

$$e'_{\omega 1} = \left(-\frac{p_1^2\psi_1}{J} + \Delta\zeta\right) i_{qs} + \left(\frac{p_1}{J} + \Delta\eta\right) T_l + \Delta\xi \tag{23}$$

where $\Delta\zeta$ and $\Delta\xi$ are the uncertain disturbances; $|\Delta\zeta| \leqslant \varsigma_1$; $|\Delta\xi| \leqslant \varsigma_2$; and ς_1 and ς_2 are bounded constants.

Here, $b(t)$ is recorded as the total disturbance factors of the system (including load). Then, (12) can be simplified as:

$$e'_{\omega 1} = -\frac{p_1^2\psi_1}{J} i_{qs} + b_1(t) \tag{24}$$

where $b_1(t) = (\Delta\zeta) i_{qs} + (\frac{p_1}{J} + \Delta\eta) T_l + \Delta\xi$; $|b_1(t)| \leqslant \varsigma$ and ς is a bounded positive constant.

Therefore, the system state equation of speed error is written as:

$$\begin{cases} e'_{\omega 1} = -\omega' = -\frac{p_1^2\psi_1}{J} i_{qs} + b_1(t) \\ e'_{\omega 2} = e_{\omega 1} = \omega^* - \omega \end{cases} \tag{25}$$

When the system enters sliding mode motion, differentiating $s = e_{\omega 1} + c e_{\omega 2} = 0$ with respect to time and the answer is computed as:

$$e_{\omega 1} = c_\Delta e^{-t/c} \tag{26}$$

As shown above, the velocity of the speed error approaching zero is $e^{-1/c}$. Moreover, it can track the speed without overshoot. Therefore, we can get the desired dynamic characteristics of the sliding mode motion by adjusting c.

Differentiating $s = e_{\omega 1} + c e_{\omega 2}$ with respect to time, s' is given as:

$$s' = e'_{\omega 1} + c e'_{\omega 2} = -\frac{p_1^2\psi_1}{J} i_{qs} + b_1(t) + c e_{\omega 1} \tag{27}$$

Using (5) and (27), it can be obtained as:

$$-\varepsilon \| e_\omega \|_1 \text{sgn}(s) - \lambda \frac{1}{1 + \alpha \| e_\omega \|_1} s = -\frac{p_1^2\psi_1}{J} i_{qs} + b_1(t) + c e_{\omega 1} \tag{28}$$

The controller can be described finally as:

$$i_{qs} = \frac{J}{p_1^2\psi_1} (\varepsilon \| e_\omega \|_1 \text{sgn}(s) + \lambda \frac{1}{1 + \alpha \| e_\omega \|_1} s + b_1(t) + c e_{\omega 1}) \tag{29}$$

The design of ASMC for BIM is defined as Theorem 2 in this paper.

Theorem 2. *For the speed error control system shown in (25), the sliding mode surface is adopted as:*

$$s_1 = e_{\omega 1} + c e_{\omega 2} \tag{30}$$

The controller is obtained as:

$$i_{qs} = \frac{J}{p_1^2\psi_1} (\varepsilon \| e_\omega \|_1 \text{sgn}(s) + \lambda \frac{1}{1 + \alpha \| e_\omega \|_1} s + b_1(t) + c e_{\omega 1}) \tag{31}$$

It can be seen that the speed error can converge to the expected place within a limited time, where λ, ε and α are system parameters and c is the integral constant.

4. Simulation and Experiment Research

4.1. Results and Analysis of the Simulation

To verify the effectiveness of the speed ASMC for BIM speed regulating system, a simulation control system was constructed by using the Matlab/Simulink toolbox. The proposed ASMC is shown in Figure 3. The specific parameters of BIM are shown in Table 1. The parameters of the ASMC are shown as follows: $\varepsilon = c = 9$; $\lambda = 2$; $\alpha = 100$; $\varsigma = 10$; and $\delta = 0.01$. The parameters of the PI are $k_p = 1$ and $k_i = 0.001$.

Table 1. Parameters of the Bearingless Induction Motor (BIM).

Parameters	Torque Winding	Suspension Winding
Rated power (Kw)	1	0.5
Rated current (A)	2.86	2.86
Stator resistance (Ω)	2.01	1.03
Rotor resistance (Ω)	11.48	0.075
Mutual inductance of stator and rotor (H)	0.15856	0.00932
Stator leakage inductance (H)	0.16310	0.01199
Rotor leakage inductance (H)	0.16778	0.01474
Rotational inertia (kg· m^2)	0.00769	0.00769
Rotor mass (kg)	2.85	2.85
Stator inner diameter (mm)	98	98
Core length (mm)	105	105
Pole pairs	1	2

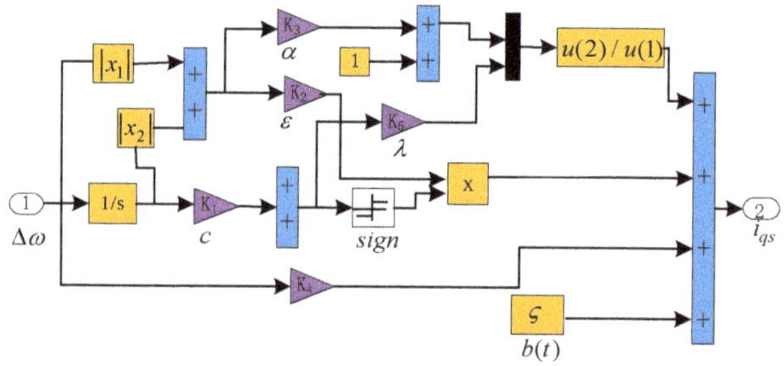

Figure 3. The proposed ASMC.

Figure 4 presents the simulation results of BIM controlled by the PI controller, SMC and ASMC under different working conditions. Figure 4a shows the simulation of the rotor operates at the speed of 6000 r/min with no-load. Figure 4b,c shows the simulation results of the system working in the condition of sudden load (5 N· m) and speed mutation (6000 r/min drops to 3000 r/min). Compared with ASMC, it can be known from Figure 4 that the system runs slowly at first under the control of PI controller and SMC. When the system is attacked by the sudden load or speed mutation, the speed has obvious fluctuation. The system needs more time to stabilize and it is unable to return to the original speed. The PI controller and SMC appear to have poor control performance.

The high speed (12,000 r/min) simulation results of BIM are shown in Figure 5. First of all, the system runs at 12,000 r/min with zero-load. Then, the load 5 N· m is given to the system. At last, the speed falls sharply from 12000 r/min to 6000 r/min. The waveforms of the system working on the conditions above separately are shown in Figure 5a–c. Suffering the external disturbances, the speed has big overshoot and operates at a steady state slowly when the system adopts the PI controller or SMC. At the same time, the value of the speed is very different from the original speed. However, the system using the ASMC responds quickly. When the system is affected by the external disturbances, the fluctuation of speed is small and it restores itself to the original value quickly. It shows that the ASMC has excellent robustness. In addition, by contrasting Figures 4 and 5, we can know that the adverse influence of disturbance to speed regulation system will be more serious if the motor runs faster. Therefore, to design a controller with strong ability to resist disturbance is very important.

Figure 6 shows the radial displacement simulation diagrams of the system operating at the speed of 12,000 r/min under the control of ASMC. It is described in the diagrams that the rotor still can approach a stable position quickly and suspend steadily even if the system is running at high speed. Good control performance of ASMC is verified.

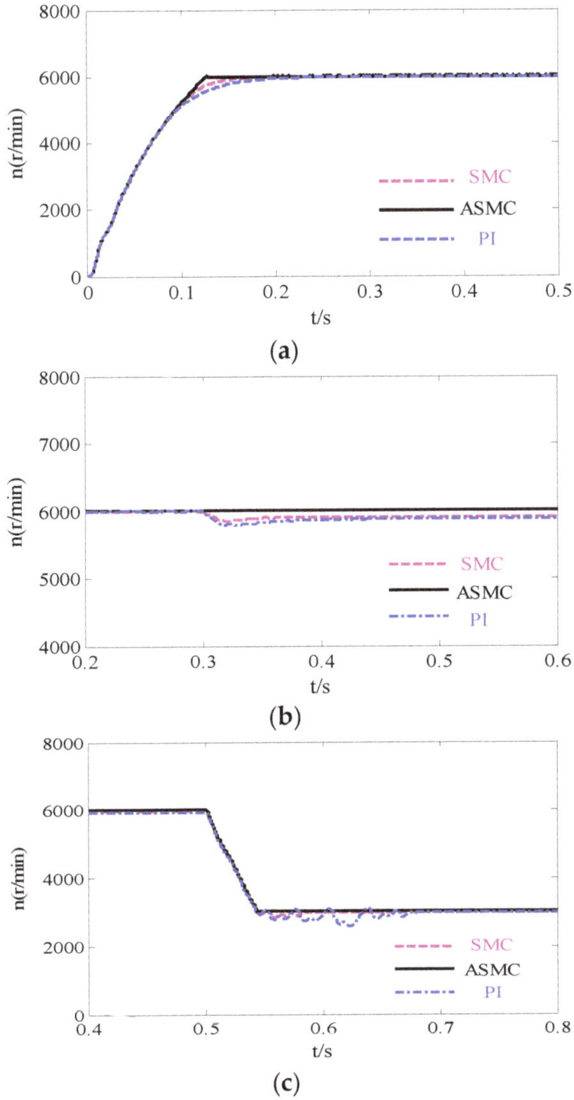

Figure 4. Waveforms of the speed dynamic response (6000 r/min): (**a**) The response of speed under no-load; (**b**) The response of speed under the sudden load (5 N· m); (**c**) The response of speed mutation under the torque disturbance (from 6000 r/min to 3000 r/min).

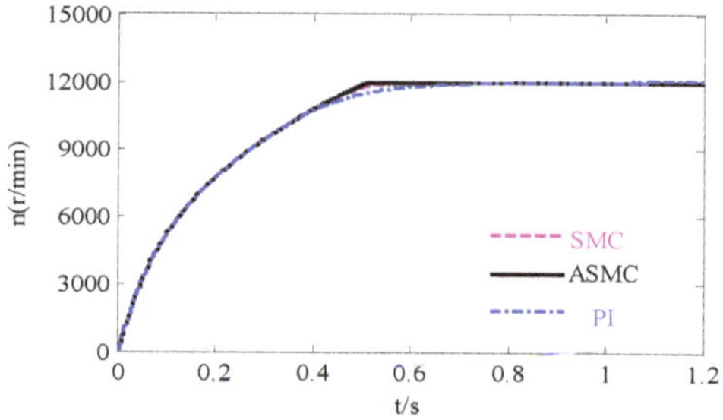

(a)

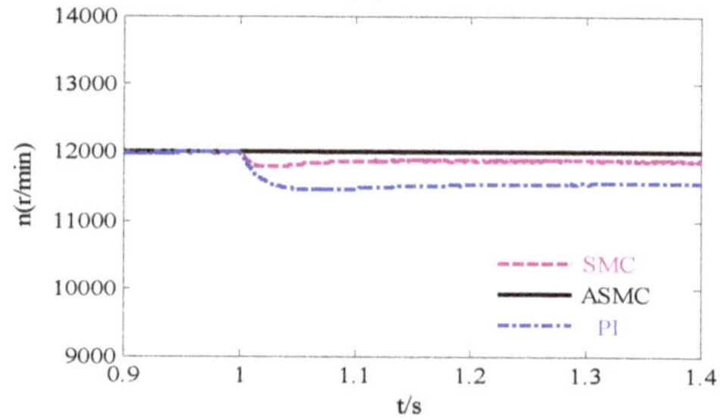

(b)

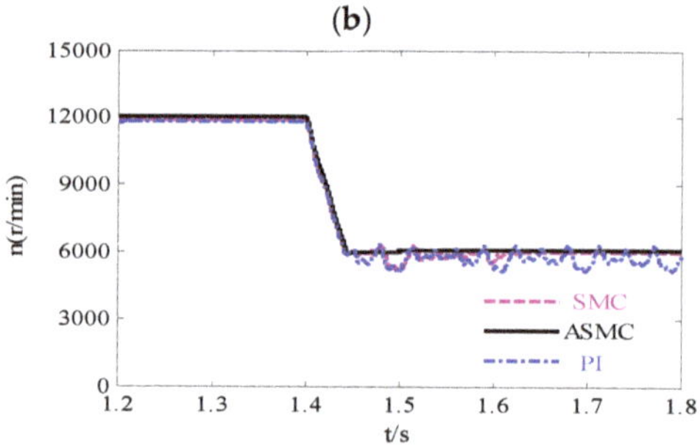

(c)

Figure 5. The simulation of system dynamic response (12,000 r/min): (**a**) The response of speed under no-load; (**b**) The response of speed under the sudden load (5 N· m); (**c**) The response of speed mutation under the torque disturbance (from 12,000 r/min to 6000 r/min).

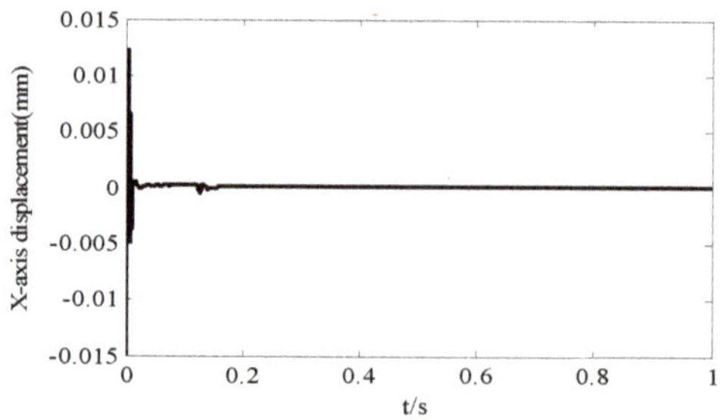

(a)

Figure 6. *Cont.*

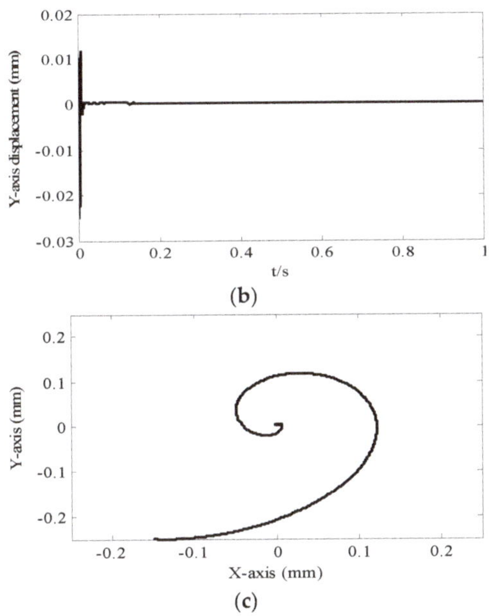

Figure 6. Waveforms of the rotor radial displacement controlled by ASMC (12,000 r/min): (**a**) The radial displacement on the x-axis; (**b**) The radial displacement on the y-axis; (**c**) The curve of rotor operation.

4.2. Results and Analysis of the Experiment

Using the experimental object of BIM as shown in Figure 7, the digital control system experimental platform is conducted to further prove the effectiveness of the proposed sliding mode control method. The experiment block diagram of the control system is shown in Figure 8. The photoelectric encoder and the eddy current displacement sensors are used to detect the speed signals and radial displacement signals of rotor, respectively. With the signals, the speed measuring module and the radial displacement detection module can acquire the actual speed and displacement of rotor. After contrasting the actual speed and the given speed (or the actual displacement and the given displacement), the drive circuit controls the two kinds of windings according to the Pulse-Width Modulation signals generated by TMS320F28335, so the rotor can rotate and be suspended. Furthermore, the sampling circuit can detect the feedback circuit from the torque windings and suspension windings. With it, the rotor can adjust the running state in time. The prototype parameters are the same as the simulation parameters. Considering the limitations of photoelectric coder measuring speed, the speed is set to 2000 r/min in the process of the experiment. The air gap of the motor auxiliary bearing is 0.4 mm. The experimental results are shown in Figure 9.

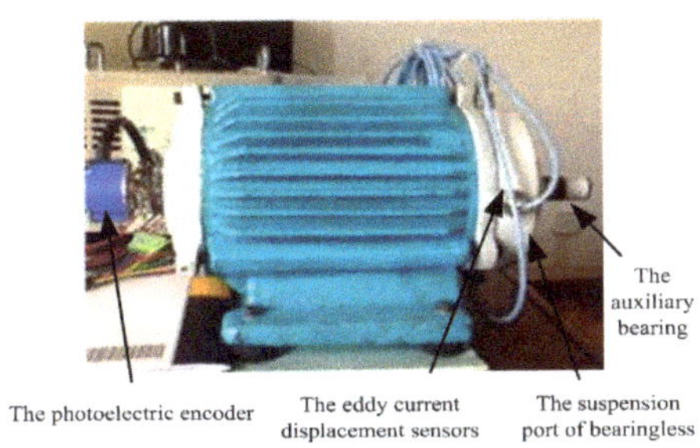

Figure 7. Prototype physical diagram.

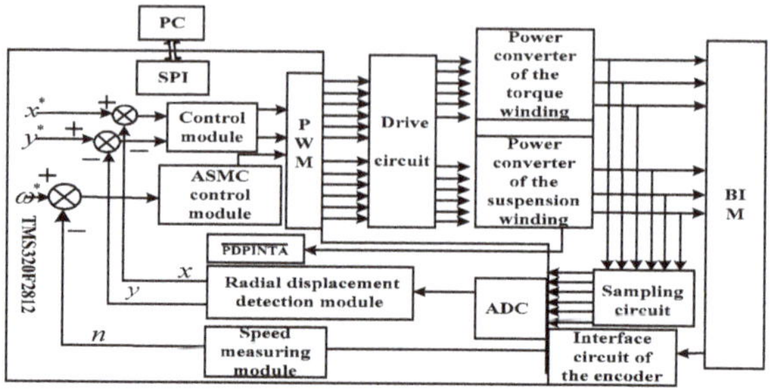

Figure 8. The experiment block diagram of the control system.

Figure 9a–c separately show the results of the system speed falling sharply from 2000 r/min to 1400 r/min under the control of PI, SMC and ASMC. From the waveforms, it can be found that the system has faster speed response, better stability and stronger robustness under the ASMC compared to the PI controller and SMC.

Figure 9d provides the curve of rotor radial displacement at the speed of 2000 r/min under the control of ASMC. It shows that the center of the rotor runs near the equilibrium position and the biggest offset value of the radial displacement is 80 μm, which is far less than the value of the air gap of motor. It proves that the rotor can be suspended steadily under the control of ASMC.

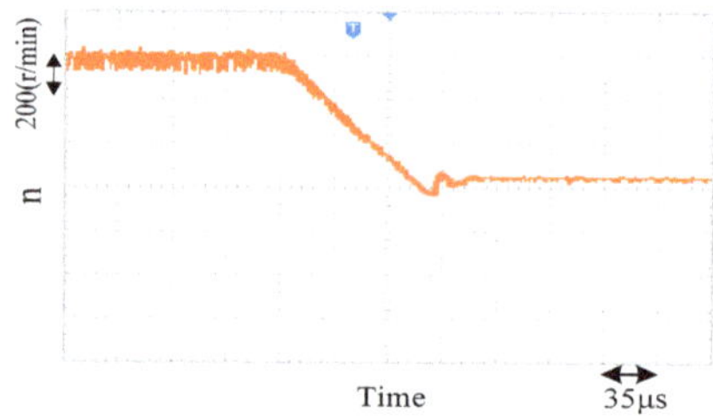

(**a**)

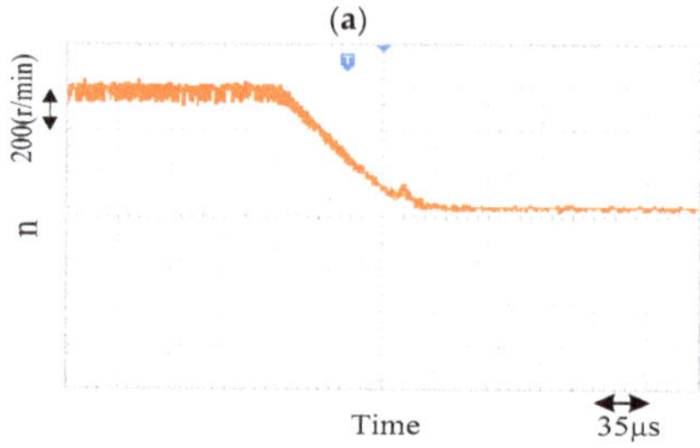

(**b**)

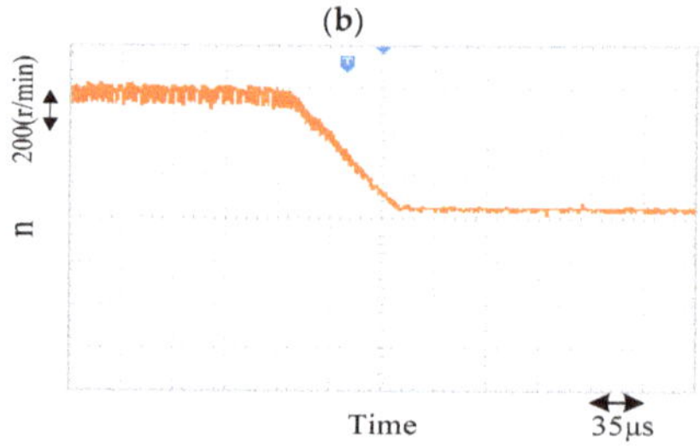

(**c**)

Figure 9. *Cont.*

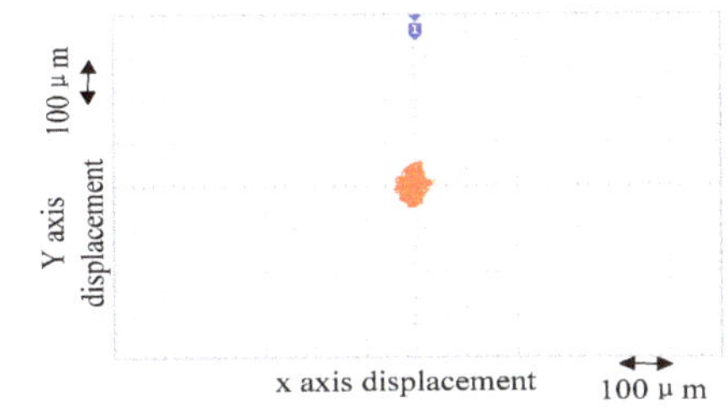

(**d**)

Figure 9. Experimental results of system dynamic response: (**a**) The diagram of speed mutation controlled by Proportional-Integral (PI); (**b**) The diagram of speed mutation controlled by SMC; (**c**) The diagram of speed mutation controlled by ASMC; (**d**) The diagram of rotor radial displacement controlled by ASMC.

5. Conclusions

In order to further optimize the operation quality of the BIM drive system, an adaptive variable-rated exponential reaching law is proposed on the basis of analyzing the conventional exponential reaching law. The proposed reaching law can adjust the approach speed adaptively according to the distances between the state variables and expected points. Moreover, it can suppress system chattering. The simulation and experimental results show that the speed controller of BIM, which is designed on the basis of the adaptive variable-rated exponential reaching law can make the system run faster, strengthen the robustness of the system and improve the operation quality of the BIM speed regulation system. In addition, the control strategy is simple and can be applied to engineering easily.

Acknowledgments: This work was supported by the National Natural Science Foundation of China under Project 51475214, 61104016 and 51305170, the China Postdoctoral Science Foundation funded project under Project 2015T80508 and 2014T70482, the Natural Science Foundation of Jiangsu Province of China under Project BK20130515, BK20141301and BK20150524, the Professional Research Foundation for Advanced Talents of Jiangsu University under Project BK20150524 and 14JDG076, and Six Categories Talent Peak of Jiangsu Province under Project ZBZZ-017, and the Priority Academic Program Development of Jiangsu Higher Education Institutions.

Author Contributions: Zebin Yang and Ling Wan proposed the new idea of the Sliding Mode Variable Structure Control of Bearingless Induction Motor Based on a Novel Reaching Law. Ling Wan derived the equations. Xiaodong Sun set up the simulation model. Fangli Li was in charge of analyzing the data. Lin Chen checked the language. All authors were involved in preparing the manuscript.

Conflicts of Interest: The authors declare no conflict of interest.

Abbreviations

BIM	Bearingless Induction Motor.
ASMC	Adaptive variable-rated sliding mode controller.
PMSM	Permanent magnet synchronous motor.
SMC	Sliding mode controller.
DC	Direct current.
PI	Proportional-Integral

References

1. De Almeida, A.T.; Ferreira, F.J.T.E.; Quintino, D.A. Technical and Economical Considerations on Super High-Efficiency Three-Phase Motors. *IEEE Trans. Ind. Appl.* **2014**, *50*, 1274–1285. [CrossRef]
2. Sun, X.D.; Chen, L.; Yang, Z.B. Overview of bearingless permanent magnet synchronous motors. *IEEE Trans. Ind. Electron.* **2013**, *60*, 5528–5538. [CrossRef]
3. Sinervo, A.; Arkkio, A. Rotor radial position control and its effect on the total efficiency of a bearingless induction motor with a cage rotor. *IEEE Trans. Magn.* **2014**, *50*, 1–9. [CrossRef]

4. Yang, Z.B.; Dong, D.W.; Gao, H.Y.; Sun, X.D.; Fan, R.; Zhu, H.Q. Rotor mass eccentricity vibration compensation control in bearingless induction motor. *Adv. Mech. Eng.* **2015**, *7*. [CrossRef]
5. Chiba, A.; Asama, J. Influence of rotor skew in induction type bearingless motor. *IEEE Trans. Magn.* **2012**, *48*, 4646–4649. [CrossRef]
6. Hou, B.J.; Gao, J.S.; Li, X.Q.; Zhou, Y.F. Study on repetitive PID control of linear motor in wafer stage of lithography. *Procedia Eng.* **2012**, *29*, 3863–3867. [CrossRef]
7. Sun, X.D.; Chen, L.; Yang, Z.B.; Zhu, H.Q. Speed-sensorless vector control of a bearingless induction motor with artificial neural network inverse speed observer. *IEEE/ASME Trans. Mechatron.* **2013**, *18*, 1357–1366. [CrossRef]
8. Sun, X.D.; Chen, L.; Jiang, H.B.; Yang, Z.B.; Chen, J.C.; Zhang, W.Y. High-performance control for a bearingless permanent magnet synchronous motor using neural network inverse scheme plus internal model controllers. *IEEE Trans. Ind. Electron.* **2016**, *63*, 3479–3488. [CrossRef]
9. Zhang, H.S.; Wang, P.; Han, B.C. Rotor Position Measurement for High-speed Permanent Magnet Synchronous Motors Based on Fuzzy PI MRAS. *Proc. Chin. Soc. Electr. Eng.* **2014**, *34*, 1889–1896. (In Chinese)
10. Mercorelli, P. An Antisaturating Adaptive Preaction and a Slide Surface to Achieve Soft Landing Control for Electromagnetic Actuators. *IEEE/ASME Trans. Mechatron.* **2012**, *17*, 76–85. [CrossRef]
11. Wang, X.D.; Liu, G. Sensorless Control of High Speed Permanent Magnet Synchronous Motor Based on Modified Sliding-Mode Observer. *Adv. Mater. Res.* **2014**, *986–987*, 1134–1137. [CrossRef]
12. Shi, T.N.; Xiao, Z.X.; Xiao, Y.W.; Xia, C. A Position Sensorless Control Strategy for BLDCM Based on an Improved Sliding Mode Observer. *Proc. Chin. Soc. Electr. Eng.* **2015**, *35*, 2043–2051.
13. Mercorelli, P. A Two-Stage Sliding-Mode High-Gain Observer to Reduce Uncertainties and Disturbances Effects for Sensorless Control in Automotive Applications. *IEEE Trans. Ind. Electron.* **2015**, *62*, 5929–5940. [CrossRef]
14. Li, S.H.; Wu, C.; Sun, Z. Design and Implementation of Clutch Control for Automotive Transmissions Using Terminal-Sliding-Mode Control and Uncertainty Observer. *IEEE Trans. Veh. Technol.* **2016**, *65*, 1890–1898. [CrossRef]
15. Serna-Garcés, S.I.; Gonzalez Montoya, D.; Ramos-Paja, C.A. Sliding-Mode Control of a Charger/Discharger DC/DC Converter for DC-Bus Regulation in Renewable Power Systems. *Energies* **2016**, *9*, 245. [CrossRef]
16. Lin, F.-J.; Shen, P.-H. Robust Fuzzy Neural Network Sliding-Mode Control for Two-Axis Motion Control System. *IEEE Trans. Ind. Electron.* **2006**, *53*, 1209–1225. [CrossRef]
17. Zhu, Y.; Cheng, M.; Hua, W.; Zhang, B.; Wang, W. Sensorless Control for Electrical Variable Transmission Based on Sliding Mode Model Reference Adaptive System. *Diangong Jishu Xuebao* **2015**, *30*, 64–72. (In Chinese)
18. Zhang, X.G.; Sun, L.; Zhao, K. Sliding mode control of PMSM based on a novel load torque sliding mode observer. *Proc. Chin. Soc. Electr. Eng.* **2012**, *32*, 111–116. (In Chinese)
19. Asad, M.; Bhatti, A.I.; Iqbal, S.; Asfia, Y. A smooth integral sliding mode controller and disturbance estimator design. *Int. J. Control Autom. Syst.* **2015**, *13*, 1326–1336. [CrossRef]
20. Guo, J.G.; Jian, X.P.; Lin, G.Y. Performance Evaluation of an Anti-Lock Braking System for Electric Vehicles with a Fuzzy Sliding Mode Controller. *Energies* **2014**, *7*, 6459–6476. [CrossRef]
21. Xu, B.; Zhu, H.Q. Adaptive nonsingular terminal sliding model control and its application to BPMSM. *Control Decis.* **2014**, *29*, 833–837.
22. Yi, B.Y.; Kang, L.Y.; Tao, S.N.; Guo, H. Design of Robust High Order Sliding Mode Observer for Permanent Magnet Synchronous Motors. *Diangong Jishu Xuebao* **2014**, *29*, 132–140. (In Chinese)
23. Gao, W.B. *Variable Structure Control Theory*; Science and Technology Press: Beijing, China, 1990; pp. 28–30.
24. Kim, M.; Song, D. A robust control of Permanent magnet synchronous Motor using Load Torque Estimation. In Proceedings of the IEEE International Symposium on Industrial Electronics, Pusan, Korea, 12–16 June 2001; pp. 1157–1162.
25. Yang, Z.B.; Sun, X.D.; Wang, M.T.; Zhu, H.Q. Decoupling control of bearingless induction motors based on least squares support vector machine inverse. *J. Comput. Theor. Nanosci.* **2014**, *11*, 1403–1409. [CrossRef]

energies

Article

Simultaneous Optimization of Topology and Component Sizes for Double Planetary Gear Hybrid Powertrains

Weichao Zhuang [1], Xiaowu Zhang [2], Huei Peng [2] and Liangmo Wang [1,*]

1 Department of Mechanical Engineering, School of Mechanical Engineering, Nanjing University of Science and Technology, 200 Xiaolingwei Street, Nanjing 210094, China; darren.wc.zhuang@gmail.com

2 Department of Mechanical Engineering, University of Michigan, G041, Walter E. Lay Autolab, 1231 Beal Ave, Ann Arbor, MI 48109, USA; xiaowuz@umich.edu (X.Z.); hpeng@umich.edu (H.P.)

* Correspondence: liangmo@njust.edu.cn; Tel.: +86-159-5186-0337

Academic Editor: Paolo Mercorelli

Received: 28 February 2016; Accepted: 20 May 2016; Published: 26 May 2016

Abstract: Hybrid powertrain technologies are successful in the passenger car market and have been actively developed in recent years. Optimal topology selection, component sizing, and controls are required for competitive hybrid vehicles, as multiple goals must be considered simultaneously: fuel efficiency, emissions, performance, and cost. Most of the previous studies explored these three design dimensions separately. In this paper, two novel frameworks combining these three design dimensions together are presented and compared. One approach is nested optimization which searches through the whole design space exhaustively. The second approach is called enhanced iterative optimization, which executes the topology optimization and component sizing alternately. A case study shows that the later method can converge to the global optimal design generated from the nested optimization, and is much more computationally efficient. In addition, we also address a known issue of optimal designs: their sensitivity to parameters, such as varying vehicle weight, which is a concern especially for the design of hybrid buses. Therefore, the iterative optimization process is applied to design a robust multi-mode hybrid electric bus under different loading scenarios as the final design challenge of this paper.

Keywords: hybrid electric vehicles (HEVs); energy management; topology optimization; optimal design methodology

1. Introduction

Tightening emissions and fuel economy standards are strong driving forces behind vehicle electrification. The future Corporate Average Fleet Average (CAFÉ) standard requires fuel economy at 54.5 miles per gallon (unadjusted) for passenger cars by 2025 [1], a significant increase from today's fleet average of about 31 miles per gallon. This fuel economy target is difficult to achieve through improving traditional powertrains using internal combustion engines. However, it can be met quite easily by hybrid electric vehicles (HEVs) and plug-in hybrid electric vehicles (PHEV).

From the perspective of power flow between the engine, the motor/generators (MGs) and the vehicle driveshaft, HEVs are generally divided into three types: series, parallel, and power-split. Among all three types, the power-split type, which uses planetary gear (PG) sets, has dominant market share [2]. This is mainly because the engine in the power-split HEV is decoupled from the vehicle speeds and can operate efficiently while much of the power flows in the mechanical path and is, thus, efficient. Therefore, about 85% of HEV and PHEV market share in the US is occupied by power-split HEVs [2].

It should be noted that the power-split powertrains are more complex than series and parallel hybrids and the control is far more complex [3]. In addition, multiple operating modes can be achieved when clutches are augmented to a power-split configuration [4]. With the introduction of clutches, the complexity of hybrid powertrains becomes unprecedented. The high system complexity provides more freedom for achieving better fuel economy, performance, cost, and comfort [5]. The multiple objectives and system complexity result in a nonlinear constrained optimization problem that is not straightforward to solve.

Researchers and engineers typically optimize the powertrain designs from three perspectives: control strategy, component sizes, and system topologies. The latter two aspects, component sizing and topology optimization, are usually integrating with control optimization problems to minimize fuel consumption. A rule-based control strategy is easy to design, but is highly dependent on the experience of engineers and cannot guarantee optimality [6]. An equivalent consumption minimization strategy (ECMS) optimizes the engine power instantaneously, but the equivalent fuel consumption factor requires tuning if the design or driving conditions are changed [7,8]. Dynamics programming (DP) guarantees global optimality, but its computation burden grows exponentially with the number of state and input variables [9], which makes it unsuitable for high-dimension systems. Some literature tries to reduce the computational burden by using a local approximation of the gridded cost-to-go [10,11]. However, it still requires long computation time that makes it unsuitable for large-scale system control or design studies. The convex optimization method computes quickly, but are not appropriate for power-split or multi-mode HEVs due to the lack of convexity [12,13]. Up to today, power-weighted efficiency analysis for rapid sizing (PEARS) is the only known method that can be used to solve multi-mode HEV control problems with demonstrated optimality and can be computed orders of magnitude faster than DP [14].

In terms of component sizing, most literature tried to identify the most fuel-efficient design in the bi-level frameworks [15–17], where the energy management strategy needs to be optimized for each component size candidate. Due to the iterations, the bi-level frameworks is time-consuming. To reduce computational load, people sometimes search the optimal designs with heuristic optimization methods to avoid looping through all designs, like using genetic algorithms (GA) [16], particle swarm optimization (PSO) [17], and self-adaptive differential evolution algorithms [18], under the constraints of performance, emissions, and cost [19]. Other studies tried to address this problem more efficiently. One method that optimizes the component sizes and control strategy simultaneously is convex optimization [20]. One study proposes a rule-based design method based on the defined hybridization ratio after analyzing the results from DP [21].

Recently, researchers have been shifting their focus to the configuration optimization that received less attention in the past, but shows great potential to improve the energy efficiency. It is known that the number of possible topologies for the parallel hybrid is very limited and can be evaluated one by one easily [22]. The power-split HEVs, however, especially when extended to multi-mode HEVs, has exponentially increasing number of designs with the number of components (PGs and clutches) and their exhaustive analysis is not easy.

The optimal topology studies started from single-PG HEVs [4]. Different modeling methods, *i.e.*, automatic modeling [23,24] and bond graphs [25], have been applied to study all possible single-PG designs. Since the design space for a single-PG hybrid is not large, the final drive and PG ratios are considered as design variables. The best single-PG design with an optimized gear ratio is identified through a systematic search [26]. In addition, Zhang *et al.* [27] develop an automated modeling technique to optimize the double-PG hybrid considering both fuel economy and launching performance. Component sizing, however, is not considered. Silvas *et al.* [28] propose a general framework to generate all possible powertrain structures by solving a constrained logic programming problem. Such frameworks can be used for HEVs with any number of PGs and even non-power-split designs, but the computation time can be a challenge.

Most of the above-mentioned studies optimize the topology without considering the component sizes and powertrain parameters that may have a significant impact on the optimization results.

However, as component sizes are also optimized, the design space grows exponentially. Therefore, it is not practical to optimize the component sizes for each topology individually. In this paper, we propose a novel optimization method, called iterative optimization, which combines the topology optimization and component sizing together. Instead of optimizing the component sizes individually (named nested optimization), the topology optimization and component sizing are executed alternately. With each iteration, the solution approaches the optimal design rapidly compared with the benchmark method, nested optimization.

The remainder of this paper is organized as follows: Section 2 describes the process for the exhaustive search of double-PG hybrid powertrains; in Section 3, the proposed iterative optimization method is introduced and the comparison between its results with the nested optimization is presented; an enhanced version of the proposed iterative optimization is also proposed. Section 4 demonstrates the performance of the proposed method using a case study that considers a set of loading scenarios. Finally, conclusions are presented in Section 5.

2. Topology Optimization via Exhaustive Search

2.1. Problem Definition

In recent years, a large number of power-split HEV powertrains were patented by original equipment manufacturers and engineers [29,30]; however, many more remain unexplored. There are two ways to generate new multi-mode power-split HEV powertrain designs: by changing the locations of the four powertrain components (engine, two MGs, and the driveline), which results in P_6^4 different combinations. In this paper, P_n^k represents k-permutations of n. The second approach adds clutches between two PG nodes, or between a PG node and the ground. The total number of possible clutches is [27]:

$$N_{\text{clutch}} = C_{3N_{\text{p}}}^2 + 3N_{\text{p}} - 2N_{\text{p}} - 1 \tag{1}$$

where C_n^k calculates the number of k-combinations from a given set of n elements, and N_{p} is the number of PGs. The first term represents the number of clutches that connect any two nodes; the second term reflects the number of clutches that can ground PG nodes; the third term stands for the redundant clutches because locking any two of the three nodes in a PG produce identical dynamics. It should be noted that the output node connected with vehicle should not be grounded, which results in the last term. Therefore, 16 clutch locations are available when composing a new design with a double-PG powertrain system, as shown in Figure 1, and the redundant clutches are marked in red.

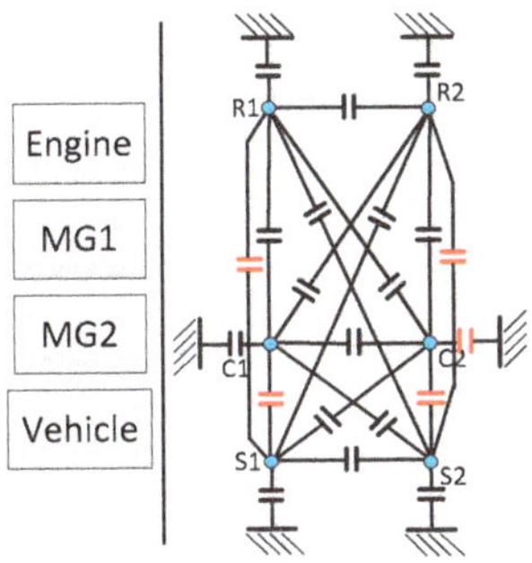

Figure 1. Powertrain components of power-split hybrid electric vehicle (HEV) and all possible clutch locations for a double planetary gear (PG) hybrid.

Each possible clutch location can be replaced by a permanent connection. A permanent connection refers to a direct mechanical connection between two nodes. Compared to a clutch, a permanent connection incurs much less cost and operation complexity. Therefore, if having a clutch does not add useful modes, it will be replaced by a permanent connection.

We define two terms below. A "design" stands for a multi-mode hybrid topology with decided clutches and permanent connections. By switching states of the clutches, multiple modes can be realized in a powertrain. A "mode" refers to a specific state of the powertrain once all clutch states are selected. It is sometimes referred as an "operating mode" in the literature.

The number of double-PG hybrid designs is enormous in theory. In reality, however, the number of clutches and permanent connections is limited because of cost and packaging. Therefore, we limit the number of clutches to no more than three in this paper. The Chevrolet Volt uses the same number of clutches [31]. Once the number of clutches is given, the number of permanent connections is only decided by degree of freedom (DoF) of the system.

A power-split HEV has three controllable components: an engine and two MGs. Since the DoF of a single PG system is two [24], the DoF of the double-PG powertrain starts from four and each effective clutch engagement reduces the DoF by one. With three controllable powertrain components, the maximum DoF of a mode is three. However, since the engine speed and torque cannot both be manipulated in modes with three DoF, only modes with two or one DoF are considered in this research.

For designs with three clutches and one permanent connection, up to six modes can be achieved for each design, even though not necessarily all of them are used. The number of possible designs with one permanent connection and three clutches is:

$$N_{\text{design}} = C^3_{N_{\text{clutch}}} \times C^1_{N_{\text{clutch}}-3} \times P^4_6 = 2,620,800 \tag{2}$$

where $C^3_{N_{\text{clutch}}}$ is the number of possible clutches pairs, $C^1_{N_{\text{clutch}}-3}$ is the number of possible selected permanent connection and P^4_6 is the number of possible powertrain locations.

2.2. System Modeling

The model of power-split HEV is generally composed by vehicle dynamics, engine model, MGs model, battery model, and power-split device model (PG system). We only optimize the powertrain of HEVs in this paper, and only longitudinal vehicle dynamics is considered. Engine and MGs are both modeled as quasi-static models because other model types, like thermodynamics-based models require extensive computation and unsuitable for large-scale design studies. In this way, the fuel consumption of the engine, the efficiency of MGs and the maximum torque under different rotational speed for both engine and MGs are calculated by employing several look-up tables. The battery model is developed by using a simple equivalent circuit where the open circuit voltage and the internal resistance are both functions of the state of charge of the battery. It should be noted that only gear ratios of PGs and final drive are optimized in component sizing, and the sizes of engine, MGs, and battery remain unchanged in this study.

In addition to the vehicle dynamics and components' models described above, we use a methodology, called automated modeling, to build up the power-split device models of different operating modes for the multi-mode HEV discussed in this paper [27]. In general, the dynamics of any specific mode is described by the characteristic matrix A^*, as shown in Equation (3). This 4×4 characteristic matrix A^* governs the relationship between the angular acceleration of powertrain devices and their corresponding torques. The detailed derivation can be found in [27].

$$\begin{bmatrix} \dot{\omega}_{\text{out}} \\ \dot{\omega}_{\text{eng}} \\ \dot{\omega}_{\text{MG1}} \\ \dot{\omega}_{\text{MG2}} \end{bmatrix} = A^* \begin{bmatrix} T_{\text{load}} \\ T_{\text{eng}} \\ T_{\text{MG1}} \\ T_{\text{MG2}} \end{bmatrix} = \begin{bmatrix} a_{11} & a_{12} & a_{13} & a_{14} \\ a_{21} & a_{22} & a_{23} & a_{24} \\ a_{31} & a_{32} & a_{31} & a_{34} \\ a_{41} & a_{42} & a_{43} & a_{44} \end{bmatrix} \begin{bmatrix} T_{\text{load}} \\ T_{\text{eng}} \\ T_{\text{MG1}} \\ T_{\text{MG2}} \end{bmatrix} \tag{3}$$

2.3. Exhaustive Search

The exhaustive search approach has been used to explore all possible designs for optimal fuel economy and launching performance [27]. The flowchart of the exhaustive search is shown in Figure 2.

With the automated modeling technology, the characteristic matrix A^* which fully describes the dynamics of the system could be generated rapidly. Then, the modes that share the same characteristic matrix (*i.e.*, identical modes) are identified, and only one of them is retained for subsequent analysis. In addition, all modes are classified into 14 different mode types based on their characteristics and functionalities. The criteria for the mode classification was proposed in [27]. After the mode analysis, winning designs are selected through a launching performance and fuel economy evaluation. Finally, a small pool of winning designs is obtained.

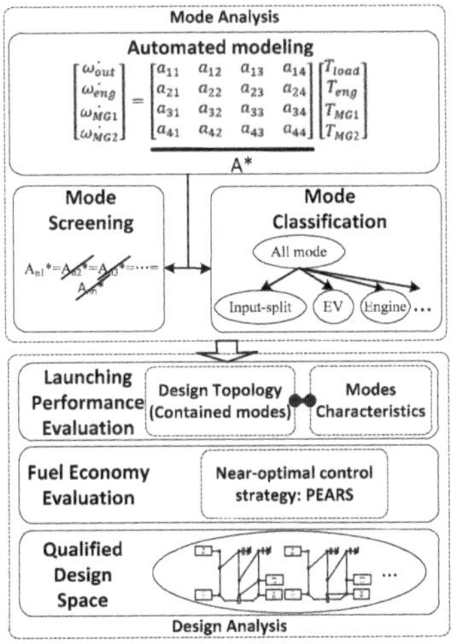

Figure 2. The proposed process for exhaustive topology search. PEARS: power-weighted efficiency analysis for rapid sizing.

2.4. Case Study

In this paper, we apply the exhaustive search method introduced above to design a plug-in hybrid bus. The benchmark is a low floor hybrid bus by Gillig that uses the GM Allison Hybrid System (AHS) as the transmission. The parameters of this bus are shown in Table 1 [32,33].

Table 1. Parameters of a Gillig plug-in hybrid electric bus with the Allison Hybrid System (AHS) powertrain [30]. PG: planetary gear.

Component	Parameter	Value
Engine	Displacement	Cummins ISB 280 6.7 L
	Maximum Power	209 kW @ 1600 RPM
	Maximum Torque	895 Nm @ 1600 RPM
	Inertia	0.82 kg·m^2
MG1, MG2	Maximum Power	80 kW
	Maximum Torque	450 Nm
	Maximum Speed	9600 RPM
$LiFePO_4$ battery	Maximum Power	60 Ah
	Voltage	900 V
All PG	Ring/Sun ratio	2
Final Drive	Gear Ratio	5.38
Vehicle	Mass	13,380 kg
	Frontal area	9.5 m^2
	Aero drag coefficient	0.62
	Rolling Resistance	0.015
	Tire radius	0.51 m

The AHS is a hybrid powertrain with two PGs and two clutches, and its lever diagram is depicted in Figure 3. By switching the states of these two clutches, an input-split mode and a compound split mode can be realized [3]. It should be noted that AHS does not contain a pure electric mode that may have a considerable effect on fuel economy, especially under a city driving cycle based on our previous study [14]. In this paper, we will also see the role of the electric mode for a hybrid electric bus.

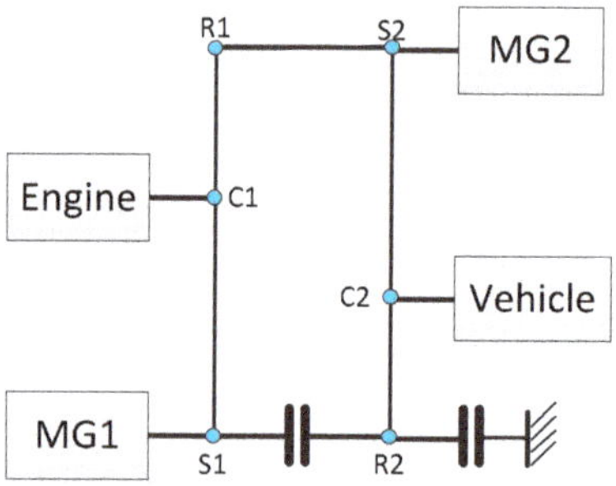

Figure 3. Lever diagram of the AHS.

Figure 4 shows the topology optimization results of this plug-in hybrid bus, in which the red point shows the performance of the original benchmark plug-in hybrid bus and blue points are survived designs whose 0–60 kph acceleration is less than 10 s. The fuel consumption in the figure refers to the total fuel consumption calculated through a near-optimal control strategy, called PEARS, under the Manhattan driving cycle. A number of designs show significant improvement on both fuel economy and launching performance. The design with the best fuel economy consumes 395 g diesel for the cycle, nearly 31.5% better than the original AHS design. Up to this point, we only explored topology optimization. Component sizes can be manipulated for even better results but this design DoF has not been explored.

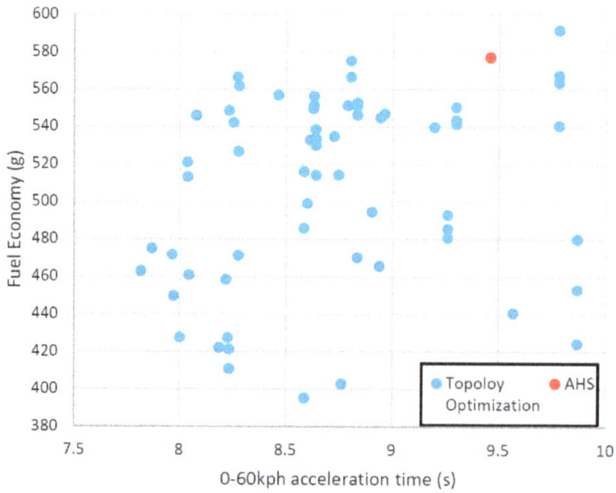

Figure 4. Fuel consumption and launching performance of survived designs and AHS hybrid.

3. Optimal Design

The method introduced in Section II solved the optimal design problem without considering the component sizes, *i.e.*, all powertrain parameters and component sizes are assumed to be given. In this section, we will introduce a methodology that combines the topology optimization and component sizing together to fully explore the fuel saving potential of hybrid electric bus.

3.1. Component Sizing

Powertrain component sizing consists of two groups: powertrain parameters, e.g., ring gear/sun gear ratio of the PGs, final drive ratio, *etc.*; and the component sizes, such as the motor power, engine power, and the energy storage capacity. In this paper, the engine is fixed. Only MG power and battery size are optimized.

The coupling between the energy management strategy and the component sizing, if fully considered, requires significant computational effort. Some researchers optimized the system with different hybridization ratios [21]. Others solved this problem heuristically, such as through GA and PSO [16,17]. In this paper, we adopt a two-loop optimization framework to solve this problem: the energy management strategy is optimized in the inner loop, while the outer loop solves optimal component sizes.

In the inner loop, the goal of the optimal energy management strategy is to minimize the cost function while avoiding frequent mode shift. The optimal control problem is defined in Equation (4) and is referred as Problem I:

Minimize:

$$J_C\left(u\left(i\right)|w\right)=\sum_{i=0}^{t_f} f_{\text{fuel}}\left(i\right)+\sum_{i=0}^{t_f} Mode\left(i\right)+\alpha\left(SOC_{\text{final}}-SOC_{\text{desired}}\right)^2 \tag{4a}$$

Subject to:

$$\begin{cases} SOC_{\min}\leqslant SOC\leqslant SOC_{\max} \\ T_{\text{e\_min}}\leqslant T_{\text{e}}\leqslant T_{\text{e\_max}} \\ T_{\text{MG1\_min}}\leqslant T_{\text{MG1}}\leqslant T_{\text{MG1\_max}} \\ T_{\text{MG2\_min}}\leqslant T_{\text{MG2}}\leqslant T_{\text{MG2\_max}} \\ \omega_{\text{e\_min}}\leqslant \omega_{\text{e}}\leqslant \omega_{\text{e\_max}} \\ \omega_{\text{MG1\_min}}\leqslant \omega_{\text{MG1}}\leqslant \omega_{\text{MG1\_max}} \\ \omega_{\text{MG2\_min}}\leqslant \omega_{\text{MG2}}\leqslant \omega_{\text{MG2\_max}} \\ Mode\in Mode_{\text{available}} \end{cases} \tag{4b}$$

where $u(i)$ is the control variables for the multi-mode HEVs, w is the driving cycle characterized by slope, velocity and time. In the cost function, $f_{\text{fuel}}(i)$ is the fuel consumption, which is determined by the engine speed ω_{e}, and engine torque T_{e}. *Mode(i)* is the mode shift penalty:

$$\begin{aligned} Mode\left(\text{i}\right)=\ &\beta_1\left[\omega_{\text{e}}\left(\text{i}+1\right)-\omega_{\text{e}}\left(\text{i}\right)\right]^2+\beta_2\left[\omega_{\text{MG1}}\left(\text{i}+1\right)-\omega_{\text{MG1}}\left(\text{i}\right)\right]^2 \\ &+\beta_3\left[\omega_{\text{MG2}}\left(\text{i}+1\right)-\omega_{\text{MG2}}\left(\text{i}\right)\right]^2 \end{aligned} \tag{5}$$

where ω_* (i) and ω_* (i + 1) are the rotational speeds of the powertrain components at current time step and next step. β_1, β_2, β_3 are the weighting factors to be tuned; the third term in the cost function is used to force the final battery state of charge SOC_{final} back to its initial value SOC_{desired}; α is a large penalty factor to form an equality constraint of the final SOC. In addition, the torque and speed of the powertrain components are restricted by their operating constraints. The available modes $Mode_{\text{available}}$ for each specific design are different and dependent on the dynamics.

In this paper, this non-linear constrained optimization problem is solved recursively, and the limitation of popular energy management strategies are discussed in Section I. Therefore, we adopt the near-optimal control algorithm, called PEARS+, whose generated results are close DP, yet computationally more efficient [14]. The procedure of the PEARS+ is shown in Figure 5 [27,34].

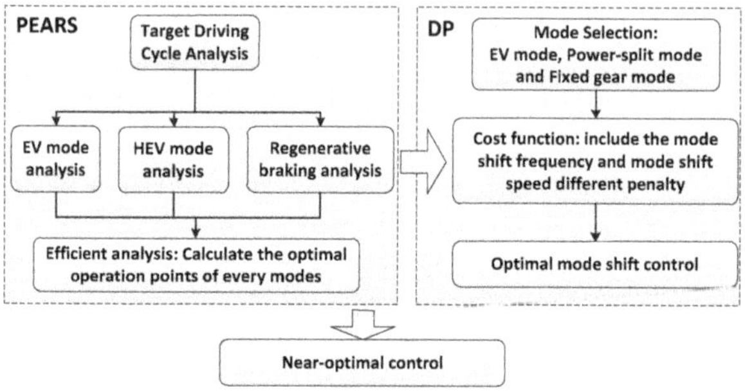

Figure 5. Procedure of the near-optimal control PEARS+.

In the outer-loop, the optimization of the component sizing, *i.e.*, Problem II in Equation (6), is solved. The target in this step is to find the best powertrain parameters and sizes with minimum fuel consumption under the required vehicle performance constraints.

Minimize:

$$F\left(x_{\text{size}} = \left[G_{\text{PG1}}, G_{\text{PG2}}, G_{\text{FR}}, P_{\text{MG1}}, P_{\text{MG2}}, C_{\text{Batt}}\right]\right) \tag{6a}$$

Subject to:

$$\begin{cases} T_{\text{acc}}\left(x_{\text{size}}\right) \leqslant T_{\text{acc\_des}} \\ \theta_{\max}\left(x_{\text{size}}\right) \geqslant \theta_{\text{max\_des}} \\ u_{\max}\left(x_{\text{size}}\right) \geqslant u_{\text{max\_des}} \end{cases} \tag{6b}$$

where the vector of the design variables x_{size} contains R/S ratio of two PGs G_{PG1}, G_{PG2}, the final drive ratio G_{FR}, power of two MGs P_{MG1}, P_{MG2} and the capacity of the battery C_{Batt}. In addition, three constraints must be met: the minimum 0–60 kph acceleration time limit $T_{\text{acc\_des}}$, the maximum gradeability limit $\theta_{\text{max\_des}}$, which refers to the maximum road slope a vehicle can ascend while maintaining a particular speed and maximum vehicle speed limit $u_{\text{max\_des}}$. The solution of this problem will be explained in the next section.

3.2. Combined Optimization

In this section, two approaches are introduced to solve the combined problem of topology optimization and component sizing.

3.2.1. Nested Optimization (Approach I)

The first approach is called nested optimization where the component sizing is embedded within the topology optimization as shown in Figure 6.

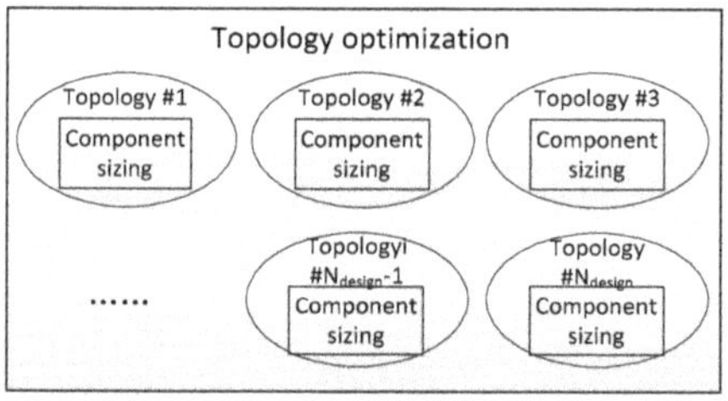

Figure 6. Schematic diagram of the nested optimization.

Optimal component sizing is calculated in the lower level for each multi-mode hybrid powertrain design obtained at the upper level, and the design with best fuel economy is identified at the upper level. This approach is brute-force by nature. The formulation of this nested optimization problem is written below and referred to as Approach I:

Minimize:

$$G\left(x_{\text{top}},\ x_{\text{size}}\right),\ x_{\text{top}} \in X_{\text{top}},\ x_{\text{size}} \in X_{\text{size}} \tag{7a}$$

Subject to:

$$\begin{cases} T_{\text{acc}}\left(x_{\text{top}}, x_{\text{size}}\right) \leqslant T_{\text{acc\_des}} \\ \theta_{\max}\left(x_{\text{top}}, x_{\text{size}}\right) \geqslant \theta_{\text{max\_des}} \\ u_{\max}\left(x_{\text{top}}, x_{\text{size}}\right) \geqslant u_{\text{max\_des}} \\ x_{\text{size}} \in \operatorname{argmin}\left\{F\left(x_{\text{top}}, y_{\text{size}}\right) : h\left(x_{\text{top}}, y_{\text{size}}\right) \leqslant 0, y_{\text{size}} \in X_{\text{size}}, j \in \{1, 2, \ldots, J\}\right\} \end{cases} \tag{7b}$$

where G represents the objective function of the upper level (topology optimization), F represents the objective function of the lower level (component sizing) and h is the set of inequality constraints at the lower level. x_{con} is the parameters that contains the powertrain locations $\overline{L_{\text{pow}}}$, clutch locations $\overline{L_{\text{clu}}}$, and permanent connection locations $\overline{L_{\text{per}}}$. In addition, vehicles should meet three vehicle performance requirements in launching, climbing, and top speed.

3.2.2. Iterative Optimization (Approach II)

The second approach is called iterative optimization, where the topology optimization and component sizing is executed sequentially and iteratively until convergence. Its process is described in Figure 7.

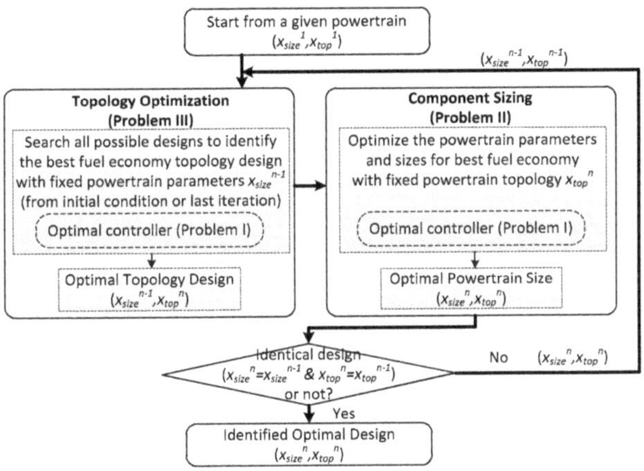

Figure 7. Flow chart of the iterative optimization.

The topology optimization problem is formulated in Equation (8) and referred to as Problem III, in which the goal is to identify the best fuel economy design under the same performance constraints as in Problem II:

Minimize:

$$G\left(x_{\text{top}} = \left[\overline{L_{\text{pow}}}, \overline{L_{\text{clu}}}, \overline{L_{\text{per}}}\right]\right) \tag{8a}$$

Subject to:

$$\begin{cases} T_{\text{acc}}\left(x_{\text{top}}\right) \leqslant T_{\text{acc\_des}} \\ \theta_{\max}\left(x_{\text{top}}\right) \geqslant \theta_{\text{max\_des}} \\ u_{\max}\left(x_{\text{top}}\right) \geqslant u_{\text{max\_des}} \end{cases} \tag{8b}$$

where x_{top} is a vector to represent each possible hybrid design with assigned powertrain locations $\overline{L_{pow}}$, clutch locations $\overline{L_{clu}}$, and permanent connection locations $\overline{L_{per}}$.

As shown in Figure 7, the loop starts from any given powertrain (x_{size}, x_{top}) that the powertrain parameters, component sizes and powertrain topology are all known, e.g., the hybrid bus as shown in Table 1, and the powertrain design from the last iteration (${x_{size}}^{n-1}$, ${x_{top}}^{n-1}$). In the first step of each iteration, the powertrain topology is optimized with powertrain parameters ${x_{size}}^{n-1}$. In the second step, the optimal powertrain parameters of the optimal topology design derived in the first step are identified by solving Problem II. Afterwards in the third step, the identified powertrain topology with optimal parameters in the second step (${x_{size}}^{n-1}$, ${x_{top}}^{n-1}$) is compared with the powertrain at the beginning of this iteration (${x_{size}}^{n}$, ${x_{top}}^{n}$). If the powertrains are the same, referring to identical powertrain topology and parameters, we denote it as the optimal topology and the optimal sizing parameters. If not, the identified powertrain (${x_{size}}^{n}$, ${x_{top}}^{n}$) requires the topology optimization and component sizing to sequentially check its optimality by comparing the powertrains at the beginning and end of each iteration. The iteration stops when the solution converges. At the end, a new hybrid design with both optimized topology and component sizes is identified.

3.3. Results

The two approaches proposed in Section 3.2 are applied to find optimal designs for a hybrid bus. The parameters described in Table 1 are used as the initial values. The nested optimization solution is taken as the benchmark to check the optimality of the iterative optimization approach. Note that some of the vehicle parameters (*i.e.*, mass, frontal area, aero drag coefficient, rolling resistance, and tire radius) are fixed. As introduced in Section II, the number of possible powertrain designs with three clutches is 2,620,800. If all powertrain parameters (R/S ratios, final drive ratio, motor sizes, and battery size) are divided into ten grids equally, we need to solve the optimal energy management problem 2.6×10^{12} times for the nested optimization platform, which takes a very long time to solve. For simplicity, several assumptions are made to define a simpler problem that can be solved in a reasonable amount of time.

(1) We fixed the powertrain component locations the same as what was shown in Figure 3.
(2) We only optimize the powertrain parameters (*i.e.*, R/S ratio of PGs and final drive) in this study.

The grids and range for the ratio of PGs and final drive is listed as follows:

$$\begin{cases} G_{PG1} = G_{PG2} = [1.2 : 0.2 : 3] \\ G_{FR} = [3 : 0.5 : 9] \end{cases} \tag{9}$$

With the above assumptions and definitions, the number of designs is three orders of magnitude smaller:

$$N_{design}' = N_{design} \times 10 \times 10 \times 13 = 3.4 \times 10^9 \tag{10}$$

Figure 8 shows the results of nested optimization, where the red triangles form the Pareto front and the X and Y axes stand for the acceleration and fuel economy performance, respectively. Design number refers to a specific design labeled before the optimization. Compared with the original AHS hybrid powertrain, Design 811 has 35.0% and 19.7% improvement on fuel economy and acceleration performance. In addition, we listed the results with only component sizing optimization and topology optimization in Table 2. It can be seen that the design with both component sizing and topology optimized achieves greater improvement than designs when only the topology or component sizing is optimized. Meanwhile, designs obtained with topology optimization perform better than the designs with only component sizing optimization. In other words, topology optimization is a more effective design exploration than tuning component sizes.

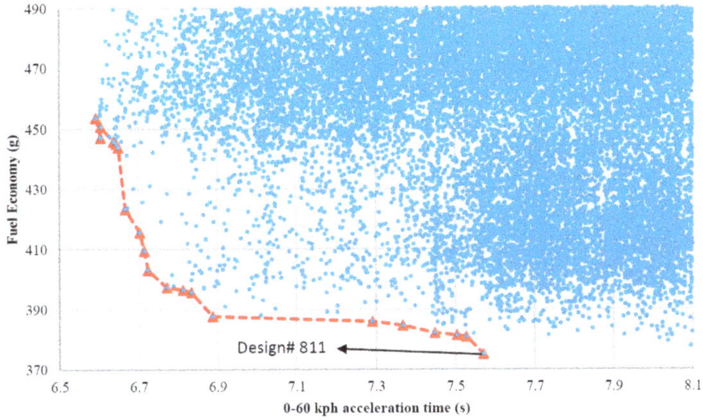

Figure 8. Results of nested optimization where red line is the Pareto-optimal.

Table 2. Performance improvement of the hybrid bus before and after nested optimization.

Optimization Method	Fuel Consumption		Acceleration Time	
	Value (g)	Improvement (%)	Value (s)	Improvement (%)
Original AHS	577	N/A	9.45	N/A
Component sizing optimization only	471.34	18.4%	8.76	7.3%
Topology optimization only	395.45	31.5%	8.58	9.2%
Both component sizing and topology optimization	375.0	35.0%	7.57	19.7%

Table 3 and Figure 9 show the optimization results of each iteration for the case study whose initial powertrain parameters, final drive, R/S ratios of two PGs equal to 6, 1.5 and 3, respectively. In Figure 9, colors indicate the performance of all possible designs for different iterations. As shown, a better design with different topologies or powertrain parameters is identified for each step until the end of the third component sizing (component sizing of Iteration III). One more iteration (Iteration IV) is executed to check the optimality of the resulted design at the end of Iteration III.

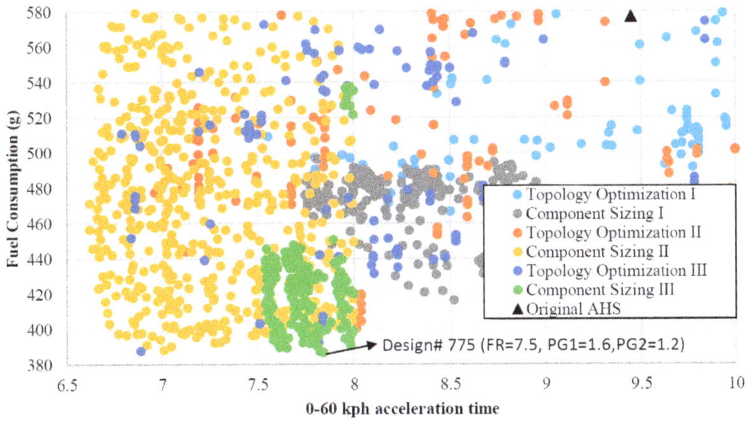

Figure 9. Results of each optimization for initial point (FR = 6, PG1 = 1.5, PG2 = 3).

Table 3. Optimal results of each iteration for initial point (FR = 6, PG1 = 1.5, PG2 = 3).

Iteration	Optimization	Optimal Topology	PG1 R/S Ratio	PG2 R/S Ratio	Final Drive Ratio	Acceleration Time (s)	Fuel Consumption (g)
Original Parameters		AHS	6	1.5	3	9.45	577
I	Topology Optimization	Design# 776	6	1.5	3	9.664	426.79
	Component Sizing	Design# 776	8	1.2	1.6	8.886	411.32
II	Topology Optimization	Design# 768	8	1.2	1.6	7.272	392.36
	Component Sizing	Design# 768	8.5	1.6	1.2	6.887	387.75
III	Topology Optimization	Design# 775	8.5	1.6	1.2	7.838	386.81
	Component Sizing	Design# 775	8.5	1.6	1.2	7.838	386.81
IV	Topology Optimization	Design# 775	8.5	1.6	1.2	7.838	386.81
	Component Sizing	Design# 775	8.5	1.6	1.2	7.838	386.81

The optimal designs of each iteration are extracted and plotted in Figure 10 to demonstrate the trend of the optimization. Four other cases with different initial conditions are also plotted in Figure 10, where the red dashed line is the fitted Pareto front from nested optimization. They all converge toward the bottom left corner. It can be seen that black and blue lines are converging to the global optimal design, Design 811, after only two iterations. Gray and yellow lines are also converging within only two iterations, but they do not converge to the global optimal design.

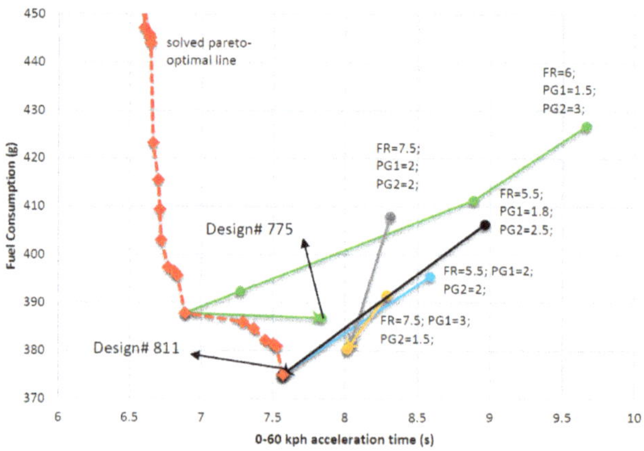

Figure 10. Sample results from iterative optimization.

In summary, the iterative optimization process successfully converges to the global optimal design that is obtained from nested optimization in last case study. In the meantime, instead of calculating the fuel economy of all 3.4×10^9 designs, as discussed in Equation (8), the iterative optimization only takes a few iterations (examining about 2500 designs in each iteration) to converge to an optimized design. Thus, the computation time is much shorter than exhaustive search. However, the iterative optimization approach is sensitive to the initial conditions (*i.e.*, powertrain parameters). In the next section, we will propose a method to enhance the iterative optimization approach and reduce of its reliance to initial conditions.

3.4. Enhanced Iterative Optimization

To overcome the sensitivity problem of the iterative optimization, a method, called enhanced iterative optimization is proposed in this part. The process of this method is presented in Figure 11. The optimization starts from a couple of random initial points. For each point, we can identify the best fuel economy design using the iterative optimization. Then, we search the design with minimum fuel consumption among the group of identified designs. In this way, the impact of the initial conditions on the final results can be reduced, while the computational efficiency is still much better than the nested optimization. In summary, Table 4 shows the performance comparison between the three approaches proposed in this paper, where m_i represents the number of initial conditions used for computation. Enhanced iterative optimization indicates better convergence performance than iterative optimization and is more efficient than nested optimization.

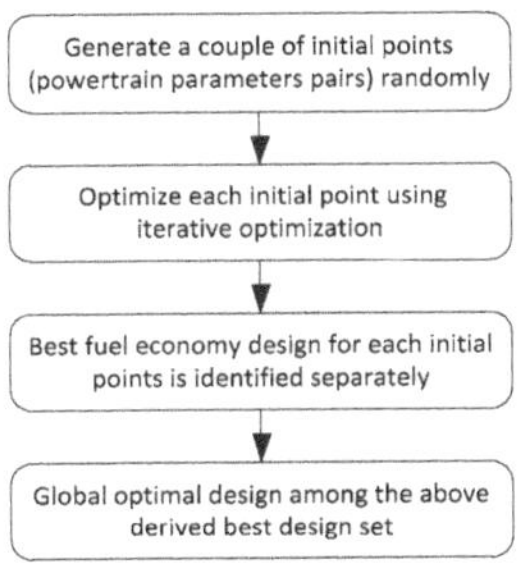

Figure 11. Process of the proposed enhanced iterative optimization.

Table 4. Comparison between three approaches proposed in this paper.

Method Name	Number of Designs Evaluated	Computational Speed	Gained Results
Nested Optimization	3.4×10^9	Very slow	Optimal design
Iterative Optimization	Up to four iterations	Fastest	Can converge to optimal design but sensitive to initial conditions
Enhanced Iterative Optimization	Up to $4m_i$ iterations	m_i times slower than iterative optimization	Better than iterative optimization

4. Payload Sensitivity Study

Many optimal sizing studies identify optimal designs for given driving cycles and some of the vehicle parameters are fixed [35]. The resulting design may be sensitive to vehicle parameter variations, such as the vehicle weight, a parameter that changes significantly (and even more for heavy trucks). A topology optimized for a specific weight may not work well when the vehicle weight changes. In this section, we apply the enhanced iterative optimization method to optimize the topology and powertrain parameters of a vehicle for fuel economy with frequently changed weight.

We assume the average passenger weight is 80 kg and the capacity of the bus is 50 persons [36], so the mass of the fully loaded bus is 13,380 + 4000 kg. In this section, we evaluate six different cases with the payload increasing in steps of 800 kg. All six cases are assumed to be equally important in the final cost function calculation. The performance requirement is defined as following: the bus needs to achieve a top speed of 100 kph and can climb a 25% slope. In addition, the minimum acceleration time of 0–60 kph should be less than 10 s at all loading conditions.

Under these conditions, we optimize the topologies and powertrain parameters of the bus for the Manhattan driving cycle using the iterative optimization method. The optimal topology and powertrain parameters are listed in Table 5 where Design 811 is the best design topology for all six scenarios as shown in Figure 12A. As we can see, all loading conditions have the same optimal power-split configuration but the optimal parameters are different.

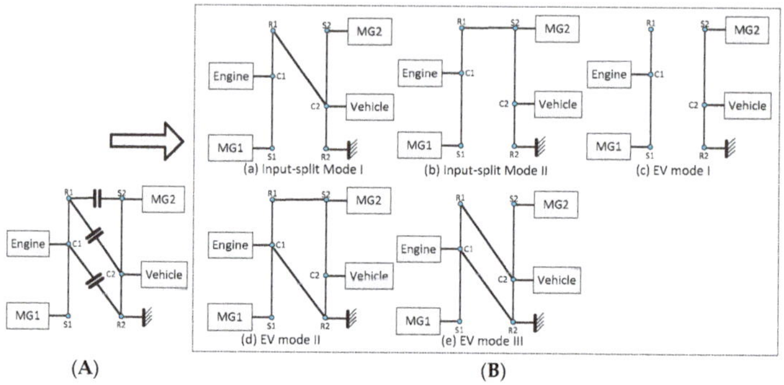

Figure 12. Identified optimal powertrain design for the target bus application and its operating modes (**A**) lever diagram of the Design 811; and (**B**) operating modes of the optimal powertrain Design 811.

Table 5. Optimal topology and powertrain parameters under a range of loading scenarios.

Vehicle Parameters	Payload (kg)	Optimal Topology	PG1 Ratio	PG2 Ratio	Final Drive Ratio
1	0	Design 811	3	1.2	8
2	800	Design 811	3	1.4	7.5
3	1600	Design 811	3	1.2	8
4	2400	Design 811	3	1.6	7
5	3200	Design 811	2.6	1.4	7.5
6	4000	Design 811	2.4	1.2	8

In order to obtain the best design under all conditions, we use the overall fuel consumption under the six payload scenarios as the cost function. The topology of the optimal design, Design 811, is shown in Figure 12. It has five operating modes—two input-split modes and three electric drive modes, also shown in Figure 12. The optimal parameters for the R/S ratios of PG1, PG2, and final drive are 2.8, 1.2, and 8, respectively. The state and control trajectories of this optimal powertrain is shown in Figure 13 where mode number 1–5 stands for input-split mode I, EV mode I, input-split mode II, EV mode II, and EV mode III. As we can see, EV modes are widely used for low-speed range and regenerative braking. That is the reason Design 811 shows better fuel economy than the original AHS. It indicates that pure electric driving modes are also essential for hybrid buses more than plug-in HEVs.

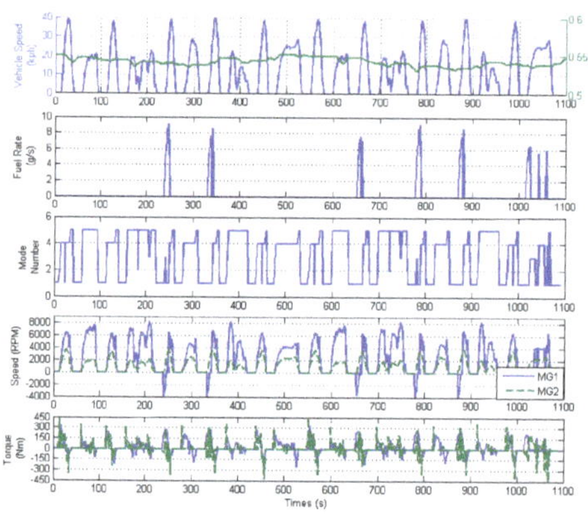

Figure 13. State and control trajectories of the simulated optimal powertrain design under the Manhattan cycle by near-optimal control strategy PEARS+.

It should be noted that the battery capacity for a hybrid bus is also important but is largely dependent on the trade-off between electric range and cost. Therefore, it is not included in this large-scale design study. We will consider this factor in the future.

5. Conclusions

The contribution of this paper is three-fold. Firstly, we optimize the multi-mode hybrid powertrain from three dimensions, *i.e.*, exploring the topologies of a power-split hybrid powertrain with double PGs and clutches, optimizing powertrain parameters, and realizing optimal control of the power split between the fuel and electricity. These three aspects are coupled together by developing an optimization framework, called iterative optimization. Instead of searching the whole design space (referred as nested optimization), iterative optimization executes topology optimization and component sizing alternately. A case study shows that iterative optimization converges to the global optimal design efficiently. Secondly, we present an enhanced iterative optimization framework to alleviate the sensitivity problem of the initial conditions. Thirdly, we develop a robust multi-mode hybrid electric bus considering a variety of loading scenarios by using the proposed enhanced iterative optimization. The resulting hybrid bus achieves better fuel economy by introducing EV modes under the Manhattan driving cycle.

In this paper, we search the optimal component size through an exhaustive search, which may be impractical if more parameters are taken into consideration in the future due to the computational burden. Instead, other heuristic methods, like GA and PSO, can be adopted to optimize the component sizes to accelerate the calculation.

Acknowledgments: This work was supported by Jiangsu Science and Technology Agency under Joint Innovation Funding (BY2014004-04 and BY2015004-02).

Author Contributions: Weichao Zhuang proposed the optimization framework and adopted it to optimize a hybrid electric bus, Xiaowu Zhang, Huei Peng and Liangmo Wang provided guidance and key suggestions.

Conflicts of Interest: The authors declare no conflict of interest.

References

1. National Highway Traffic Safety Administration (NHTSA). Obama Administration Finalizes Historic 54.5 mpg Fuel Efficiency Standards. Available online: http://www.nhtsa.gov/About+NHTSA/Press+Releases/2012/Obama+Administration+Finalizes+Historic+54.5+mpg+Fuel+Efficiency+Standards (accessed on 25 December 2015).
2. Alternative Fuels Data Center: Maps and Data. U.S. HEV Sales by Model. 2014. Available online: http://www.afdc.energy.gov/data/10301 (accessed on 25 December 2015).
3. Miller, J.M. Hybrid electric vehicle propulsion system architectures of the e-CVT type. *IEEE Trans. Power Electron.* **2006**, *21*, 756–767. [CrossRef]
4. Zhang, X.W.; Li, C.T.; Kum, D.; Peng, H. Prius$^+$ and Volt$^-$: Configuration analysis of power-split hybrid vehicles with a single planetary gear. *IEEE Trans. Veh. Technol.* **2012**, *61*, 3544–3552. [CrossRef]
5. Xu, L.F.; Ouyang, M.G.; Li, J.Q.; Yang, F.Y.; Lu, L.G.; Hua, J.F. Optimal sizing of plug-in fuel cell electric vehicles using models of vehicle performance and system cost. *Appl. Energy* **2013**, *103*, 477–487. [CrossRef]
6. Jalil, N.; Kheir, N.; Salman, M. A Rule-Based Energy Management Strategy for a Series Hybrid Vehicle. In Proceedings of the 1997 American Control Conference, Albuquerque, NM, USA, 4–6 June 1997; pp. 689–693.
7. Musardo, C.; Rizzoni, G.; Guezennec, Y.; Staccia, B. A-ECMS: An adaptive algorithm for hybrid electric vehicle energy management. *Eur. J. Control* **2005**, *11*, 509–524. [CrossRef]
8. Sciarretta, A.; Back, M.; Guzzella, L. Optimal control of parallel hybrid electric vehicles. *IEEE Trans. Control Syst. Technol.* **2004**, *12*, 352–363. [CrossRef]
9. Lin, C.C.; Peng, H.; Grizzle, J.W.; Kang, J.-M. Power management strategy for a parallel hybrid electric truck. *IEEE Trans. Control Syst. Technol.* **2003**, *11*, 839–849.

10. Larsson, V.; Lars, J.; Egardt, B. Analytic solutions to the dynamic programming subproblem in hybrid vehicle energy management. *IEEE Trans. Veh. Technol.* **2015**, *64*, 1458–1467. [CrossRef]
11. Vinot, E. Time Reduction of the Dynamic Programming Computation in the Case of Hybrid Vehicle. In Proceedings of the 13th International Workshop on OIPE 2014—Optimization and Inverse Problems in Electromagnetism, Delft, The Netherlands, 10–12 September 2014.
12. Hu, X.; Murgovski, N.; Johannesson, L.M.; Egardt, B. Optimal dimensioning and power management of a fuel cell/battery hybrid bus via convex programming. *IEEE/ASME Trans. Mechatron.* **2015**, *20*, 457–468. [CrossRef]
13. Nüesch, T.; Elbert, P.; Flankl, M.; Onder, C.; Guzzella, L. Convex optimization for the energy management of hybrid electric vehicles considering engine start and gearshift costs. *Energies* **2014**, *7*, 834–856. [CrossRef]
14. Zhang, X.; Peng, H.; Sun, J. A near-optimal power management strategy for rapid component sizing of multimode power split hybrid vehicles. *IEEE Trans. Control Syst. Technol.* **2015**, *23*, 609–618. [CrossRef]
15. Zou, Y.; Li, D.G.; Hu, X.S. Optimal sizing and control strategy design for heavy hybrid electric truck. *Math. Probl. Eng.* **2012**, *2012*. [CrossRef]
16. Fang, L.C.; Qin, S.Y.; Xu, G.; Li, T.L.; Zhu, K.M. Simultaneous optimization for hybrid electric vehicle parameters based on multi-objective genetic algorithms. *Energies* **2011**, *4*, 532–544. [CrossRef]
17. Ebbesen, S.; Doenitz, C.; Guzzella, L. Particle swarm optimisation for hybrid electric drive-train sizing. *Int. J. Veh. Des.* **2012**, *58*, 181–199. [CrossRef]
18. Wu, L.H.; Wang, Y.N.; Yuan, X.F.; Chen, Z.L. Multiobjective optimization of HEV fuel economy and emissions using the self-adaptive differential evolution algorithm. *IEEE Trans. Veh. Technol.* **2011**, *60*, 2458–2470. [CrossRef]
19. Shankar, R.; Marco, J.; Assadian, F. The novel application of optimization and charge blended energy management control for component downsizing within a plug-in hybrid electric vehicle. *Energies* **2012**, *5*, 4892–4923. [CrossRef]
20. Murgovski, N.; Johannesson, L.; Sjoberg, J.; Egardt, B. Component sizing of a plug-in hybrid electric powertrain via convex optimization. *Mechatronics* **2012**, *22*, 106–120. [CrossRef]
21. Sundström, O.; Guzzella, L.; Soltic, P. Torque-assist hybrid electric powertrain sizing: From optimal control towards a sizing law. *IEEE Trans. Control Syst. Technol.* **2010**, *18*, 837–849. [CrossRef]
22. Nuesch, T.; Ott, T.; Ebbesen, S.; Guzzella, L. Cost and Fuel-Optimal Selection of HEV Topologies Using Particle Swarm Optimization and Dynamic Programming. In Proceedings of the 2012 American Control Conference (ACC), Montreal, QC, Canada, 27–29 June 2012; pp. 1302–1307.
23. Zhang, X.W.; Peng, H.; Sun, J. A Near-Optimal Power Management Strategy for Rapid Component Sizing of Power Split Hybrid Vehicles with Multiple Operating Modes. In Proceedings of the 2013 American Control Conference (ACC), Washington, DC, USA, 17–19 June 2013; pp. 5972–5977.
24. Liu, J.M.; Peng, H. A systematic design approach for two planetary gear split hybrid vehicles. *Veh. Syst. Dyn.* **2010**, *48*, 1395–1412. [CrossRef]
25. Bayrak, A.E.; Ren, Y.; Papalambros, P.Y. Design of Hybrid-Electric Vehicle Architectures Using Auto-Generation of Feasible Driving Modes. In Proceedings of the ASME 2013 International Design Engineering Technical Conferences and Computers and Information in Engineering Conference: American Society of Mechanical Engineers, Portland, OR, USA, 4–7 August 2013; p. V001T01A005.
26. Kang, M.; Kim, H.; Kum, D. Systematic Configuration Selection Methodology of Power-Split Hybrid Electric Vehicles with a Single Planetary Gear. In Proceedings of the ASME 2014 Dynamic Systems and Control Conference: American Society of Mechanical Engineers, San Antonio, TX, USA, 22–24 October 2014; p. V001T15A001.
27. Zhang, X.; Li, S.; Peng, H.; Sun, J. Efficient exhaustive search of power split hybrid powertrains with multiple planetary gears and clutches. *J. Dyn. Sys. Meas. Control* **2015**, *137*. [CrossRef]
28. Silvas, E.; Hofman, T.; Serebrenik, A.; Steinbuch, M. Functional and cost-based automatic generator for hybrid vehicles topologies. *IEEE/ASME Trans. Mechatron.* **2015**, *20*, 1561–1572. [CrossRef]
29. Klemen, D.; Schmidt, M.R. Two-Mode, Compound-Split, Electro-Mechanical Vehicular Transmission Having Significantly Reduced Vibrations. U.S. Patent 6,358,173, 19 March 2002.
30. Holmes, A.G.; Schmidt, M.R. Hybrid Electric Powertrain Including a Two-Mode Electrically Variable Transmission. U.S. Patent 6,478,705, 12 November 2002.

31. Conlon, B.M.; Blohm, T.; Harpster, M.; Holmes, A.; Palardy, M.; Tarnowsky, S. The next generation "Voltec" extended range EV propulsion system. *SAE Int. J. Altern. Powertrains* **2015**, *4*, 248–259. [CrossRef]
32. Hallmark, S.; Wang, B.; Qiu, Y.; Sperry, R. Evaluation of In-Use Fuel Economy for Hybrid and Regular Transit Buses. *J. Transp. Technol.* **2013**, *3*, 52–57. [CrossRef]
33. Allison Hybrid H40 EP/H50 EP. Available online: http://www.allisontransmission.com/docs/default-source/marketing-materials/sa5983en-h40-50-ep1BCB31AC06C2F2B94ACCEED0.pdf?sfvrsn=4 (accessed on 25 December 2015).
34. Zhuang, W.; Zhang, X.; Zhao, D.; Peng, H.; Wang, L. Optimal design of three-planetary-gear power-split hybrid powertrains. *Int. J. Automot. Technol.* **2016**, *17*, 299–309. [CrossRef]
35. Bayrak, A.E.; Ren, Y.; Papalambros, P.Y. Optimal Dual-Mode Hybrid Electric Vehicle Powertrain Architecture Design for a Variety of Loading Scenarios. In Proceedings of the ASME 2014 International Design Engineering Technical Conferences and Computers and Information in Engineering Conference: American Society of Mechanical Engineers, Buffalo, NY, USA, 17–20 August 2014; p. V003T01A005.
36. Mendes, E. In U.S., Self-Reported Weight Up Nearly 20 Pounds Since 1990. Available online: http://www.gallup.com/poll/150947/self-reported-weight-nearly-pounds-1990.aspx (accessed on 25 December 2015).

energies

MDPI

Article

Aerodynamic Analysis and Three-Dimensional Redesign of a Multi-Stage Axial Flow Compressor

Tao Ning [1,2,*], Chun-Wei Gu [1,2], Wei-Dou Ni [1,2], Xiao-Tang Li [3] and Tai-Qiu Liu [3]

1. Key Laboratory for Thermal Science and Power Engineering of Ministry of Education, Department of Thermal Engineering, Tsinghua University, Beijing 100084, China; gcw@mail.tsinghua.edu.cn (C.-W.G.); niwd@tsinghua.edu.cn (W.-D.N.)
2. Collaborative Innovation Center of Advanced Aero-Engine, Beijing 100191, China
3. Shenyang Engine Design and Research Institute (SEDRI), Shenyang 110000, China; xiaotang_li@126.com (X.-T.L.); taiqiuliu@163.com (T.-Q.L.)

* Correspondence: nt12@mails.tsinghua.edu.cn; Tel.: +86-10-6278-1739; Fax: +86-10-6277-1209

Academic Editor: Paolo Mercorelli
Received: 14 September 2015; Accepted: 11 April 2016; Published: 16 April 2016

Abstract: This paper describes the introduction of three-dimension (3-D) blade designs into a 5-stage axial compressor with multi-stage computational fluid dynamic (CFD) methods. Prior to a redesign, a validation study is conducted for the overall performance and flow details based on full-scale test data, proving that the multi-stage CFD applied is a relatively reliable tool for the analysis of the follow-up redesign. Furthermore, at the near stall point, the aerodynamic analysis demonstrates that significant separation exists in the last stator, leading to the aerodynamic redesign, which is the focus of the last stator. Multi-stage CFD methods are applied throughout the three-dimensional redesign process for the last stator to explore their aerodynamic improvement potential. An unconventional asymmetric bow configuration incorporated with leading edge re-camber and re-solidity is employed to reduce the high loss region dominated by the mainstream. The final redesigned version produces a 13% increase in the stall margin while maintaining the efficiency at the design point.

Keywords: axial compressor; multi-stage; stator; bow; re-camber; re-solidity

1. Introduction

High performance requirements and cost reduction were proposed for the development of modern aero engines and heavy-duty gas turbines, which improve the compressor design with higher loading, fewer stage counts and a shorter axial distance. Consequently, modern compressors have challenging aerodynamic designs because the larger adverse pressure gradient and stronger blade row interaction lead to internal turbulent flow with high three-dimensionality and nonlinearity. The highly three-dimensional flow in modern compressors also makes it difficult to evaluate the aerodynamic performance, resulting in the high cost and risk of compressor development.

Two main approaches are employed in the aerodynamic analysis and design of compressors. The first is the through-flow code, which is a traditional quasi-three-dimension (quasi-3D) method based on streamline curvature equations. The through-flow code was the main tool for compressor performance predictions two decades ago, with the advantage of quickly and effectively predicting performance [1–3]. However, the through-flow code typically requires the input of many empirical parameters, such as blockage, loss and deviation, which are dependent on the design history. The larger number of inputs limits the application of the code in compressor design beyond the state of the art and suppresses the improvement potential of modern design performance because proper correlation models and empirical data are unavailable. However, through-flow is very important for determining the general compressor design and assembly, especially for one-dimension (1-D) and two-dimension (2-D) designs.

With the recent rapid development of computer science and computational fluid dynamic (CFD) technology, a fully viscous, 3-D CFD method has played a key role in the development of modern compressors, especially for the advanced 3-D design, by providing detailed flow fields without an empirical parameter input. In particular, after the introduction of steady multi-stage CFD approaches in the early 1990s, a breakthrough was achieved for the aerodynamic analysis and design philosophy of turbomachinery. Two main blade row-coupling models exist for modeling and simulating turbomachinery in a multi-stage environment: the average-passage approach and the mixing plane approach. The average-passage approach was developed by Adamczyk [4,5] and accounts for the unsteady deterministic flow and blade-row interaction as the body force and source terms of the passage-averaged equations. The CFD code APNASA by National Aeronautics and Space Administration (NASA) is a typical representation of the average-passage approach, which was used in enormous multi-stage compressor analysis and design [6–11]. In the mixing plane approach, the flow variables of the interface planes between the stationary and rotating blade rows are averaged circumferentially and exchanged as the boundary between neighboring blade rows. This approach is used in the commercial CFD codes NUMECA-Fine/Turbo and ANSYS-CFX and has been widely employed in multi-stage compressor analyses in recent years [11–15].

Many researchers, including Adamczyk [7], Wellborn [10], Brilliant [16], Mansour [11], and Ikeguchi [15], successfully developed multi-stage CFD methods to reduce development costs and gain performance improvement. In the last two decades, with the extensive application of multi-stage CFD methods and the development of advanced high-fidelity CFD methods [17–19], CFD technologies have improved compressor aerodynamics and revolutionized the philosophy of compressor design. Indeed, the development of large novel 3-D blading design concepts, including bow [20,21], sweep [22,23], re-camber [23], and blended blade and endwall (BBEW) [24], can be mainly attributed to the application of CFD technology, which leads to improved and more highly three-dimensional design philosophies for compressor aerodynamics. In turn, this revolution produces many benefits for increasing the machine's loading capacity beyond the current state-of-the-art technology.

Nevertheless, most of the 3-D blading techniques researched by CFD methods mainly only concentrated on the isolated or part-stage blade row environments and were relatively limited in the entire multi-stage environment. The effects of 3-D blading are significantly determined by 3-D internal flow structures on modern compressors with high three dimensionality and strong blade row interactions, which may be rather different between the isolated and multi-stage environment. Therefore, it is worth applying multi-stage CFD methods to capture the realistic physics and understand the mechanism of 3-D blading in the multi-stage environment. However, multi-stage CFD methods are not without limits. As noted by Denton [25], the main issues facing CFD users include uncertainties of the reliability of CFD due to the numerical errors, modeling errors, steady flow assumption, and so on. Therefore, it is essential to conduct validation research on multi-stage CFD methods based on test data to guarantee the validity and reliability of CFD methods applied in the design process.

Currently, CFD is frequently integrated into a state-of-the-art blade design system coupled with automatic multi-objective optimization. The optimization system typically includes geometry generation tools, optimization algorithms, CFD and finite element method (FEM) solvers, as well as artificial neural networks (ANNs) to speed up the optimization process, which involves mechanical, thermal, and aerodynamic analyses [26–28]. This paper focuses on the 3-D aerodynamic redesign of a 5-stage compressor with multi-stage CFD methods, with the purpose of developing the next upgrade version of this type of compressor.

2. Compressor Overview

The highly loaded 5-stage compressor is designed for developing a modern industrial gas turbine; the overall compressor configuration is illustrated in Figure 1. A high loading design strategy is applied in the design process. MCA airfoils are used for the transonic rotor of the first stage, whereas CDA profiles are employed in the design of the remainder of the subsonic blades. The inlet guided vane (IGV) and first two stators are variable for surge control at partial speed. However, other detailed compressor design specifications cannot be provided in this work because of the proprietary nature of the information. A full-scale rig test was conducted at the Shenyang Engine Design and Research Institute (SEDRI). The overall performance data, the measured casing static pressure and the flow profiles at the leading edge of the stators were obtained for both design and off-design conditions.

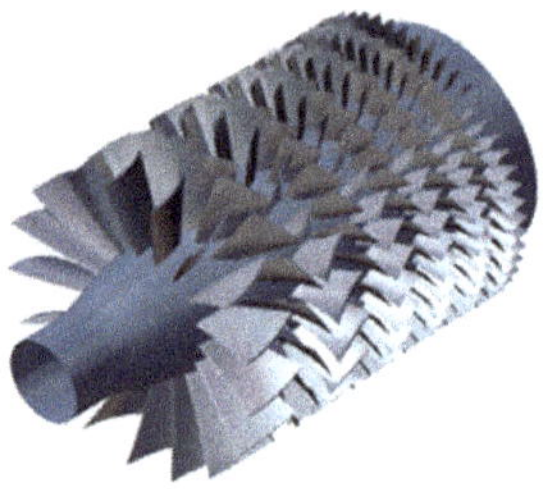

Figure 1. Overall configuration of the compressor.

3. CFD Computational Method

3.1. Numerical Scheme

Multistage CFD methods are applied in the aerodynamic analysis and redesign process with the commercial CFD software "Fine/Turbo" [29], where the 3D Reynolds-Averaged Navier-Stokes (RANS) equations are solved using a cell-centered finite volume approach. After calibration of several turbulence models, single passage steady simulations are conducted using the Spalart-Allmaras turbulence model [30]. The second-order upwind scheme for spatial discretization and the four-stage Runge-Kutta scheme for time discretization are employed for the solver. Moreover, a multigrid approach is applied to accelerate the convergence. In particular, the mixing plane approach is used as the rotor-stator interface strategy. Air (perfect gas) is applied as the simulation fluid.

3.2. Computational Mesh

The rotor tip clearances and variable stator clearances are considered to be simple constant span gaps. The mesh is generated using structured hex elements, employing a HOH (inlet, O, and outlet blocks) mesh topology, with an O-grid covering the blade surface. The rotor tips and variable stator clearances are filled with butterfly mesh blocks (HO mesh topology) using a non-matching mesh interface with a mainstream domain, as demonstrated in Figure 2.

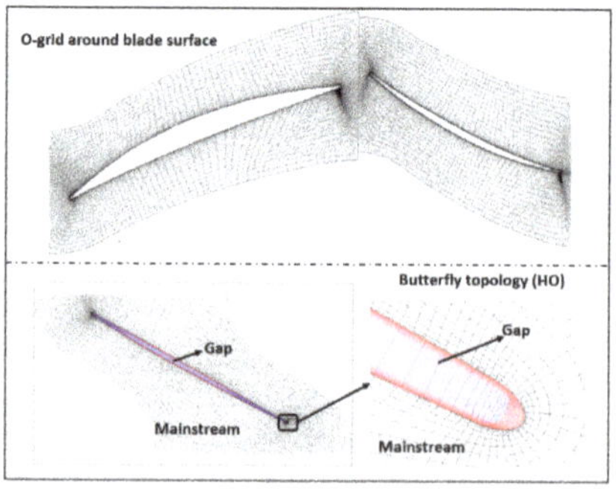

Figure 2. Typical mesh topology.

A mesh-dependence test is carried on three mesh sizes of four million, six million and eight million. A total mesh size of six million for the 11 blade passage, including IGV and five stages (rotor and stator), is adopted. To guarantee the mesh quality, the minimum grid skewness is approximately 18 degrees. The distance between the first grid line and the solid wall is set to $y+ < 10$. An overview of the mesh for the 5-stage compressor is presented in Figure 3.

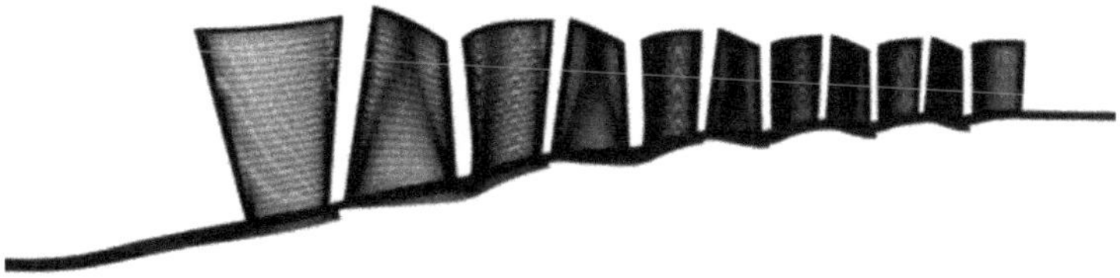

Figure 3. Mesh for the CFD computations.

3.3. Boundary Conditions

A total pressure profile and inlet flow angle profile are specified at the inlet based on the test data, while a radial pressure gradient to match the radial equilibrium condition is specified at the outlet. Moreover, one-blade passage flows are simulated with the periodic conditions applied on the pitchwise boundaries of the flow domain. Nonslip and adiabatic conditions are adopted for all of the solid walls. In particular, rotors are calculated in the rotating frame of reference, while the hub surface is defined as a stationary wall; the casing surface is defined as a counter-rotating wall in the rotating frame of reference.

4. Aerodynamic Analysis

Multi-stage CFD methods are used for the detailed aerodynamic analysis of the baseline compressor to put forward recommendations for redesign efforts. The entire redesign process is conducted in the multi-stage environment, even for a single blade. Therefore, it is essential to carefully conduct validation research on the multi-stage CFD methods applied in this paper prior to aerodynamic redesign.

Test data from the full-scale rig are used for validation. The overall performance is compared with the experimental data, and details of the flow fields, such as the casing static pressure and total pressure profiles at the leading edge of stator 4 and stator 5, are presented for validation at the design and off-design operating points (near the stall). The experimental and computational performance data presented below are normalized using the same normalization factors taken from the rig conditions at the design point. The measured and predicted overall performances of the 5-stage compressor at design speed are presented in Figure 4. Compared with the test data, it is clear that the shape of the speed line predicted by the multi-stage CFD is similar to the experimental line, where the error bars were derived based on one standard deviation. The deviation of the peak adiabatic efficiency (total to total) between the CFD calculation and experiment is within 0.5%. The CFD slightly over-predicts the mass flow rate by approximately 0.4%. Above all, the overall performance of this 5-stage compressor is well predicted.

The predicted static pressure on the casing is compared with test data at the design and near stall points as shown in Figure 5. The relative axial distance position 0 corresponds to the IGV inlet, while position 1 represents the position of the stator 5 outlet. The casing static pressure of the CFD model and experiment is non-dimensionalized by the outlet casing static pressure of the CFD model at the design point. The maximum difference between the predicted casing static pressure and the test data is approximately 5%.

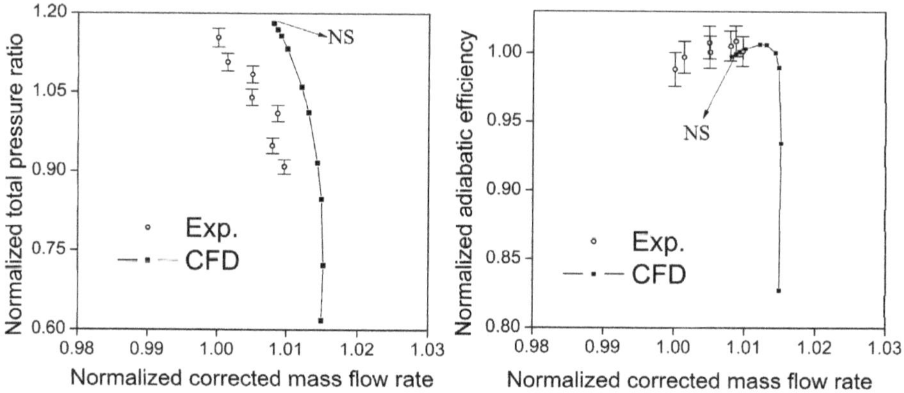

Figure 4. Measured and predicted overall performance. NS: Near Stall.

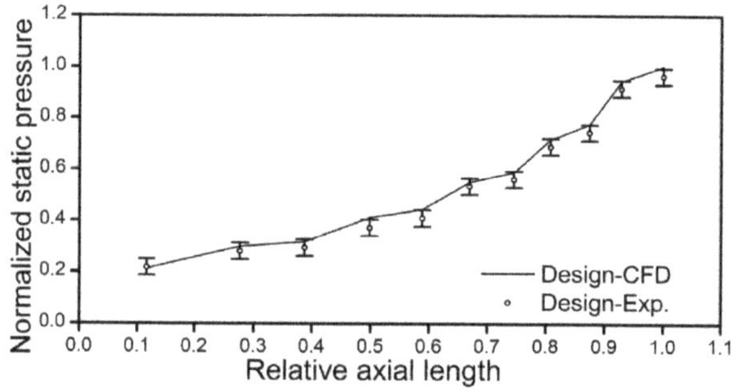

(**a**)

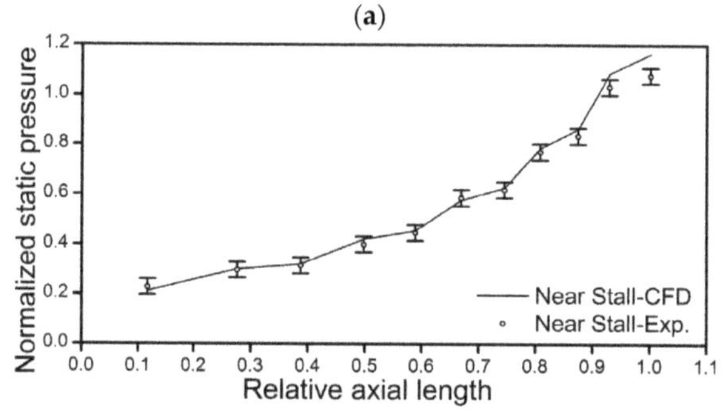

(**b**)

Figure 5. Measured and predicted casing static pressure. (**a**) Design point; (**b**) Near stall point. Exp.: Experiment.

Figure 6 shows the measured and predicted total pressure profiles at the leading edges of stator 4 and stator 5, where the total pressure is non-dimensionalized by the averaged outlet total pressure of the test rig at the design point. The agreement between the experimental and CFD results is satisfactory, except for a deviation at the tip region.

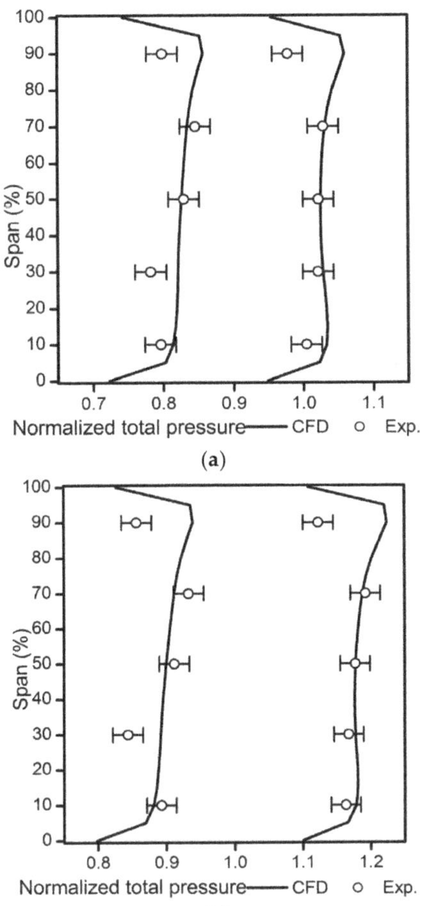

Figure 6. Measured and predicted total pressure profiles at the leading edges of stator 4 and stator 5. (**a**) Design point of stator 4 and stator 5; (**b**) Near stall point of stator 4 and stator 5.

To summarize, the multistage CFD methods applied in the present research can predict the overall compressor performance well. Furthermore, the prediction of the casing static pressure and spanwise total pressure profiles is good. CFD methods are applied for further aerodynamic analysis and aerodynamic redesign below.

The calculation results are analyzed to further examine the flow fields of the baseline compressor to propose the aerodynamic redesign strategies in the next section. In Figure 7, the limiting streamlines on the suction side of the compressor near the stall point indicate that no remarkable flow separation occurs, except in stator 5. It is obvious that stator 5 suffers separations on the suction side across the span, which causes the machine to stall. Figure 8 shows the streamwise distributions of the diffusion factor at the 20%, 50% and 80% spans for the design and near stall points. It is clear that the diffusion factors of the last two stages are much higher than in the other stages, especially for the stators. Additionally, the operating condition of the last stator is the most serious, especially for the hub and middle span regions, where the diffusion factors exceed 0.6. Furthermore, Figure 9 shows the surface static pressure rise coefficient distribution at the 20%, 50% and 80% spans of stator 5 for the design point and near the stall point. These results indicate that stator 5 operates in a high positive incidence even at the design point, which is obviously not a satisfied operating condition. At the near stall point, the incidence of the middle span is larger than that of the hub and tip, which is in agreement with Figure 8. According to the above analysis, the aerodynamic redesign of a baseline compressor should be the focus of the last stator to improve the performance and operating stability of the compressor.

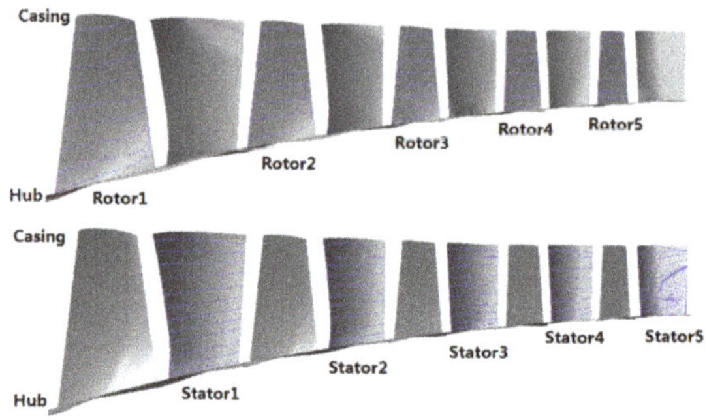

Figure 7. Limiting streamlines on the suction side of each blade at the near stall point.

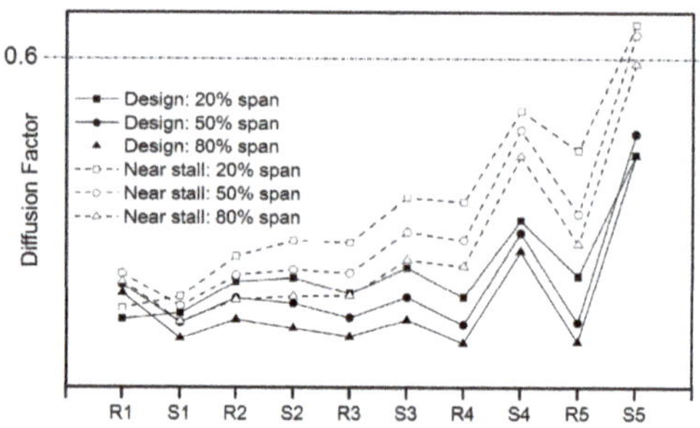

Figure 8. Streamwise diffusion factor distributions of each row.

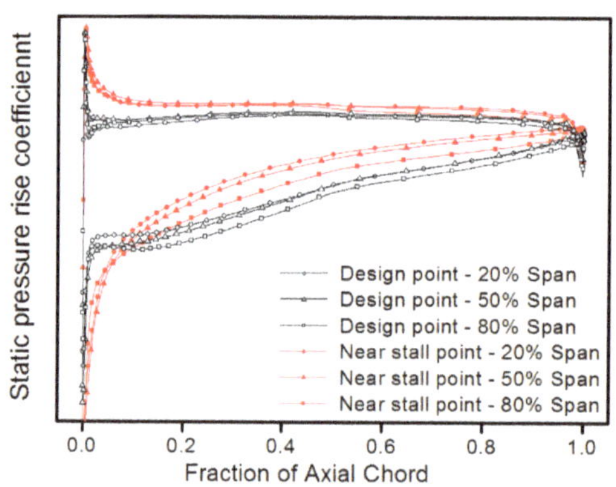

Figure 9. Surface static pressure rise coefficient distributions at the 20%, 50% and 80% spans of stator 5.

5. Three-Dimensional Redesign

In this section, three-dimensional redesign strategies are introduced to gain aerodynamic performance improvements, which will be implemented in the next upgrade version of this type of compressor.

The previously shown results indicate that stator 5 is the main contradiction for aerodynamic improvement. The total pressure ratio of the last stage is approximately 1.25. The blade loading diffusion factor of stator 5 is approximately 0.46, with a typical subsonic inlet Mach number of 0.48. In this paper, upgrades are mainly made at the last stator. Various three-dimensional blading techniques, including bow, leading edge re-camber (LER) and re-solidity, are adopted in the aerodynamic redesign of stator 5.

To verify the effect of the redesign on the entire machine, the complete redesign process of stator 5 is incorporated with the multi-stage CFD methods, which is different from many redesigns of isolated or part-stage blade rows. Moreover, the newly developed in-house 3-D blade toolkit "NT-Blade" is employed to modify the redesigned blade geometry.

5.1. Bow

To explore the potential of the aerodynamic improvements of a radially stacked straight blade of the baseline, the 3-D blade bow design technique is first employed as the redesign strategy of stator 5. A positive bow is defined as a positive lean angle at the hub and a negative lean angle at the shroud in comparison with a normal, radially stacked straight blade [18]. It is well acknowledged that the bow is broadly adopted to eliminate corner stall in compressor vanes [17,18]. The bow induces radial forces on the fluids and results in shifting the fluids from the end walls to the midspan, which can reduce endwall loading and postpone the tendency towards corner stall. However, there are still no recognized, well-established bow criteria in the 3-D blade design of compressors that lead to performance benefits, mainly because different compressors are dominated by different flow controlling mechanisms between the endwall and mainstream regions [21].

It is found that the symmetric positive hub and tip bow type applied by many researchers to eliminate endwall corner stall does not lead to a performance improvement for the compressor in this research. This is mainly because the separation in stator 5 of this research is mainstream dominated instead of endwall dominated. The iso-surface of the axial velocity (-1 m/s) of stator 5 is shown in Figure 10, and the blue zones represent the low-momentum fluids, which demonstrate that the flow separation in stator 5 of the baseline is mainstream dominated. The symmetric bow types off-load the endwall region and also increase the secondary flow to the mainstream, which may migrate the low-momentum fluids to a larger region along the span, resulting in poorer flow conditions. Additionally, the results of Figure 10 correspond to the near stall point marked in Figure 4.

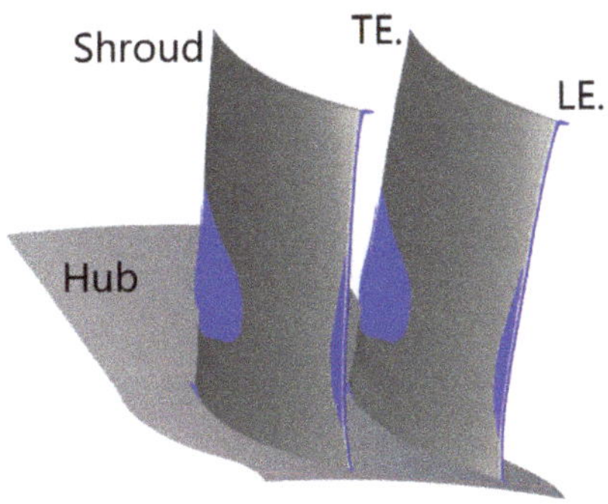

Figure 10. Iso-surface of the axial velocity (-1 m/s) for stator 5 of the baseline (near the stall points). TE.: trailing edge; LE.: leading edge.

Considering that the loss mechanism is mainstream dominated, to suppress the migration of low-momentum fluids and improve the high loss region with high-momentum fluids, a compromise is applied as the redesign principle to implement the bow strategy. A dramatically asymmetric bow configuration is applied in this paper, as shown in Figure 11. In the 50%–100% span of the tip region, a circular arc stacking line is adopted. The parameters of bow height H_b and bow angle α_b are applied to describe the bowed configuration. Providing enough radial force in the tip, a large circumferential displacement and 50% spanwise range of stacking (H_b = 0.5span, α_b = 40°) are applied in the tip region. These are expected to improve the low-momentum fluid region (in the 15%–60% span) with the upper high-momentum fluids.

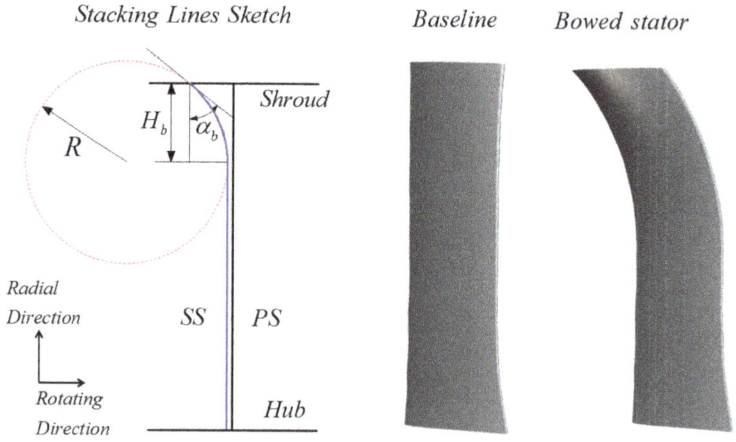

Figure 11. Stacking lines sketch and blade shapes of stator 5. *SS*: suction side; *PS*: pressure side.

The flows shifting from the tip to the hub along the span is evident in Figure 12, where the pitch-averaged inlet axial velocity profile of the bowed stator 5 becomes much more uniform compared with the baseline.

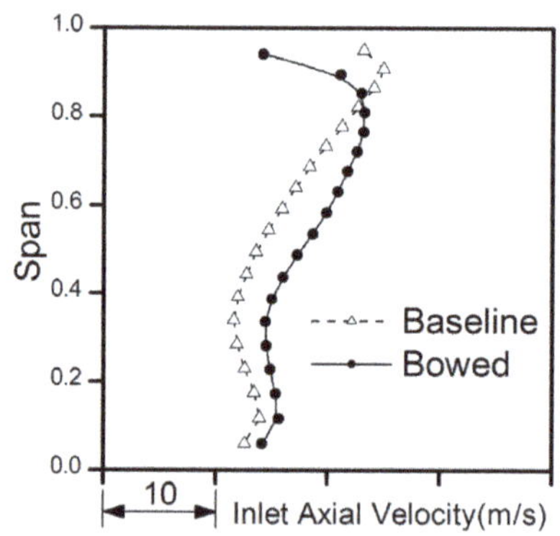

Figure 12. Spanwise distributions of the inlet axial velocity (near the stall points).

The iso-surface of the axial velocity (−1 m/s) of bowed stator 5 in Figure 13 shows that the low-momentum fluid zone near the trailing edge at the middle span disappeared compared to that of the baseline, as shown in Figure 10.

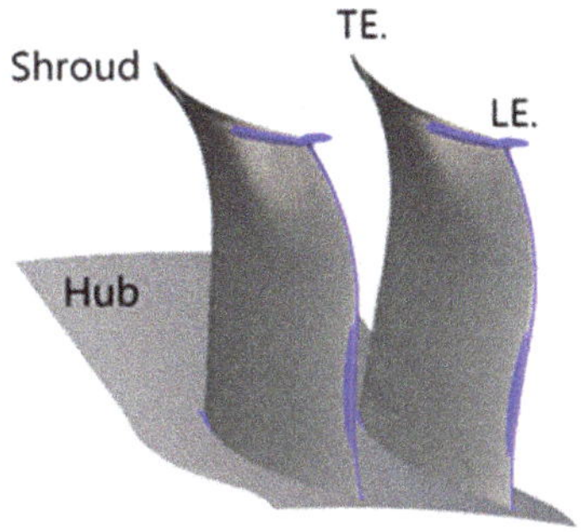

Figure 13. Iso-surface of the axial velocity (−1 m/s) for the bowed stator 5 (near the stall points).

Figure 14 demonstrates the relative Mach number and streamlines at the 35% span of the baseline and bowed stator 5. Compared with the baseline, the separations on the suction side completely vanished. The previously shown results show that the bow configuration helped to effectively improve the mainstream dominated low-momentum region.

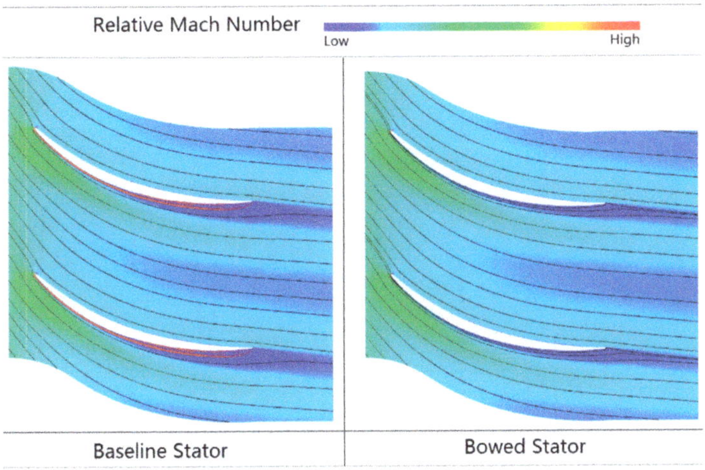

Figure 14. Relative Mach number contours and streamlines at the 35% span of stator 5 (near the stall points).

5.2. LER & Re-Solidity

According to previous aerodynamic analyses, the last stator of the baseline compressor operates at much higher unsatisfied incidences along the full span. Moreover, a larger incidence at the tip region was introduced by the dramatically asymmetric tip bow, and the high loss region in the middle span could not be completely eliminated by only applying the bow, as shown in Figure 15. To gain more performance improvement, other 3-D design methods, including LER and re-solidity, were added to the final redesigned strategy incorporated with the bow configuration.

LER alters the inlet blade angle in the partial chordwise range while maintaining the outlet blade angle to align the leading edge towards the flow direction. An increased incidence is introduced by the strong tip bow and high loss middle region, so the middle span and tip span regions are provided with larger inlet blade angles to reduce the high incidence, as demonstrated in Figure 16. Simultaneously, the pressure rise capability depends on both the incidence and solidity, especially for the endwall regions. Re-solidity is defined in this research as altering the solidity by changing the chord length while maintaining the thickness distribution. To prevent the tip and hub region flows from separation, the solidities in the endwall regions are increased, as illustrated in Figure 16. Additionally, the spanwise distribution of the solidity and inlet metal angle are parameterized using cubic spline curves.

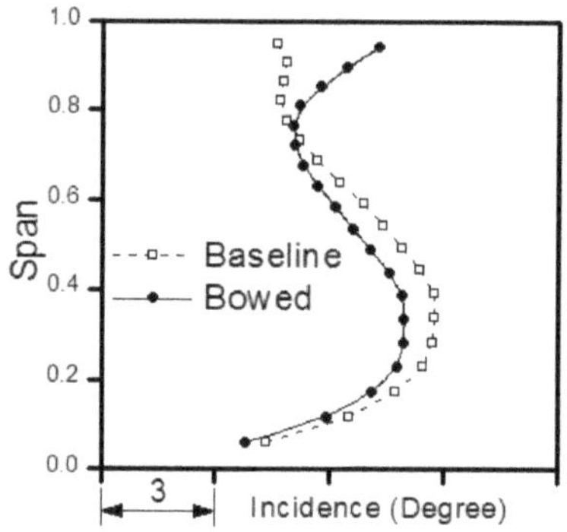

Figure 15. Spanwise distributions of the incidence at the design point.

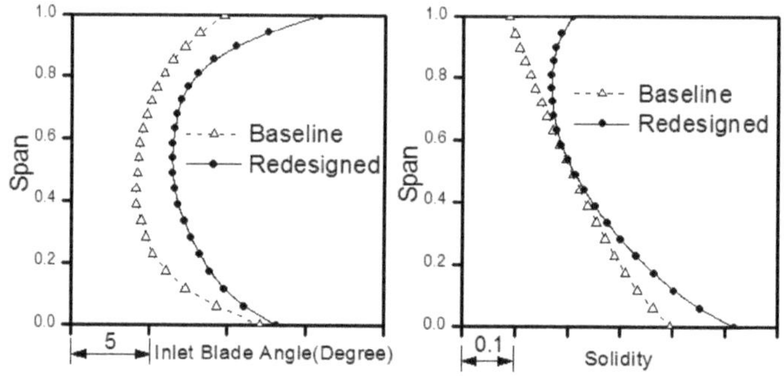

Figure 16. Inlet blade angle and solidity profiles of the baseline and redesigned stator 5.

Multi-stage CFD computations were conducted to evaluate the redesigned version, which applied the strong tip bow incorporated with LER and re-solidity. Figure 17 compares the incidence profiles between the baseline and redesigned stator 5 at the near stall points. The high incidence at the middle span was reduced by approximately three degrees, and the large inlet blade angle that increased in the tip region and that was introduced by re-camber prevented the strong tip bow from developing high incidence. A more reasonable spanwise incidence distribution and uniform loading, as well as the solidity redistribution, led to a significant loss reduction and performance improvement, as depicted in Figure 18. The total pressure loss of each S1 stream surface is defined as:

$$Loss = \frac{Pt_{in} - Pt_{out}}{Pt_{in} - Ps_{in}} \tag{1}$$

where Pt_{in} and Ps_{in} respectively denote the pitch-average total pressure and static pressure of each S1 stream surface leading edge (a short distance upstream blade). Pt_{out} denotes the pitch-average total pressure of each S1 stream surface trailing edge (a short distance downstream blade).

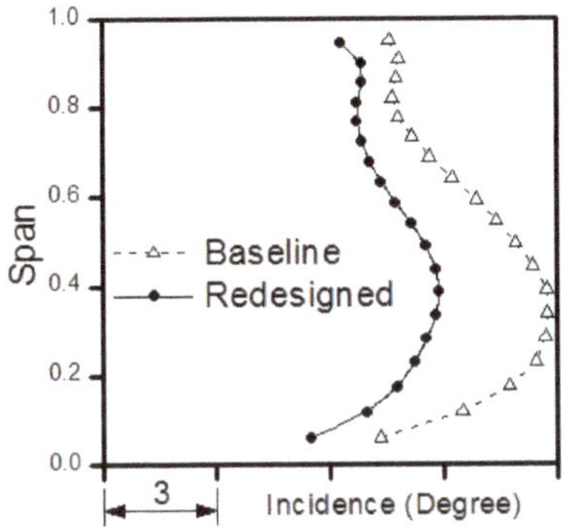

Figure 17. Spanwise distributions of the incidence for the baseline and redesigned stator 5 (near the stall points).

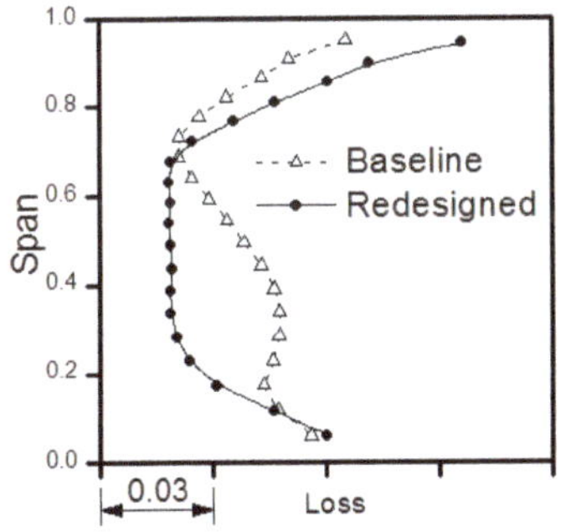

Figure 18. Loss profiles of the baseline and redesigned stator 5 (near the stall points).

To further substantiate the redesign effects, the total pressure contours downstream of the baseline and redesigned stator 5 at the near stall points are shown in Figure 19. Compared with the baseline, the high loss in the middle and lower spans are significantly reduced in the redesigned version. The iso-surface of the axial velocity (-1 m/s) of stator 5 in Figure 20 shows that the low-momentum fluid zone is reduced significantly compared with that of the baseline. The previously shown results illustrate that the bow configuration and the LER and re-solidity applied in the redesigned version help to effectively improve the low-momentum region.

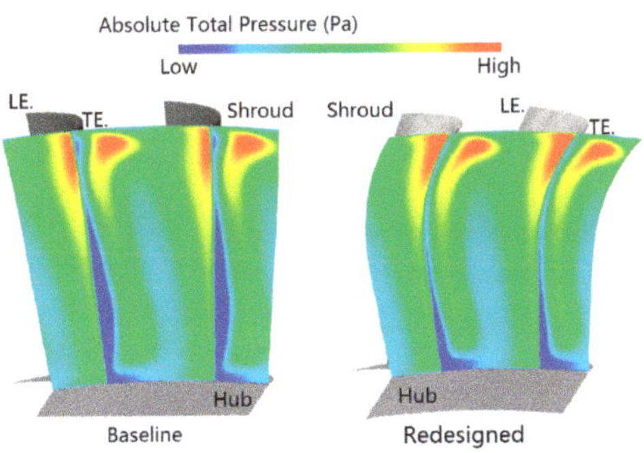

Figure 19. Total pressure contours downstream of the baseline and redesigned stator 5 (near the stall points).

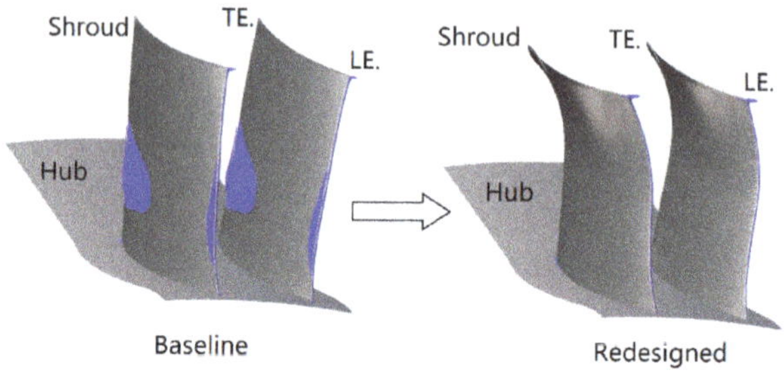

Figure 20. Iso-surface of the axial velocity (−1 m/s) for the baseline and redesigned stator 5 (near the stall points).

Finally, the overall performances are identified, as presented in Figure 21. Compared with the baseline, the three-dimensional redesign contributes an approximate 13% stall margin increase to the redesigned version. The stall margin is defined as follows:

$$Stall\ Margin = \frac{\varepsilon_{NS}}{\varepsilon_D} \times \frac{m_D}{m_{NS}} \tag{2}$$

where ε_D and ε_{NS} denote the total pressure ratio at the design point and at the near stall point, respectively, and m_D and m_{NS} denote the corrected mass flow rate at the design point and at the near stall point, respectively.

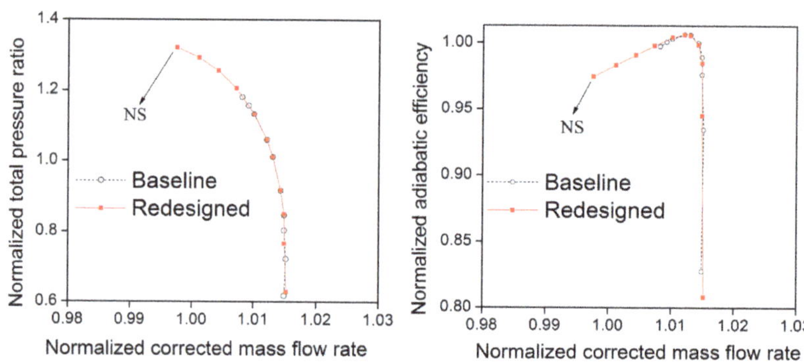

Figure 21. Overall performances.

Furthermore, the redesigned version also maintains the efficiency at the design point while providing a sufficient operating range, which proves that the redesign strategies are effective.

6. Conclusions

In this paper, the three-dimensional redesign of a stator blade is conducted in a 5-stage axial compressor with multi-stage CFD methods. The conclusions are as follows:

1. Validation with the full-scale test data is conducted for the multi-stage CFD methods prior to redesign. The multi-stage CFD methods applied in this research predict the overall performance quite well and also accurately capture the casing static pressure and stator leading edge total pressure profiles, both under design and off-design operating conditions.
2. The detailed aerodynamic analysis indicates that a large separation exists in the suction side of the last stator, resulting in stall. Thus, the follow-up redesign focuses on the last stator.

3. The mainstream flow separation dominates the loss, so a dramatically asymmetric positive bow incorporated with leading edge re-camber and re-solidity is applied for the redesigned version. The most significant achievement of the three-dimensional blade redesign in this research is the stall margin improvement, with a total 13% stall margin increase compared with the baseline, while maintaining efficiency at the design points.

This research explains the power of multi-stage CFD methods applied in the aerodynamic analysis and three-dimensional redesign of a modern highly loaded compressor in the multi-stage environment. The redesigning exploitation by bow, re-camber and re-solidity techniques provides various beneficial design philosophies of 3-D blading for the development of modern highly loaded compressors.

Acknowledgments: This work is supported by the Shenyang Engine Design and Research Institute (SEDRI), the Collaborative Innovation Center of Advanced Aero-Engine and the National Natural Science Foundation of China (Grant No. 51136003). The authors would like to thank SEDRI for permission to publish this paper.

Author Contributions: In this paper, Tao Ning developed the in-house 3-D blade toolkit "NT-Blade", proposed the blade optimization method, performed the aerodynamic analysis and wrote the manuscript; Chun-Wei Gu and Wei-Dou Ni provided guidance for the blade optimization and polished the manuscript; Xiao-Tang Li and Tai-Qiu Liu conducted part of the experiments of the full-scale test rig.

Conflicts of Interest: The authors declare no conflict of interest.

Nomenclature

ANNs	Artificial Neural Networks
ANSYS-CFX	CFD Code by ANSYS
APNASA	CFD Code by NASA
BBEW	Blended blade and endwall
CDA	Controlled Diffusion Airfoil
CFD	Computational Fluid Dynamics
FEM	Finite Element Method
IGV	Inlet Guided Vane
LE	Leading Edge
LER	Leading Edge Re-camber
MCA	Multiple Circular Arc
NASA	National Aeronautics and Space Administration
Numeca-Fine/Turbo	CFD Code by Numeca
PS	Pressure Surface
RANS	Reynolds-Averaged Navier-Stokes
SEDRI	Shenyang Engine Design and Research Institute
SS	Suction Side
TE	Trailing Edge
m_D	Corrected mass flow rate at the design point
m_{NS}	Corrected mass flow rate at the near stall point
Pt_{in}	Inlet pitch-average total pressure
Pt_{out}	Outlet pitch-average total pressure
Ps_{in}	Inlet pitch-average static pressure
$y+$	Dimensionless distance of the first grid node off the wall, $y+ = \rho u_\tau y_{wall}/\mu$
ε_D	Total pressure ratio at the design point
ε_{NS}	Total pressure ratio at the near stall point

References

1. Denton, J.D. Throughflow calculations for transonic axial flow turbines. *J. Eng. Power* **1978**, *100*, 212–218. [CrossRef]
2. Hirsch, C.; Denton, J.D. Through-flow calculations in axial turbomachines. Advisory Group for Aeronau-tical Research and Development (AGARD), Advisory Report, No.175. 1981.

3. Hale, A.A.; Davis, M.W.; Kneile, K.R. Turbine engine analysis compressor code: TEACC—Part I: Technical approach and steady results. In Proceedings of the AIAA 32nd Aerospace Sciences Meeting, Reno, NV, USA, 10–13 January 1994.
4. Adamczyk, J.J. Model Equation for Simulating Flows in Multistage Turbomachinery. In Proceedings of the 30th International Gas Turbine Conference and Exhibit, Houston, TX, USA, 1985; American Society of Mechanical Engineers (ASME): New York, NY, USA, 1985.
5. Adamczyk, J.J.; Mulac, R.A.; Celestina, M.L. A model for closing the inviscid form of the average-passage equation system. *ASME J. Turbomach.* **1986**, *108*, 180–186. [CrossRef]
6. Adamczyk, J.J.; Celestina, M.L.; Beach, T.A.; Barnett, M. Simulation of three-dimensional viscous flow within a multistage turbine. *ASME J. Turbomach.* **1990**, *112*, 370–376. [CrossRef]
7. Adamczyk, J.J. Aerodynamic analysis of multistage turbomachinery flows in support of aerodynamic design. *ASME J. Turbomach.* **2000**, *122*, 189–217. [CrossRef]
8. Mansour, M.; Hingorani, S.; Dong, Y. A new multistage axial compressor designed with the APNASA multistage CFD code: Part 1—Code calibration. In Proceedings of the ASME Turbo 2001: Power for Land, Sea, and Air, New Orleans, LA, USA, 4–7 June 2001; American Society of Mechanical Engineers (ASME): New York, NY, USA, 2001.
9. Dong, Y.; Mansour, M.; Hingorani, S.; Hayes, J. A new multistage axial compressor designed with the APNASA multistage CFD code: Part 2—Application to a new compressor design. In Proceedings of the ASME Turbo Expo 2001: Power for Land, Sea, and Air, New Orleans, LA, USA, 4–7 June 2001; American Society of Mechanical Engineers (ASME): New York, NY, USA, 2001.
10. Wellborn, S.R.; Delaney, R.A. Redesign of a 12-stage axial-flow compressor using multistage CFD. In Proceedings of the ASME Turbo Expo 2001: Power for Land, Sea, and Air, New Orleans, LA, USA, 4–7 June 2001; American Society of Mechanical Engineers (ASME): New York, NY, USA, 2001.
11. Mansour, M.L.; Gunaraj, J.; Goswami, S. Validation of steady average-passage and mixing plane CFD approaches for the performance prediction of a modern gas turbine multistage axial compressor. In Proceedings of the ASME Turbo Expo 2008: Power for Land, Sea, and Air, Berlin, Germany, 9–13 June 2008; American Society of Mechanical Engineers (ASME): New York, NY, USA, 2008; Volume 6, pp. 393–402.
12. Belamri, T.; Galpin, P.; Braune, A.; Cornelius, C. CFD analysis of a 15 stage axial compressor: Part I—Methods. In Proceedings of the ASME Turbo Expo 2005: Power for Land, Sea, and Air, Reno-Tahoe, NV, USA, 6–9 June 2005; American Society of Mechanical Engineers (ASME): New York, NY, USA, 2005; Volume 6 (Part B), pp. 1001–1008.
13. Belamri, T.; Galpin, P.; Braune, A.; Cornelius, C. CFD analysis of a 15 stage axial compressor: Part II—Results. In Proceedings of the ASME Turbo Expo 2005: Power for Land, Sea, and Air, Reno-Tahoe, NV, USA, 6–9 June 2005; American Society of Mechanical Engineers (ASME): New York, NY, USA, 2005; Volume 6 (Part B), pp. 1009–1017.
14. Terauchi, K.; Kariya, D.; Maeda, S.; Yoshiura, K. Redesign of an 11-stage axial compressor for industrial gas turbine. In Proceedings of the ASME Turbo Expo 2005—Gas Turbine Technology: Focus for the Future, Reno-Tahoe, NV, USA, 6–9 June 2005; American Society of Mechanical Engineers (ASME): New York, NY, USA, 2005; Volume 6 (Part A), pp. 261–267.
15. Ikeguchi, T.; Matsuoka, A.; Sakai, Y.; Sakano, Y.; Yoshiura, K. Design and development of a 14-stage axial compressor for industrial gas turbine. In Proceedings of the ASME Turbo Expo 2012: Turbine Technical Conference and Exposition, Copenhagen, Denmark, 11–15 June 2012; American Society of Mechanical Engineers (ASME): New York, NY, USA, 2012; Volume 8, pp. 125–134.
16. Brilliant, L.; Burger, G.; Balamucki, S.; Dong, Y.; Lejambre, C. Application of multistage CFD analysis to low pressure compressor design. In Proceedings of the ASME Turbo Expo 2004, Vienna, Austria, 14–17 June 2004; American Society of Mechanical Engineers (ASME): New York, NY, USA, 2004; Volume 5, pp. 645–653.
17. Ren, X.D.; Xu, K.; Shyy, W.; Gu, C.W. A multi-dimensional high-order discontinuous Galerkin method based on gas kinetic. *J. Comput. Phys.* **2015**, *292*, 176–193. [CrossRef]
18. Ren, X.D.; Gu, C.W. Investigation of compressor tip clearance flow based on the discontinuous galerkin methods. In Proceedings of the ASME Turbo Expo 2013: Turbine Technical Conference and Exposition, San Antonio, TX, USA, 3–7 June 2013; American Society of Mechanical Engineers (ASME): New York, NY, USA, 2013.

19. Ren, X.D.; Gu, C.W. Application of a discontinuous Galerkin method on the compressible flow in the transonic axial compressor. *Appl. Therm. Eng.* **2015**, *93*, 707–717. [CrossRef]
20. Weingold, H.; Neubert, R.; Behlke, R.; Potter, G. Bowed stators: An example of CFD applied to improve multistage compressor efficiency. *ASME J. Turbomach.* **1997**, *119*, 161–168. [CrossRef]
21. Fischer, A.; Riess, W.; Seume, J.R. Performance of strongly bowed stators in a four-stage high-speed compressor. *ASME J. Turbomach.* **2004**, *126*, 333–338. [CrossRef]
22. Denton, J.; Xu, L. The exploitation of three-dimensional flow in turbomachinery design. *Proc. Inst. Mech. Eng. C* **1998**, *213*, 125–137. [CrossRef]
23. Gallimore, S.J.; Bolger, J.J.; Cumpsty, N.A. The use of sweep and dihedral in multistage axial flow compressor blading, Part 2: Low and high speed designs and test verification. *ASME J. Turbomach.* **2002**, *124*, 533–541. [CrossRef]
24. Ji, L.C.; Tian, Y.; Li, W.; Yi, W.; Wen, Q. Numerical studies on improving performance of rotor-67 by blended blade and endwall technique. In Proceedings of the ASME Turbo Expo 2012: Turbine Technical Conference and Exposition, Copenhagen, Denmark, 11–15 June 2012; American Society of Mechanical Engineers (ASME): New York, NY, USA, 2012; Volume 8, pp. 135–145.
25. Denton, J.D. Some limitations of turbomachinery CFD. In Proceedings of the ASME Turbo 2010: Power for Land, Sea, and Air, Glasgow, UK, 14–18 June 2010; American Society of Mechanical Engineers (ASME): New York, NY, USA, 2010; Volume 7, pp. 735–745.
26. Pierret, S.; Coelho, R.F.; Kato, H. Multidisciplinary and multiple operating points shape optimization of three-dimensional compressor blades. *Struct. Multidiscip. Optim.* **2007**, *33*, 61–70. [CrossRef]
27. Zhao, Q.J.; Zhou, X.Y.; Xiang, X.Y. Multi-Objective optimization of groove casing treatment in a transonic compressor. *Proc. Inst. Mech. Eng. A* **2014**, *228*, 626–637. [CrossRef]
28. Zhang, Y.Y.; Huang, D.G.; Sun, X.J.; Wu, G.Q. Exploration in optimal design of an airfoil with a leading edge rotating cylinder. *J. Therm. Sci.* **2010**, *19*, 318–325. [CrossRef]
29. NUMECA International. Available online: http: //www.numeca.com (accessed on 6 June 2015).
30. Spalart, P.R.; Allmaras, S.R. A one equation turbulence model for aerodynamic flows. *Rech. Aerosp.* **1994**, *1*, 5–21.

energies

Article

Control and Optimization of a Variable-Pitch Quadrotor with Minimum Power Consumption

Shouzhao Sheng * and Chenwu Sun

College of Automation Engineering, Nanjing University of Aeronautics and Astronautics, 29 YuDao St., Nanjing 210016, China; sunchenwu@nuaa.edu.cn

* Correspondence: shengsz@nuaa.edu.cn; Tel.: +86-25-8489-2305

Academic Editor: K.T. Chau

Received: 15 February 2016; Accepted: 18 March 2016; Published: 24 March 2016

Abstract: Recently, there has been a rapid growth of interest in quadrotors with electric variable-pitch propellers. The control and optimization of such propellers are important factors for improving the flight performance of the vehicles. Therefore, the steady-state identification method to estimate the parameters of the mathematical model of the electric variable-pitch propeller is developed. The steady-state control and optimization scheme with minimum power consumption and the adaptive compensation scheme for the variable-pitch propeller are then proposed, based on which the response performance of the lift force produced by the variable-pitch propeller can be greatly improved by using a cascade compensation scheme. Furthermore, the direct lift-based flight control strategy is presented, which can significantly contribute to the improvement of the flight performance, precisely because the roll, pitch, yaw and vertical channels of the variable-pitch quadrotor are approximately linearized and completely decoupled from each other in this case. The experimental results demonstrate that both the endurance performance and the positioning accuracy of the variable-pitch quadrotor are improved simultaneously by using the proposed method with minimum power consumption.

Keywords: minimum power consumption; steady-state optimization; steady-state identification; adaptive compensation; variable-pitch quadrotor

1. Introduction

With the development of small and micro Unmanned Aerial Vehicles (UAVs) in recent years, small, multi-rotor UAVs equipped with electric propellers have gained considerable momentum, and are widely used as experimental and hobby platforms because of their simple mechanical structure, good operability, maneuverability and agility. Therefore, multi-rotor UAVs have recently attracted great interest. Also, considerable work exists on various modeling, design, control, and optimization schemes for multi-rotor UAVs [1–9].

Fixed-pitch multi-rotor designs are mechanically simple. The stability and flight control of fixed-pitch multi-rotors are well established [10–17]. However, the only way to change the lift force produced by a fixed-pitch propeller is by changing the voltage to the motor, which restricts the aggressive and aerobatic maneuvers that multi-rotor UAVs can perform, therefore limiting the applicability of fixed-pitch multi-rotor UAVs in agility-intensive missions [18]. Variable-pitch multi-rotor UAVs can largely overcome the limitations resulting from the fixed-pitch flight.

The benefits of variable-pitch propellers over fixed-pitch propellers for a quadrotor have been analyzed in [19]. The addition of variable-pitch propellers to a quadrotor platform results in an additional degree of freedom for changing the lift produced by each motor-propeller combination. With a variable-pitch propeller, the lift can be changed by either changing the propeller pitch or by changing the rotational speed. These two actuators, to a large extent, overlap, and there are many combinations of rotational speed and propeller pitch that yield an identical lift. The number of possible combinations is mainly limited by aerodynamic constraints, the maximum propeller pitch and the maximum available motor power. It can be seen from the discussion thus far that on the one hand, the combination could be adjusted to more power efficient settings as the desired lift increases or decreases, and on the other hand, only one combination can yield a desired lift with minimum power consumption. Thus, the control allocation problem of which actuator to use, propeller pitch or rotational speed or combination of the two, needs to be explored systematically under the precondition of ensuring the flight performance, which can certainly contribute to the improvement of the endurance performance of quadrotors.

The modeling and control of the variable-pitch quadrotors have been explored only recently, yet little literature or research is available until now. The design, development, and control of a variable-pitch quadrotor have been studied in [20]. The problem of characterizing the dynamics of a variable-pitch quadrotor from data has been discussed, and black-box versus grey-box models have been analyzed in [21]. The variable-pitch model and the nonlinear proportional squared control algorithm for a quadrotor have also been implemented in the quaternion space [22]. The design of a variable-pitch quadrotor with constant motor speed has been investigated, which has also proven to be most effective in increasing the maneuverability of the quadrotor while largely maintaining its mechanical simplicity [23]. Control and trajectory generation algorithms for a variable-pitch quadrotor have been presented both theoretically and experimentally. The control law is not based on near-hover assumptions, allowing for large attitude deviations from hover [24]. Nevertheless, that research work mainly focused on the aerobatic maneuvers of electric variable-pitch quadrotors. The control and optimization of quadrotors with minimum power consumption has not, to date, been addressed in detail under the precondition of ensuring the flight performance.

The purpose of this study is therefore to develop a proper control and optimization strategy with minimum power consumption for the variable-pitch quadrotor, based on which the endurance performance and the positioning accuracy can be improved. The efficiency and superiority of the proposed method is to be verified by a series of experimental tests.

The structure of the paper is as follows: first, the control and optimization problems are addressed in Section 2, and the steady-state identification method for the variable-pitch propeller is developed in Section 3. Then, the control and optimization strategy for the variable-pitch propeller is presented in Section 4, followed by the direct lift-based flight control strategy for the variable-pitch quadrotor in Section 5. Experimental results are shown in Section 6. Finally, conclusions are drawn in Section 7.

2. The Control and Optimization Problem Statement of the Variable-Pitch Quadrotor

The prototype quadrotor vehicle chosen for this study, 0.5 m in height and 1.45 m in opposite axial distance, is depicted in Figure 1. A maximum take-off mass of 15 kg is assumed including 7 kg of maximum load. The maximum climb speed is approximately 8 m/s while the maximum descent speed is about 13 m/s. The vehicle is used specifically because of the simple mechanical structure, good operability, maneuverability and agility. As the core of the vehicle, the propulsion system mainly consists of an advanced lithium battery, four brushless DC motors, four variable-pitch propellers, and torque-transmitting mechanisms. The variable-pitch propellers driven by the electric motors through the torque-transmitting mechanisms can provide sufficient lift force to lift the vehicle weighing from 8 kg for no load to 15 kg for full load according to the mission equipment.

Figure 1. The variable-pitch quadrotor.

The quadrotor has six degrees of freedom. Let ϕ,θ,ψ represent the roll, pitch and yaw angles of the quadrotor in the inertial frame. Let v_x, v_y, v_z represent the flight velocities of the quadrotor in the body frame. Considering that the flight velocities v_x, v_y are dependent on the attitude θ, ϕ in flight control system design, for simplicity, only the control schemes for ϕ,θ,ψ and v_z are investigated in this study.

The four propellers driven by each motor constitute the propulsion system of the quadrotor. Therefore, the quadrotor has only four control inputs to produce the desired roll, pitch and yaw moments, and the desired vertical movement, as shown in Figure 2, where α_j, ω_j and L_j denote the propeller pitch, the rotational speed and the lift force. Specially, the numerical subscript j indicates different variable-pitch propellers in this paper.

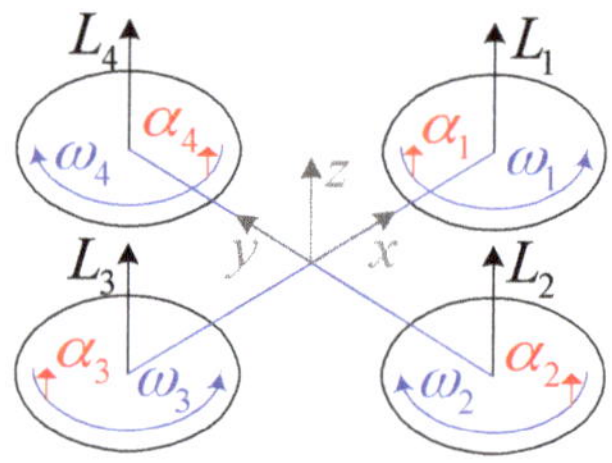

Figure 2. The control inputs and the frames of reference for the variable-pitch quadrotor.

As for the jth control input, the combination of the rotational speed ω_j and the propeller pitch α_j produces the lift force L_j in the direction of the propeller axis as follows [25]:

$$L_j = b_L \alpha_j \omega_j^2 \tag{1}$$

where b_L is an aerodynamic constant, whose estimate, $\hat{b}_L$, can be obtained by using least-squares regression. The total lift force required to produce the movement along the body z axis is then given by:

$$T = \sum_{j=1}^{4} L_j \tag{2}$$

Assuming the variable-pitch quadrotor is near hover, the drag produced by the motor-propeller combination can then be approximately modeled by [25,26]:

$$Q_j = b_{D_1} \alpha_j^2 \omega_j^2 + b_{D_2} \omega_j^2 + b_{D_3} \alpha_j \omega_j \tag{3}$$

where b_{D1}, b_{D2} and b_{D3} can be considered as aerodynamic constants. The torques used to control the roll, pitch and yaw moments can then be described by:

$$T_\phi = l\,(L_4 - L_2) \tag{4a}$$

$$T_\theta = l\,(L_3 - L_1) \tag{4b}$$

$$T_\psi = Q_1 + Q_3 - Q_4 - Q_2 \tag{4c}$$

where l is the horizontal distance from the propeller center to the center of gravity of the quadrotor.

The use of control and optimization of variable-pitch propellers is desirable for a high-performance quadrotor, mainly because on the one hand, the characteristics of the variable-pitch propeller strongly affect the flight dynamics, stability and endurance performance of the vehicle; while on the other hand, any great changes in the gross flight weight with the loads and the flight speed also require the control system to be more complex and more exact. Obviously each motor and propeller has the same characteristics and can be controlled independently. To avoid repetition, the present study is focused on any pre-specified motor and propeller to demonstrate the proposed control algorithm, which, without loss of generality, can apply to others. For simplicity, the subscript j, denoting the jth motor and propeller, will be omitted in the following two sections, except when there is ambiguity.

3. The Steady-state Identification Method of the Electric Variable-pitch Propeller

The analysis of the basic principle model is crucial for the development of a proper control and optimization scheme to improve the performance of the variable-pitch quadrotor.

The motor is normally modeled by a circuit containing a resistor, inductor, and voltage generator. With regard to the aim of this study, the inductance of a small, brushless hobby motor is negligible when compared to the physical response of the system and so can be ignored. The motor terminal voltage is then approximately given by [27]:

$$u = Ri + \frac{\omega}{K_V} \tag{5}$$

where u is the armature voltage, i is the armature current, R is the armature coil resistance, K_V is the voltage constant, ω is the rotational speed of the motor.

Generally, the motor torque, Q_{m}, is assumed proportional to the difference between the armature current, i, and the no-load current, i_0, through the torque constant, K_Q:

$$Q_{\mathrm{m}} = K_Q\left(i - i_0\right) \tag{6}$$

The dynamics of the variable-pitch propeller can then be modeled as a simple first order differential equation:

$$I\dot{\omega} = Q_{\mathrm{m}} - Q \tag{7}$$

where I is the inertia including the motor and propeller.

Substituting Equations (3), (5) and (6) into Equation (7) yields the following nonlinear differential equation for the electric variable-pitch propeller:

$$I\dot{\omega} + \left(b_{D_1}\alpha^2 + b_{D_2}\right)\omega^2 + \left(b_{D_3}\alpha + b_{D_4}\right)\omega = bu + b_0 \tag{8}$$

where $b_{D_4} = \dfrac{K_Q}{RK_V}$, $b = \dfrac{K_Q}{R}$ and $b_0 = -K_Q i_0$. Both the armature voltage, u, and the propeller pitch, α, have a substantial effect on the rotational speed of the propeller, thus having a large impact on the lift force of the propeller, which will be discussed in sufficient details in the following section.

However, the control and optimization of the electric variable-pitch propeller is a difficult and often not very precise task, mainly because the coefficients of the differential equation are unknown. As for the steady-state operation, we have:

$$\gamma\left(\alpha, u\right)\Theta = bu \tag{9}$$

where $\gamma\left(\alpha, u\right) = \left[\alpha^2\varpi^2\left(\alpha, u\right), \varpi^2\left(\alpha, u\right), \alpha\varpi\left(\alpha, u\right), \varpi\left(\alpha, u\right), -1\right]$, $\Theta = \left[b_{D_1}, b_{D_2}, b_{D_3}, b_{D_4}, b_0\right]^{\mathrm{T}}$, and $\varpi\left(\alpha, u\right) = \lim_{t\to\infty} \omega\left(t\right)\Big|_{\substack{\alpha = \text{constant},\\ u = \text{constant}}}$ is the steady-state response to constant inputs.

Considering that the motor rotates in a single direction, which provides sufficient thrust force to lift the quadrotor, the commutation interval angle can be expressed as follows:

$$\vartheta\left(t_k\right) - \vartheta\left(t_{k-1}\right) = 60/N_{\mathrm{p}} \tag{10}$$

where ϑ denotes the rotational angle; t_{k-1}, t_k denote the $(k-1)$th and $k$thcommutation instants, respectively, both of which are available from the electric speed controller (ESC) of the motor; N_p is the number of pole pairs. The steady-state output, ϖ, can therefore be accurately estimated by:

$$\varpi \approx \frac{60\ell}{N_p\left(t_k - t_{k-\ell}\right)} \tag{11}$$

where ℓ is an optional integral number. On the basis of the discussion made thus far, we can present the steady-state identification algorithm to estimate the coefficients in Equation (8) as follows:

The Steady-state Identification Algorithm:

Input:

The steady-state sample set $S = \{(\alpha_k, u_k, \varpi(\alpha_k, u_k))\}_{k=1}^{n}$, *where the subscript,* k , *denotes the sample number, and* $n \geqslant 5$. *Moreover, The chosen* S *satisfies the constraint condition* $\text{cond}\left(\mathbf{R}^{\mathrm{T}}\mathbf{R}\right) < c_0$, *where* $\mathbf{R} = \begin{bmatrix} \gamma(\alpha_1, u_1) \\ \vdots \\ \gamma(\alpha_n, u_n) \end{bmatrix}$, *and* c_0 *is an optional positive value.*

Output:

$\hat{\Theta} = \left[\hat{b}_{D_1}, \hat{b}_{D_2}, \hat{b}_{D_3}, \hat{b}_{D_4}, \hat{b}_0\right]^{\mathrm{T}}$ *and* $\hat{b}$, *which are the estimates of* Θ *and* b, *respectively.*

Steps:

(1) *Calculate* $\hat{\Theta}_b = \left(\mathbf{R}^{\mathrm{T}}\mathbf{R}\right)^{-1}\mathbf{R}^{\mathrm{T}}\mathbf{U}$, *where* $\mathbf{U} = \begin{bmatrix} u_1 \\ \vdots \\ u_n \end{bmatrix}$;

(2) *Calculate* $\hat{b} = \frac{1}{n}\sum_{k=1}^{n}\frac{\gamma(\alpha_k, u_k)\hat{\Theta}_b}{u_k}$;

(3) *Calculate* $\hat{\Theta} = \hat{\Theta}_b\hat{b}$

As for the DUALSKY® XM5060EA-9 (Shanghai Dualsky Models Co., Ltd., Shanghai, China) brushless motor and the 16 inch diameter propeller used in this study, the estimates of the coefficients of the nonlinear differential equation are shown in Table 1 for reference.

Table 1. The estimates of the coefficients of the electric variable-pitch propeller.

Configuration	Variable	Value
XM5060EA-9 motor and 16 inch diameter propeller	$\hat{b}_{D_1}$	2.73×10^{-8}
	$\hat{b}_{D_2}$	1.13×10^{-6}
	$\hat{b}_{D_3}$	2.93×10^{-5}
	$\hat{b}_{D_4}$	2.37×10^{-2}
	$\hat{b}_0$	-2.97×10^{-2}
	$\hat{b}$	0.78

4. The Control and Optimization Strategy for the Electric Variable-Pitch Propeller

4.1. The Steady-State Control and Optimization Scheme with Minimum Power Consumption

As shown in Equation (1), the lift force can be changed by either changing the rotational speed or by changing the propeller pitch. There are many rotational speed and propeller pitch combinations that yield the same lift force values. The optimization of the motor power consumption among the combinations is therefore essential to improve the endurance of the variable-pitch quadrotor. The motor power, P, can be given by:

$$P \approx i^2 R + Q_p \omega \tag{12}$$

The online optimization of the motor power is often an undecidable problem, which, however, can be approximately converted into the minimization of the input voltage to the motor and propeller under the steady-state condition [18]. The solution to the optimization problem can therefore be formulated as follows:

$$(\alpha_c, \omega_c) \approx \underset{L=L_c}{\arg\min}\, u \tag{13}$$

where α_c, ω_c are the sub-optimized propeller pitch command and the sub-optimized rotational speed command of the electric variable-pitch propeller, respectively, and L_c is the lift force command from the flight controller discussed in Section 5.

Not considering the estimation error of the coefficients of the nonlinear differential in Equation (8), the steady-state component of the input voltage can be derived as follows:

$$u_0 = \hat{b}^{-1}\left\{L_c\hat{b}_L^{-1}\left(\hat{b}_{D_1}\alpha_c + \hat{b}_{D_2}\alpha_c^{-1}\right) + L_c^{0.5}\hat{b}_L^{-0.5}\left(\hat{b}_{D_3}\alpha_c^{0.5} + \hat{b}_{D_4}\alpha_c^{-0.5}\right) - \hat{b}_0\right\} \tag{14}$$

According to Equations (1) and (14), the steady-state optimization problem can then be formulated as:

$$\alpha_c \approx \underset{L=L_c}{\arg\min}\left\{L_c^{0.5}\left(\hat{b}_{D_1}\alpha_c + \hat{b}_{D_2}\alpha_c^{-1}\right) + \hat{b}_L^{0.5}\left(\hat{b}_{D_3}\alpha_c^{0.5} + \hat{b}_{D_4}\alpha_c^{-0.5}\right)\right\} \tag{15a}$$

$$\omega_c = \sqrt{\frac{L_c}{\hat{b}_L\alpha_c}} \tag{15b}$$

We therefore conclude that the sub-optimized propeller pitch command, α_c, satisfies the following condition:

$$2\ell_c\left(\hat{b}_{D_1}\beta_c^4 - \hat{b}_{D_2}\right) = \hat{b}_\ell\left(\hat{b}_{D_4}\beta_c - \hat{b}_{D_3}\beta_c^3\right) \tag{16}$$

where $\ell_c = L_c^{0.5}, \hat{b}_\ell = \hat{b}_L^{0.5}$ and $\beta_c = \alpha_c^{0.5}$. It is worth noting that α_c can be obtained online by using the iterative algorithm discussed below. Figure 3 shows the change of the sub-optimized propeller pitch command α_c with the lift force command L_c.

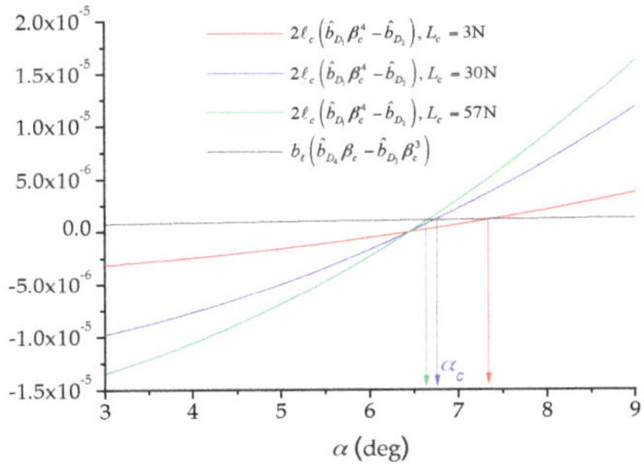

Figure 3. The variation of the sub-optimized propeller pitch command with the lift force command.

The steady-state estimates of the lift force L and the motor power P, as functions of the rotational speed ω and the pitch angle α, can be obtained by using numerical analyses, as shown in Figure 4. The plots show how the lift can be increased by either increasing the rotational speed, increasing pitch, or by increasing both. Given a curve of constant lift force, only one curve of minimum constant motor power remains tangent to it. The black curve indicates the optimal $\alpha - \omega$ trajectory with minimum power consumption, and the sub-optimized propeller pitch command α_c displayed with "X" is comparatively shown in Figure 4. We can conclude that the difference between the sub-optimized propeller pitch and the optimal propeller pitch is insignificant.

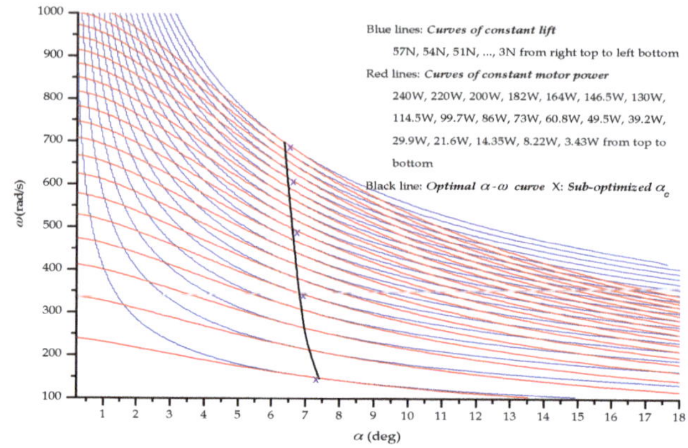

Figure 4. The lift force and the motor power dependences on the rotational speed and the propeller pitch.

Therefore, the motor power consumption is approximately the lowest while changing the propeller pitch and the rotational speed according to Equation (15). Referring to Equation (16) and Figure 3, we then present the online iterative algorithm to optimize the propeller pitch command and the rotational speed command quickly:

The Online Iterative Algorithm:

Input:

$L_c(t_k)$, $\alpha_c(t_{k-1})$ and $\Delta\alpha_c$, where t_{k-1}, t_k denote the $(k-1)$ th and kth control instants, respectively; $\Delta\alpha_c$ is the optional minimum increment of α_c.

Output:

$\alpha_c(t_k), \omega_c(t_k)$.

Steps:(1)

(1) $\alpha_c(t_k) \leftarrow \alpha_c(t_{k-1})$, and $i_+, i_- = 0$;
(2) If $i_+ > 0$ and $i_- > 0$, then
Go to Step 4;
Else
Go to Step 3;
End
(3) If $2\ell_c\left(\hat{b}_{D_1}\beta_c^4 - \hat{b}_{D_2}\right) > \hat{b}_\ell\left(\hat{b}_{D_4}\beta_c - \hat{b}_{D_3}\beta_c^3\right)$, then
$\alpha_c(t_k) \leftarrow \alpha_c(t_k) - \Delta\alpha_c$, $i_- \leftarrow i_- + 1$;
Else
$\alpha_c(t_k) \leftarrow \alpha_c(t_k) + \Delta\alpha_c$, $i_+ \leftarrow i_+ + 1$;
End
Return to Step 2;
(4) Calculate $\omega_c(t_k)$.

4.2. The Adaptive Compensation Scheme

We now assume that:

$$u = u_0 + u_a \tag{17}$$

where the adaptive compensation component of the input voltage, u_a, is given by:

$$u_a = \hat{b}^{-1}\left(\hat{b}_{D_1}\alpha_c^2 + \hat{b}_{D_2}\right)\left(\omega^2 - \omega_c^2\right) + \hat{b}^{-1}\left(\hat{b}_{D_3}\alpha_c + \hat{b}_{D_4} - k\right)(\omega - \omega_c) - \hat{b}^{-1}\tilde{k}\omega \tag{18}$$

where k and $\tilde{k}$ are adjustable parameters. Considering the estimates of the coefficients in Table 1 and the small change of α_c in Figure 3, u_a should be insignificant, as long as $k \approx \hat{b}_{D_4}, \tilde{k} \approx 0$, and ω is close to ω_c within the normal operating rotational speed range, thus having little effect on the optimization of the electric variable-pitch propeller. Substituting Equations (14), (17) and (18) into Equation (8) yields:

$$\dot{\omega} + \left(k_a + \tilde{k}_a + \varepsilon_b\right)\omega = k_a\omega_c \tag{19}$$

where $k_a = \hat{I}_b^{-1}k$, $\tilde{k}_a = \hat{I}_b^{-1}\tilde{k}$, $\varepsilon_b = \left\{\left(\hat{b}_{D_1}\alpha_c^2 + \hat{b}_{D_2}\right)\omega + \left(\hat{b}_{D_3}\alpha_c + \hat{b}_{D_4}\right) - b\omega^{-1}\right\}\left(I^{-1} - \hat{I}_b^{-1}\right)$, and $\hat{I}_b = I\hat{b}b^{-1}$. Note that ε_b is also insignificant.

Considering that u_a should be insignificant within the normal operating rotational speed range, the ideal model is therefore defined as:

$$\dot{\omega}_m + \hat{k}_m\omega_m = \hat{k}_m\omega_c \tag{20}$$

where $\hat{k}_m = \hat{I}^{-1}\hat{b}_{D_4}$, and $\hat{I}$ is the estimate of I.

The tracking error between the outputs of the plant and the model is then defined as:

$$e = \omega_m - \omega \tag{21}$$

Now, we consider the Lyapunov function candidate:

$$V = e^2 + \lambda_a\delta_a^2 + \tilde{\lambda}_a\tilde{\delta}_a^2 \tag{22}$$

where $\delta_a = \hat{k}_m - k_a, \tilde{\delta}_a = \tilde{k}_a + \varepsilon_b$, and $\lambda_a, \tilde{\lambda}_a$ are optional positive values. The derivative of the Lyapunov function candidate is then negative definite as the adaptive laws are presented as follows:

$$\begin{cases} \dot{k} = \gamma_a\,(e + \mu)\,e_0 \\ \dot{\tilde{k}} = -\tilde{\gamma}_a\,(e + \mu)\,\omega \end{cases} \tag{23}$$

where $\gamma_a = \lambda_a^{-1}\hat{I}_b, \tilde{\gamma}_a = \tilde{\lambda}_a^{-1}\hat{I}_b, e_0 = \omega_c - \omega$, and μ satisfies the constraint condition:

$$\operatorname{sgn}(\mu) = \operatorname{sgn}\left(\dot{e} + \hat{k}_m e\right) \tag{24}$$

Note that the term μ can greatly contribute to the convergence rates of adjustable parameters, as can be seen from:

$$\dot{V} = 0.5\left\{-\hat{k}_m e^2 - \left(\dot{e} + \hat{k}_m e\right)\mu\right\} \tag{25}$$

The adaptive compensation given by Equation (23) can reduce the tracking error such that ω is, as far as possible, close to ω_m, thus improving the response performance of the rotational speed.

4.3. The Cascade Compensation Scheme

As described thus far, referring to Equations (1) and (20), we approximately have:

$$\frac{L(s)}{L_c(s)} \approx \frac{a_m}{s + a_m} \tag{26}$$

where $a_m = 2\hat{k}_m$. We can further improve the response performance of the variable-pitch propeller through a single cascade compensation element:

$$\frac{L_c(s)}{L_p(s)} \approx \frac{a_m^{-1}s + 1}{a_p^{-1}s + 1} \tag{27}$$

where a_p is an optional positive value, and L_p is the control signal from the flight controller discussed in Section 5. Therefore:

$$\frac{L(s)}{L_p(s)} \approx \frac{a_p}{s+a_p} \tag{28}$$

It is finally noted that both the adaptive compensation and the cascade compensation can greatly contribute to the improvement of the control quality of the variable-pitch propeller. The control scheme of the variable-pitch propeller, including the steady-state optimization, the adaptive compensation and the cascade compensation, is shown in Figure 5.

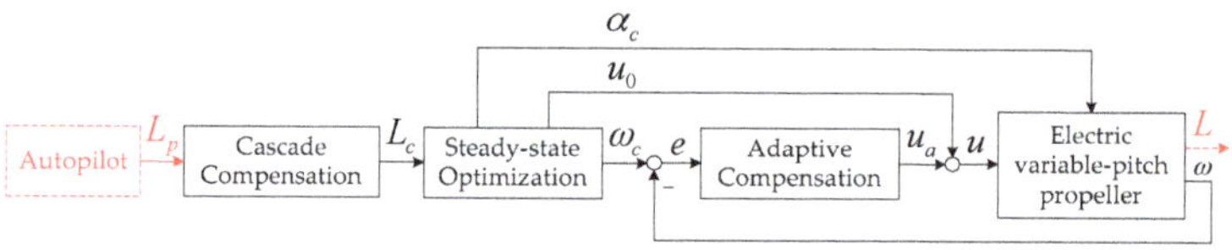

Figure 5. The schematic diagram of the electric variable-pitch propeller.

5. The Direct Lift-based Flight Control Strategy for the Variable-Pitch Quadrotor

5.1. The Direct Lift-based Height Control Scheme

Considering that each motor and propeller has the same input-output properties, according to Equations (2) and (28), the total lift produced by the combined effect of the forces produced by the four propellers can also be expressed by:

$$\frac{T(s)}{T_p(s)} \approx \frac{a_p}{s+a_p} \tag{29}$$

where:

$$T_p = \sum_{j=1}^{4} L_{pj} \tag{30}$$

and L_{pj} denotes the lift command signal corresponding to the jth propeller, as shown in Figure 5.

As for the near-hover flight, we suppose that:

$$T_{ph} = T_p - mg \tag{31a}$$

$$T_h = T - mg \tag{31b}$$

where T_{ph} denotes the control signal from the height controller discussed below, m denotes the mass of the quadrotor, and g denotes the gravitational acceleration. After offsetting the gravity acting on the variable-pitch quadrotor, Equation (29) can then be rewritten by:

$$\frac{T_h(s)}{T_{ph}(s)} \approx \frac{a_p}{s+a_p} \tag{32}$$

Let z, z_c be the actual height and the height command signal, respectively. We propose the direct lift-based height control law as follows:

$$T_{ph} = k_{zp}\varepsilon_z + k_{zi}\int \varepsilon_z \mathrm{d}t - m a_p v_z - T_h \tag{33}$$

where k_{zp}, k_{zi} are optional parameters, and:

$$\varepsilon_z = ma_p(z_i - z) + m(v_{zi} - v_z) \tag{34}$$

where z_i is the output of a cascade compensation element in response to z_c; v_{zi} is correspondingly the derivative of z_i.

It is worthy to note that the direct-lift feedback term in Equation (33) can contribute to the stability of the height. The transfer function, from the input z_i to the output z, can then be derived by using the knowledge of rigid-body kinematics and by substituting Equations (33) and (34) into Equation (32) as follows:

$$\frac{z(s)}{z_i(s)} \approx \frac{a_p k_{zp} s + a_p k_{zi}}{s^3 + a_p s^2 + a_p k_{zp} s + a_p k_{zi}} \tag{35}$$

The cascade compensation element is further chosen as:

$$\frac{z_i(s)}{z_c(s)} \approx \frac{\omega_z^2 s^3 + \omega_z^2 a_p s^2 + \omega_z^2 a_p k_{zp} s + \omega_z^2 a_p k_{zi}}{a_p k_{zp} s^3 + \left(2\xi_z \omega_z a_p k_{zp} + a_p k_{zi}\right) s^2 + \left(\omega_z^2 a_p k_{zp} + 2\xi_z \omega_z a_p k_{zi}\right) s + \omega_z^2 a_p k_{zi}} \tag{36}$$

where ξ_z and ω_z are optional parameters. Thus:

$$\frac{z(s)}{z_c(s)} \approx \frac{\omega_z^2}{s^2 + 2\xi_z \omega_z s + \omega_z^2} \tag{37}$$

where ξ_z and ω_z can be considered as the damping and natural frequency. The dynamics of the height is therefore determined by the choices of $k_{zp}, k_{zi}, \xi_z, \omega_z$ and a_p.

5.2. The Direct Lift-based Attitude Control Scheme

As described in Section 2, the moments are produced by generating a differential lift across the two propellers on the same arm of the quadrotor, thus providing the input necessary for the quadrotor to move longitudinally or laterally; for making the quadrotor yaw to a particular orientation, the lifts of a set of opposite propellers are changed simultaneously and by the same amount.

It can be seen from Table 1 that the second term on the right-hand side of Equation (3) remains dominant in the drag produced by the motor-propeller combination. We therefore suppose that:

$$T_{\phi c} = L_{p4} - L_{p2} \tag{38a}$$

$$T_{\theta c} = L_{p3} - L_{p1} \tag{38b}$$

$$T_{\psi c} = L_{p1} + L_{p3} - L_{p4} - L_{p2} \tag{38c}$$

The roll, pitch and yaw channels can then be considered to be approximately linear and decoupled as $T_{\phi c}$, $T_{\theta c}$ and $T_{\psi c}$ are chosen as the control input signals without considering other aerodynamic effects.

We then propose the direct lift-based attitude control laws as follows:

$$T_{\phi c} = k_{\phi p} e_\phi + k_{\phi i} \int e_\phi \mathrm{d}t + k_{\phi d} \dot{e}_\phi - k_{\dot{\phi}} \dot{\phi} \tag{39a}$$

$$T_{\theta c} = k_{\theta p} e_\theta + k_{\theta i} \int e_\theta \mathrm{d}t + k_{\theta d} \dot{e}_\theta - k_{\dot{\theta}} \dot{\theta} \tag{39b}$$

$$T_{\psi c} = k_{\psi p} e_\psi + k_{\psi i} \int e_\psi \mathrm{d}t + k_{\psi d} \dot{e}_\psi - k_{\dot{\psi}} \dot{\psi} \tag{39c}$$

where $e_\phi = \phi_i - \phi$, $e_\theta = \theta_i - \theta$, $e_\psi = \psi_i - \psi$; ϕ_i, θ_i, ψ_i are the roll, pitch and yaw command signals, respectively; $k_{\phi p}, k_{\phi i}, k_{\phi d}, k_{\dot{\phi}}, k_{\theta p}, k_{\theta i}, k_{\theta d}, k_{\dot{\theta}}, k_{\psi p}, k_{\psi i}, k_{\psi d}, k_{\dot{\psi}}$ are optional parameters.

Compared with conventional flight control methods for variable-pitch quadrotors, the direct lift-based flight control method can improve the flight performance, precisely because all channels are approximately linearized and completely decoupled from each other in this case. It is finally noted that the lift command signal to each motor and propeller can be obtained from Equations (30), (31), (33), (38) and (39).

The longitudinal movement is constituted by the forward velocity caused by the pitch angle θ produced by the moment T_θ, and so does the lateral movement. The longitudinal and lateral trajectory controllers are commonly referred to as the outer-loop controllers and can also be easily designed by using conventional linear control techniques, which have been studied in many literatures, such as [4–6], so won't be covered here.

6. Experimental Tests

In this section, we first describe the details of the experimental setup used for verification purposes. Finally, extensive experimental tests are carried out along with necessary discussions and evaluations.

6.1. The Experimental Setup and Description

Apart from the variable-pitch propeller described in Section 2, the component parts of the control system of the quadrotor include a DSP-based hardware platform and a high integrity navigation unit. The software of the proposed control scheme runs at 200 Hz on the DSP-based hardware platform that can provide reliable support for high precision timer and synchronization operations. The flight controller, performing the closed-loop attitude and altitude control module, is employed to generate the command as inputs to the control system of the variable-pitch propeller. The high integrity navigation system, based on the combined use of the Global Positioning System (GPS) and an inertial measurement unit (IMU), is used to provide the flight controller with the accurate attitude, position and velocity information. The discrete estimate of the rotational speed can be obtained by the use of the commutation instant from the ESC since the commutation interval angle is constant, shown in Equation (10), and the rotational speed is high within the normal operating rotational speed range.

To test and verify the efficiency of the proposed control method, the prototype quadrotor is first conducted to repeatedly perform climb and descent flights in attitude hold mode. Figure 6 shows the experimental test scenes for the present study. For simplicity, only the experimental results related to the pre-specified variable-pitch propeller are given to demonstrate the proposed method under the aerodynamic constraints and the power constraints.

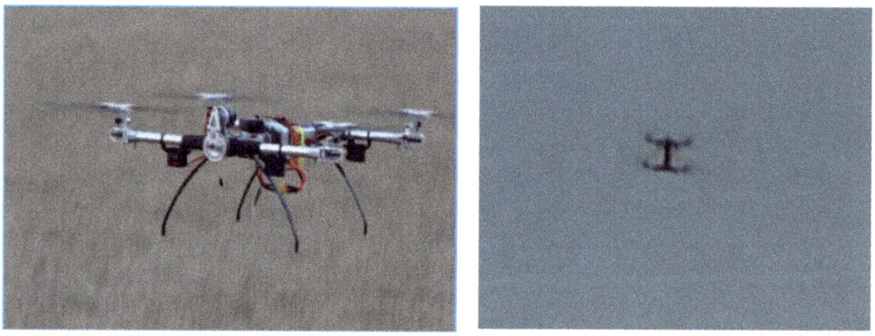

(a) (b)

Figure 6. Experimental test scenes: (**a**) take-off flight; (**b**) climb and descent flights.

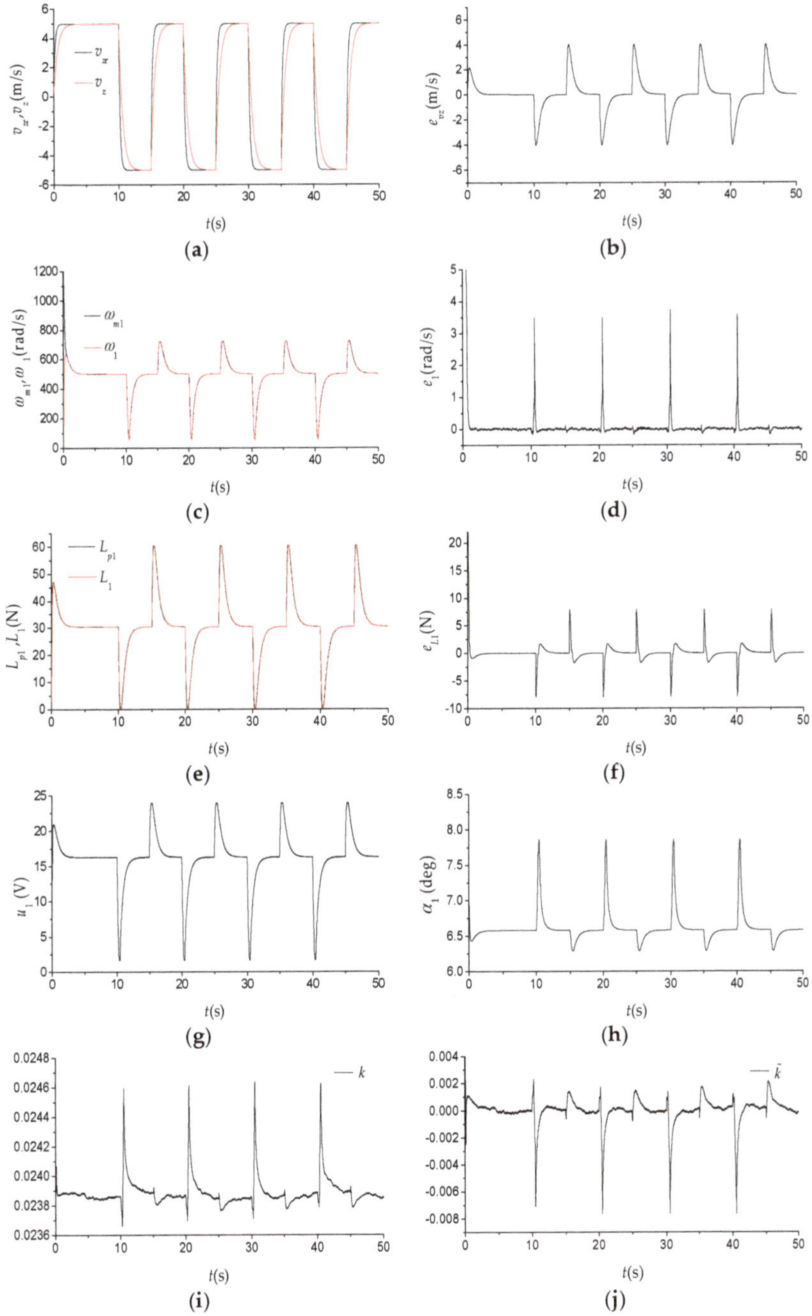

Figure 7. Experimental results of the proposed method in the case of the vertical flight with constant velocity: (**a**) v_{zc} and v_z ; (**b**) e_{vz} ; (**c**) ω_{m1} and ω_1 ; (**d**) e_1 ; (**e**) L_{p1} and L_1 ; (**f**) e_{L1} ; (**g**) u_1 ; (**h**) α_1 ; (**i**) k ; (**j**) $\tilde{k}$. Other propellers have similar characteristics.

6.2. Vertical Flight and Endurance Tests

6.2.1. Experimental Tests of the Proposed Method

The vertical flight tests are conducted to comparatively assess the endurance performance of the variable-pitch quadrotor by using the proposed method and the conventional method. All predefined parameters of the flight control system of the variable-pitch quadrotor are first described as follows:

(1) The estimates of aerodynamic parameters: $\hat{b}_L = 1.75 \times 10^{-5}, \hat{I} = 1.37 \times 10^{-3}$;
(2) Adaptive compensation in the rotational speed loop: $\gamma_a = 1.0 \times 10^{-3}$, $\tilde{\gamma}_a = 1.0 \times 10^{-5}$, $\mu = \dot{e} + \hat{k}_m e$;
(3) Cascade compensation in the lift force loop: $a_p = 5$;
(4) Direct lift-based height control scheme: $k_{zp} = 2.0, k_{zi} = 1.6, \xi_z = 2.0, \omega_z = 0.75$;
(5) Direct lift-based attitude control scheme: $k_{\phi p}, k_{\theta p} = 2.1, k_{\phi i}, k_{\theta i} = 0, k_{\phi d}, k_{\theta d} = 0.5, k_{\dot{\phi}}, k_{\dot{\theta}} = 0.1$, $k_{\psi p} = 9.0, k_{\psi i} = 0.5, k_{\psi d} = 3.0, k_{\dot{\psi}} = 0.5$;
(6) Longitudinal position control scheme: $\theta_i = k_p e_x + k_i \int e_x \mathrm{d}t + k_d \dot{e}_x - k_v v_x$, where e_x is the position error along the body x axis, $k_p = 1.0$, $k_i = 0.1$, $k_d = 0.2$, $k_v = 1.0$, and so does the lateral position control scheme. A series of real-time experimental tests are then conducted based on the above assumptions.

Figure 7 shows the experimental results of the proposed method in the case of the vertical flight with constant velocity, where $z_c(t) = \int_0^t v_{zc}(\tau)\,\mathrm{d}\tau$; v_{zc} is the vertical velocity command signal; e_{vz} is the error between v_{zc} and v_z; e_{L1} is the error between L_{p1} and L_1.

For simplicity, only the results related to the first propeller are shown in the figure. It is seen from Figure 7a–f that the resulting controlled system can achieve good tracking performances, even though the voltage and the propeller pitch exhibit significant variations in accordance with the vertical velocity command signal shown in Figure 7g,h. Meanwhile, the adjustable parameters of the adaptive compensation element also have good convergence properties as shown in Figure 7i,j during the maneuver flights. It is, however, to note that the rotational speed tracking error demonstrates a remarkable difference during the climb flight mode and the descent flight mode due to the nonlinear characteristics of the propeller.

To further demonstrate the positive effect of the propeller pitch on the endurance performance and comparatively assess the efficiency and superiority of the proposed method, the propeller pitch α involved in the proposed method is set equal to α_c and different constant values, respectively, and simultaneously, the variable-pitch quadrotor is particularly equipped with the same type of Lithium battery with a capacity of 5000 mAh for each experimental test. Table 2 and Figure 8 show the endurance performance versus propeller pitch angles using the proposed method.

Table 2. The endurance performance *versus* propeller pitch angles.

Battery Capacity	α(deg)	Endurance (s)
5000 mAh	Sub-optimized value	267.30
	3	155.36
	4	197.50
	5	226.61
	6	246.72
	7	251.65
	8	249.22
	9	246.77
	10	243.27
	11	239.70
	12	234.91
	13	229.19
	14	220.52
	15	212.58

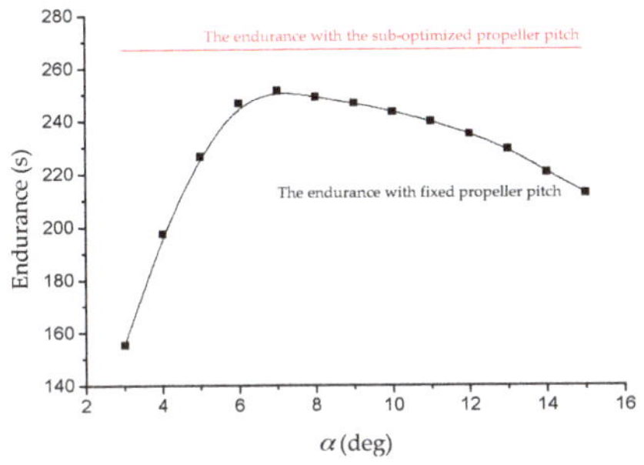

Figure 8. The endurance performance *versus* propeller pitch angles

The proposed method with minimum power consumption has the most significant improving effect on the endurance performance when compared to the fixed propeller pitches. According to the test results, the improvement of the endurance performance of the quadrotor ranges from 6.2% to 72.1%.

6.2.2. Experimental Tests of the Conventional Method with Constant Rotational Speed

In order to make a fair comparison between the proposed method and the conventional method with constant rotational speed implemented through the widely used Pixhawk® autopilot module, the experimental tests of the conventional method are then conducted for this purpose, where the autopilot employs one optimal proportional-integral controller to obtain the armature voltage according to the rotational speed tracking error, and another optimal proportional-integral- differential controller to determine the propeller pitch according to the vertical velocity tracking error.

The predefined parameters are described as follows:

(1) Rotational speed control scheme (optimal proportional-integral control): proportional and integral coefficients equal to 0.85 and 1.95;
(2) Vertical velocity control scheme (optimal proportional-integral-differential control): proportional, integral and differential coefficients equal to 2.5, 1.0 and 0.2.

Figure 9 shows the experimental results with the rotational speed command signal $\omega_{cj} = 400$ (j = 1, 2, 3, 4). The propeller pitch exhibits a significant variation in accordance with the vertical velocity command signal during the maneuver flights, as shown in Figure 9f, although the vertical velocity tracking performance deteriorates slightly and is basically acceptable when compared to the proposed method.

Similarly, to demonstrate the effect of the rotational speed on the endurance performance in this test case, the rotational speed command signal is set equal to different constant values. The experimental tests are conducted with the same equipment as mentioned above, and the resulting endurance performance *versus* rotational speeds is shown in Table 3 and Figure 10.

Table 3. The endurance performance *versus* rotational speeds.

Battery Capacity	ω_c(rad/s)	Endurance (s)
	350	230.95
	400	240.09
	450	239.89
5000 mAh	500	232.52
	550	220.43
	600	205.31
	650	189.06

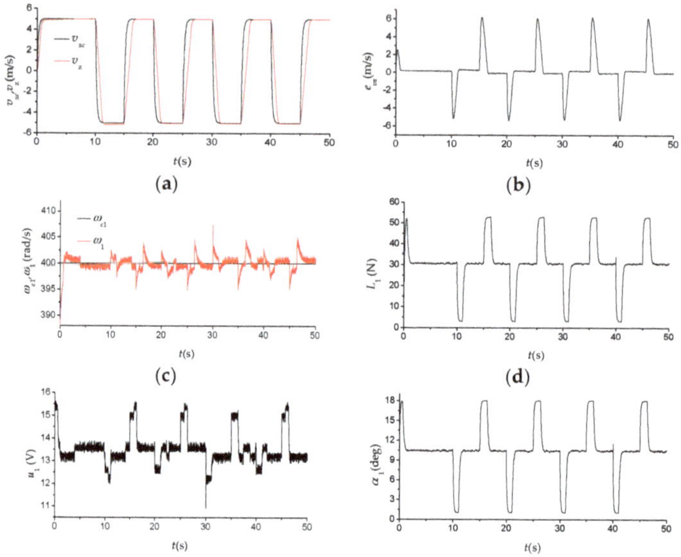

Figure 9. Experimental results of the conventional method with $\omega_{cj} = 400$ ($j = 1,2,3,4$): (**a**) v_{zc} and v_z ; (**b**) e_{vz} ; (**c**) ω_{c1} and ω_1 ; (**d**) L_1 ; (**e**) u_1 ; (**f**) α_1 . Other propellers have similar characteristics.

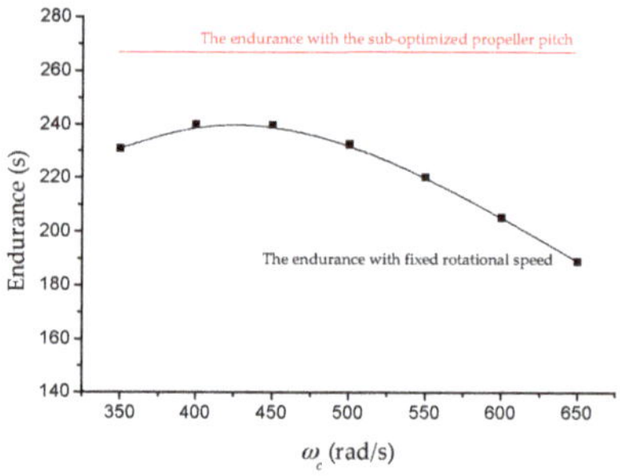

Figure 10. The endurance performance versus rotational speeds.

In contrast to what has been shown in Table 2 and Figure 8, the endurance with fixed rotational speed decreases distinctly, although the control quality is basically satisfactory in practice, which also comparatively demonstrates that the proposed method can contribute to the significant improvement of the endurance performance.

6.3. Positioning Accuracy Tests

All predefined parameters of the proposed algorithm are defined in Section 6.2.1. Figure 11 shows the experimental results of the proposed algorithm during near-hover operations, where v_n, v_e, v_h are the northward, eastward and vertical velocities in the Earth frame. The proposed method can achieve a good trajectory tracking performance when the variable-pitch quadrotor is conducted to perform a low-velocity trajectory tracking task, as shown in Figure 11a–c. Furthermore, according to the experimental data, the deviation distance is less than 0.08 m during the straight line flight, and the height tracking error is less than 0.12 m in the presence of disturbances and uncertainties.

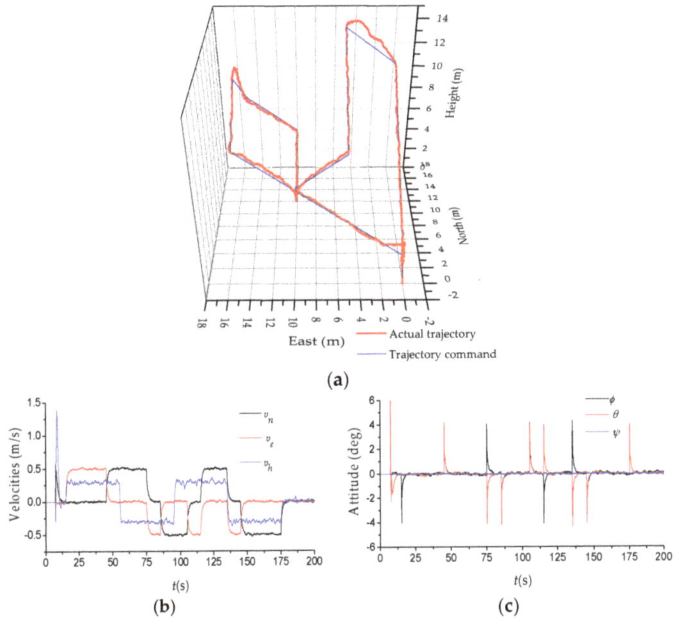

Figure 11. The experimental results of the proposed method during near-hover operations: (**a**) 3D flight trajectory tracking; (**b**) flight velocities; and (**c**) attitude.

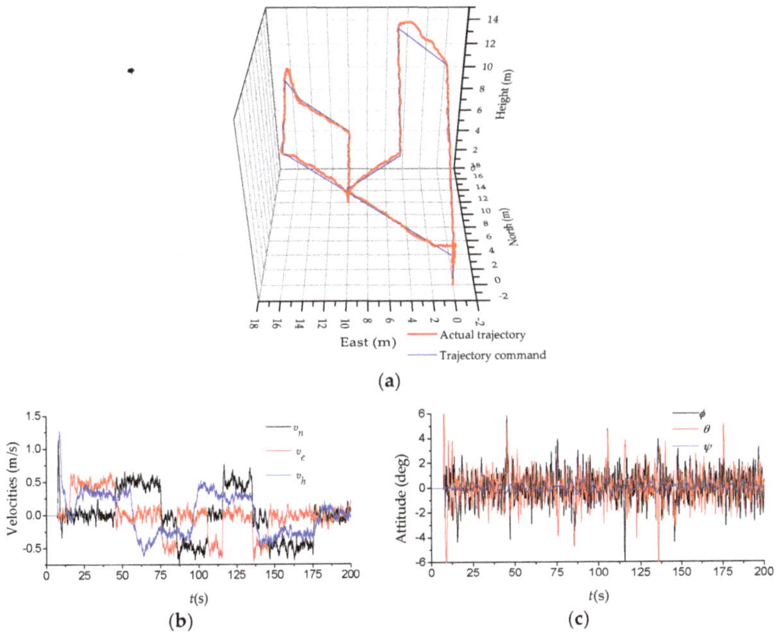

Figure 12. The experimental results of the conventional method during near-hover operations: (**a**) 3D flight trajectory tracking; (**b**) flight velocities; and (**c**) attitude.

Figure 12 shows the experimental results of the conventional method during near-hover operations. According to the experimental data, the deviation distance is more than 0.23 m during the straight line flight, and the height tracking error is more than 1.3 m, as shown in Figure 12a–c. The experimental results of the conventional method also demonstrate that the horizontal position error reaches a maximum value of 0.20 m, and height error also reaches a maximum value of 0.55 m in the presence of disturbances and uncertainties during the hovering flight.

Compared with the results of the proposed method, the resulting performance of the conventional method deteriorates significantly precisely because of the nonlinearity and the coupling interactions mentioned above, as demonstrated in [3,21]. From the comparison of the experimental results of two flight control methods, we can therefore conclude that the proposed method contributes to the significant improvement of the flight performance of the variable-pitch quadrotor, such as the endurance performance and the positioning accuracy.

7. Conclusions

To address the control and optimization problems of the electric variable-pitch quadrotor, a steady-state identification method to estimate the parameters of the mathematical model of the electric variable-pitch propeller has been first developed. A control and optimization strategy for the variable-pitch propeller, mainly including the steady-state control and optimization scheme with minimum power consumption, the adaptive compensation scheme and the cascade compensation scheme, is then proposed, which greatly improves the response performance of the lift force produced by the variable-pitch propeller. Furthermore, the direct lift-based flight control strategy is presented, which contributes to the improvement of the flight performance of the variable-pitch quadrotor, precisely because the roll, pitch, yaw and vertical channels are approximately linearized and completely decoupled from each other in this case. The experimental test results have comparatively demonstrated the efficiency and superiority of the proposed method. These achievements can also apply to other micro air vehicles with electric variable-pitch propellers.

Acknowledgments: This study was supported in part by National Natural Science Foundation of China (NSFC) (Under Grant No. 61374188), Natural Science Foundation of Jiangsu Province of China (Under Grant No. BK20141412), Applied Basic Research Programs of Natural Science Foundation of Jiangsu Province, China (Under Grant No. BY2015003-10). The authors would like to thank Professor Ben M. Chen in National University of Singapore for providing facilities and assistance.

Author Contributions: All authors discussed the contents of the manuscript. Shouzhao Sheng contributed to the research idea and the framework of this study. Chenwu Sun performed the experiments.

Conflicts of Interest: The authors declare no conflict of interest.

References

1. Altug, E.; Ostrowski, J.P.; Mahony, R. Control of a Quadrotor Helicopter Using Visual Feedback. In Proceedings of the 2002 IEEE International Conference on Robotics and Automation (ICRA), Washington, DC, USA, 10–17 May 2002; Volume 1, pp. 72–77.
2. Tayebi, A.; McGilvray, S. Attitude stabilization of a VTOL quadrotor aircraft. *IEEE Trans. Control Syst. Technol.* **2006**, *14*, 562–571. [CrossRef]
3. Madani, T.; Benallegue, A. Control of a Quadrotor Mini-Helicopter via Full State Backstepping Technique. In Proceedings of the 2006 45th IEEE Conference on Decision and Control, San Diego, CA, USA, 13–15 December 2006; pp. 1515–1520.
4. How, J.P.; Bethke, B.; Frank, A.; Dale, D.; Vian, J. Real-time indoor autonomous vehicle test environment. *IEEE Trans. Control Syst. Mag.* **2008**, *28*, 51–64. [CrossRef]
5. Pounds, P.; Mahony, R.; Corke, P. Modeling and control of a large quadrotor robot. *Control Eng. Pract.* **2010**, *18*, 691–699. [CrossRef]
6. Ryll, M.; Bulthoff, H.H.; Giordano, P.R. First Flight Tests for a Quadrotor UAV with Tilting Propellers. In Proceedings of the 2013 IEEE International Conference on Robotics and Automation (ICRA), Karlsruhe, Germany, 6–10 May 2013; pp. 295–302.

7. Sheng, S.; Sun, C. Design of a Stability Augmentation System for an Unmanned Helicopter Based on Adaptive Control Techniques. *Appl. Sci.* **2015**, *5*, 575–586. [CrossRef]
8. Sheng, S.; Sun, C. A Near-Hover Adaptive Attitude Control Strategy of a Ducted Fan Micro Aerial Vehicle with Actuator Dynamics. *Appl. Sci.* **2015**, *5*, 666–681. [CrossRef]
9. Sheng, S.; Sun, C.; Zhao, H. Indirect Adaptive Attitude Control for a Ducted Fan Vertical Takeoff and Landing Microaerial Vehicle. *Math. Probl. Eng.* **2015**, *2015*. [CrossRef]
10. Bouabdallah, S.; Murrieri, P.; Siegwart, R. Towards Autonomous Indoor Micro VTOL. *Auton. Robots* **2005**, *18*, 171–183. [CrossRef]
11. Bresciani, T. Modeling, Identification and Control of a Quadrotor Helicopter. Master's Thesis, Lund University, Lund, Sweden, 1 January 2008.
12. Shirsat, A.R. Modeling and Control of a Quadrotor UAV. Master's Thesis, Arizona State University, Tempe, AZ, USA, May 2015.
13. Erginer, B.; Altug, E. Modeling and PD control of a quadrotor VTOL vehicle. In Proceedings of the 2007 IEEE Intelligent Vehicles Symposium, Istanbul, Turkey, 13–15 June 2007; pp. 894–899.
14. Walid, M.; Slaheddine, N.; Mohamed, A.; Lamjed, B. Modeling and Control of a Quadrotor UAV. In Proceeding of the 2014 15th International Conference on Sciences and Techniques of Automatic Control and Computer Engineering (STA), Hammamet, Tunisia, 21–23 December 2014; pp. 343–348.
15. Alexis, K.; Nikolakopoulos, G.; Tzes, A. Model Predictive Quadrotor Control: Attitude, Altitude and Position Experimental Studies. *IET Control Theory Appl.* **2012**, *6*, 1812–1827. [CrossRef]
16. Kim, J.; Kang, M.S.; Park, S. Accurate Modeling and Robust Hovering Control for a Quad–Rotor VTOL Aircraft. In Proceeding of the Selected Papers from the 2nd International Symposium on UAVs, Reno, NV, USA, 8–10 June 2009; pp. 9–26.
17. Hoffmann, G.M.; Huang, H.; Waslander, S.L.; Tomlin, C.J. Quadrotor Helicopter Flight Dynamics and Control: Theory and Experiment. In Proceeding of the AIAA Guidance, Navigation, and Control Conference, Hilton Head, SC, USA, 21–22 August 2007; Volume 2.
18. Bouabdallah, S. Design and Control of Quadrotors with Application to Autonomous Flying. Ph.D. Thesis, Swiss Federal Institute of Technology, Lausanne, Switzerland, 10 January 2007.
19. Cutler, M.; Ure, N.K.; Michini, B.; How, J.P. Comparison of Fixed and Variable Pitch Actuators for Agile Quadrotors. In Proceeding of the AIAA Conference on Guidance, Navigation and Control and Co-Located Conferences (GNC), Portland, OR, USA, 8–11 August 2011. [CrossRef]
20. Cutler, M. Design and Control of an Autonomous Variable-Pitch Quadrotor Helicopter. Ph.D. Thesis, Massachusetts Institute of Technology, Cambridge, MA, USA, 23 August 2012.
21. Riccardi, F.; Panizza, P.; Lovera, M. Identification of the Attitude Dynamics for A Variable-Pitch Quadrotor UAV. In Proceeding of the European Rotorcraft Forum (ERF) 2014, Southampton, UK, 2–5 September 2014.
22. Fresk, E. Modeling, Control and Experimentation of a Variable Pitch Quadrotor. Master's Thesis, Luleå University of Technology, Luleå, Sweden, 2013.
23. Michini, B.; Redding, J.; Ure, N.K.; Cutler, M.; How, J.P. Design and Flight Testing of an Autonomous Variable-pitch Quadrotor. In Proceeding of the 2011 IEEE International Conference on Robotics and Automation (ICRA), Shanghai, China, 9–13 May 2011; pp. 2978–2979.
24. Cutler, M.; How, J.P. Actuator Constrained Trajectory Generation and Control for Variable-Pitch Quadrotors. In Proceeding of the AIAA Guidance, Navigation, and Control Conference (GNC), Minneapolis, MN, USA, 13–16 August 2012.
25. Bristeau, P.J.; Martin, P.; Salaun, E.; Petit, N. The Role of Propeller Aerodynamics in the Model of a Quadrotor UAV. In Proceeding of the European Control Conference, Budapest, Hungary, 23–26 August 2009.
26. Drela, M. QPROP Users Guide. 2009. Available online: http://web.mit.edu/drela/Public/web/qprop/ (accessed on 10 November 2015).
27. Hanselman, D.C. *Brushless Permanent Magnet Motor Design*, 2nd ed.; The Writers' Collective: Cranston, RI, USA, 2003.

energies

MDPI

Article

A Multi-Objective Optimization Framework for Offshore Wind Farm Layouts and Electric Infrastructures

Silvio Rodrigues [1,*], Carlos Restrepo [2], George Katsouris [1], Rodrigo Teixeira Pinto [3], Maryam Soleimanzadeh [4], Peter Bosman [5] and Pavol Bauer [1]

1 DC systems, Energy conversion & Storage Group, Delft University of Technology, Mekelweg 4, Delft 2628 CD, The Netherlands; G.Katsouris@student.tudelft.nl (G.K.); P.Bauer@tudelft.nl (P.B.)
2 Department of Industrial Technologies, Universidad de Talca, Talca 3340000, Chile; restrepo.cr@gmail.com
3 CITCEA-UPC, Carrer de Jordi Girona, 31, Barcelona 08034, Spain; rodrigo.teixeira@citcea.upc.edu
4 ECN, Westerduinweg 3, Petten 1755 LE, The Netherlands; soleimanzadeh@ecn.nl
5 Centrum Wiskunde & Informatica, Science Park 123, Amsterdam 1098 XG, The Netherlands; Peter.Bosman@cwi.nl
* Correspondence: s.m.fragosorodrigues@tudelft.nl; Tel.: +31-1-52-78-18-98; Fax: +31-6-53-65-07-14

Academic Editor: Frede Blaabjerg
Received: 2 November 2015; Accepted: 4 March 2016; Published: 18 March 2016

Abstract: Current offshore wind farms (OWFs) design processes are based on a sequential approach which does not guarantee system optimality because it oversimplifies the problem by discarding important interdependencies between design aspects. This article presents a framework to integrate, automate and optimize the design of OWF layouts and the respective electrical infrastructures. The proposed framework optimizes simultaneously different goals (e.g., annual energy delivered and investment cost) which leads to efficient trade-offs during the design phase, e.g., reduction of wake losses *vs* collection system length. Furthermore, the proposed framework is independent of economic assumptions, meaning that no *a priori* values such as the interest rate or energy price, are needed. The proposed framework was applied to the Dutch Borssele areas I and II. A wide range of OWF layouts were obtained through the optimization framework. OWFs with similar energy production and investment cost as layouts designed with standard sequential strategies were obtained through the framework, meaning that the proposed framework has the capability to create different OWF layouts that would have been missed by the designers. In conclusion, the proposed multi-objective optimization framework represents a mind shift in design tools for OWFs which allows cost savings in the design and operation phases.

Keywords: design parameters; economic functions; multi-objective optimization; offshore wind farms; trade-offs; wind farm designers

1. Introduction

Currently the offshore wind industry is aiming at reducing its cost of energy (COE) (M€/MWh) to breach the 100 €/MWh barrier as soon as 2020 [1–6] from the current 163 €/MWh [7]. Although the technologies used in offshore wind farms (OWFs) have greatly improved, the COE generated offshore is yet not competitive [8]. In fact, electricity generated offshore is currently approximately 50% more expensive when compared to onshore wind generation [9]. Figure 1a demonstrates that the cost of power (COP) (M€/MW) installed of OWFs has increased since the initial project and has not reduced in the last years [10,11].

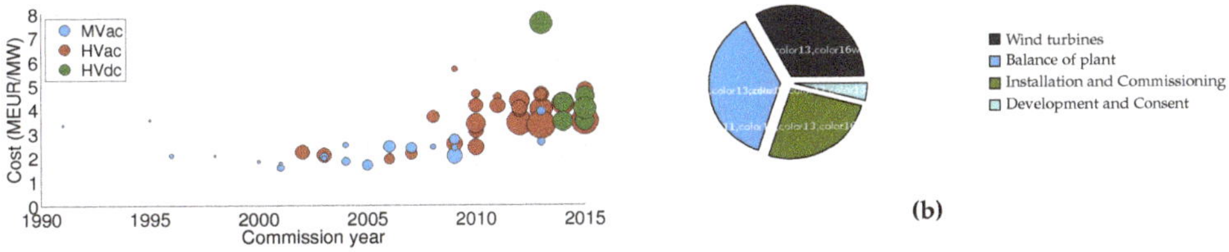

(a)

(b)

Figure 1. Cost of power (COP) and capital expenditure (CAPEX) breakdown of offshore wind farms (OWFs). (**a**) COP of the European OWFs composed of five or more turbines [10,11]. Circle size represents the installed capacity of the OWFs. The monetary values were updated considering a Eurozone inflation of 1.85% [12]; (**b**) Typical CAPEX breakdown of an OWF [13].

There are different measures that can decrease the costs of the energy generated offshore [2–6]. Key factors are, for example, exploitation of economies of scale and greater standardization, introduction of turbines with higher rated power and reliability and greater activity at the front end engineering and design (FEED) phase.

The design phase of an OWF is performed during the FEED phase (see Figure 2), after the initial feasibility studies have been done and permission has been granted, and before final investment decisions are made [2,14]. FEED studies allow wind farm developers to make a pre-selection of economically viable design concepts and the respective key components [3]. During the FEED phase, decisions have not yet been made regarding the number of turbines [15,16], the support structures that will be used or the number of substations that will be built [15,16]. In this phase, several layout concepts are preliminary designed, and although the final wind farm layout will be based on these designs, it may still differ considerably [15].

Figure 2. Lifecycle of an OWF and location of the FEED phase [14,17].

Several aspects have increased the need for a broader activity at the FEED stage [3]. The development phase of OWFs is time consuming due to the time needed to manually create several designs and the necessary cable routing [18–20]. In fact, circa 4% of the total capital expenditure (CAPEX) of an OWF are allocated to the development phase (see Figure 1b), in which all the components and technologies that lead to an optimized and feasible system must be assessed [13,21]. Recent OWFs occupy larger areas, which often have variable water depth and seabed conditions [22] and are situated further from shore [22], leading to more complex constraints and design challenges on the grid connection. Finally, the large number of wind turbines leads to complex collection systems, which need to be carefully assessed to achieve wind farm layouts with higher efficiencies [2]. Figure 3 shows the difference in complexity between the first OWF, Vindeby, and the recent British Gwynt y Môr.

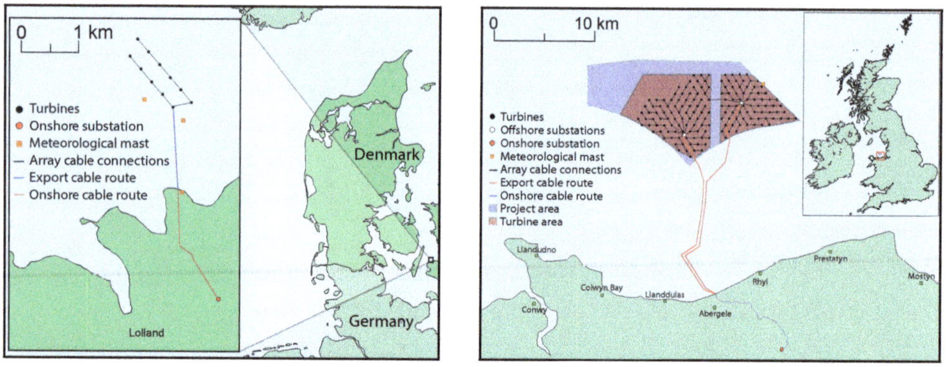

(a) (b)

Figure 3. Differences in topology and design complexity between two OWFs. (**a**) Vindeby wind farm [10]; (**b**) Gwynt y Môr wind farm [23].

Additionally, current design processes are based on a sequential approach (or decoupled strategy) due to the complexity of designing an OWF [19,24]. Such strategy does not guarantee system optimality because the interactions between different system components are disregarded. Moreover, early project decisions may become constraints in later stages [9]. Automated optimization is crucial to optimize the wind farm layout, since the design of OWFs with standard tools is highly complex and time consuming [19]. A reduction of up to 10% in the cost of energy is possible through more integrated design methods [9].

The increased difficulty in the design of modern OWFs comes from the fact that, mostly, all the design aspects of an OWF influence both its energy production and its investment and operational costs [2]. For example, the energy production is increased by placing more turbines in the OWF area, however this also makes the costs rise. Also, interactions between turbines reduce the increase in energy production that results from more turbines being closed together. Hence, these design goals are conflicting, meaning that there is not a single solution for the problem but a set of solutions which represent the trade-off. In the multi-objective (MO) space, a layout is optimal if there is no other layout which is better in all objectives.

Although more than 150 research articles on the wind farm layout optimization problem (WFLOP) may be found in literature, few studies have investigated the inherent trade-offs of designing an OWF [25]. For example, the trade-off between the wind farm capacity factor and the power density within the project area was assessed in [26]. The authors analyzed the conflict between increasing the spacing between the turbines to increase energy production (via a decrease of wake losses) and the need for larger project areas.

Comprehensive studies that explicitly consider multiple goals during the optimization process are rare [27]. Table 1 presents the characteristics of the MO WFLOP (MOWFLOP) from previous studies. The work carried out in [28] optimized the annual energy production (AEP) considering the problem constraints (minimum proximity constraint between wind turbines and area constraint which guarantees that all turbines are placed within the wind farm area) as a second objective function. The AEP and the turbine noise were optimized in [29,30]. Similarly, the AEP was maximized and the sum of the wind farm area and number of turbines were treated as a second objective function in [31]. Three simultaneous optimization goals were used in [32]: AEP, area used and collection system length. The work presented in [33] studied the different features that a MO algorithm should have to efficiently solve the MOWFLOP. The optimization goals were the AEP and the system efficiency, while the optimization variables were the location and number of turbines.

Table 1. Existing approaches for the MOWFLOP.

References	Optimization Variables	Design Goals
Kusiak *et al.* [28]	Turbine locations	Energy generation, Problem constraints
Zhang *et al.* [29,30], Sorkhabi *et al.* [34]	Turbine locations	Energy generation, Noise level
Veeramachaneni *et al.* [31]	Turbine locations	Energy generation, Cost
Tran *et al.* [32]	Turbine locations	Energy generation, Collection system length, Wind farm area
Sisbot *et al.* [35]	Turbine locations and quantity	Energy generation, Cost
Rodrigues *et al.* [33]	Turbine locations and quantity	Energy generation, System efficiency

A framework able to efficiently optimize, at once, both the wind farm layout and its respective electrical infrastructure for large OWFs is a highly desired tool by wind farm designers [20,29]. However, none of the optimization frameworks displayed in Table 1 captured all the key aspects pertaining to the development of OWFs. In other words, as far as the authors knowledge a MO optimization framework—which is able to give general recommendations and trade-offs insight to OWF developers—has not yet been established.

To bridge this existing research gap, this article proposes a MO optimization framework to *integrate*, *automate* and *optimize* the design of OWF layouts and their electrical infrastructure. The most suitable and relevant optimization goals and design variables for the MOWFLOP will be identified. The optimization framework is then applied to the design of an OWF in a case study to demonstrate the advantages and differences of the proposed method.

The article is organized as follows: Section 2 provides an overview of the relevant commercial and academic optimization methods for OWFs. Furthermore the most common economic functions used to assess the profitability of wind projects are introduced and explained. Thereafter, Section 3 presents the MO optimization framework for the design of OWF layouts and respective electrical infrastructures and its boundaries and selection criteria. Section 4 introduces the optimization variables considered in the framework as well as their boundaries, constraints and influences over the energy production and expenditures. The industrial trends of the different components of an OWF are also investigated. Section 5 then describes the case study used to demonstrate the usefulness of the proposed framework, followed by Section 6 in which the results obtained are presented and discussed. Section 7 presents the main conclusions of the article.

2. Current Wind Farm Optimization Tools

2.1. Commercially Available Software

Currently, there are several software commercially available to optimize the design of wind farms. For example, WindPRO [36], WindFarmer [37] and WAsP [38] are among the most famous optimization tools. Table 2 presents the main features and limitations of the best-known available commercial software.

Table 2. Wind Farm Design Software Commercially Available.

Software	Optimization Goals	Design Parameters	Design Constraints	Design Considerations	Limitations
OpenWind (AWS Truepower) [39]	Cost of Energy	Wind turbine coordinates	Site constraints for turbine placement, noise levels	Wake and turbulence losses, shadow flicker, visual impact, uncertainty analysis	Electrical losses are not considered
WAsP (DTU) [38]	None	Pre-defined wind turbine coordinates	None	Wake losses, different turbine models may be used simultaneously, estimation of loads in complex terrains	There is no optimization algorithm, electrical losses are not considered
WindFarmer (DNV Garrad Hassan's) [37]	Energy production	Wind turbine coordinates	Environmental constraints, wind farm boundaries, exclusion zones and set-back distances from boundaries, soil topography	Turbulence intensity, environmental and visual impacts, noise levels, shadow flicker, uncertainty analysis, wake loss, turbine loading, electrical losses calculation, reactive power production	No optimization of the collection system layout
WindPRO [36]	Energy production or Minimum production loss	Wind turbine coordinates	Minimum distances between turbines, site constraints for turbine placement	Noise levels, visual impact, shadow effects, Wake and turbulence losses, different turbine models may be used simultaneously, electrical losses calculation, uncertainty analysis	Restricted to geometrical layouts, addition of turbines incrementally
WindSim [40]	Wind project profit	Wind turbine coordinates	Environmental constraints, wind farm boundaries, exclusion zones for turbine placement, IEC Constraints	Optimum number of turbines, terrain features effects on the wind, wake and turbulence losses	No collection system design, electrical losses are not considered
WindFarm (ReSoft) [41]	Energy production or cost of energy	Wind turbine coordinates	Turbine separation, exclusion zones for turbine placement	Wake losses, visual impact, shadow flicker, noise levels, different turbine models may be used simultaneously	No collection system design, electrical losses are not considered

All the commercial wind farm design tools were specifically built for onshore environments and, therefore, consider some irrelevant design aspects for offshore areas such as visual impact (irrelevant for far offshore), shadow flickering, noise levels [34] and complex terrain elevations [42,43]. Although it is possible to use them to design OWFs, none of them consider some of the important offshore aspects such as number of turbines, collection and transmission systems design, number and location of offshore substations and transmission technology. Furthermore, no commercial tool uses a MO algorithm to optimize the trade-offs between the chosen goals.

2.2. Academic Studies

On the other hand, several academic studies were specifically tailored to design OWFs. A brief description and the main drawbacks of some of these works are given next.

2.2.1. Offshore Wind Farm Layout Optimization (OWFLO) (2005)

The OWFLO project proposed a framework to design OWFs which uses the levelized cost of energy (LCOE) (€/MWh) as the optimization goal. It includes turbine availability, wake and cable losses and cost models for the turbines, support structures, cables, operation expenditure (OPEX) (bn €), installation and decommissioning [44]. With these models, the OWFLO tool captures the key factors that influence the cost of energy: turbine size, water depth, distance to shore, soil types and wind and wave conditions.

The main goal of the OWFLO project was the development of an optimization software to provide insights into the trade-offs between cost and energy. However, even though both energy and cost models are used, wind farm developers are only presented with one solution at the end of the optimization routine. Hence, no information regarding the existing trade-offs is obtained. Furthermore, the design of the wind farm collection system, which is a key design aspect, was not considered.

2.2.2. Lackner and Elkington (2007)

The trade-offs inherent to OWFs design problems are identified in [45]. These trade-offs are captured by considering the COE as the optimization goal. However, by optimizing directly the COE, the wind farm designers are only presented with one final solution, hence, once again, no information regarding the trade-offs is obtained. Furthermore, the water depth and the wind speed are considered as functions of the distance to shore.

2.2.3. Gribben *et al.* (2010)

This study presents an OWF layout design tool based on an engineering approach [46]. The authors also identify the multi-disciplinary nature of the problem and, although it is stated that it is important to make a selection of the significant design considerations since it is a highly complex problem, no selection criteria are provided. Moveover, there is no optimization routine implemented in the design approach and only a variation of the design parameters is carried out.

Trade-off insights are presented between, for example, energy yield and foundation costs. Nonetheless, similar to standard design strategies, the method is based on a sequential approach. Moreover, only geometrical layouts are considered and the turbine type is set beforehand.

2.2.4. Topology Optimization of Wind Farms (TopFarm) (2011)

The main goal of this EU-funded project, TopFarm, was the design of an optimization tool for wind farm developers [47]. The financial balance was used as the optimization goal, whereas the turbine coordinates were chosen as design variables. Relevance and relative cost basis were the selection criteria for the cost models in the framework. TopFarm only considered costs which depend on the wind farm topology, since only these provide useful information for the optimization algorithm to guide the search.

Although TopFarm has a comprehensive number of modules, it still lacks key aspects for the optimization of OWF layouts. For example, the collection system routing was determined by solving the auxiliary road problem, *i.e.*, interconnecting all turbines and assuming that the cables could transport all the power connected to them. Furthermore, the TopFarm tool does not consider offshore substations or export cables, which are key factors for OWF developers.

2.3. Economic Functions for Offshore Wind Projects

Investing capital in one project means that the same capital cannot be applied to another investment that is equally or more attractive. Therefore, before final investment decisions are made, wind farm designers create several wind farm designs during the FEED phase to evaluate them economically [2,14].

The choice of which economic function to use is based on different factors, e.g., risk associated, financial structure, regulation, project size [48]. Due to the variety of factors, depending on the case some economic functions are more suitable than others. Therefore, different functions should be used to evaluate possible investments [49]. Table 3 presents the objective functions commonly used in the existing optimization frameworks as well as other economic functions which may be used to assess the profitability of OWFs. Next, a brief description of each function is given and their respective advantages and disadvantage are also discussed.

2.3.1. Annual Energy Delivered

Maximizing the annual energy production (AEP) (GWh/year) is among the most common goals [25] and aims at maximizing the AEP of the wind turbines. Another more comprehensive goal is to maximize the annual energy delivered (AED) (GWh/year) to the onshore electrical network. This goal covers the AEP but also the electrical losses through the entire system, e.g., transformers and cabling systems. Nonetheless, these goals do not take into consideration costs associated with the project or the efficiency of the project.

2.3.2. Utilization Factor

The utilization factor (UF) [–], or capacity factor, of an OWF is the ratio between its AEP and its power production if working constantly at rated power, for the same period of time. This measure is also captured by the wind farm efficiency which is the ratio between the energy delivered over the energy produced without losses. The UF is maximized considering the AED and the efficiency of the system. However, the costs of the project are still neglected if designers use this measure.

2.3.3. Cost of Power

A simple approach to also considerer the CAPEX is to minimize the COP installed which is the ratio between in the CAPEX and the rated capacity of the project. In this way, the optimization tries to maximize the AED and reduce the cost simultaneously. However, and specially in offshore environments, the OPEX also plays an important role in the total costs of the project. Hence, a better approach is to use economic functions which relate the AED, CAPEX and OPEX of the project through economic factors.

2.3.4. Levelized Cost of Energy

The LCOE is also one of the most common economic functions used to evaluate wind farm layouts [25]. The LCOE represents the minimum energy price that meets the desired interest rate by the designers [17]. The LCOE function requires the interest rate and the wind farm lifetime. However, it may not be straightforward to set a value for the interest rate since it is associated with the risk perceived by the investor and it also may change throughout the wind farm lifetime [50]. Furthermore, the LCOE is not suitable to choose between mutually exclusive options for OWF layouts since it does not capture the difference between investment sizes [48].

2.3.5. Net Present Value

The net present value (NPV) (bn €) defines the total profit of the OWF. It considers both the cash outflows (costs) and inflows (revenues) [48]. Therefore, the NPV requires all the economic factors that the LCOE function requires as well as the price of energy. A positive NPV represents a good investment whereas a negative NPV value indicates that the income (with the interest rate chosen by the designers) is lower than the costs [48].

The NPV is suitable to evaluate and compare mutually exclusive projects because it can distinguish the difference in size of different wind farms [48]. Moreover, the NPV should also be used as an alternative measure to further validate the investment [48]. On the other hand, the NPV does not consider, for example, the time required to cover the investment costs (see Section 2.3.7), which may lead the designers to choose a layout that will require a long time to generate profit to the investors.

2.3.6. Internal Rate of Return

The internal rate of return (IRR) (%) is the interest rate that sets the NPV function equal to zero. The IRR function is usually used to approve or disapprove wind farm designs and allows to check if a minimum rate of return set by the designers is met [48]. Nonetheless, the IRR is not suitable to choose between exclusive alternative designs since it does not differentiate wind farm project sizes. Furthermore, the IRR may over predict the profitability of a project because it assumes that the interim revenues are reinvested at a similar rate to the IRR [48]. Finally the IRR does not accurately distinguish the profitability of projects with different lifetimes.

2.3.7. Discounted Payback Time

The discounted payback time (DPT) (years) determines the time, in years, required to cover the initial investment while taking into account the time value of the money. The use of the DPT is recommended when risk is an issue since it represents the period of time in which the capital invested was not recovered, and is therefore, still at risk [48].

The DPT is also not recommended to select one wind farm layout among a set of alternatives because it also does not discern wind farm project sizes. Furthermore, the DPT is not suitable to rank the different possibilities since it ignores the cash revenues after the payback time [48].

2.3.8. Return on Investment

The return on investment (ROI) [–] represents the revenues per unit invested. It is calculated by calculating the ratio between the cash revenue and the initial investment. The ROI is recommended to accepting or rejecting single projects [48]. The ROI is not suitable to choose among mutually exclusive wind farm designs for the same reasons presented before.

2.3.9. Benefit to Cost Ratio

The benefit to cost ratio (BCR) [–] is similar to the NPV function but instead of subtracting the CAPEX to the revenues, it computes a ratio. Higher values for the BCR function represent better wind farm designs. Similarly to other functions the BCR function is recommended to accepting or rejecting single projects but not for selecting among mutually exclusive alternatives. However, it can be used to differentiate the size of the projects if the BCR ratio is obtained with incremental revenues and expenses [48].

To calculate the incremental ratio, the different projects are sorted by increasing CAPEX values and the BCR of the first project is calculated. Next, the incremental differences between the revenues and CAPEX of the second project to the first are obtained. The ratio of these values is then multiplied by the BCR of the first project. If the result is higher than one, the second project becomes the reference. The procedure is repeated until a ratio lower than one is obtained [48].

2.3.10. Annualized Value

The annualized value (AV) (bn €) represents the annual revenues. The AV is constant if no escalation of the price of energy or the OPEX are considered [48].

Table 3. Economic functions for the wind farm layout optimization problem (WFLOP) [48].

Function	Equation	Parameters
AED (GWh/year)	$8760 \sum_{i=1}^{angle\ bins} \sum_{j=1} f_i^j \cdot \left(\left(\sum_{k=1}^{turbines} T_k \left(\overline{wind_i^j} \right) \right) - \text{losses} \left(\overline{wind_i^j} \right) \right)$	f_i^j—wind frequency in direction i and wind bin j, $T_k \left(\overline{wind_i^j} \right)$—power production of the k-th turbine, losses $\left(\overline{wind_i^j} \right)$—system electrical losses
UF [-]	$\frac{AED}{8760 \sum_{j=1}^{turbines} P_{rated}^j}$	P_{rated}^j—rated power of the j-th turbine
COP (M€/MWh)	$\frac{CAPEX}{\sum_{j=1}^{turbines} P_{rated}^j}$	
LCOE (€/MWh)	$\frac{1}{AED} \left(\frac{CAPEX}{a} + OPEX \right)$	a—annuity factor ($a = \frac{1-(1+r)^{-n}}{r}$), r—interest rate, n— wind farm lifetime, OPEX—annual operational expenditure
NPV (bn €)	$(AED \cdot p_{kWh} - OPEX)\, a - CAPEX$	p_{kWh}— market energy price
IRR (%)	$(AED \cdot p_{kWh} - OPEX) \frac{1-(1+r_{IRR})^{-n}}{r_{IRR}} - CAPEX$	r_{IRR}—interest rate that zeroes the NPV equation
DPT (years)	$\frac{n \cdot CAPEX}{(AED \cdot p_{kWh} - OPEX) \cdot a}$	
ROI [-]	$\frac{(AED \cdot p_{kWh} - OPEX) \cdot a}{CAPEX}$	
BCR (bn €)	$\frac{AED \cdot p_{kWh} \cdot a}{CAPEX + OPEX \cdot a}$	
AV (bn €)	$AED \cdot p_{kWh} - OPEX$	

2.4. Current Optimization Frameworks

Figure 4 shows the current optimization strategy used by wind farm designers. The main disadvantage of this approach is that, by varying the economic factors, e.g., interest rate and price of energy, of the objective functions, it is not guaranteed that the layout solution remains the most advantageous. In fact, optimization strategies that optimize economic functions such as the LCOE or the NPV, may be thought as weighted optimization procedures, in which a MO problem is simplified into a single problem one.

Current single-objective optimization frameworks only output one layout which is, often, pre-constrained due to decision taken before the optimization. An example of this practice is the choice of the wind turbine model or the number of turbines in the initial phase of the design process (see Section 1). However, the designers should not limit the design themselves beforehand but rather should explore all the designs and select the most suitable design only after a careful inspection of economic factors and high-level constraints.

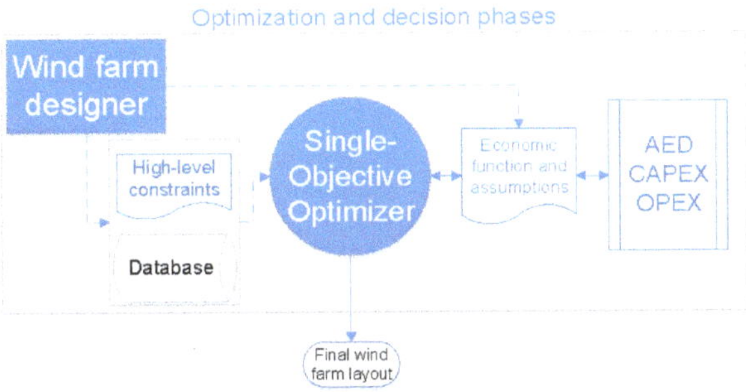

Figure 4. Flowchart of current single-objective optimization strategies. Dotted-line arrows represent input from the designer, solid-line arrows represent algorithm flow and dotted-point-line arrows represent component use.

3. Multi-Objective Optimization Framework

3.1. Optimization Goals

All economic functions used to evaluate the profitability of OWFs have the same backbone variables: the AED, CAPEX and OPEX. They were the chosen as optimization goals because they retain their original function direction when optimizing any economic function. In other words, to optimize an economic function $f\,(AED, CAPEX, OPEX)$, the AED must be maximized and both the CAPEX and OPEX have to be decreased. In this way, it is possible to obtain optimized wind farm designs without setting *a priori* economic factors, e.g., interest rate or price of energy.

Although MO optimization is generally slower and computationally more demanding than single-objective optimization due to the larger problem search space (optimizing conflicting goals is more challenging than optimizing only one function), the former only requires one optimization routine because it optimizes all the goals simultaneously and therefore the optimization and decision phases are decoupled (see Figure 5). In this way, time is saved since only one optimization routine is run. Afterwards, all the possible results are presented to the designers. Nonetheless, the proposed framework still depends on component costs. The optimization strategy would be more flexible if it differentiated wind farm layouts solely on raw comparisons (e.g., cable length, number of turbines or support structures height) without relying on any costs. However, as for any optimization problem, adding many optimization goals is counter-productive as it raises the problem complexity and largely increases the required computational cost.

The optimization framework targets the FEED phase of an OWF. At this stage the farm layout is not yet decided, but the wind farm location is already defined [15]. Therefore, the macro siting optimization of the wind farm is out of scope. It is considered that the project location resulted from several factors that minimized environmental impacts such as bird and marine life, military activity, nature conservation, shipping and navigation, sub-sea cabling and pipelines, existing offshore industries and aviation [22].

Furthermore, it is assumed that the location has suitable wave and tidal conditions, soil properties, feasible water depths, as well as an economically viable wind resource. Although transportation and installation of wind farm components are not directly considered, part of these costs are captured in the framework by avoiding the most challenging seabed areas and by including an installation cost share in the price of turbines.

Finally, a pre-selection of components, e.g., wind turbines and cables, has been performed beforehand. In this way, the wind farm designers provide a database with information on these components to the optimization phase (see Figure 5).

3.1.1. Selection Criteria

The following criteria were used to select the design aspects to be incorporated in the framework:

- Topic relevance—only design considerations that are relevant for OWFs should be included. For example, as mentioned in Section 2, visual impact, shadow flickering and noise are not relevant in offshore environments [15,19,51] although they are highly relevant when designing onshore wind farms.
- Impact—the optimization framework should include design features that play an important role in the wind farm layout. This is because the optimization algorithm needs to capture the difference between the wind farm layouts [47].

The trends of the different components will be assessed in the next section to assure that the components selected in the case study of the paper are representative of the current offshore wind industry. Furthermore the trends of the components will also identify the important design characteristics, e.g., number, location and type of offshore substations, that need to be implemented in the optimization framework.

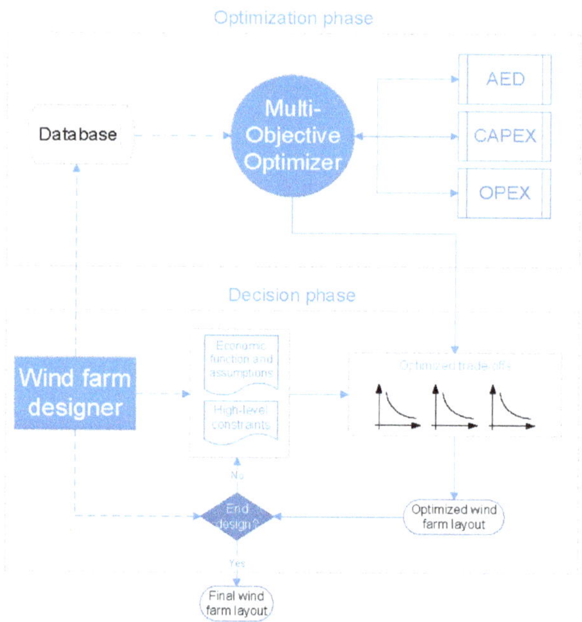

Figure 5. Proposed MO optimization framework for the design of OWFs and their electrical infrastructure. Dotted-line arrows represent input from the designer, solid-line arrows represent algorithm flow and dotted-point-line arrows represent component use.

4. Selected Design Aspects and Optimization Variables

This section presents the optimization variables and respective design aspects which, based on the aforementioned selection criteria, should be considered in the optimization of OWF layouts and their electrical infrastructure. The optimization variables and design aspects are presented separately for the main components of an OWF: turbines, offshore substations and cables (shown in Figure 6). Table 4 summarizes all the design aspects with their respective boundaries and constraints. Furthermore, it also shows the direct influences of the design variables over the AED and CAPEX. Next, the industrial trends of each optimization variable are given to assure that the optimization framework here presented embraces all the important points of current state-of-the-art OWFs.

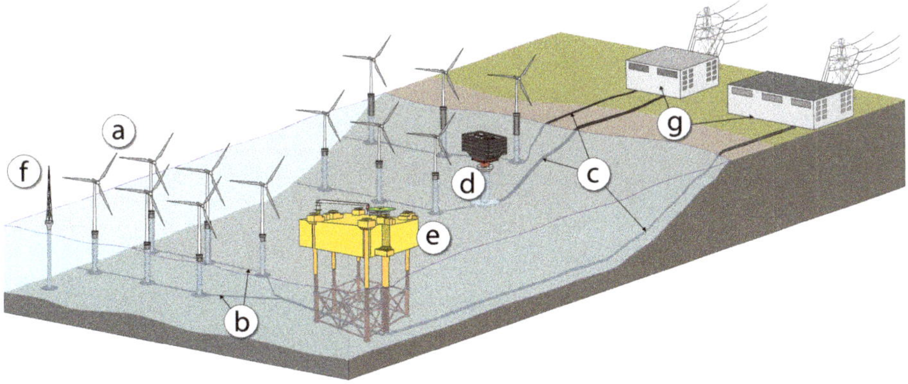

Figure 6. Main components of an OWF: (**a**) Wind turbines; (**b**) Collection cables; (**c**) Export cables; (**d**) Transformer station; (**e**) Converter station; (**f**) Meteorological mast; (**g**) Onshore stations.

Table 4. Design variables: influences, boundaries and constraints.

Design Variables	AED	Costs	Boundaries	Constraints
Wind Turbines				
Number	Energy conversion, wakes losses	Turbine cost, collection cables cost	Maximum stated in the development consent order (DCO) or maximum turbine packing	Wind farm efficiency, capacity factor, installed capacity
Location	Wake losses	Support structure costs	Wind project area	Turbine separation, water depth, natural and man-made constraints [15]
Model	Energy conversion, wakes losses	Turbine costs	Pre-selected turbine models	Commercial and consistency reasons [15]
Offshore Substations				
Number	Electrical losses	Components cost, substation cost, cabling costs	Maximum stated in the DCO	—
Location	Electrical losses	Support structure costs, cabling costs	Wind project area	Same as for the turbine locations [52]
Type	Electrical losses	Components cost, support structure costs, substation cost, cabling costs	Technologies allowed in the DCO	Same as for the turbine models
Collection Cables				
Rated power	Electrical losses	Cabling costs	Pre-selected cables, Voltages allowed in the DCO	Turbines rated power
Rated voltage	Electrical losses	Cabling costs, reactive power compensation costs	Pre-selected cables	—
Transmission Cables				
Rated power	Electrical losses	Cabling costs	Pre-selected cables	Project rated power (overplanting)
Rated voltage	Electrical losses	Cabling costs, reactive power compensation costs	Pre-selected cables, Voltages allowed in the DCO	—

4.1. Wind Turbines

Figure 7 shows that the rated power, rotor diameter and hub height of offshore wind turbines is increasing since the initial projects. The average turbine rated power installed between 2012–2014 rounded the 4 MW mark (see Figure 8a) [53]. This means that, generally, the offshore wind industry prefers larger wind turbines.

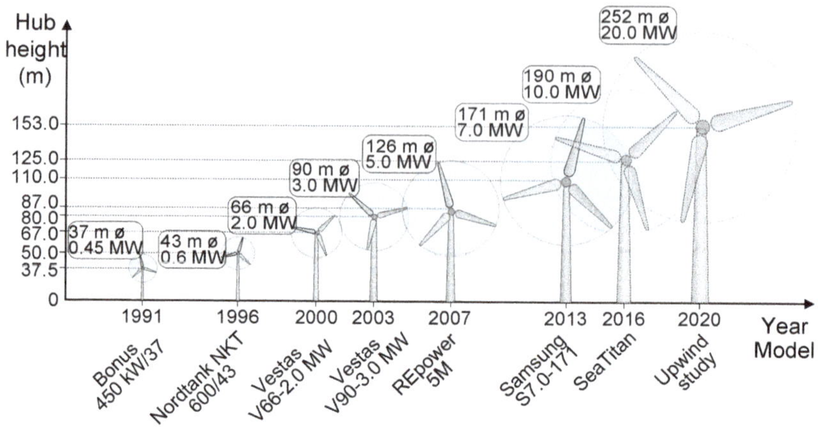

Figure 7. Hub height, rated power and rotor diameter of several wind turbine models and their commission year [10,11].

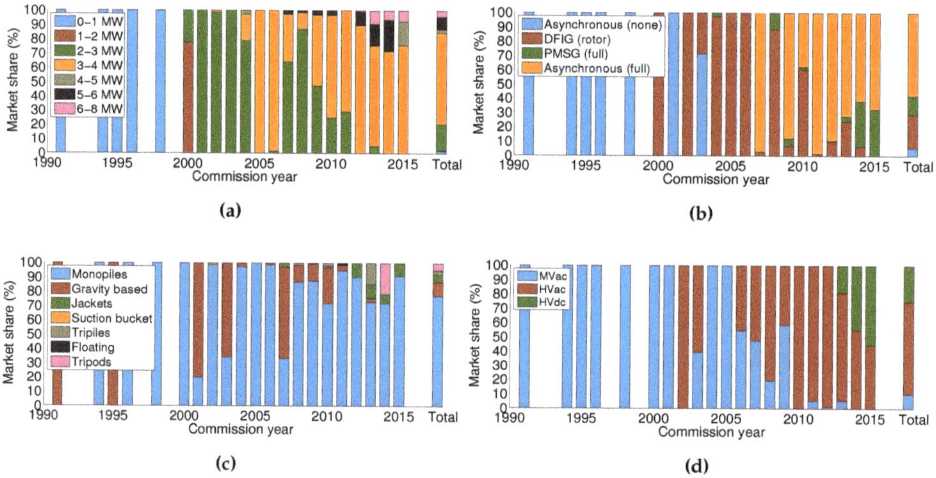

Figure 8. Early and total statistics for commissioned and under construction European OWFs [10,11,54]; (**a**) Turbine rated power; (**b**) Turbine electrical configuration; (**c**) Turbine support structure; (**d**) Transmission technology.

The electrical systems of wind turbines have also evolved. Figure 8b shows that the first turbines used simple fixed-speed systems with asynchronous generators and with no power converters [55]. In a second technological step, pitch-controlled variable speed wind turbines technologies emerged as the preferred technologies and became the dominating type of wind turbines in the following years. Wind turbines equipped with doubly-fed induction generators (DFIGs) constitute approximately 25% of the offshore market [10]. Nowadays, most of the offshore wind turbines are equipped with fully-rated voltage source converters (VSCs) in a back-to-back configuration. These converters allow for enhanced controllability which helps to meet the challenging grid code requirements and the reduction in mechanical loads achieved with variable-speed operation [56].

Regarding the support structures, Figure 8c shows that the monopiles have the highest market share, although different grounded support designs, e.g., tripods, gravity based and jackets, have been tested. Figure 9 shows the most common support structures and two innovative designs: the suction bucket and the twisted jacket [57,58]. Nonetheless, monopiles are expected to maintain their market share for the coming years since efforts are being done to enhance the design of monopiles to be able to support 6–8 MW turbines in water depths up to 40 m [59].

In the far future, floating support structures are expected to obtain a fair share of the offshore wind market due to their use in deep waters [60]. Figure 9 shows several floating solutions which have been tested in the last years. However, much has to be done before floating solutions become economically more viable than monopiles [60].

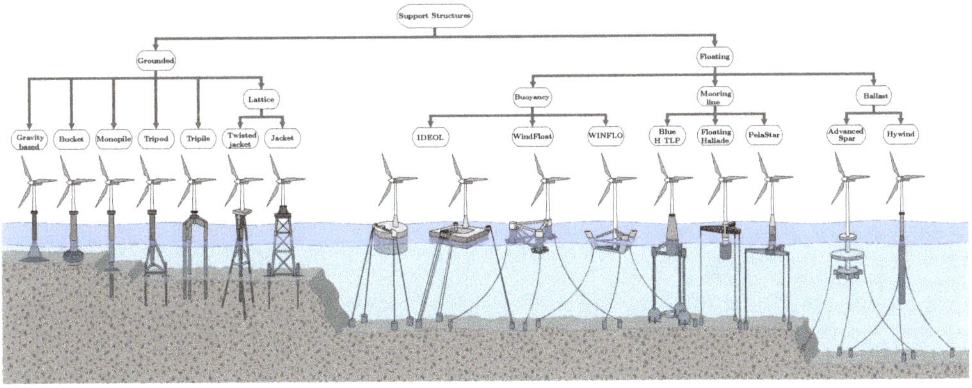

Figure 9. Most common grounded turbine support structures and several existing floating solutions [61–64].

4.1.1. Design Variables, Constraints and Influences

The number, model and location of the turbines have to be simultaneously optimized to guarantee that enhanced layouts are obtained. The locations of the wind turbines have a strong impact on the overall efficiency of the wind farm. The turbines energy production is directly related to the turbines power curve and to the wind resource specific from the wind farm area. Installing wind turbines close to each other causes interferences such as shadowing effects, which lead to reduced power production and, therefore, lower wind farm efficiencies. For example, the Danish Horns Rev OWF produces 11% less due to wake effects [65].

Different turbine models have distinct market prices, hence, it is important to evaluate the influence of the number of turbines and respective types in the CAPEX. Also the support structures have to be assessed since the turbine model and water depth and soil properties play a major role. Support structure costs are typically dominated by the steel price and the influence of the water depth and ground conditions on the structure design [13]. In fact, the support structures investment cost does not scale linearly with their height. Therefore, costly support structures may be prevented if deep locations of the wind farm area are avoided. The proposed optimization framework needs to address the dependency of the cost of the turbine support structures with the seabed water depth.

4.2. Offshore Substations

Since 2002 most of the OWFs make use of offshore substations to increase the voltage levels of the transmission system (see Figure 8d). Currently, there are two types of offshore substations: transformer substations, that simply increase the ac voltage to suitable transmission levels; and converter substations which also perform ac to dc conversion (and viceversa).

4.2.1. Design Variables, Constraints and Influences

The number and location of substations is an important design aspect as it directly impacts the total length of the collection system [66]. The number of substations has also a strong impact in the wind project CAPEX, as well as type of substation. Converter substations are much more expensive due to their larger size and weight, but also due to the extra components required to rectify ac into dc. Furthermore, the type of substation also defines the transmission technology, which plays an important role in the overall system cost and electrical losses [67].

Figure 10a shows that initially one external substation was commonly used, *i.e.*, the substation was placed at the edge of the projects. In a second step, the substation started to be inside the wind farm to minimize the collection system length. Nowadays, large OWFs usually make use of two offshore substations as shown in Figure 3b [10,11]. In this way, an optimization framework has to be able to place offshore substations both inside and outside the wind turbines as well as handle multiple offshore substations.

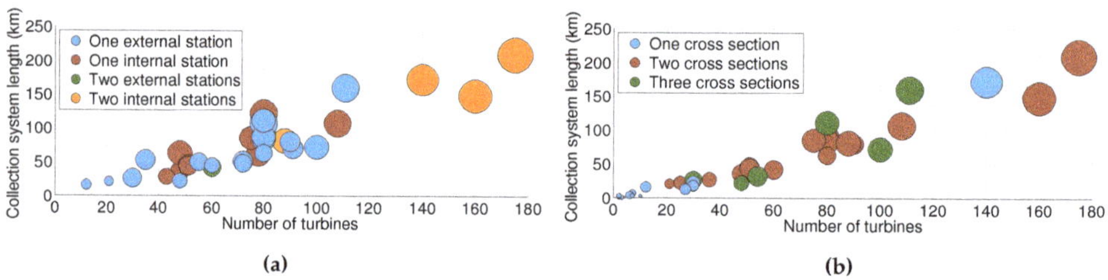

Figure 10. Statistics for commissioned and under construction European OWFs [10,11,54]. Circle size represents the installed capacity of the wind farms; (**a**) Number and location of offshore substations; (**b**) Number of different collection cables.

4.3. Collection Cables

The collection cables interconnect different wind turbines and transport the energy generated to an offshore substation or to shore if no substation is present. More precisely, the collection cables connect to the switchgear inside the turbines to the switchgears of the substation [19,68].

4.3.1. Design Variables, Constraints and Influences

The main design variables of the collection cables are the rated voltage and their cross section which is related to their rated power. So far, 33 kV is the highest and most common voltage level in wind farm collection systems [10]. If a 66 kV collection system was used, the transmittable power would double but the costs would increase by 12% [69]. Therefore, the voltage level of the collection system has to be assessed to assure that optimized layouts are obtained.

The cable layout design is performed during FEED studies and several topologies may be used, e.g., radial, looped or branched arrays [15,70–73]. The design of the collection system plays an important role in the electrical losses. For example, the use of branching may lead to shorter collection systems (Figure 3b shows an example of a branched collection system) and, hence, lower installation costs [74].

The usage of cables with different cross sections is another strategy to reduce the investment costs related to the cabling system. Figure 10b shows that, hitherto, the offshore industry has opted to use array cables with different cross sections only if the total array cable length was above 25 km (an exception is the British Greater Gabbard wind farm) [52]. This strategy allows for costs reduction because cables that carry the power of only a few turbines have lower cross sections and, therefore, need less raw material [75].

4.4. Transmission Cables

Initial OWFs exported their energy production via medium-voltage ac (MVac) transmission systems since the transmission losses were low due to their low installed capacities (up to 100 MW) and close proximity to shore (shorter than 15 km) [22]. The need for an offshore transformer came from the high installed capacities and higher distances to shore [22]. In 2002, the Danish Horns Rev 1 became the first project to make use of an offshore transformer station [10].

Thus far, the industry has used the high-voltage dc (HVdc) technology when distances higher than 50 km and installed capacities above 100 MW were considered [22].

Although HVac technology is used to transport 65% of the energy generated offshore and the HVdc technology accounts for 25% of the total installed capacity (see Figure 8d), currently it is still unclear if one technology will obtain a large share of the market or if several technologies will be used [22]. Therefore, it is desired to obtain an optimization framework that accounts for the different transmission technologies.

4.4.1. Design Variables, Constraints and Influences

The type of cable has a direct impact in the overall system cost as well as in the transmission losses. Furthermore also the cross section of the cables is directly linked with their power transfer capability. Finally, the rated voltage of the transmission system also relates with the power losses and with the obtainable transmission distances. This is of high importance when HVac systems are considered due to the large reactive power compensation needs [76–78].

5. Case Study

The key variables during the design of an OWF, the optimization goals and the need for MO optimization were introduced in the previous sections. Now, the proposed framework is applied to a case study to demonstrate the advantages and the design capabilities given to the designers.

5.1. Borssele Wind Farm

In 2013, the Netherlands defined an Energy Agreement which set as target to increase its renewable energy production by 14% until 2020 [79]. To achieve this target the government has allocated up to € 18 billion to subsidize the offshore wind industry. The Energy Agreement identified the need for an additional offshore capacity of 3.5 GW by 2023. For this purpose, 700 MW licences and respective subsidies will be tendered annually between 2015 and 2019 [80].

Several areas, shown in Figure 11, have been identified for the development of OWFs. The areas Borssele, IJmuiden Ver, Hollandse kust and Ten Noorden van de Waddeneilanden cover approximately 2900 km^2 (representing 5% of the Dutch EZZ [81]) and have a potential offshore wind capacity of 17.4 GW (assuming an average of 6 MW/km^2) [82].

The first two zones of the Borssele wind farm area are considered for the case study. The Dutch Borssele wind farm zone is situated 0.5 km away from the Belgian exclusive economic zone [83] and has a total area of 344 km^2 which is divided into four sites with a maximum installed power of 350 MW each (see Figure 12a) [83].

Two 700 MW platforms will be built in the Borssele area [80]. Each platform will connect two sites to the onshore grid via 220 kV ac cables [83,84]. The developers have a maximum bid value of 0.124 €/kWh for the Borssele zones I and II [83]. The collection system is planned to have a rated voltage of 66 kV [83,85].

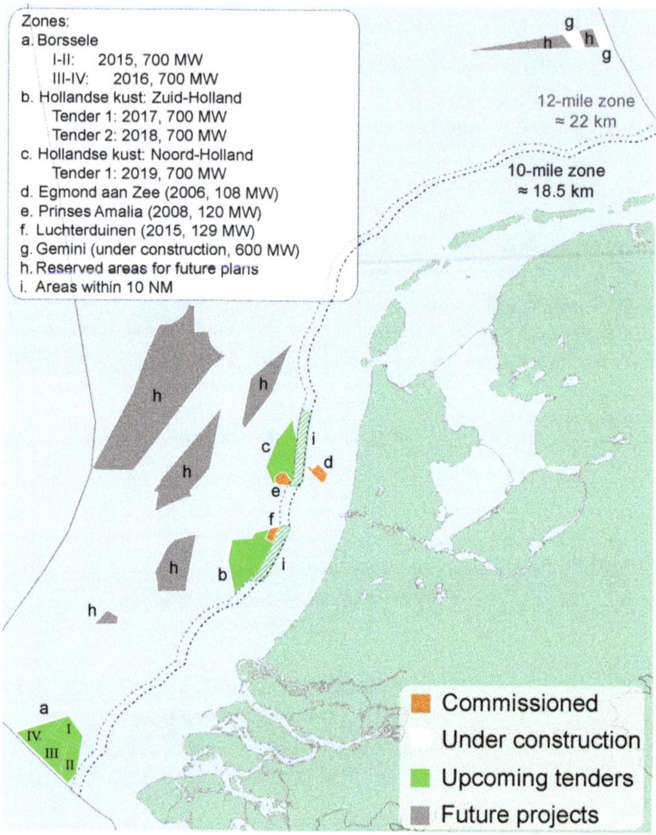

Figure 11. Dutch commissioned, under construction, planned and future OWF areas [81,82].

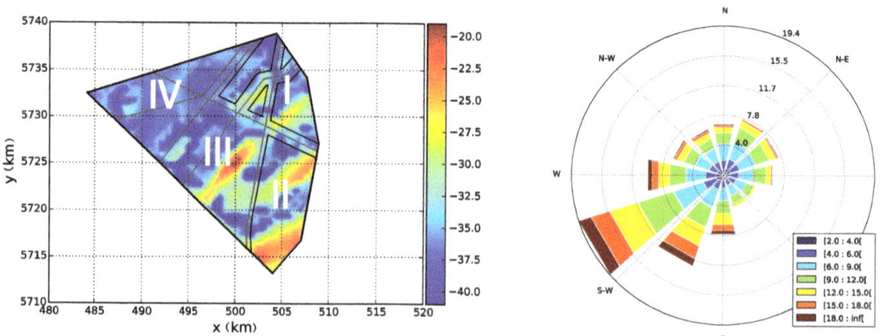

(a) (b)

Figure 12. Description of the Borssele wind farm area, existing pipeline and telecom cables and water depth [84]. The color bar of the left figure represents the water depth in meters and the legend of the right figure presents the wind bins in m/s. (**a**) Wind farm water depth [84]; (**b**) Annual wind rose at height of 90 m [86].

The wind farm area, mean water depth and existing pipelines or telecom cables are shown in Figure 12a. The water depth ranges between approximately 16 and 38 m [83]. The presence of unexploded ordnances (UXOs), which are likely to be present at the site, are neglected since they will be removed if necessary by governmental agencies without additional costs to the wind farm developers [87,88].

Although, most of the existing wind farm projects make use of a standard grid layout [10,11], this is not the case for the Borssele wind farm area in which there are no restrictions regarding the placing of the turbines [89,90].

The wind resource data (see Figure 12b) was measured at four different locations [91]. The measurement points are located between 15 to 82 km of the offshore wind farm and the measurement height varied between 21 and 315 m above sea level [91]. The measurement campaigns were carried between 5.5 months to 12 years [91]. The long-term mean wind speed at a hub height of 100 m above sea level at the center of the Borssele zone is 9.6 m/s [91]. The wind speeds were extrapolated to the turbine hub heights to account for its vertical profile [92]. The mean wind speed at hub height of the Vestas turbine (107 m) is 9.67 m/s, whereas for the Darwind turbine (hub height of 92.5 m) a value of 9.52 m/s is obtained. Lastly, it was considered that the wind turbines could operate during 80% of the time for wind speeds higher than 18 m/s, *i.e.*, during 80% of that time the wind speeds were lower than 25 m/s and, in this way, the turbines operate. For wind speeds higher than 25 m/s the turbines have to be shut down for safety reasons (see Figure 13).

Lastly, the point of common coupling (PCC), situated in Borssele, is approximately 50 km away from the wind farm area [93] and the length of the export cables are considered to be 15% longer than the distance to shore to account for any necessary rerouting of the transmission cables (see Figure 3b).

5.1.1. Component Data

Turbines

The turbines considered in the case study are the Vestas 8 MW with a rotor diameter (RD) of 164 m and hub height of 107 m [94] and the Darwind 5 MW with a 115 m rotor diameter and a hub height of 92.5 m [95]. Figure 13 presents the power and thrust curves of both turbines. Since these curves represent the power delivered at the output of the turbine, it is not necessary to calculate the power losses in the turbine electrical system [25]. Both turbines makes use of full-rated power converters connected in back to back. A minimum separation between two wind turbines has to be respected [96]. In this case study a minimum distance of 4RD was used [84].

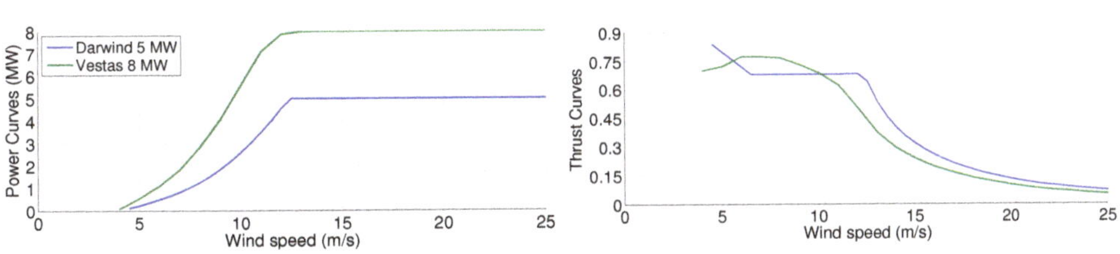

(a) (b)

Figure 13. Power and thrust curves of the turbines used in the case study. (**a**) Power curves; (**b**) Thrust curves.

Cables

All cables are modeled with a π-model which takes into consideration the type of soil surrounding the cables and the geometric dimensions of the cable according to the IEC standards 60228 and 60287 [97,98]. For a thorough description of the cable model the reader is referred to [76,99]. Table 5 presents the properties of the all the cables used in the case study. The 33 kV cables (240 or 500 mm^2

of cross section) are able to carry 5 or 8 Darwind turbines; and 3 or 5 Vestas turbines, respectively. Regarding the 66 kV cables, they could carry the power of 11 or 16 Darwind turbines; and 6 or 10 Vestas turbines, respectively.

Table 5. Parameters of the cables used in the case study [97,98,100].

	MVac				HVac						HVdc							
Rated Voltage (kV)	33		66		132			220			±150				±320			
Cross section (mm^2)	240	500	240	630	500	800	1000	500	800	1000	240	630	1200	2000	240	630	1200	2000
Rated Power (MVA or MW)	27.7	40	54.9	81.7	150	178	189	250	295	330	209	374	537	722	446	797	1147	1540
Current Rating (Copper) (A)	485	700	480	715	655	775	825	655	775	866	697	1246	1791	2406	697	1246	1791	2406
Resistance @20°C (mΩ/km)	80.4	44.7	85	41.4	51.3	37.4	34	39.1	24.6	21	75.4	28.3	15.1	9	75.4	27.3	15.1	9
Capacitance per phase (nF/km)	280	300	220	320	140	170	190	140	170	190	—	—	—	—	—	—	—	—
Inductance per phase (mH/km)	0.34	0.31	0.38	0.33	0.43	0.4	0.38	0.43	0.4	0.39	—	—	—	—	—	—	—	—
Cost (k€/km)	243	373	283	437	598	782	886	771	1009	1301	502	664	824	1005	502	677	1019	1405

Multi-Level Modular Converters (MMCs)

A MMC steady-state model recently introduced in the literature is used in this work [76,101,102]. The model captures the conduction and switching losses of the semiconductors, the conduction losses of the arm reactors, the cooling system losses and the influence of the temperature of operation. The MMC is based on Infineon IGBTs and its properties are shown in Table 6 [103]. The MMC is considered to balance the voltage of the capacitors of the submodules through the reduced switching control strategy which lowers the switching losses and allows the efficiency of the MMC to be close to 99.5% [101].

Transformers

The transformers are modeled with a T-model which considers the core and ohmic losses (see Figure 14) [76,99]. The transformer parameters (shown in Figure 14) are based on the three-phase transformer of the SimPowerSystems toolbox of MATLAB Simulink [76].

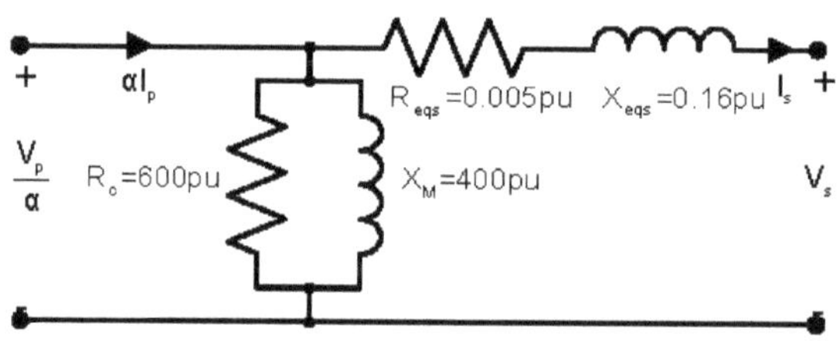

Figure 14. Transformer model [76,99].

Table 6. Multi-Level Modular Converter (MMC) parameters [76,77,101].

Parameter	Value
Rated power	>500 MVA
Rated dc voltage	±150 kV or ±320 kV
Rated ac voltage	380 kV
System frequency	50 Hz
Carrier frequency	100 Hz
Number of arm SMs	58
Arm resistance	0.07 Ω
Reactance phase reactor	0.1 pu
Resistance phase reactor	0.01 pu

5.2. Optimization Goals

5.2.1. AED

Energy Production

The energy generated by the turbines is calculated based on their power curves. The availability of the turbines is considered to be 97% which is a value envisioned to be obtained between 2015 and 2035 [104].

Wake Losses

The Jensen model was used to calculate the wake interaction between turbines. The Jensen model, one of the most common models used, is a simplified and fast manner of calculating the wind speed inside the wake of a turbine [25]. More precisely it was used a later version of the model, which was developed by Katic *et al.* [105]. All the MO approaches presented in Table 1 used the Katic-Jensen model (except [29,30] which used the original Jensen model). For a description of the wake model the reader is referred to [105–107].

Collection System Design

Finding the optimal branched infield cable topology for OWFs corresponds to the capacitated minimum spanning tree problem [108]. A savings heuristic approach is used since the optimization framework requires fast models. It is a hybrid between the heuristics Planar Open Savings [72] and Esau and Williams are the possibility for multiple cable capacities, the minimization of pipeline/cable crossings and the option to select the maximum number of cables branches per turbine [75].

It is assumed that the collection system connects every turbine to one substation through one distinct cable. The inter-array cables cannot cross each other or the transmission lines of their substation. The maximum number of cable connections per turbine is considered to be four due to space limitations inside the turbine tower [68,85]. Finally, the crossings with existing pipelines and telecom cables should be minimized [66,109]. For a more detailed explanation of the heuristic the reader is referred to [75].

Electric Infrastructures

Three different transmission technologies (TT) are considered in this work: MVac, HVac and HVdc (see Figure 15). The MVac interconnects directly the turbines to the PCC, whereas the other two options make use of offshore substations.

Recent studies investigated the advantages of using an overplanting strategy. Such design strategy breaks the common rule used up to now of using a transmission system with the same power capacity of the wind farm. The main idea is to harvest energy from more turbines for low wind speeds and apply curtailment strategies when the wind speed is too high [110].

This idea is implemented in the HVdc scenario by setting the MMC rated power as the sum of the turbines rated power that are interconnected to it. Since the number of cables per MMC is fixed, it is assumed that if the chosen cables have not enough energy transport capacity, the algorithm upgrades them to have larger cross sections. If the cable with the highest cross section available is not able to carry all the power, the MMC caps the power that flows through it to its rated power and the rest is curtailed.

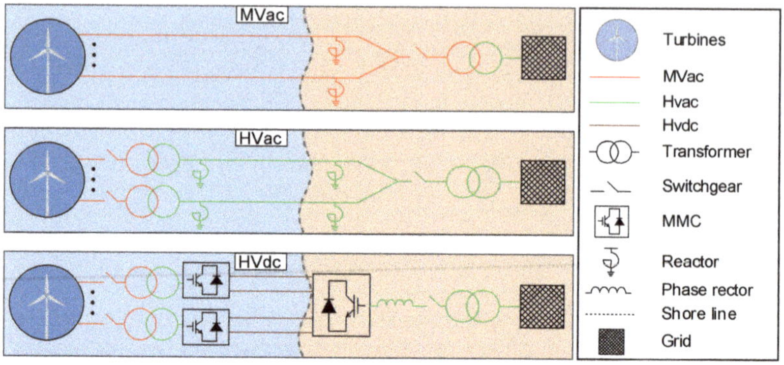

Figure 15. Electrical infrastructures considered in the optimization framework.

For the case of HVac technology if there is at least one power flow in which the chosen cable is not able to transport all the power, the number of cables of that substation is increased by one and the power flows are run once again. Lastly, the number of turbines that could be connected to each MVac cable is calculated assuming the turbines to operate with a power factor of 0.95 to allow for some reactive power transport [85].

Power Flows

The power flows from the turbines to the PCC are calculated using a power flow and optimal power flow solver library developed in the Python programming language called PyPower [111]. There are three possible types of nodes in a network: PQ, PV and slack nodes. A node PQ is a node in which the active and reactive power are known and the state variables are defined as the load angle (δ) and voltage. On a PV node the active power and voltage level are known but the reactive power and δ are unknown [77]. Finally, on slack nodes a reference angle and the voltage are set but the active and reactive powers are not determined before solving the load flow. Table 7 shows a comparison between the different types of network nodes.

The turbines are considered to be nodes of type PQ, *i.e.*, the active power is set to the values obtained with the wake losses model and it is considered that the turbines are not producing any reactive power. In this way the power flow equations determine the voltage and angle at the turbine nodes. Alternatively, the turbines could be set as PV nodes to help in the reactive power support. However, this falls under the wind farm power control which is out of the scope of this work.

The ac side of the MMCs and the onshore network nodes are defined as slack buses, *i.e.*, the voltage is equal to 1 pu and the angle δ is set to zero degrees. The transformer nodes are considered to be of type PQ and with no net injection of active or reactive power flowing into their nodes. The sizing of the transformers is initially done by summing the turbines rated power which are connected to it. After running the power flows if the rated power of the transformers is surpassed, these become the new values and the power flow scenarios are repeated.

The reactors nodes (see Figure 15) are considered to be of type PV (see Table 7). It is considered that no active power flows to the reactors and that the voltage is equal to 1 pu. In this way, the reactors provide total reactive power compensation. The reactor sizing is based on the scenario in which no power is produced at the turbines, since it represents the most demanding case for the reactive power compensation units [77].

Table 7. Comparison between the network node types [77].

Node Type	Known Variables	Unknown Variables
PQ	Active power and Reactive power	Voltage and δ
PV	Active power and Voltage	Reactive power and δ
Slack node	Voltage and δ	Active power and Reactive power

5.2.2. CAPEX

The cost models used in this case study are based on several literature works [49,74,85,112–115] and it is not the scope of this work to further develop or validate them. The cost values used are updated considering an annual Eurozone inflation of 1.85% [12].

Turbines

The price for the turbines is obtained using the data in [74]. The cost of the Darwind turbine is 4.85 M€ and the Vestas 8 MW costs 6.39 M€. The values include 10% extra for the transport and installation.

If the 66 kV collection system is used, the cost of the turbines increases by 20% due to the extra costs of the turbine transformers and switchgears [85].

The cost to decommissioning the turbines at the end of the wind farm lifetime is set to 1.05 M€ for the Darwind turbine and 1.68 M€ for the Vestas 8 MW [116].

Support Structures

The cost of the monopile support structures and their transport and installation is given by [74]:

$$C_{ss} = 0.54 \cdot P_{\text{rated}}^{turb} \left(1 + 0.02\left(w_d - 8\right)\right) \left(1 + 0.8 \cdot 10^{-6} \left(h\left(\frac{RD}{2}\right)^2 - 10^5\right)\right) \tag{1}$$

where C_{ss} is the support structure cost in M€, w_d is the water depth in meters, h is the turbine hub height in meters, RD is the rotor diameter in meters and P_{rated}^{turb} is the turbine rated power in MW.

Cables

The cost values of the 33 kV collection cables are obtained from [74] and are shown in Table 5. The cost of the 66 kV cables is considered to be 15% more expensive than the 33 kV cables for similar cross sections [85]. The cost to install the MVac cables is considered to be 365 k€/km independently of the voltage level or the cable cross section [74,85].

The costs of the HVac cables are based on values taken from [114] and are shown in Table 5. An additional cost of 720 k€/km is added to account for the installation of the cable [74]. The costs of the HVdc cables are based on [115] and are also shown in Table 5. Similarly to the HVac cables, the installation had an extra cost of 720 k€/km [74].

The cost to cross an existing pipeline or telecom cable is considered to be 100 k€ [75].

Switchgears

The switchgears for the 33 kV collection system cost 73.25 k€ and the 66 kV ones cost 101.25 k€ [114]. One MV switchgear is installed in each collection cable as shown in Figure 15. The HV switchgears cost 1.57 M€ for a rated voltage of 132 kV and 2.60 M€ for 220 kV transmission systems [115].

Offshore Substations

The cost of the HVac offshore substations is given by [114]:

$$C_{ss}^{HVac} = 2.8286\text{ M€} + 0.099 P_{ss} \quad (2)$$

where P_{ss} is the sum of the substation turbines rated powers (in MW) and C_{ss}^{HVac} is the cost M€.

The same expression is used for the HVdc substations but with an 85% increase to account for the larger and heavier structure needed to house the extra components [117].

The cost of reserve diesel generators to run the essential equipment of the substation is given by [74]:

$$C_{gen} = 0.0237\text{ M€} + 0.0023 P_{ss} \quad (3)$$

where P_{ss} is the sum of the rated powers of the turbines that interconnect to the substation (in MW) and C_{gen} is the cost of the diesel generator in M€.

Transformers

The transformer costs are expressed as [74]:

$$C_{trafo} = 0.0477 P_{trafo}^{0.7513} \quad (4)$$

where P_{trafo} is the rated power of the transformer in MVA and C_{trafo} is the cost in M€. The minimum transformer rated capacity is considered to be 50 MW.

MMCs

The cost of the MMCs is given by the following expression [115]:

$$C_{MMC} = 61.3777\text{ M€} + 0.0657 P_{MMC} \quad (5)$$

where P_{MMC} is the rated power of the MMC in MVA and C_{MMC} is the cost of the MMC in M€. The minimum MMC rated capacity is consider to be 500 MVA.

Reactive Power Compensation

The 33 kV reactors are considered to have an efficiency of 98% [118] and cost 66% of what a transformer of the same rating would cost [114]. A 40% extra cost is added to the 66 kV reactors due to their higher volume and weight [85].

SCADA, Project Development, Insurance

The SCADA system is considered to cost 0.75 M€/turbine [74], whereas the project development is estimated as 0.468 M€/MW [74]. The insurance and contingencies during the project phase are considered to be 0.7 M€ per Darwind turbine and 1.12 M€ for the Vestas turbines [116].

5.2.3. OPEX

The operational costs are an important component of OWFs and, hence, should be considered during the optimization phase. Inasmuch to use a more comprehensive OPEX calculation model, such as the OPEX Cost Estimator from ECN [119], a considerable amount of data is required, e.g., available vessels, crew members and detailed weather information.

Simpler approaches can be used to avoid this drawback. The OPEX may be considered as a percentage of the investment costs [44,45] or monetized through a cost value per MWh produced by the farm [49,120,121]. Because the value of the OPEX is directly proportional to either the AED or the

CAPEX, it is not included in the optimization process and is only calculated afterwards. In this way, the problem becomes bi-objective with the AED and CAPEX as optimization goals.

5.3. Multi-Objective Gene-Pool Optimal Mixing Evolutionary Algorithm (MOGOMEA)

The Multi-objective Gene-pool Optimal Mixing Evolutionary Algorithm (MOGOMEA) was used in this work. The MOGOMEA, recently introduced in the literature [122], is a metaheuristic discrete MO optimization algorithm which demonstrated to outperform, *i.e.*, obtain better optimized trade-offs, other state-of-the-art well-known MO optimization algorithms on standard combinatorial benchmark functions [122], as well as in real-world applications [33,123] for a given budget of function evaluations. Although the MOGOMEA is capable of obtaining high-quality results, it does not guarantee global optimal solutions are found given a finite amount of evaluations of the optimization functions [122].

Since the MOGOMEA only uses discrete variables, it is straightforward to optimize the number and locations of turbines in simultaneous. Constraints on the location of turbines, such as minimum distance to existing power cables and shipwrecks, within the OWF area are automatically respected by not considering constrained places in the possible set of locations for the turbines [47]. Furthermore, the limited precision offered by the models used, such as the wake model, do not require the accuracy of continuous variables [33]. The location of a turbine may be described by a single parameter in a discrete approach, whereas a continuous domain demands two variables. Lastly, given the same number of problem variables and the same inherent underlying problem complexity, continuous optimization problems are typically slower to solve than discrete problems because of the larger variable-domain size of continuous variables [33].

MOGOMEA is started by initializing n solutions (the definition of a solution is later introduced in Equation (6)) which are evaluated to obtain their objective values. The algorithm then alters solutions into offspring by exchanging variables between different solutions. The MOGOMEA clusters solutions that are in the same objective space vicinity and only performs variation within each cluster since solutions tend to be very dissimilar for different areas of the objective space [122].

The general variant of the algorithm uses the mutual information as measure to build a hierarchical linkage model to identify group of variables that are important to be copied together under black-box optimization, *i.e.*, nothing is known about the optimization problems [124,125]. However, it is known that turbines influence the energy production of neighboring turbines and that this influence might be considered negligible for turbines situated far enough apart [126,127]. For this reason, in this work the distance between the possible locations is used as a measure of dependency instead of the standard statistical analysis used in MOGOMEA [33].

The population size used by the algorithm is an important internal parameter that should be adjusted according to the instance of the WFLOP being solved [127,128]. A population-free sizing scheme of MOGOMEA is used, avoiding the need for the right population size to be assessed [33]. In this way, the population size is incrementally increased throughout the optimization [33]. This is done to allow new genetic material into the population since the algorithm performs a very intensive local-search strategy and hence, it may fail to layouts that cannot be reached from the initial solutions (only through mutation) [122].

Although the MOGOMEA was initially designed to have a population of solutions that are sequentially altered to generate offsprings [122], the version of MOGOMEA employed evaluates the solutions in an asynchronous parallel manner. Currently, it is becoming even more important to move towards a parallel environment in which several CPU cores may be used simultaneously [129]. This is of high importance especially because the evaluation time of a solution varies with the complexity of the OWF layout it represents. For example, the wake loss evaluation highly depends on the number of turbines [126,127] and wind directions [107], whereas the collection system design also depends on the number of turbines of the project [72,75].

5.3.1. Optimization Variables

A solution is composed as follows:

$$x = \left[\begin{array}{ccccccc} t_1 & \cdots & t_i & V_{HVac} & C_{HVac} & V_{HVdc} & C_{HVdc} & V_{MVac} \end{array}\right] \tag{6}$$

where $t_1 \ldots t_i$ determine what is placed in each available position of the wind farm area. The following representation is used: 0—Empty, 1—Darwind 5 MW turbine, 2—Vestas 8 MW, 3—Transformer substation, 4—Converter substation. The variables V_{HVac}, V_{HVdc} and V_{MVac} determine the voltage level of the HVac, HVdc and MVac cables, respectively. The variables C_{HVac} and C_{HVdc} identify the cross section of the cables.

5.3.2. Random Wind Farm Layouts

An heuristic that aims at maximizing the distance between turbines is used to create random layouts. Firstly, the number of turbines of the wind farm, m, is randomly generated between one and the number of locations of the grid. The location of the initial turbine is randomly chosen from all the possible locations. After, the distances of the remaining possible locations are computed to the first turbine and the most distant location is chosen for the second turbine. The distances f the remaining locations are updated by checking whether the distance to the new turbine is smaller than the currently stored distance (the shorter distance is kept). The procedure is repeated m times or until a turbine violates the proximity constraint. In this way, it is guaranteed that feasible wind farm layouts are generated, similarly to previous works [130–132].

Whenever a turbine is placed in the layout, the algorithm randomly selects a turbine model from the ones considered. At the end, the turbine model used in the OWF is the one that was assigned the most since, generally speaking, OWFs are composed of only one turbine type [10,11]. Nonetheless, the framework was designed as is so that this constraint could be removed in the future, if desired.

A random number of substations is sampled between zero and three. Next, the k-means clustering algorithm is run with that same number [133]. The closest locations of the grid to the cluster centroids are chosen to represent the substations. Thereafter, another random number is sampled for each substation to decide if it represents either a converter or a transformer substation. In case there is at least one converter station, the algorithm considers that all substations perform ac-dc conversion, *i.e.*, the transmission system uses the HVdc technology.

5.3.3. Multi-Resolution

The results obtained in [33] indicate that it could be beneficial to use a multi-resolution scheme in which a large grid step is employed in the first optimization iteration to allow the algorithm to find highly packed wind farms easier. Thereafter, these solutions would serve as initial population for a second optimization routine with a smaller grid step. Therefore, the algorithm starts the optimization routine with the 4RD grid step and advances to smaller step sizes of 2RD and 1RD. The layouts are then mapped onto the finer grid.

5.3.4. Internal Parameters

The initial population size, n, is set to 32 since it corresponds to the number of CPU cores available. The algorithm groups the wind farm layouts into five clusters during the optimization process since this setting has proved to already provide improved results over using a single cluster in recent literature [122].

6. Results

The case study was run in Python on a server with 32 cores (Intel Xeon ES-2690@2.9 GHz) running the 64-bit version of Ubuntu 12.04. The optimization framework was given in total 30 days to run.

The largest grid step (4RD) size was run for 6 days, whereas 12 days were given to each of the two remaining more refined grid steps (2RD and 1RD).

6.1. Optimized Trade-off

Figure 16 shows the optimized trade-off obtained with the MO optimization algorithm (and for the two layouts designed by hand, which are introduced next) between the AED and CAPEX. The trade-off, composed of 358 different OWF layouts, shows that three different transmission technologies were used. For small wind farms the 66 kV interconnection to shore was the best option. On the other hand, for AEDs higher than 4000 GWh/year, the HVdc transmission technology was used. In the right-hand extreme part of the trade-off curve the HVac technology was used. However, it came with much higher investment costs and did not increase the AED significantly.

Although the optimized trade-off shows a linear relationship between the goals up to an AED of approximately 7000 GWh/year, it is important to note that nothing guarantees that the layouts that are in a close vicinity in the objective space, are also similar in variables space, *i.e.*, the layouts and technologies may vary considerably.

The optimization case study had also the possibility to use the Darwind 5 MW wind turbine; however, no layout that is present in the trade-off used it. The results for the standard layouts also corroborate this result: although the wind farm layout composed of Vestas 8 MW turbines is situated in the trade-off, the OWF with Darwind 5 MW turbines performs much worse than the other layouts. The Darwind 5 MW wind turbines produced less energy and, even though they are cheaper, this difference did not suffice for profitability.

The data gathered for this specific case study hinted that the best turbine was the Vestas one. However, this result is specific for the case study of this paper and should not be generalized since turbines with lower rated powers could be the best option under different circumstances. For example, the support structure costs, insurance costs, among others, do weight in the decision making. Although, one turbine demonstrated to be clearly superior for this case, for other scenarios, e.g., different wind farm area or distance to shore, the outcome could differ.

A more difficult situation would be when one could not have a direct hint from the specifications of the wind turbines. Under such scenario, the advantages of the proposed optimization framework would clearly stand out. Furthermore, the optimization framework is also able to select from several wind turbine types (more than two), which is a very difficult selection process to be manually performed.

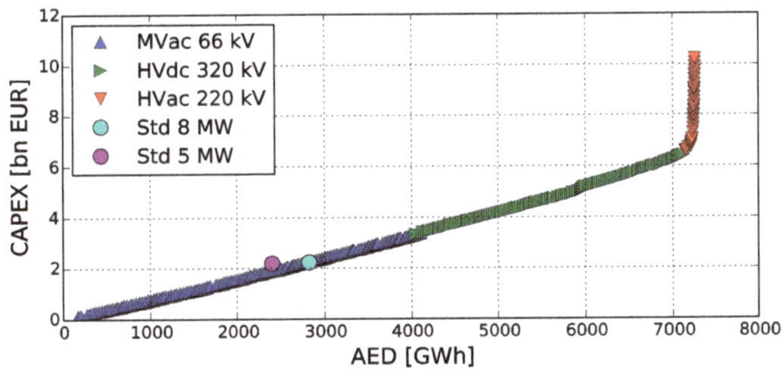

Figure 16. Optimized trade-off obtained between the optimization goals for areas I and II and standard layouts. The optimized layouts will be numbered from left to right, *i.e.*, with increasing annual energy delivered (AED) and CAPEX.

6.2. Wind Farm Layouts Designed with Standard Approaches

Two OWF layouts were hand-designed by taking into consideration the guidelines given to these areas [134]. In this way, 350 MW (352 MW for the 8 MW turbines) were placed in each wind farm area and all the turbines were interconnected to one 220 kV HVac offshore substation placed in the exact location where it will be built [83]. It is important to note that some level of optimization was used during the design of these layouts. The wind turbines were placed in the grid that maximized the distance between them. Furthermore, as shown in Figure 17, the turbines have a larger distance between them in the main wind direction (see Figure 12b). The collection systems were designed with the heuristic algorithm presented in [75].

The grid step sizes used in the standard layouts are different from the ones that the optimization framework had access to. This difference may explain why no layouts that export the power to shore via HVac technology were found in the vicinity of the trade-off (see Section 3.1) shown in Figure 16. Nonetheless, the layout with 8 MW wind turbines did not dominate (was not superior in any way) the layouts of the optimized trade-off.

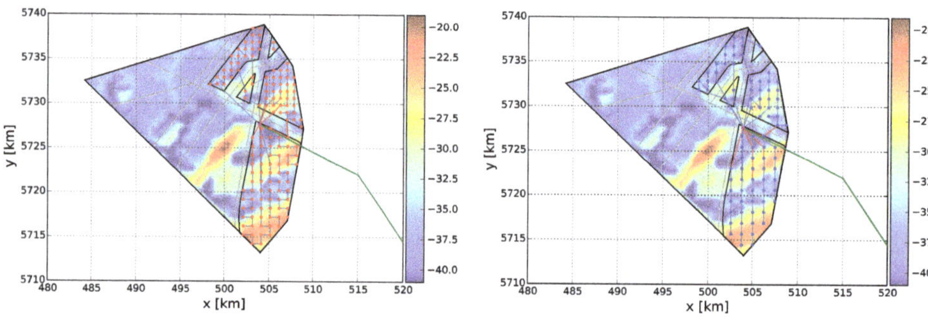

(a) (b)

Figure 17. Layouts obtained with standard design philosophies and restricted to the design specifications [134]. Red and blue circles represent the 5 MW and 8 MW turbines, respectively. Gray lines are existing pipelines and telecom cables, blue and red lines are the collection system cables and green lines are the exporting HVac cables. The purple circles represent the offshore substations. (**a**) Layout with 140 turbines (5 MW); (**b**) Layout with 88 turbines (8 MW).

6.3. Economic Functions

Figure 18 shows how the layouts of the optimized trade-off perform according to the economic functions presented in Table 3. Table 8 presents the economic values and general characteristics of the layouts (shown in Figure 19) that demonstrated to be the best according to the different economic functions. The results were obtained with the following assumptions: interest rate of 7% [70], wind farm lifetime of 20 years, annual OPEX of 2% of the respective CAPEX [44,45,135] and price of energy of 0.124 €/kWh, which is the highest bid for the Borssele areas I and II [83]. In Section 6.4 the impact of these parameters will be assessed. Although fixed annual values were used, distinct annual values could have been applied in a straightforward manner to simulate, for example, the financial incentives that the energy generated at OWFs obtain in the first years of exploration [136]. However, these incentives are country specific and, therefore, were not considered.

Table 8. Economic values and general characteristics of the best layouts.

Layout nr.	NPV (bn €)	LCOE (€/MWh)	COP (M€/MW)	UF [–]	IRR (%)	DPT (years)	ROI [-]	AV (bn €)	BCR [–]	AED (GWh/year)	CAPEX (bn €)	IC (MW)	TT	CSL (km)	NSS
3	0.10	88.22	3.32	**0.49**	12.80	13.41	1.49	0.03	1.41	0.28	0.21	64	MVac	76.13	0
5	0.14	83.91	3.11	0.48	13.78	12.67	1.57	0.04	1.48	0.34	0.25	80	MVac	74.53	0
18	0.28	**83.79**	**3.08**	0.48	**13.81**	**12.64**	**1.58**	0.07	**1.48**	0.67	0.49	160	MVac	182.45	0
248	**1.73**	96.18	3.17	0.43	11.23	14.81	1.35	0.63	1.29	5858	4.93	1552	HVdc	281.54	1
322	1.28	107.20	3.39	0.41	9.35	16.80	1.19	**0.76**	0.76	7210	6.76	1992	HVac	181.30	8
358	-2.86	161.21	5.13	0.42	3.12	27.77	0.72	0.69	0.77	**7256**	10.23	1992	HVac	170.65	28
Std 8 MW	1.01	90.19	3.16	0.45	12.55	13.75	1.45	0.31	1.37	2822	2225	704	HVac	138.31	1
Std 5 MW	0.50	104.24	3.12	0.46	10.04	16.26	1.23	0.25	1.19	2399	2187	700	HVac	163.79	1

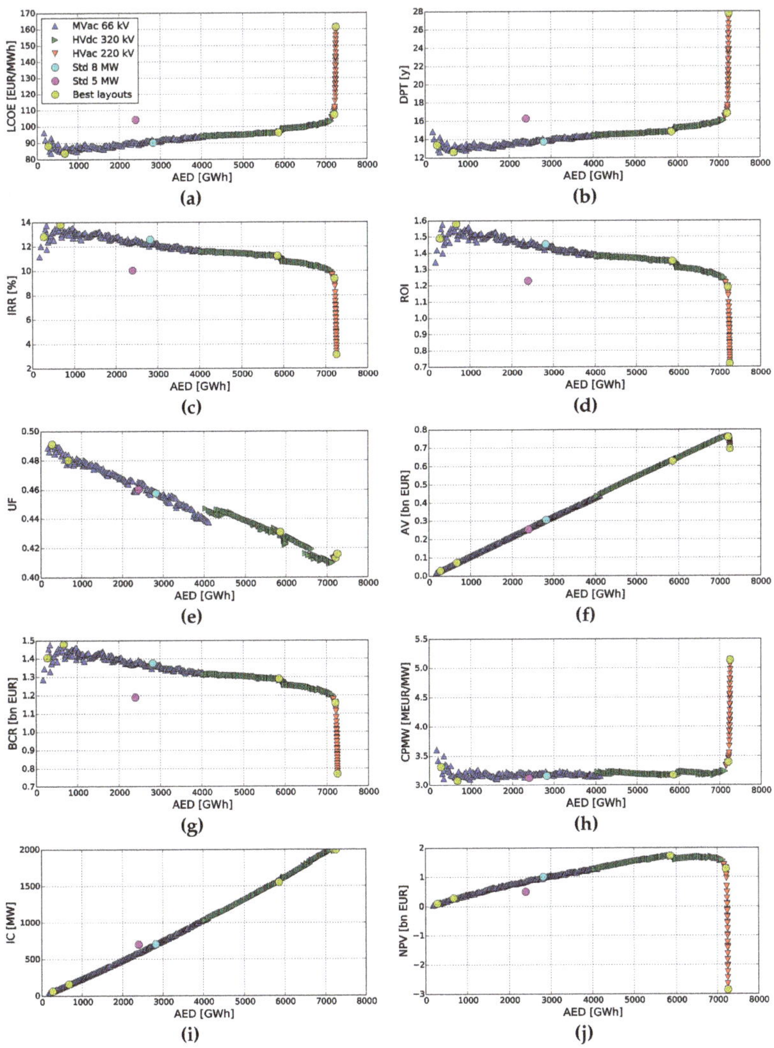

Figure 18. Results obtained for the different economic functions of the layouts of the optimized trade-off. The values obtained with the standard layouts are also shown. (**a**) levelized cost of energy (LCOE); (**b**) discounted payback time (DPT); (**c**) Internal rate of return (IRR); (**d**) return on investment (ROI); (**e**) utilization factor (UF); (**f**) annualized value (AV); (**g**) benefit to cost ratio (BCR); (**h**) cost of power (COP); (**i**) Installed capacity; (**j**) net present value (NPV).

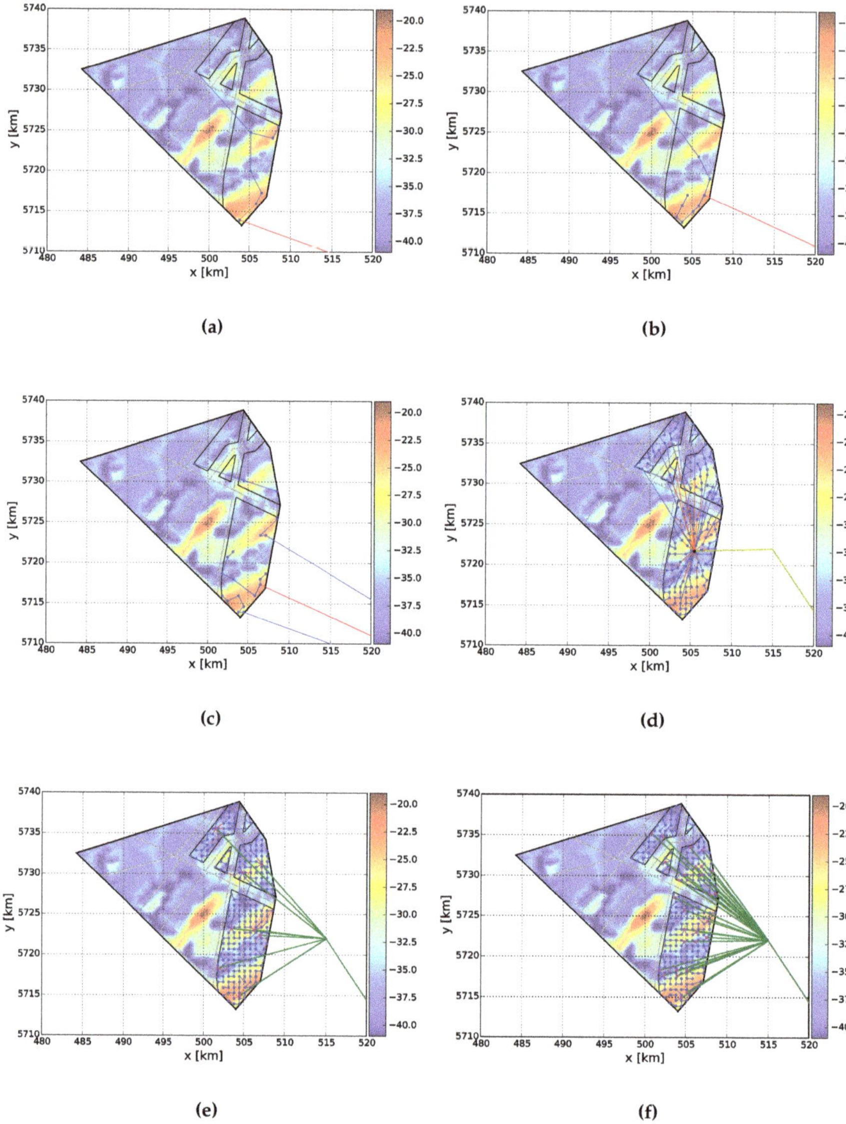

Figure 19. Layouts obtained with proposed optimization framework. Blue, purple and black circles represent, respectively, the 8 MW turbines, HVac and HVdc substations. Gray lines are existing pipelines and telecom cables, blue and red lines are the collection system cables and green and yellow lines are the exporting HVac and HVdc cables, respectively. (**a**) Layout number 3; (**b**) Layout number 5; (**c**) Layout number 18; (**d**) Layout number 248; (**e**) Layout number 322; (**f**) Layout number 358.

6.3.1. AED

Figure 19f shows the layout with the highest AED, whereas some of its characteristics are given in Table 8. The layout uses the concept of distributed substations in which several small offshore substations housing a single transformer and one reactor for compensation of the export cable are used [137]. This concept, as shown in Figure 19f and Table 8, leads to a significant reduction in the total collection system length (CSL) since this layout has 110 km less array cables than layout number 248, which has 55 turbines less. The electrical losses are reduced and the usage of array cables with high cross sections is also minimized in layout number 358. Furthermore, this concept leads to fewer crossings with existing pipelines and telecom cables as shown in Figure 19f. On average, there are approximately nine turbines (72 MW) connected to each substation.

The AED function is fully biased towards energy production maximization, hence, designers should be cautious when applying it in case there are no constraints over the installed power or a maximum project CAPEX. Moreover, the economic values obtained for this layout clearly demonstrated that. This layout obtained a negative NPV, a DPT of almost 28 years and a ROI value lower than one, meaning that the investment would not be able to generate profit for the wind farm developers, despite its very high AED.

6.3.2. Utilization Factor

The layout which presented the highest UF value, shown in Figure 19a, was composed of only eight 8-MW turbines. Since the objective is to minimize the losses one might initially think that a layout with only one turbine would provide the best results. However not only the wake losses are being considered but all the power losses until the PCC. In this way, the layout with eight turbines demonstrated to also make a better use of the MVac cable. The turbines were placed far apart from each other and in shallow areas. The same layout would present the best result if the efficiency of the system would have been used instead of the UF [33].

The standard layout with 5-MW turbines presented worse results than its counterpart composed of Vestas turbines in all function but the UF. This means that despite the lower AED, the 5-MW turbines are slightly better used and it would be in the preferred layout if the UF was the most important decision factor.

6.3.3. LCOE

The LCOE values of the optimized layouts are shown in Figure 18a. Values between approximately 80 to 105 €/MWh were found for MVac and HVdc transmission technologies. On the other hand, the layouts based on HVac technology presented much higher LCOE values due to their higher investment costs than the other layouts which used HVdc and MVac transmission systems.

The layout which presented the lowest LCOE (83.79 €/MWh) was composed of 20 turbines as shown in Figure 19c. Although the turbines were placed far apart to minimize wake losses, they were also placed in regions of shallow waters to minimize the cost of support structures. Furthermore, the turbines were also placed in the wind farm area closest to shore, minimizing the cabling system cost and the electrical losses.

6.3.4. IRR, ROI, COP, BCR and DPT

The same layout that had the lowest LCOE also presented the best values for the IRR, DPT, COP, BCR and ROI functions. Although these economic indicators present different values, they did not alter the ranking between layouts, *i.e.*, if a certain layout performs best at one of these economic functions it also turns out to be the best according to the remaining ones.

6.3.5. AV

The AV equation is biased towards larger wind farms since it measures the annual revenue. The layout that maximized the AV equation is shown in Figure 19e and it is very similar to layout number 358, shown in Figure 19f. The only difference is that it makes use of eight offshore HVac substations to interconnect the 249 turbines to shore.

6.3.6. Incremental BCR

According to the incremental BCR economic analysis, the wind farm layout number 5 (see Figure 19b) was the preferred choice. It is composed of ten 8 MW turbines and, therefore, makes full use of the 66 kV cable with the largest cross section (see Table 5), thus minimizing the ratio between the AED and the investment costs.

6.3.7. NPV

The wind farm layout which presented the highest NPV value (Figure 19d) has one HVdc offshore substation that interconnects 194 turbines to shore.

6.4. *The Influence of Economic Factors*

In this section, the influence of the economic factors over the wind farm layout design is investigated. To this end, different parameters, such as the interest rate, wind farm lifetime and price of energy, were altered to analyze the resulting implications. Initially the LCOE will be analyzed, followed by the NPV function.

6.4.1. LCOE

Assuming the OPEX to be a percentage of the investment costs [44,45], p_{CAPEX}, the LCOE clearly becomes a ratio between the CAPEX and the AED:

$$LCOE = \frac{CAPEX}{AED}\left(\frac{1}{a} + p_{CAPEX}\right) \tag{7}$$

This equation shows that both an alteration to the interest rate and wind farm lifetime or a different percentage of the CAPEX for the OPEX, will affect in the same way all the wind farm layouts since it only shifts the ratio between the CAPEX and the AED.

Another possibility, in the absence of a more refined OPEX model, is to monetize the OPEX through a cost value per MWh, c, delivered by the OWF [49,120]:

$$LCOE = \frac{1}{AED}\left(\frac{CAPEX}{a} + c \cdot AED\right) = \frac{CAPEX}{a \cdot AED} + c \tag{8}$$

Once again, all the layouts will be affected in the same way if the value c is changed. This means that, if the OPEX is somewhat directly dependent on either the CAPEX or the AED, the LCOE function is insensitive to the variation of its economic factors. Figure 20a shows that the order of the layouts of the optimized trade-off remained unaltered, hence, the wind layout number 18 remained the one with the lowest LCOE value.

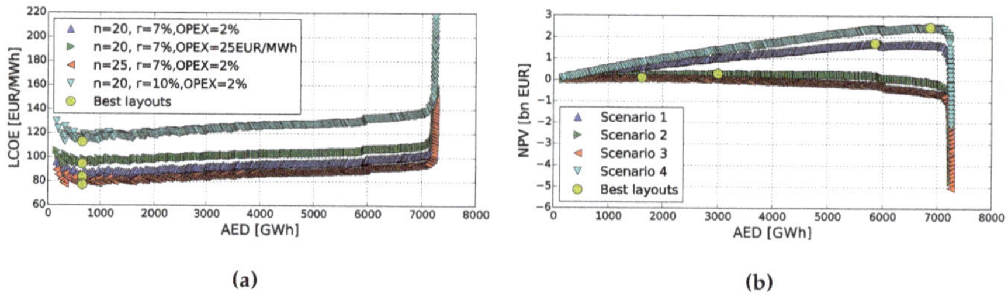

Figure 20. LCOE and NPV values for different economic factors. (**a**) LCOE; (**b**) NPV.

6.4.2. NPV

Different from the LCOE function, the variation of the economic parameters will influence differently the cash flows of the NPV equation even if the OPEX is proportional to one of the other goals (see Figure 20b). In this way, the economic parameters play an important role as weighting factors in the outcome of the NPV function. Table 9 presents the characteristics of the four layouts (shown in Figure 21) with the highest NPV value for different economic parameters and shows the layouts.

If the designers used an energy price of 0.1 €/kWh, which is the target for energy generated offshore by 2020 (see Section 1), a layout with an IC of 752 MW presents the highest NPV (see Figure 21b). This is the only layout of the optimized trade-off curve that has a similar capacity to the 700-MW guideline which demonstrated to be advantageous at some point.

If the lifetime operation of the OWF is considered to be 25 years [138,139], it is more beneficial, according to the NPV function, to install more turbines to increase the AED and take advantage of the extra time of operation (Figure 21d).

Figure 20b demonstrates that the higher the revenues, either through a higher price of energy or via a longer exploration of the site, the NPV function is maximized with OWFs with higher installed capacities (IC). If the revenues are lower, wind farms with less turbines present the highest NPV values.

Table 9. Influence of the economic factors on the NPV function.

Scenario nr.	Lifetime (Years)	Interest Rate (%)	Energy Price (€/kWh)	NPV (bn €)	Layout nr.	AED (GWh/Year)	CAPEX (bn €)	IC (MW)	TT	CSL (km)	NSS
1	20	7	0.124	1.726	**248**	5858	4.93	1552	HVdc	281.54	1
2	20	7	**0.1**	0.29	**115**	2998	2.38	752	MVac	821.98	0
3	20	**12**	0.124	0.097	**57**	1617	1.22	392	MVac	418.91	0
4	**25**	7	0.124	2.468	**301**	6860	6.04	1904	HVdc	230.63	2

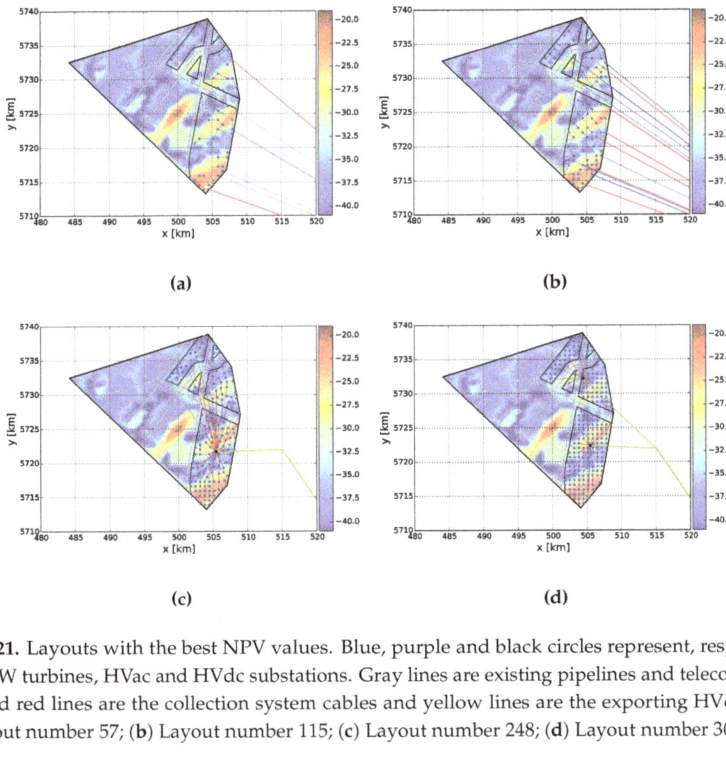

Figure 21. Layouts with the best NPV values. Blue, purple and black circles represent, respectively, the 8 MW turbines, HVac and HVdc substations. Gray lines are existing pipelines and telecom cables, blue and red lines are the collection system cables and yellow lines are the exporting HVdc cables. (**a**) Layout number 57; (**b**) Layout number 115; (**c**) Layout number 248; (**d**) Layout number 301.

6.5. Discussion

The optimized trade-off curve was obtained in 30 days with the proposed framework, which is relatively fast if compared to the design phase of a state-of-the-art OWF which takes several months [140]. The optimization strategy also demonstrates that designers would have missed other wind farm designs that could lead to similar AED and CAPEX values. For example, Figure 22 shows a layout with 88 Vestas 8 MW turbines connected to shore via 66-kV cables, which has similar values for the AED and CAPEX as the standard layout of Figure 17b. If, for example, there would be a risk associated with building structures offshore (e.g., substations, or reliability values for the different components of the system), the layout obtained with the algorithm would be preferable to the standard HVac-based one.

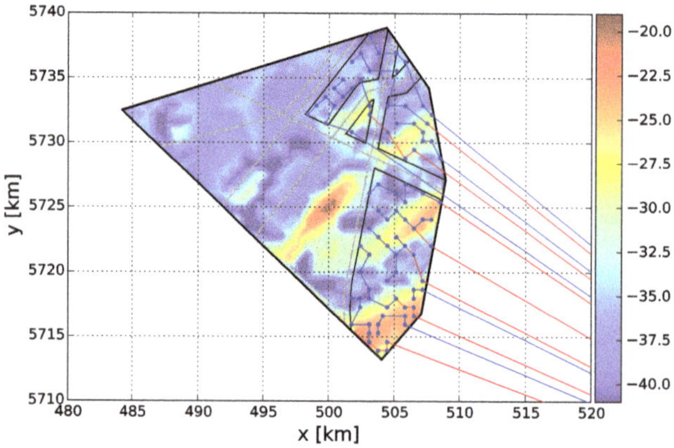

Figure 22. Layout with similar annual energy delivered (AED) and CAPEX values as the standard layout with Vestas 8 MW turbines.

No high-level constraints were applied to the case study, e.g., maximum installed capacity. Although these will have an influence on the choice of the layout, the basis of the framework holds since the designers are only required to firstly discard the layouts which do not meet the high-level constraints and then apply the economic functions over the remaining ones.

The optimization phase itself can still be an iterative process of refining the optimization goals and constraints because the MO algorithm is powerful enough to exploit peculiarities in an optimization goal that actually still gives undesired results. For example, in the case study there was no restriction regarding the maximum number of export cables or a preference towards layouts which used offshore substations. In this way, even though the layouts shown in Figures 17b and 22 have similar AED and CAPEX values, the latter, found with the optimization framework, may not be desired by the wind farm designers due to the high amount of cables connected to shore. On the other hand, the layout of Figure 22 does not require offshore substations, which if a certain level of risk is associated to them, may represent an advantage over the layout shown in Figure 17b.

The proposed MO optimization framework provides the designers with important information that could not easily be obtained with single-objective alternatives. For example, the optimized trade-off shown in Figure 16 shows to the designers the best transmission technology and voltage level for the different range of installed capacities. Furthermore the trade-off also informs the designers to any possible important point of transition in the problem design space, *i.e.*, it shows when increasing the installed capacity does not lead to better designs but solely to more expensive layouts. For example, this is the case for the right part of the optimized trade-off of Figure 16. Even though the layouts have higher AED, this comes with a dramatic increase of the CAPEX. Therefore, this alerts the designers that such layouts do not represent the best layouts.

The discovery of important points of transition in the problem search space (Figure 16) is also directly applicable in other search spaces, such as the economic functions shown in Figure 18. The designers are informed when the layouts do not present interesting economic figures in a straightforward manner.

The multi-objective optimization approach also provides the designers with wind farm designers that were optimized without using economic parameters. This allows the designers to iterate over the different layouts with different economic parameters without the need to rerun the optimization (see Section 6.4).

7. Conclusions

A review of existing commercial and academic optimization tools showed their drawbacks and gaps to be twofold: first, the optimization processes are based on a sequential approach and, second, a priori economic assumptions need to be made for the optimization phase. To solve these issues, a MO optimization framework to automate, integrate and optimize the design of OWFs was presented in this work.

The design aspects of existing offshore wind farms, e.g., number and location of offshore substations and collection system design, were analyzed. Furthermore, also the industrial trends of the components was investigated. In this way, it was assured that the optimization framework captures all the important design aspects and the industrial trends of the components used in state-of-the-art offshore wind farm. Also, the important optimization variables of the wind farm layout optimization problem were assessed.

A state-of-the-art multi-objective optimization algorithm, so-called MOGOMEA, was used in this work. More specifically, a variant of the algorithm tailored for the wind farm layout optimization problem was employed. The algorithm and the optimization framework were then applied to a case study in the Dutch offshore Borssele area. The framework found several wind farm layouts that represent an optimized trade-off of between the design goals, the annual energy delivered to the onshore electrical network and the investment costs.

Distinct wind farm layouts demonstrated to be the most profitable choice depending on the economic function used. The economic factors, e.g., interest rate and price of energy, demonstrated to play an important role in the outcome of the best layout. These results show that wind farm designers should not decide the layout of the wind farm based solely on one economic function such as the NPV or LCOE. Instead, designers should carefully inspect the performance of the layouts at hand using several economic functions.

The proposed framework helps the designers to perform such analysis since all efficient trade-offs are presented to them. This way, designers may iterate over the optimized trade-offs to select which layout is the most suitable without rerunning the optimization tool, accelerating the decision phase and reducing costs. Moreover, the framework designed layouts that performed similarly to layouts based on existing design strategies, e.g., placement of turbines in a standard grid. This result indicates that designers may be missing other design concepts that lead to similar performances. Nonetheless, the selection of the wind farm layout has to undergo a selection to assure that it is the one that most meets the desire of the designers.

Acknowledgments: The work was done within the project "Far and Large Offshore Wind (FLOW)", which is supported by the Ministry of Economic Affairs, Agriculture and Innovation of the Netherlands within the EOS-LT program of Agentschap-NL (P201101-005-ECN). The opinion expressed by the authors does not necessarily reflect the position of the Ministry of Economic affairs, nor does it involve any responsibility on its part. The work was also performed under the Funding Agency Grant CONICYT-FONDECYT 11150069. The authors would like to thank the Het Collectief company for providing the computational resources used to obtain the results presented in this paper.

Author Contributions: All authors were involved during the development of the manuscript. S. Rodrigues wrote the article and was in charge of most of the modeling. G. Katsouris wrote the heuristic to design the collection systems, R. Teixeira Pinto wrote the code to compute the power flows and P. Bosman designed the multi-objective optimization algorithm. C. Restrepo contributed with the illustrations present in the work and provided general input during the writing of this work. M. Soleimanzadeh and P. Bauer provided valuable insights throughout the research and writing processes.

Conflicts of Interest: The authors declare no conflict of interest.

Abbreviations

AED (GWh/year)	Annual Energy Delivered
AEP (GWh/year)	Annual Energy Production
AV (bn €)	Annualized Value
BCR [–]	Benefit to Cost Ratio
CAPEX (bn €)	Capital Expenditure
COE (M€/MWh)	Cost of Energy
COP (M€/MW)	Cost of Power
CSL (km)	Collection System Length
DCO	Development Consent Order
DFIG	Doubly-Fed Induction Generator
DPT (years)	Discounted Payback Time
FEED	Front-End Engineering and Design
HVac	High-Voltage alternate current
HVdc	High-Voltage direct current
IC (MW)	Installed Capacity
IRR (%)	Internal Rate of Return
LCOE (€/MWh)	Levelized Cost of Energy
MMC	Modular Multi-level Converter
MO	Multi-Objective
MOGOMEA	Multi-Objective Gene-pool Optimal Mixing Evolutionary Algorithm
MVac	Medium-Voltage alternate current
NPV (bn €)	Net Present Value
NSS	Number of Substations
OPEX (bn €)	Operational Expenditure
OWF	Offshore Wind Farm
PCC	Point of Common Coupling
PMSG	Permanent-Magnet Synchronous Generator
RD (m)	Rotor Diameter
ROI [–]	Return on Investment
TT	Transmission Technology
UXO	UneXploded Ordnance
UF [–]	Utilization Factor
VSC	Voltage-Source Converter
WFLOP	Wind Farm Layout Optimization Problem

References

1. Christensen, B. The road to below 10 ct EUR/kWh—Siemens response to the challenge. In Proceedings of the EWEA (European Wind Energy Association) Offshore 2015, Copenhagen, Denmark, 10–12 March 2015.
2. BVG Associates. *Future Renewable Energy Costs: Offshore Wind—How Technology Innovation Is Anticipated to Reduce the Cost of Energy from European Offshore Wind Farms*; Technical Report; KIC InnoEnergy: Eindhoven, The Netherlands, 2014.
3. The Crown Estate. *Offshore Wind Cost Reduction—Pathways Study*; Technical Report; Available online: http://www.thecrownestate.co.uk/energy-and-infrastructure/offshore-wind-energy/working-with-us/strategic-workstreams/cost-reduction-study/ (accessed on 15 March 2016).
4. Offshore Wind Cost Reduction Task Force. *Offshore Wind Cost Reduction Task Force Report*; Technical Report; Available online: https://www.gov.uk/government/groups/offshore-wind-cost-reduction-task-force (accessed on 15 March 2016).
5. Offshore Renewable Energy Catapult. *Cost Reduction Monitoring Framework*; Technical Report; Available online: https://ore.catapult.org.uk/documents/10619/168655/pdf/a8c73f4e-ba84-493c-8562-acc87b0c2d76 (accessed on 15 March 2016).
6. Offshore Wind Programme Board. *Sharing Good Practice in the Supply Chain to Facilitate Cost Reduction*; Technical Report Rev 03; The Crown Estate: London, UK, 2012.
7. Milborrow, D. Onshore wind is more competitive than ever. Available online: http://www.windpowermonthly.com/article/1330525/onshore-wind-competitive-ever (accessed on 15 March 2016).
8. The Wind Power. EWEA Offshore: Offshore to be cost competitive by 2023. Available online: http://www.windpowermonthly.com/article/1337352/ewea-offshore-offshore-cost-competitive-2023 (accessed on 1 June 2015).
9. DNV GL. *Project FORCE—Offshore Wind Cost Reduction through Integrated Design*; Technical Report; DNV GL: Hovik, Norway, 2014.
10. Lindoe Offshore Renewables Center. List of Offshore Wind Farms. Available online: http://www.lorc.dk/offshore-wind-farms-map/list (accessed on 1 March 2015).
11. 4C Offshore. Offshore Wind Farms. Available online: www.4coffshore.com/windfarms (accessed on 23 March 2015).
12. Trading Economics. Available online: http://www.tradingeconomics.com/euro-area/inflation-cpi (accessed on 8 August 2015).
13. BVG Associates. *Value Breakdown for the Offshore Wind Sector*; Technical Report; Renewables Advisory Board: Eindhoven, The Netherlands, 2010.
14. Scottish Enterprise. *Offshore Wind Fact Overview of a Wind Farm Project*; Technical Report; Scottish Enterprise: Glasgow, Scotland, 2010.
15. Galloper Wind Farm Limited. *Galloper Wind Farm Project Environmental Statement—Chapter 5 Project Details*; Technical Report; RWE Innogy and SSE Group: Wiltshire, UK, 2011.
16. Galloper Wind Farm Limited. *Gwynt y Mor Offshore Wind Farm—Environmental Statement (Non Technical Summary)*; Technical Report; RWE Innogy and SSE Group: Wiltshire, UK, 2005.
17. Ebbesen, M.C. Seeing the big picture—Finding the optimal conceptual design. In Proceedings of the EWEA (European Wind Energy Association) Offshore 2015, Copenhagen, Denmark, 10–12 March 2015.
18. Esteban, M.D.; Lopez-Gutierrez, J.S.; Diez, J.J.; Negro, V. Methodology for the design of offshore wind farms. In Proceedings of the 11th International Costal Symposium, Szczecin, Poland, 9–14 May 2011.
19. European Wind Energy Association. *Wind Energy—The Facts Part I Technology*; Technical Report; Wind Energy: Brussels, Belgium, 2009.
20. Clausen, R.S. Optimising wind farm layouts—Weighing costs against benefits. In Proceedings of the EWEA (European Wind Energy Association) Offshore 2015, Copenhagen, Denmark, 10–12 March 2015.
21. Wyatt, S.; Govindji, A.K.; James, R.; Duan, N.; Xie, B.; Liu, M.; Yi, G. *Detailed Appraisal of the Offshore Wind Industry in China*; Technical Report; Carbon Trust: London, UK, 2014.
22. Rodrigues, S.; Restrepo, C.; Kontos, E.; Pinto, R.T.; Bauer, P. Trends of offshore wind projects. *Renew. Sustain. Energy Rev.* **2015**, *49*, 1114–1135.

23. Rheinisch-Westfalisches Elektrizitatswerk. Gwynt y Mor Offshore Wind Farm Key Statistics. Available online: https://www.rwe.com/web/cms/en/1252456/rwe-innogy/sites/wind-offshore/under-construction/gwynt-y-mr/tech-and-spec/ (accessed on 31 March 2015).
24. Lumbreras, S.; Ramos, A. Offshore wind farm electrical design: A review. *Wind Energy* **2013**, *16*, 459–473.
25. Herbert-Acero, J.F.; Probst, O.; Rethore, P.E.; Larsen, G.C.; Castillo-Villar, K.K. A Review of Methodological Approaches for the Design and Optimization of Wind Farms. *Energies* **2014**, *7*, 6930–7016.
26. Tong, W.; Chowdhury, S.; Mehmani, A.; Messac, A. Multi-Objective wind farm design—Exploring the trade-off between capacity factor and land use. In Proceedings of the 10th World Congress on Structural and Multidisciplinary Optimization, Orlando, FL, USA, 19–24 May 2013.
27. Khan, S.; Rehman, S. Computational intelligence techniques for placement of wind turbines: A brief plan of research in Saudi Arabian perspective. In Proceedings of the 2010 IEEE International Energy Conference and Exhibition (EnergyCon), Manama, Bahrain, 18–22 December 2010; pp. 519–523.
28. Kusiak, A.; Song, Z. Design of wind farm layout for maximum wind energy capture. *Renew. Energy* **2010**, *35*, 685–694.
29. Zhang, P.Y. Topics in Wind Farm Layout Optimization: Analytical Wake Models, Noise Propagation, and Energy Production. Master's Thesis, University of Toronto, Toronto, ON, Canada, 2013.
30. Kwong, W.Y.; Zhang, P.Y.; Romero, D.; Moran, J.; Morgenroth, M.; Amon, C. Wind farm layout optimization considering energy generation and noise propagation. In Proceedings of the ASME 2012 International Design Engineering Technical Conferences & Computers and Information in Engineering Conference, Chicago, IL, USA, 12–15 August 2012.
31. Veeramachaneni, K.; Wagner, M.; O'Reilly, U.M.; Neumann, F. Optimizing energy output and layout costs for large wind farms using particle swarm optimization. In Proceedings of the IEEE Congress on Evolutionary Computation, Brisbane, Australia, 10–15 June 2012; pp. 1–7.
32. Tran, R.; Wu, J.; Denison, C.; Ackling, T.; Wagner, M.; Neumann, F. Fast and effective multi-objective optimisation of wind turbine placement. In Proceedings of the 15th Annual Conference on Genetic and Evolutionary Computation, Amsterdam, The Netherlands, 6–10 July 2013; pp. 1381–1388.
33. Rodrigues, S.; Bauer, P.; Bosman, P.A. Multi-Objective Optimization of Wind Farm Layouts—Complexity, Constraint Handling and Scalability. *Renew. Sustain. Energy Rev.* **2016**, submitted.
34. Sorkhabi, S.; Romero, D.; Beck, J.; Amon, C. Constrained multi-objective wind farm layout optimization—Introducing a novel constraint handling approach based on constraint programming. In Proceedings of the International Design Engineering Technical Conference, Boston, MA, USA, 2–5 August 2015.
35. Sisbot, S.; Turgut, O.; Tunc, M.; Camdali, U. Optimal positioning of wind turbines on Gokceada using multi-objective genetic algorithm. *Wind Energy* **2010**, *13*, 297–306.
36. *WindPRO*. EMD International A/S: Aalborg, Denmark. Available online: http://www.emd.dk/WindPRO (accessed on 20 April 2014).
37. *GHWindFarmer*. Available online: http://www.gl-garradhassan.com/en/software/GHWindFarmer.php (accessed on 20 April 2014).
38. Wind Resources for Wind Turbine Production. Wind resources for energy production of wind turbines. Available online: http://www.wasp.dk/wasp (accessed on 20 April 2014).
39. *OpenWind*. AWS Truepower: Albany, NY, USA. Available online: http://www.awstruepower.com/solutions/products/openwind/ (accessed on 20 April 2014).
40. *WindSim*. Available online: http://www.windsim.com/ (accessed on 20 April 2014).
41. *ReSoft*. Wind Farm Analysis, Design and Optimisation. Available online: http://www.resoft.co.uk/ (accessed on 20 April 2014).
42. Kallioras, N.A.; Lagaros, N.D.; Karlaftis, M.G.; Pachy, P. Optimum layout design of onshore wind farms considering stochastic loading. *Adv. Eng. Softw.* **2015**, *88*, 8–20.
43. Song, M.; Chen, K.; Zhang, X.; Wang, J. Optimization of wind turbine micro-siting for reducing the sensitivity of power generation to wind direction. *Renew. Energy* **2016**, *85*, 57–65.
44. Elkinton, C.N.; Manwell, J.F.; McGowan, J.G. Offshore wind farm layout optimization (OWFLO) project: An introduction. In Proceedings of the European Offshore Wind Conference & Exhibition, Copenhagen, Denmark, 26–28 October 2005.

45. Lackner, M.A.; Elkinton, C.N. An Analytical Framework for Offshore Wind Farm Layout Optimization. *Wind Eng.* **2007**, *31*, 17–31.
46. Gribben, B.J.; Williams, N.; Ranford, D. Offshore wind farm layout design—A systems engineering approach. In Proceedings of the Ocean Power Fluid Machinery, London, UK, 19 October 2010.
47. Rethore, P.E.; Fuglsang, P.; Larsen, G.; Buhl, T.; Larsen, T.; Madsen, H. TopFarm: Multi-fidelity optimization of offshore wind farm. In Proceedings of the ISOPE (International Society of Offshore and Polar Engineers) Conference, Maui, HI, USA, 19–24 June 2011.
48. Short, W.; Packey, D.J.; Holt, T. *A Manual for the Economic Evaluation of Energy Efficiency and Renewable Energy Technologies*; Technical Report NREL/TP-462-5173; National Renewable Energy Laboratory: Golden, CO, USA, 1995.
49. Elkinton, C.N. Offshore Wind Farm Layout Optimization. Ph.D. Thesis, University of Massachusetts, Amherst, MA, USA, 2007.
50. de Oliveira, W.S.; Fernandes, A.J. Investment Analysis For Wind Energy Projects. *Rev. Bras. Energ.* **2013**, *19*, 239–285.
51. Erlich, I.; Shewarega, F.; Feltes, C.; Koch, F.; Fortmann, J. Offshore Wind Power Generation Technologies. *Proc. IEEE* **2013**, *101*, 891–905.
52. Project Management Support Services Ltd (PMSS). *Greater Gabbard Offshore Wind Farm Non Technical Summary*. Technical Report; Greater Gabbard Offshore Winds Ltd.: Wiltshire, UK, 2005.
53. European Wind Energy Association. *The European Offshore Wind Industry—Key Trends and Statistics 2013*. Technical Report; Wind Energy: Brussels, Belgium, 2014.
54. Countries list, The Wind Power. Available online: http://www.thewindpower.net/windfarms_offshore_en.php (accessed on 1 March 2015).
55. Polinder, H.; Ferreira, J.; Jensen, B.; Abrahamsen, A.; Atallah, K.; McMahon, R. Trends in Wind Turbine Generator Systems. *IEEE J. Emerg. Sel. Top. Power Electron.* **2013**, *1*, 174–185.
56. Teixeira Pinto, R.; Rodrigues, S.; Bauer, P.; Pierik, J. Grid code compliance of VSC-HVDC in offshore multi-terminal DC networks. In Proceedings of the 39th Annual Conference of the IEEE Industrial Electronics Society, Vienna, Austria, 10–13 November 2013.
57. Ibsen, L.B.; Liingaard, M.; Nielsen, S.A. Bucket Foundation, a status. In Proceedings of the Copenhagen Offshore Wind, Copenhagen, Danmark, 26–28 October 2005.
58. de Villiers, P. *Demonstrating Keystone Engineering's Innovative Inward Battered Guide Structure (IBGS) Offshore Foundation Concept at Hornsea—Best Practice for Private-Public Cooperation*; Technical Report; Carbon Trust: London, UK, 2012.
59. Gengenbach, J.; Mikkelsen, K.; Rudinger, F.; Brommundt, M.; Gretlund, J.S. Design challenges of XL monopiles. In Proceedings of the EWEA (European Wind Energy Association) Offshore 2015, Copenhagen, Denmark, 10–12 March 2015.
60. Govindji, A.K.; James, R.; Carvallo, A. *Appraisal of the Offshore Wind Industry in Japan*; Technical Report; Carbon Trust: London, UK, 9 October 2014.
61. European Wind Energy Assoctiation. *Deep Water—The Next Step for Offshore Wind Energy*; Technical Report; European Wind Energy Association: Brussels, Belgium, 2013.
62. Roddier, D.; Cermelli, C.; Aubault, A.; Weinstein, A. WindFloat: A floating foundation for offshore wind turbines. *Renew. Sustain. Energy* **2010**, *2*, 1–34.
63. *Japan's floating offshore wind projects: An overview*; Technical Report; Main(e) International Consulting LLC: Bremen, ME, USA, 2013.
64. *Floating Offshore Wind Foundations: Industry Consortia and Projects in the United States, Europe and Japan—An Overview*; Technical Report; Main(e) International Consulting LLC: Bremen, ME, USA, 2013.
65. Sorensen, T.; Thogersen, M.L.; Nielsen, P. *Adapting and Calibration of Existing Wake Models to Meet the Conditions Inside Offshore Wind Farms*; Technical Report; EMD International A/S: Aalborg, Denmark, 2008.
66. The Crown Estate. *Submarine Cables and Offshore Renewable Energy Installations—Proximity Study*; Technical Report; The Crown Estate: London, UK, 2012.
67. Rodrigues, S.; Bauer, P.; Pierik, J. Comparison of offshore power transmission technologies: A multi-objective optimization approach. In Proceedings of the 15th International Power Electronics and Motion Control Conference (EPE/PEMC), Novi Sad, Serbia, 4–6 September 2012.

68. Schachner, J. Power Connection for Offshore Wind Farms. Master's Thesis, Delft University of Technology, Delft, The Netherlands, 2004.
69. Carbon Trust. Offshore Wind Accelerator—Driving Down the Cost of offshore Wind. Available online: http://carbontrust.com/media/105322/electrical_presentation_jm_-dec_2011.pdf (accessed on 1 May 2014).
70. Pierik, J.; Pavlovsky, M.; Bozelie, J.; Bauer, P.; de Haan, S. *Dowec Electrical System Baseline Design*; Technical Report DOWEC 045 Rev. 2; Energy Research Centre of the Netherlands (ECN): Petten, The Netherlands, 2002.
71. Berzan, C.; Veeramachaneni, K.; McDermott, J.; O'Reilly, U.M. *Algorithms for Cable Network Design on Large-Scale Wind Farms*; Technical Report; Tufts University: Medford, MA, USA, 2011.
72. Bauer, J.; Lysgaard, J. The offshore wind farm array cable layout problem: A planar open vehicle routing problem. *J. Op. Res. Soc.* **2014**, *66*, 360–368.
73. Tesauro, A.; Rethore, P.E.; Larsen, G. State of the art of wind farm optimization. In Proceedings of EWEA—European Wind Energy Conference & Exhibition, Copenhagen, Denmark, 16–19 April 2012.
74. Dicorato, M.; Forte, G.; Pisani, M.; Trovato, M. Guidelines for assessment of investment cost for offshore wind generation. *Renew. Energy* **2011**, *36*, 2043–2051.
75. Katsouris, G. Infield Cable Topology Optimization of Offshore Wind Farms. Master's Thesis, Delft University of Technology, Delft, The Netherlands, 2015.
76. Papadopoulos, A. Modeling of Collection and Transmission Losses of Offshore Wind Farms for Optimization Purposes. Master's Thesis, Delft University of Technology, Delft, The Netherlands, 2015.
77. Teixeira Pinto, R. Multi-Terminal dc Networks—System Integration, Dynamics and Control. Ph.D. Thesis, Delft University of Technology, Delft, The Netherlands, 2014.
78. Rodrigues, S. Dynamic Modeling and Control of VSC-based Multi-Terminal DC Networks. Master's Thesis, Instituto Superior Técnico, Lisbon, Portugal, 2011.
79. Donat, L.; Eberle, A.; Velten, E.K.; Prahl, A.; Wevers, M. *Assessment of Climate Change Policies in the Context of the European Semester—Country Report The Netherlands*; Technical Report; Ecologic Institute and Eclareon: Berlin, Germany, 2014.
80. Loyens and Loeff. *North Sea offshore wind—Developments in Belgium and the Netherlands*; Technical Report; Rotterdam, The Netherlands, 2014.
81. Netherlands Enterprise Agency. *North Sea 2050 Spatial Agenda*; Technical Report; Ministerie van Economische Zaken: The Hague, The Netherlands, 2014.
82. Netherlands Enterprise Agency. *Rijksstructuurvisie Windenergie op Zee (in Dutch)*; Technical Report; Ministerie van Economische Zaken: The Hague, The Netherlands, 2014.
83. Netherlands Enterprise Agency. *Borssele Wind Farm Zone Project & Site Description*; Technical Report 2; Ministry of Economic Affairs: The Hague, The Netherlands, 2015.
84. Netherlands Enterprise Agency. *Borssele Wind Farm Zone—Project & Site Description*; Technical Report; Ministry of Economic Affairs: The Hague, The Netherlands, 2014.
85. DNV GL. *66 kV Systems for Offshore Wind Farms*; Technical Report 113799-UKBR-R02, Rev. 2; TenneT: Arnhem, The Netherlands, 2015.
86. *Metocean Study for the Borssele Wind Farm Zone*; Technical Report; Deltares: Delft, The Netherlands, 2015.
87. Netherlands Enterprise Agency. UXO removal/clearance. Available online: http://offshorewind.rvo.nl/questions/view/41293422/uxo-removalclearance-09-12-2015 (accessed on 28 December 2015).
88. *Borssele Wind Farm Zone Wind Farm Sites I and II Project and Site Description*; Technical Report; Netherlands Enterprise Agency: The Hague, The Netherlands, 2015.
89. Netherlands Enterprise Agency. Wind farm layout-location turbines. Available online: http://offshorewind.rvo.nl/questions/view/32317252/wind-farm-layout-location-turbines-22-05-2015 (accessed on 23 August 2015).
90. Netherlands Enterprise Agency. Wind farm layout. Available online: http://offshorewind.rvo.nl/questions/view/33205612/wind-farm-layout-07-07-2015 (accessed on 23 August 2015).
91. *Site Studies Wind Farm Zone Borssele—Wind Resource Assessment*; Technical Report; Ecofys: Utrecht, The Netherlands, 2015.
92. Hsu, S.A.; Meindl, E.A.; Gilhousen, D.B. Determining the power-law wind-profile exponent under near-neutral stability conditions at sea. *J. Appl. Meteor.* **1994**, *33*, 757–765.

93. *Offshore Grid Connection System Borssele—Scoping Document (Notitie Reikwijdte en Detailniveau)*; Technical Report; Tennet: Den Haag, The Netherlands, 2015.
94. Vestas Wind Systems A/S. Offshore v164-8.0 mw v112-3.3 mw. *Tech. Rep.* 2013. Available online: http://nozebra.ipapercms.dk/Vestas/Communication/Productbrochure/OffshoreProductBrochure/Offshore ProductBrochure/ (accessed on 23 August 2015).
95. XEMC Darwind. XD115-5MW. Available online: www.darwind.nl/Wind-turbines/XD115-5MW (accessed on 20 March 2015).
96. Esteban, M.D.; Diez, J.J.; Lopez, J.S.; Negro, V. Why offshore wind energy? *Renew. Energy* **2011**, *36*, 444–450.
97. International Electrotechnical Commission. *Electric Cables—Conductors of Insulated Cables*; International Electrotechnical Commission: Geneva, Switzerland, 2004.
98. International Electrotechnical Commission. *Electric Cables—Calculation of the Current Rating*; International Electrotechnical Commission: Geneva, Switzerland, 2006.
99. Papadopoulos, A.; Rodrigues, S.; Kontos, E.; Todorcevic, T.; Pinto, R.T.; Bauer, P. Collection and Transmission Losses of Offshore Wind Farms for Optimization Purposes. In Proceedings of the IEEE ECCE (Energy Conversion Congress and Exposition) 2015, Montreal, QC, Canada, 20–24 September 2015.
100. Asea Brown Boveri (ABB). *XLPE Submarine Cable Systems—Attachment to XLPE Land Cable Systems—User's Guide*; Lyckeby, Sweden, 2010.
101. Papadopoulos, A.; Rodrigues, S.; Kontos, E.; Todorcevic, T.; Bauer, P. A Fast Steady-State Loss Model of a Modular Multilevel Converter for Optimization Purposes. In Proceedings of the 2015 IEEE ECCE Asia, Seoul, Korea, 1–5 June 2015.
102. Rodrigues, S.; Papadopoulos, A.; Kontos, E.; Todorcevic, T.; Bauer, P. Steady-State Loss Model of Half-Bridge Modular Multilevel Converters. *IEEE Trans. Ind. Appl.* **2016**, *1*, doi:10.1109/TIA.2016.2519510.
103. *Technical Information of IGBT Module FZ1000R33HL3*; Infineon: Rotterdam, The Netherlands, 2013.
104. Levitt, A.C.; Kempton, W.; Smith, A.P.; Musial, W.; Firestone, J. Pricing offshore wind power. *Energy Policy* **2011**, *39*, 6408–6421.
105. Katic, I.; Hojstrup, J.; Jensen, N. A Simple Model for Cluster Efficiency. In Proceedings of the EWEC'86, Rome, Italy, 7–9 October 1986; Volume 1, pp. 407–410.
106. Jensen, N. *A Note on Wind Generator Interaction*; Technical Report Riso-M-2411; Riso National Laboratory: Roskilde, Denmark, 1983.
107. Rodrigues, S.; Bauer, P.; Pierik, J. Modular approach for the optimal wind turbine micro siting problem through CMA-ES algorithm. In Proceeding of the Fifteenth Annual Conference Companion on Genetic and Evolutionary Computation Conference Companion, Amsterdam, The Netherlands, 6–10 July 2013.
108. Amberg, A.; Domschke, W.; Voß, S. Capacitated Minimum Spanning Trees: Algorithms Using Intelligent Search. *Citeseer* **1996**. Available online: http://citeseerx.ist.psu.edu/viewdoc/download?doi=10.1.1.47.3749&rep=rep1&type=pdf (accessed on 18 March 2016).
109. *Subsea Cables UK Guideline No 6 the Proximity of Offshore Renewable Energy Installations & Submarine Cable Infrastructure in UK Waters*; Technical Report; The Crown Estate: London, UK, 2012.
110. Henderson, A.; Baldock, N.; Yendole, H.; Parker, G.; Greig, G.P.E. Money does grow on turbines—Overplanting offshore windfarms. In Proceedings of the Global Offshore Wind Conference, Glasgow, UK, 11–12 June 2014.
111. PYPOWER. Available online: http://rwl.github.io/PYPOWER/index.html (accessed on 1 May 2015).
112. Barten, O. What is the Ideal Powercurve? Master's Thesis, Delft University of Technology, Delft, The Netherlands, 2012.
113. Georgios, S. Techno-Economical Analysis of DC Collection Grid for Offshore Wind Parks. Master's Thesis, University of Nottingham, Nottingham, UK, 2010.
114. Lundberg, S. Performance Comparison of Wind Park Configuration. Master's Thesis, Chalmers University of Technology, Goteburg, Sweden, 2003.
115. Lluch, J.R. Power Transmission Systems for Offshore Wind Farms Technical-Economic Analysis. Master's Thesis, Universitat Politcnica de Catalunya, Barcelona, Spain, 2015.
116. Flow Offshore Wind Cost Model. Available online: http://flow-offshore.nl/page/flow-offshore-wind-cost-model (accessed on 28 August 2015).
117. Van Eeckhout, B. The Economic Value of VSC HVDC Compared to HVAC for Offshore Wind Farms. Master's Thesis, Katholieke Universiteit Leuven, Leuven, Belgium, 2007.

118. Gajic, Z.; Hillstrom, B.; Mekic, F. HV shunt reactor secrets for protection engineers. In Proceedings of the 30th Western Protective Relaying Conference, Spokane, WA, USA, 21–23 October 2003.
119. Braam, H.; Obdam, T.; van de Pieterman, R.; Rademakers, L. *Properties of the O & M Cost Estimator (OMCE)*; Technical Report ECN-E-11-045; ECN: Petten, The Netherlands, 2011.
120. Deloitte. *Establishing the Investment Case Wind Power*; Technical Report; Copenhagen, Denmark, 2014.
121. Cory, K.; Schwabe, P. *Wind Levelized Cost of Energy—A Comparison of Technical and Financing Input Variables*. Technical Report NREL/TP-6A2-46671; National Renewable Energy Laboratory: Golden, CO, USA, 2009.
122. Luong, N.H.; La Poutré, H.; Bosman, P.A. Multi-objective gene-pool optimal mixing evolutionary algorithms. In Proceedings of the 2014 Conference on Genetic and Evolutionary Computation, Vancouver, BC, Canada, 2014; pp. 357–364.
123. Luong, N.H.; Grond, M.; La Poutré, H.; Bosman, P.A. Scalable and practical multi-objective distribution network expansion planning. In Proceedings of the IEEE PES General Meeting, Denver, CO, USA, 26–30 July 2015.
124. Bosman, P.; Thierens, D. On measures to build linkage trees in LTGA. In *Parallel Problem Solving from Nature—PPSN XII*; Springer: Berlin, Germany, 2012; Volume 7491, pp. 276–285.
125. Thierens, D.; Bosman, P.A. Optimal mixing evolutionary algorithms. In Proceedings of the 13th Annual Conference on Genetic and Evolutionary Computation, Dublin, Ireland, 12–16 July 2011; pp. 617–624.
126. Rodrigues, S.; Bauer, P.; Pierik, J. A clustering approach for the wind turbine micro siting problem through genetic algorithm. In Proceedings of the 39th Annual Conference of the IEEE Industrial Electronics Society, Vienna, Austria, 10–13 November 2013.
127. Gonzalez, J.S.; Burgos Payan, M.; Riquelme Santos, J. A New and Efficient Method for Optimal Design of Large Offshore Wind Power Plants. *IEEE Trans. Power Syst.* **2013**, *28*, 3075–3084.
128. Gonzalez, J.S.; Payan, M.B.; Santos, J.M.R.; Gonzalez-Longatt, F. A review and recent developments in the optimal wind-turbine micro-siting problem. *Renew. Sustain. Energy Rev.* **2014**, *30*, 133–144.
129. de Bokx, R. Parallelizing the Linkage Tree Genetic Algorithm and Searching for the Optimal Replacement for the Linkage Tree. Master's Thesis, Delft University of Technology, Delft, The Netherlands, 2015.
130. Saavedra-Moreno, B.; Salcedo-Sanz, S.; Paniagua-Tineo, A.; Prieto, L.; Portilla-Figueras, A. Seeding evolutionary algorithms with heuristics for optimal wind turbines positioning in wind farms. *Renew. Energy* **2011**, *36*, 2838–2844.
131. Wagner, M.; Veeramachaneni, K.; Neumann, F.; O'Reilly, U.M. Optimizing the layout of 1000 wind turbines. In Proceedings of the European Wind Energy Association Annual Event, Brussels, Belgium, 14–17 March 2011; pp. 1–10.
132. Wagner, M.; Day, J.; Neumann, F. A fast and effective local search algorithm for optimizing the placement of wind turbines. *Renew. Energy* **2013**, *51*, 64–70.
133. MacQueen, J. Some methods for classification and analysis of multivariate observations. In Proceedings of the Fifth Berkeley Symposium on Mathematical Statistics and Probability, Berkeley, USA, 27 December 1965–7 January 1966; University of California Press: Berkeley, CA, USA, 1967; Volume 1, pp. 281–297.
134. Netherlands Enterprise Agency. *Borssele Wind Farm Zone Project & Site Description*; Technical Report; Ministry of Economic Affairs: The Hague, The Netherlands, 2014.
135. Bulder, B.; Hagg, F.; van Bussel, G.; Zaaijer, M. *Dutch Offshore Wind Energy Converter Task 12—Cost Comparison of the Selected Concepts*; Technical Report ECN-C-01-080; Energy Research Centre of the Netherlands (ECN), Petten, The Netherlands, 2000.
136. *Offshore Wind Power: Big Challenge, Big Opportunity Maximising the Environmental, Economic and Security Benefits*; Technical Report; Carbon Trust: London, UK, 2008.
137. Ahrenfeldt, S.F. Distributed Substations—A cost-efficient multi-platform topology. In Proceedings of the EWEA (European Wind Energy Association) Offshore, Copenhagen, Denmark, 10–12 March 2015.
138. Smith, P. Question of the Week: Are offshore projects built to last? Available online: http://windpowermonthly.com/article/1320109/question-week-offshore-projects-built-last (accessed on 5 September 2015).

139. *The Economics of Wind Energy*; Technical Report; European Wind Energy Association: Brussels, Belgium, 2009.
140. Thomsen, K.E. Chapter Three—Project Planning. In *Offshore Wind*; Thomsen, K.E., Ed.; Elsevier: Boston, MA, USA, 2012; pp. 27–49.

MDPI AG
Grosspeteranlage 5
4052 Basel
Switzerland
Tel.: +41 61 683 77 34

Energies Editorial Office
E-mail: energies@mdpi.com
www.mdpi.com/journal/energies

www.ingramcontent.com/pod-product-compliance
Lightning Source LLC
Chambersburg PA
CBHW061609220926
51850CB00005B/458
* 9 7 8 3 7 2 5 8 4 9 7 7 2 *